Rusnanotech International Forum on Nanotechnology 2010

AA002776

Journal of Physics: Conference Series Volume 291

Moscow, Russia
1-3 November 2010

Editors:

Anatoly Dvurechenskii
Vasily Klimov
P Kashkarov
Eduard Son
Valentin Parmon
E Obraztsova
Marat Gallamov
Denis Logunov
Konstantin Skryabin
Rem Petrov
Sergey Deev

ISBN: 978-1-61782-766-2
ISSN: 1742-6588

Rusnanotech International Forum on Nanotechnology 2010

Journal of Physics: Conference Series Volume 291

Moscow, Russia
1-3 November 2010

ISBN: 978-1-61782-766-2
ISSN: 1742-6588

IOP Institute of Physics

Printed from e-media with permission by:

Curran Associates, Inc.
57 Morehouse Lane
Red Hook, NY 12571

Some format issues inherent in the e-media version may also appear in this print version.

Printed by Curran Associates, Inc. (2011)

For permission requests, please contact the Institute of Physics
at the address below.

Institute of Physics
Dirac House, Temple Back
Bristol BS1 6BE UK

Phone: 44 1 17 929 7481
Fax: 44 1 17 920 0979

techtracking@iop.org

Additional copies of this publication are available from:

Curran Associates, Inc.
57 Morehouse Lane
Red Hook, NY 12571 USA
Phone: 845-758-0400
Fax: 845-758-2634
Email: curran@proceedings.com
Web: www.proceedings.com

TABLE OF CONTENTS

CATALYSIS AND CHEMICAL INDUSTRY

NANOBIOTECHNOLOGY

NANODIAGNOSTICS

NANOELECTRONICS

NANOMATERIALS

NANOPHOTONICS

NANOTECHNOLOGY IN THE ENERGY INDUSTRY

NANOTECHNOLOGY IN MEDICINE

PREFACE

The Rusnanotech 2010 International Forum on Nanotechnology was held from November 1–3, 2010, in Moscow, Russia. It was the third forum organized by RUSNANO (Russian Corporation of Nanotechnologies) since 2008. In March 2011 RUSNANO was established as an open joint-stock company through the reorganization of the state corporation Russian Corporation of Nanotechnologies. RUSNANO's mission is to develop the Russian nanotechnology industry through co-investment in nanotechnology projects with substantial economic potential or social benefit. Within the framework of the Forum Science and Technology Program, presentations on key trends of nanotechnology development were given by foreign and Russian scientists, R&D officers of leading international companies, universities and scientific centers.

The science and technology program of the Forum was divided into eight sections as follows (by following hyperlinks you may find each section's program including videos of all oral presentations):

- Catalysis and Chemical Industry
- Nanobiotechnology
- Nanodiagnostics
- Nanoelectronics
- Nanomaterials
- Nanophotonics
- Nanotechnolgy In The Energy Industry
- Nanotechnology in Medicine

The scientific program of the forum included 115 oral presentations by leading scientists from 15 countries. Among them in the "Nanomaterials" section was the lecture by Dr Konstantin Novoselov, winner of the Nobel Prize in Physics 2010. The poster session consisted of over 500 presentations, 300 of which were presented in the framework of the young scientists' nanotechnology papers competition. This volume of the *Journal of Physics: Conference Series* includes a selection of 57 submissions.

The scientific program committee:

Prof Zhores Alferov, Academician
Vice-president of Russian Academy of Sciences, Nobel Prize winner, Russia, *Chairman of the Program Committee*
Prof Sergey Deev, Corresponding Member of Russian Academy of Sciences
Head of the Laboratory of Molecular Immunology, M M Shemyakin and Yu A Ovchinnikov Institute of Bioorganic Chemistry, Russian Academy of Sciences, Russia, *Deputy Chairman of the Program Committee*
Prof Alexander Aseev, Academician
Vice-president of Russian Academy of Sciences Director, A V Rzhanov-Institute of Semiconductor Physics, Siberian Branch of Russian Academy of Sciences, Russia
Prof Sergey Bagaev, Academician

UC Berkeley, Director of NSF Nano-scale Science and Engineering Center (NSEC), USA
Prof Andrei Zvyagin
Professor, Macquarie University, Australia
Prof Sergey Kalyuzhny
Director of the Scientific and Technological Expertise Department, RUSNANO, Russia
Konstantin Kazaryan, PhD
Expert of the Scientific and Technological Expertise Department, RUSNANO, Russia, *Program Committee Secretary*
Simeon Zhavoronkov
Head of Nanotechnology Programs Development Office, Rusnanotech Forum Fund for the Nanotechnology Development, Russia

Editors of the proceedings:

Section "Nanoelectronics" – Corresponding Member of Russian Academy of Sciences, Professor Anatoly Dvurechenskii (Institute of Semiconductor Physics, RAS).
Section "Nanophotonics" – Professor Vasily Klimov (Institute of Physics, RAS).
Section "Nanodiagnostics" – Professor P Kashkarov (Russian Scientific Center, Kurchatov Institute).
Section "Nanotechnology for power engineering" – Corresponding Member of Russian Academy of Sciences, Professor Eduard Son (Joint Institute for High Temperatures, RAS).
Section "Catalysis and chemical industry" – Member of Russian Academy of Sciences, Professor Valentin Parmon (Institute of Catalysis SB RAS).
Section "Nanomaterials" – E Obraztsova, PhD (Institute of Physics, RAS), Marat Gallamov PhD (Moscow State University).
Section "Nanotechnology in medicine" – Denis Logunov, PhD (Gamaleya Research Institute of Epidemiology and Microbiology, RAMS).
Section "Nanobiotechnology" – Member of Russian Academy of Sciences, Professor Konstantin Skryabin (Bioengineering Center, RAS), Member of Russian Academy of Sciences, Professor Rem Petrov (RAS), Corresponding Member of Russian Academy of Sciences, Professor Sergey Deev (Institute of Bioorganic Chemistry).

All papers published in this volume of *Journal of Physics: Conference Series* have been peer reviewed through processes administered by the proceedings Editors. Reviews were conducted by expert referees to the professional and scientific standards expected of a proceedings journal published by IOP Publishing.

Journal of Physics: Conference Series **291** (2011) 012001 doi:10.1088/1742-6596/291/1/012001

New prospects of low-scale gas chemistry

Vladimir Arutyunov[1]

Semenov Institute of Chemical Physics, Russ. Acad. Sci., Kosygina, 4, Moscow, 119991, Russia

E-mail: arutyunov@center.chph.ras.ru

Abstract. The large-scale industrial sources of conventional natural gas are limited and can be exhausted in the nearest future. But there are still a significant number of low deposit, remote, low-pressure fields, and abundant resources of unconventional gas including coal-bed methane, shale gas, gas hydrates. However, usually these sources are less intensive, short living and spread through a large area. So gas industry very likely will have to develop new low-scale technologies to explore and process such resources into more valuable and easily transportable liquid products to satisfy consumers in any point of the world as we do it now with oil and petrochemicals.

1. The problem

Due to all reliable forecasts, Natural Gas will be the main sours of energy at any rate to the end of this century. Oil resources become more and more limited and expensive. “Clean coal” proved to be too expensive for practical use. Thermonuclear synthesis still needs dozens of years of thorough and expensive investigations before possible industrial application. So call “clean alternative sources” including biofuels, sun, wind, etc. in principal can’t meet the growing need in industrial power supply [1].

On the contrary, the resources of Natural Gas are large enough, and it is the cleanest fossil fuel. The basis of the modern gas industry is the conventional gas with total amount of world resources of about 200 trillion m^3 and predicted supply period of about 30 years (figure 1).

Meanwhile, the annual consumption of conventional natural gas has already exceeded 3 trillion m^3 and rapidly increasing (figure 1). To the beginning of this century it exceeded the volume of annually discovered resources (figure 2) which provoked the tension with gas supply on the world market and the increase of its price. The similar event in the oil industry occurred about 25 years ago, approximately in 1985, with well known consequences for oil market and the whole world economy.

[1] To whom any correspondence should be addressed.

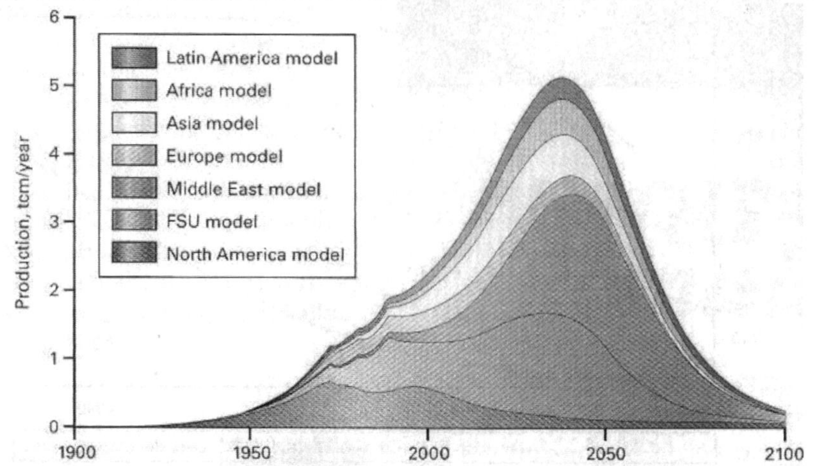

Figure 1. Predictions of the world supply of conventional natural gas [2].

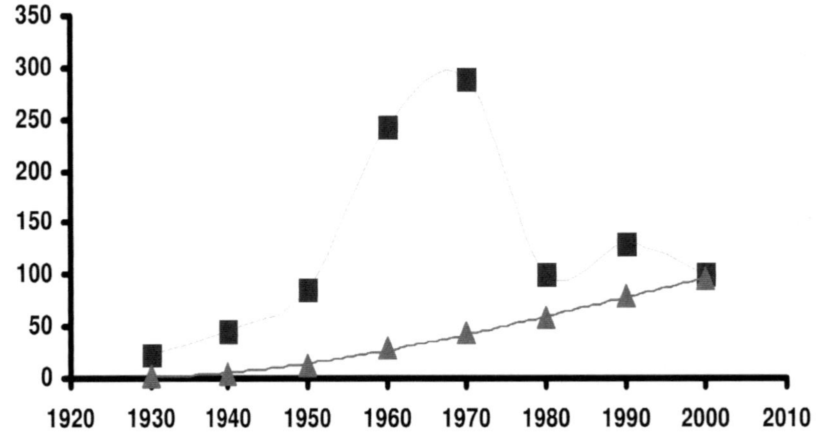

Figure 2. World Natural Gas production and discoveries by decades.

Although it is difficult to expect a discovery of a significant number of new giant conventional gas fields which can feed long-distance gas pipe systems, large gas liquefying or gas chemical plants of the type that were built in recent years, there are huge amount of gas hydrates and other unconventional sources. Therefore, in the nearest future, we will have to more and more intensively exploit low deposit, remote, low-pressure fields, and unconventional gas including coal-bed methane and shale gas. The recent industrial development of shale gas in the USA which now covers about 20% of demand and guarantee at least 30 years of supply (figure 3), evidently showed the tendency to less intensive, short living and spread sources.

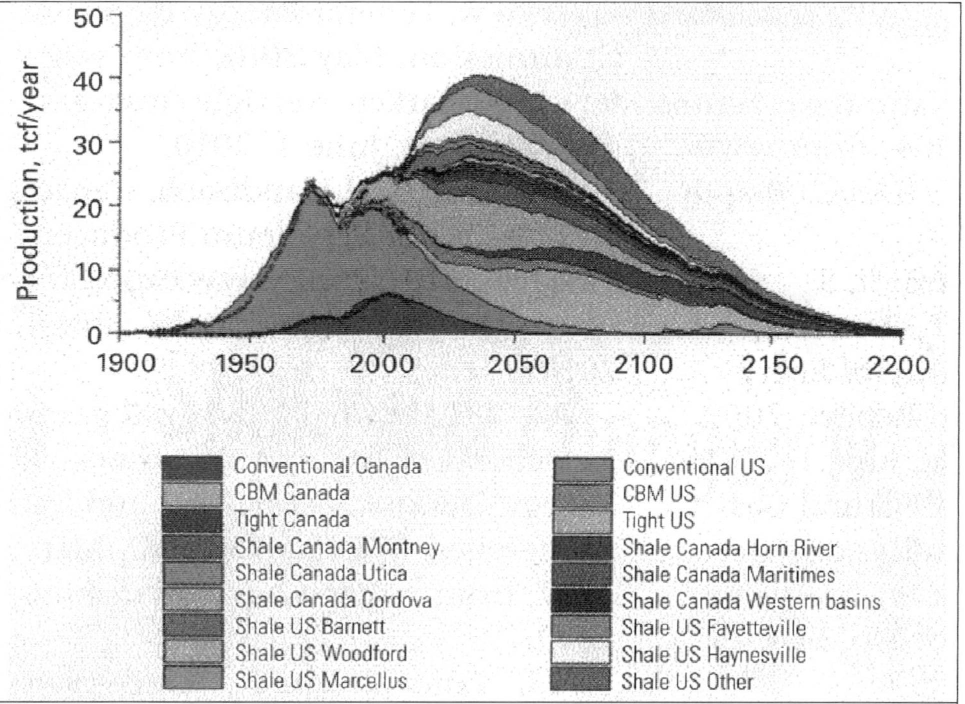

Figure 3. Resources of Natural Gas in the USA and Canada including unconventional gas [3].

Although unconventional sources are yet more expensive than conventional, this difference is not dramatic (figure 4) and rapidly diminishes. The production cost of shale gas has already reached \$120-140 per 1000 m^3 of gas which makes it quite competitive in comparison with conventional gas from shelf or arctic deposits.

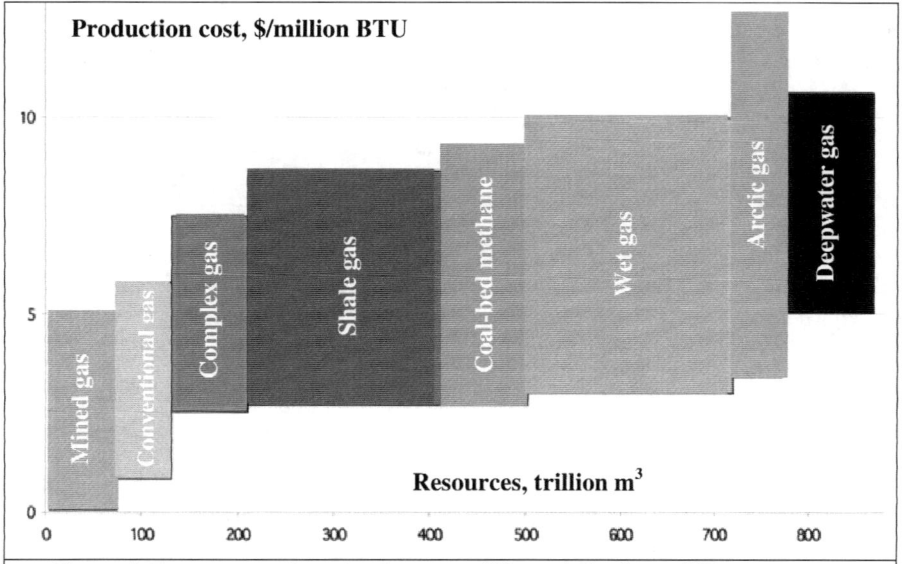

Figure 4. Resources and production cost of Natural Gas from different sources.

One of the most difficult problems of gas industry is a high cost of gas transportation which significantly exceeds the cost of transportation of oil and in many cases significantly exceeds the cost of gas production (figure 5). Now we relay mostly on a very expensive long distant pipelines and large-scale liquefying LNG plants to make gas accessible for remote consumers.

The other problem is a high cost of gas chemical processing, especially in comparison with processing of oil. It is very desirable to convert gas into more valuable and easily transportable chemicals and liquid fuels, but modern large-scale gas chemical GTL plants demonstrate only marginal profitability and need huge gas resources to provide at any rate several decades of operation to compensate billion dollars of capital costs.

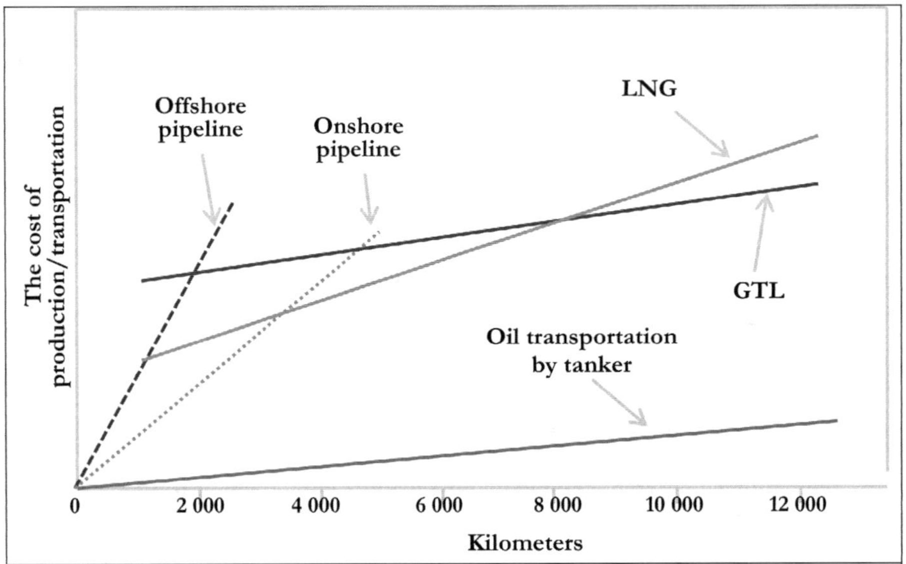

Figure 5. The cost for different methods of gas transportation.

So to meet new potential sources gas industry will have to develop new technologies to explore, distribute and process such resources into more valuable and easily transportable products to satisfy consumers in any point of the world as we do it now with oil and petrochemicals.

2. Principals and problems of low-scale gas chemistry

One of the most evident ways to develop less intensive, short living and spatially extended future unconventional gas sources is their on-sight processing into any liquid products. But as it can be seen from figure 5, the cost of such transformation is very high. Present-day complex and very expensive gas chemical technologies are unprofitable at low-scale realization (figure 6). And as it can be seen from figure 6, in many cases these technologies are inapplicable at all, especially in remote gas producing regions of Russia with severe climate and absence of transport and industrial infrastructure.

For Russia it is especially actual problem. Due to the absence of low-scale gas processing technologies it annually loses up to 50 billion m^3 of flared oil associated gases and about 50 million tons of natural gas liquids (NGL) which can't be separated from pipeline gas. Besides, there is more than 600 low deposit gas fields with total resources of about 6 trillion m^3 which up to now do not involved into industrial development.

To develop and transport these resources gas industry needs simpler low-scale technologies to convert them just on site into more valuable and easily transportable liquid chemical or fuels. It must be principally new processes initially designed for low-scale applications. And it worth to note, that in some cases even the well-known solutions that proved to be unacceptable for large-scale applications can be suitable enough for such low-scale technologies.

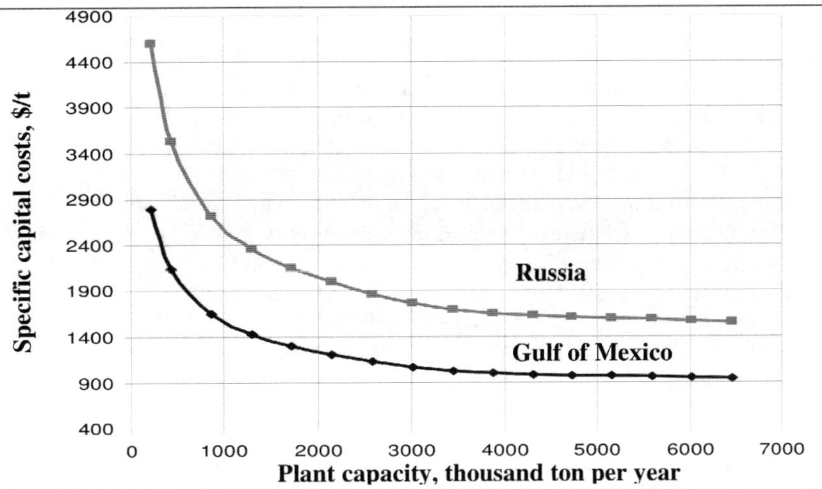

Figure 6. Specific capital costs for traditional GTL production vs. plant capacity [4].

Some evident t basic principals for such low-scale gas chemical technologies are listed below:

- Simplicity and less tonnage in comparison with existing technologies;
- Modular construction that makes it possible easily increase or decrease capacity or move the equipment into other place;
- Completely factorial production of equipment without complex constructing-and-mounting works on place;
- Universality with respect to gas composition and possible fluctuation in its debit;
- Possibility to combine different modules from different producers;
- Autonomous power supply just from producing gas, etc.

Today there are no such technologies of industrial level in Russia or anywhere else. However, the results of some Russian research engineering studies that were accomplished during the last years have created a strong base for such new technologies. There are two main directions of investigations:

- Alternative technologies for production of syngas and subsequent synthesis of chemicals or fuels using syngas (Fischer-Tropsch synthesis, methanol, DME, etc.);
- Direct conversion of hydrocarbon gases into chemicals (ethylene, methanol, etc.).

2.1. Alternative technologies for production of syngas

The economical attractiveness of indirect methods via syngas is connected with high conversion of initial gas carbon into final products and the maturity of these well developed technologies. But high capital investment and high energy consumption for syngas production make questionable their low-scale application. The well known fact that this first and most energy and capital consuming stage of all modern gas chemical plants – the conversion of natural gas into syngas - consumes up to 70% of all costs makes it evident that only the principal improvement in syngas production can give the acceptable solution [5].

Below are listed the most intensively studied alternative technologies for conversion of hydrocarbons into syngas:

- Methane oxidation with the use of oxygen permeable ceramic membranes;
- Microchannel reactors;
- Short-time catalysis;

- Production of syngas on the base of power engines;
- Production of syngas and hydrogen in radiation burners;
- Production of syngas at filtration combustion;
- Electrochemical conversion of methane to syngas.

As an example of alternative method of syngas production potentially applicable for low-scale gas processing we can mention our recent development of compact and highly productive generator of syngas and hydrogen based on gas-phase combustion of hydrocarbons in volumetric permeable matrix [6] (figure 7). Flameless combustion near inner surface of permeable volumetric matrix proceeds with much lover temperature due to intensive heat exchange with surface. And partial lock of IR radiation of combustion front in deep volumetric matrix lets to significantly widen combustion limits. This auto-thermal process let to burn very rich hydrocarbon-oxidant mixtures with oxygen excess coefficient $\alpha = [O_2]_0/2[CH_4]_0$ up to 0.36 for methane-air mixtures and to use practically any compositions of gas-phase hydrocarbons. At that the yield of hydrogen and CO is very close to thermodynamically equilibrium values. This device is very simple in operation and has high specific thermal power up to 30 Wt/cm^2. It uses no catalyst and demonstrates no soot formation in cavity. As oxidant air, enriched air or oxygen can be used.

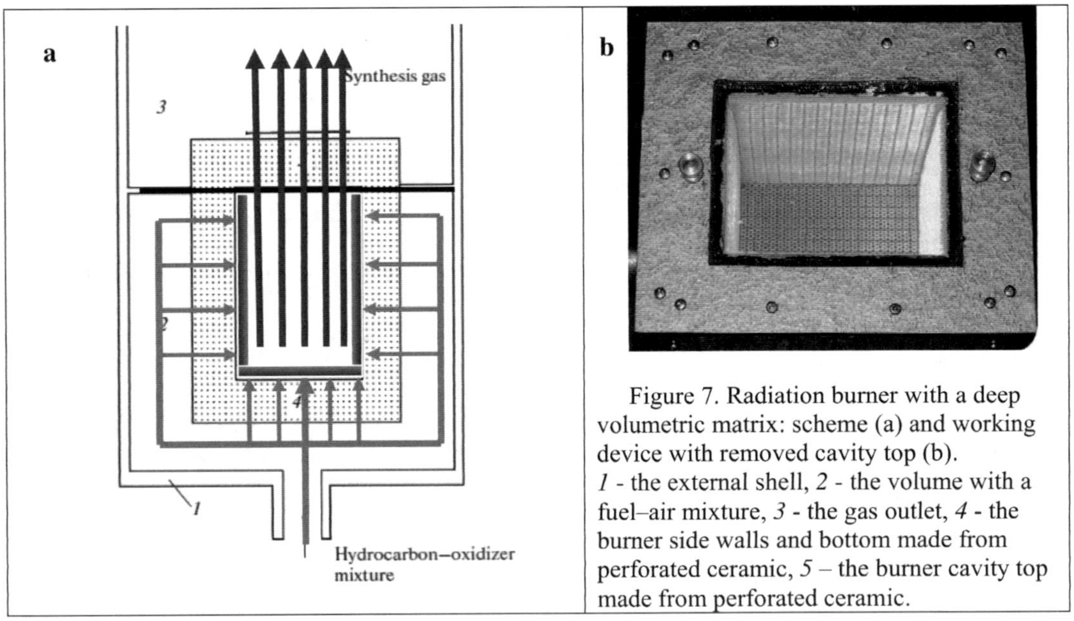

Figure 7. Radiation burner with a deep volumetric matrix: scheme (a) and working device with removed cavity top (b).
1 - the external shell, *2* - the volume with a fuel–air mixture, *3* - the gas outlet, *4* - the burner side walls and bottom made from perforated ceramic, *5* – the burner cavity top made from perforated ceramic.

2.2. Direct processes of natural gas conversion into chemicals

Direct processes potentially are simpler and so better adaptable for low-scale applications. They usually well compatible with power and heat production which is important for remote producing sites. But their common and very serious disadvantage is significantly lower selectivity and conversion per pass, as well as the absence of industrial experience. Among the others there can be mentioned such long-studied direct processes as:

- Oxidative coupling of methane to ethylene (OCM) [7];
- Direct oxidation of methane to methanol (DMTM) and connected processes [5, 8];
- Catalytic aromatization of methane;
- Halogenation and oxy-halogenation of methane.

As it was mentioned above, the low-scale processes needn't to copy large-scale technologies. For example, the use of OCM for production of polymerization grade ethylene needs its thorough separation from reaction gases and complex purification which hardly can be profitable at low-scale. But there are often can be offered more simple solutions, e.g. ethylene oligomerization to liquid fuels without its preliminary separation from reaction gases [7].

The same is true for other direct processes. The direct oxidation of methane to methanol (DMTM) although unacceptable for big multi-tonnage processes using expensive pipeline gas due to low selectivity and yield has its own vast niche where nowadays there are practically no other concurrent technologies (figure 8) [8, 9].

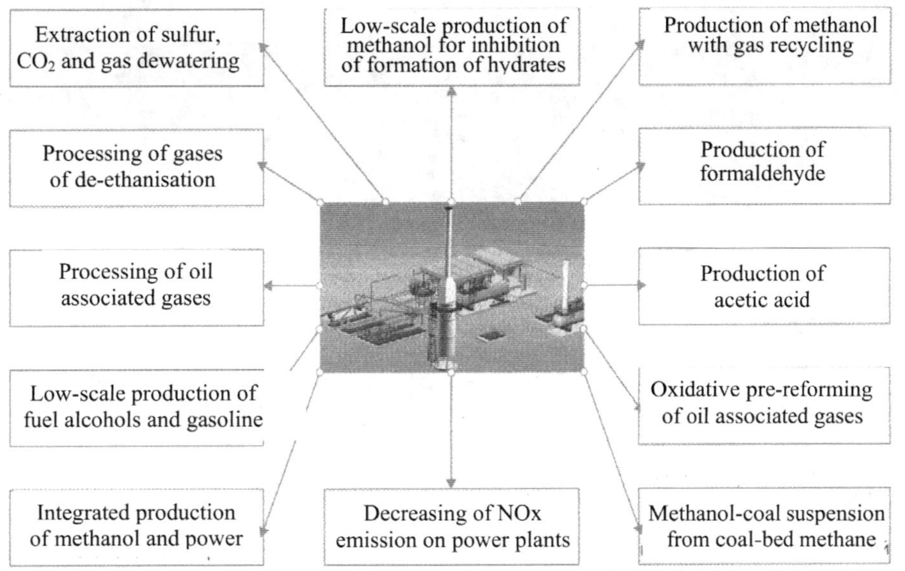

Figure 8. Processes on the base of Direct oxidation of Methane To Methanol (DMTM) [9]

E.g., the integrated production of methanol by DMTM process and power on power plants not only provide the simplest and cheapest way to produce methanol without need in syngas and oxygen but as well lets to significantly reduce and even completely eliminate NOx emission by power plant practically without addition costs [8, 9].

3. Russian specificity

Today Russia severely depends on mineral raw staff export that provides about 20% of its GDP and up to 70% of its foreign currency inflow. Such situation dictates the permanent tendency of further expansion of hydrocarbon export. That is why practically all main Russian projects in gas industry are connected with developing of remote Arctic and offshore fields and building of new long distant exporting pipelines. The total cost of these projects estimates in several hundred billion dollars. Taking into account the real perspective of rapid development of unconventional gas in Europe and very possibly in China, and rapid formation of spot LNG market, it may be foreseen the further decrease of gas prices and the volume of European gas import from Russia. So this policy doesn't seem very reliable. Meanwhile, the chemical processing of Natural Gas tenfold increases its added cost (figure 9) [10], facilitates it transportation to domestic and world markets and provide the diversification of export. It is possible to produce and process only about 10% of modern gas production to provide the same in volume and more reliable currency inflow.

Now the share of Russian petrochemistry in the world is only 2% which doesn't correspond to Russian share in production of hydrocarbons (about 15%). The share of petrochemistry in Russian GDP is less than 2% which is incommensurable to that in other big countries (India -12%, USA - 25%, China – 30%). So it is severe need in rapidly developing of Russian gas chemistry to adjust national economic demands and possibilities with changing gas and petrochemical markets. But this gas chemistry must rely on new technologies appropriate for specific Russian conditions.

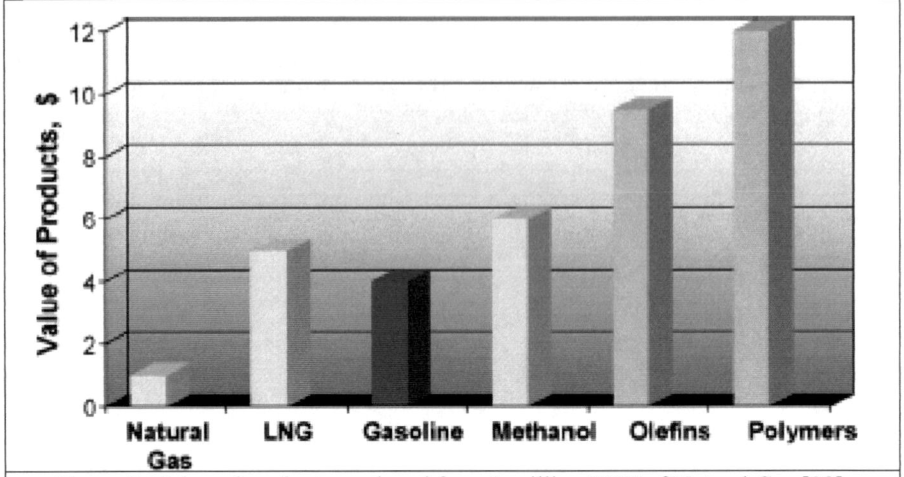

Figure 9. Value of products produced from 1 million BTU of Natural Gas [10].

The interest to low scale alternative gas chemical processes in Russia was initiated about twenty years ago by a severe need in processing of flared oil associated gases from remote oil fields, the need to utilize gas from low debit remote gas fields and necessity to provide remote areas with motor and domestic fuels and necessary chemicals. As a result, Russia has the advanced positions in low scale gas chemistry which gives it a chance to be among the first countries that can reach the industrial level in manufacturing of necessary equipment, and not only satisfy its own requirements but become a leader in this innovative and high-tech market with potential annually capacity of dozens billion dollars.

References

[1] Arutyunov V S 2007 Biofuels: pro et contra *Ross. Khim. Zhurn. (Zhurn. Ross. Khim. ob-va im. D.I. Mendeleeva)* **51** №6 94-99 (in Russian)

[2] Mohr S H and Evans G M 2007 Model proposed for world conventional, unconventional gas *Oil &Gas J.* Dec. 17 46-51

[3] Mohr S H and Evans G M 2010 Shale gas changes N. American gas production projections *Oil &Gas J.* July 26 60-64

[4] Miroshnichenko D A and Kessel I B 2007 In: Procidings of conference "Gas chemistry-2007" VNIIGAS (in Russian)

[5] Arutyunov V S and Krylov O V 1998 *Oxidative Conversion of Methane* (Moscow: Nauka) (in Russian).

[6] Arutyunov V S, Shmelev V M, Lobanov I N and Politenkova G G 2010 A generator of synthesis gas and hydrogen based on a radiation burner *Theor. Found. Chem. Eng.* **44** 20-29

[7] Sinev M Yu, Fattakhova Z T, Lomonosov V I and Gordienko Yu A 2009 Kinetics of oxidative coupling of methane: bridging the gap between comprehension and description *J. Nat. Gas Chem.* **18** 273-287

[8] Arutyunov V S 2004 Recent results on fast flow gas-phase partial oxidation of lower alkanes *J. Nat. Gas Chem.* **13** 10-22

[9] Arutyunov V S 2009 Prospective technologies based on partial oxidation of hydrocarbon gases *Gasokhimiya* №2 (6) 50-61 (in Russian)

[10] Vora B, Chen J Q, Bozzano A, Glover B and Barger P 2009 Various routes to methane utilization - SAPO-34 catalysis offers the best option *Catalysis Today* **141** 77-83

Journal of Physics: Conference Series **291** (2011) 012002 doi:10.1088/1742-6596/291/1/012002

The nanosize catalysts role in the modern hydroprocesses

K N Irisova[1] ,V K Smirnov and E L Talisman
Catachem Company Ltd., 20 Narodnaya st.,Moscow,117152,Russia

E-mail:catachem@mtu-net.ru

Abstract. Introduction of the modern technological procedures operating the catalytic systems with different nanosized characteristics is the only way to fabricate components of commercial oils that meet the current requirements. Specifications to the individual catalysts, which form a catalytic system, differ both in nanostructural features of the support porosity and in distribution of nanosized active site. These specifications are related to the purpose of the process and the role of the catalyst in the process.

1. Introdaction

The efficiency of catalytic technologies in oil processing and oil chemistry is related to the functional properties of the catalysts such as activity, selectivity and stability. The progress in scientific understanding of the catalytic action, the use of advanced technologies and modern devices for preparing the catalysts observed in the last decade are associated with increasing demands to the quality of the products of hydroprocessing. On developing catalytic processes it is desirable not only to use the potentialities of the relevant catalysts but to control the process variables that affect their performance including hydrodynamics (the ideal displacement, the ideal mixing or an intermediate version) and kinetics (the shift of equilibrium by varying the product concentration).

2. Experimental details

In terms of the volume of the catalysts loaded in the industrial reactors of hydroprocessing, hydrodynamic characteristics of the feed stream and the mixture of gaseous products are close to a certain extent to the conditions of ideal displacement. With the aim to approach the conditions of ideal displacement, the methods and facilities are developed for a uniform ("dense") introduction of the catalyst charge in the reactor that makes it possible to divide the downright stream of the feed stream into microstreams with the lowest transverse gradient. At the same period of time the methods and facilities are developed that serve to organize introduction of the feed stream onto the catalyst charge in the piston –bed mode: the trays of different design for the high dispersive distribution[1], "baskets" and "protective" catalysts arranged as a single layer or a number of layers. These constructions facilitate the mixing of feed to form the filter for mechanical admixtures and to trap asphaltenes and metals from heavy and recycled feeds. In some cases both the trays for high dispersive distribution and protective catalysts are used.

[1] To whom any correspondence should be addressed.

Published under licence by IOP Publishing Ltd

Journal of Physics: Conference Series **291** (2011) 012002 doi:10.1088/1742-6596/291/1/012002

However, the operation in the piston-bed mode (ideal displacement) at the inlet of the catalyst bed cannot ensure the maintenance of this mode over the entire catalyst body. When the desired reaction is accompanied by gaseous reaction products (H_2S and NH_3 in our case) gaseous counterflows are formed, liquid flow is mixed and the role of the mixing increases. At the same time product accumulation is increasingly constrained by kinetic limitations due to the decreasing constants of equilibrium $RS+2H_2 \leftrightarrow RH_2+H_2S$. To counterbalance the negative influence of the increasing concentrations of H_2S and NH_3 an additional amount of hydrogen (quench) is introduced in the space between the fixed catalyst beds operated in the reactor. When a deeper desulphurization is desired the hydrogen consumption increases 5-8 times compared to that needed by the reaction chemistry [2,3;figure 1]. It is the hydrogen deficiency that limits the yield of products with the lowest sulfur content at the units currently operated tor hydroprocess crude oil fractions.

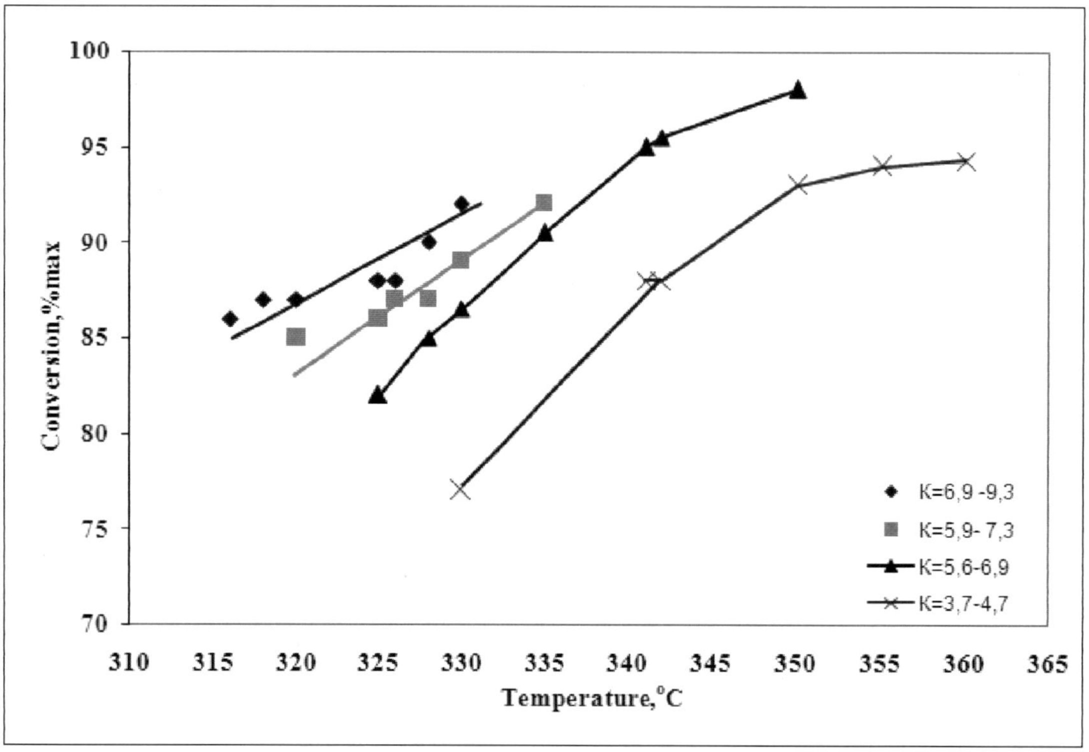

Figure 1. The influence of the process parameters on the conversion of the sulphur in disel fraction with industrial plant on the catalysts RK-242+RK-231Co Company Catachem ltd.
(«K» is ratio of hydrogen expense to chemical consumption)

3.Results and discussions

Catalytic distillation, that becomes increasingly important last time, integrates reaction and separation into liquid and gaseous products in a single vessel (reactor/column) charged with the fixed catalyst bed[4]. In this case, the expense of excessive hydrogen is negligible. However under these conditions the level of hydrodesulfurization is relatively small and the manufacture of the product with the lowest sulfur content requires a second stage of the process in the presence of a very efficient catalyst in excess hydrogen. Note that the excess is related to low sulfur content in the feed processed at the second stage.

Hydroprocessing in reactors with the fixed catalyst bed can also be intensified by mixing the gas–containing hydrogen, the feed and a fraction of the product (isotherming) [5]. In this case, the level of hydrodesulfurization is also relatively small and a second step including a very efficient catalyst in excess hydrogen is needed but to base content of the sulphur in this stage. However total hydrogen expense is lower than in the one-stage process.

Currently, a widespread technique is the control of the process along the height of the reactor using catalyst layers endowed with different functional properties. The layers are separated through the cavities in which the feed stream is mixed in the vicinity of the next layer. Facilities for an additional dosage of hydrogen are also installed. The catalysts are expected to meet severe requirements of high conversions at high flow rates. This can be ensured by combining a high initial activity of the nanosized active sites with their high density on the catalyst surface and the negligible influence of the diffusion that can be achieved by the trend to use the extrudates with the lowest diameter (0.8-1.2mm). The latter condition means severe requirements of the mechanical strength of the extrudates needed to maintain the integrity when subjected to overloads. When this scheme is implemented the hydrogen consumption increases many times compared to that needed by the reaction chemistry.

The transfer to two-stage processes such as catalytic distillation or isotherming considerably changes the requirements to the catalysts operating at different steps. From the catalyst for the first-stage of hydroprocessing (hydrotreatment) the highest level of conversion of sulfur containing compounds like mercaptanes, sulfides and disulfides is expected that would approache the complete conversion. The feed for the second stage includes sulfur containing compounds like thiophene and benzothiophene homologues, as well as nitrogen– containing compounds competing with sulfur containing compounds for the active sites on the catalyst surface. Accordingly, the requirements to the catalysts can be formulated.

At the first stage, the residual sulfur content in the diesel fraction cannot exceed 0.2 weight % at a high space velocity ranging from 4 to 5 h^{-1}. These requirements are as a rule fulfilled by nearly each catalytic system used in one-stage processes free of hydrogen deficiency. When a two-stage operation is realized, the demands to the first-stage catalyst operated in the mode of catalytic distillation are complemented by the need to optimize the pore nanostructure.

4.Conclusions

To the catalyst used at the second step of hydroprocessing of diesel oil, a need to ensure the lowest residual sulfur content (≤10 ppm) at a reasonable space velocity (1-2h^{-1}) is complemented by additional requirements. Among these is either high density of active sites responsible for hydrogenolysis of sulfur -containing and nitrogen -containing compounds or the presence of sites of different nature active in the hydrogenolysis of S- and N- containing compounds. The latter demand can be met either by a nanostructural design of the support or by specific conditions of introducing active components. A control of the combination of active sites on the catalyst surface is also possible through the periodical introduction of a special reagent to suppress excessive active sites.

References

[1] Ленчевский В В, Пинаев А С, Шевченко Г В, Галиев Р Ф 2006 *Нефтепереработка и нефтехимия*, научно-информационный сборник **5** 35-40.
[2] Талисман Е Л , Трофимов О В, Смирнов В К, Ирисова К Н 2008 *Нефтепереработка и нефтехимия, научно-информационный сборник* **11** 13-7.
[3] Huve L G, Creyghton E J, Ouwehand C and al. 2004 EMEA.
[4] Song C 2003 *Catalyst Today* **86** 211-63.
[5] Тайс Э З 2009 *Мир нефтепродуктов.Вестник нефтяных компаний* **4** 6-11.

Investigation of catalytic vacuum black in the electrodes low-temperature electrochemical devices

S.I. Nefedkin [1], N.G. Sedelnikov [2], A.M. Fatyushin [2], O.V.Holichev[1], A.S. Bogomolova [1], I.V. Kiselev [1,3]
[1] Moscow Power Engineering Institute (Technical University), Russia
[2] GINALMAZZOLTO, Moscow, Russia
[3] JSC «Gazprom VNIIGAZ», Moscow, Russia
E-mail: snefedkin@mail.ru

ABSTRACT

Research of catalytic compositions based on noble metal black, received by magnetron deposition of metal in vacuum thickness-control, disclosed. Such blacks have unique bush-like structure formed by thin filaments of platinum with diameter near 10 nm and 200-300 nm long. This structure compared with structure of platinized platinum, received by electrochemical deposition from solution. On basis of electrochemical research disclosed that vacuum platinum blacks (VPB) based on titan, nickel and carbon, provide high activity in anode hydrogen oxidation reaction and cathode hydrogen restoration in low temperature electrochemical devices.

1. INTRODUCTION

Amount of platinum for commercial use of power modules based on hydrogen-air proton exchange membrane fuel cells (PEFC) must be lowered to 0.2 mg/cm^2 and specific power value should be no less than 0.6 W/cm^2 [1]. Similar objectives for minimization of noble metals in electrode stand in catalyst development for other low temperature electrochemical devices that uses platinum group metals. This applies to polymer electrolyte electrolysis units, alkaline hydrogen-air fuel cells. Also using of platinum catalysts in alkaline hydrogen generator electrodes could be reasonable for noticeable lowering of power inputs in water electrolysis.

In connection with this, scientific centers provide large number of researches devoted to development of active electrode with low noble metal content. Different methods are used for catalytic layer formation. Among them such methods as rolling, screen printing, aerosol catalyst deposition, platinum restoration from saline, etc. could be separated. Some publications contain alternative methods such as using of ion beam for deposition of platinum particles on E-TEK paper [2-4]. Using of thin film deposition vacuum technology allows effective preparation of electrode surface, application of intermediate layer and receiving of catalytic compositions that could be controlled by constitution, structure and thickness of noble metal particle layer. It's been showed that aerosolizing method provides effective usage and high platinum activity [5-8]. Method of cathode aerosolizing provides significant lowering of platinum in hydrogen-air PEFC and keeps platinum and iridium catalytic activity at the same level [9-10]. Such iridium deposition technology on anode of polymer membrane electrolyzer allows reaching 10-fold catalyst lowering [11-12].

Described method allows to receive unique nanodispersed structure of vacuum catalytic black (VCB) received by aerosolizing of Pt and PtIr cell. This work assigned task for detailed research of such dispersed structures by methods of structural, elemental, electrochemical analysis and catalysis activity research.

2. RESEARCH METHODS

2.1. Electrodes and catalyst preparation

Application of catalyst on substrate front surface was provided by using of magnetron. Pt cell

Published under licence by IOP Publishing Ltd

Journal of Physics: Conference Series **291** (2011) 012003 doi:10.1088/1742-6596/291/1/012003

was sputtered in vacuum. Developed technology allows controlling structure of deposited layers from quasiamorphous condition to high grade texture. Using of preliminary surface ionic cleaning and deposition of intermediate layers allows receiving adhesion coating durability much higher than film putted by vacuum evaporation or chemical deposition. In addition, method of magnetron deposition has high productivity. Sputtering speed was (1-3)×10^{-4} g/min·cm^2 Coating adhesion durability was provided by high condensing particles energy (5-10 eV). In this research was used specially developed method that allows receiving dispersed catalytic compositions with controlled by constitution and thickness of noble metal blacks. Black matted surface of dispersed structure is characteristic of this method.

For structure and activity of catalytic composition based on plain substrate research, titan foil (100 μm thickness), preliminary treated in 50% solution of H_2SO_4 and temperature 100°C was used. For development of PEFC anode catalyst deposited on the microporous layer of carbon paper Sigracet 10bb (0.42 mm thickness, 84% porosity). Also for expansion of electrochemical reaction area on fuel cell anode, emulsion of Nafion ionomer was sprayed (in air environment). Scheme of gas diffusing electrode is showed on Fig.1.

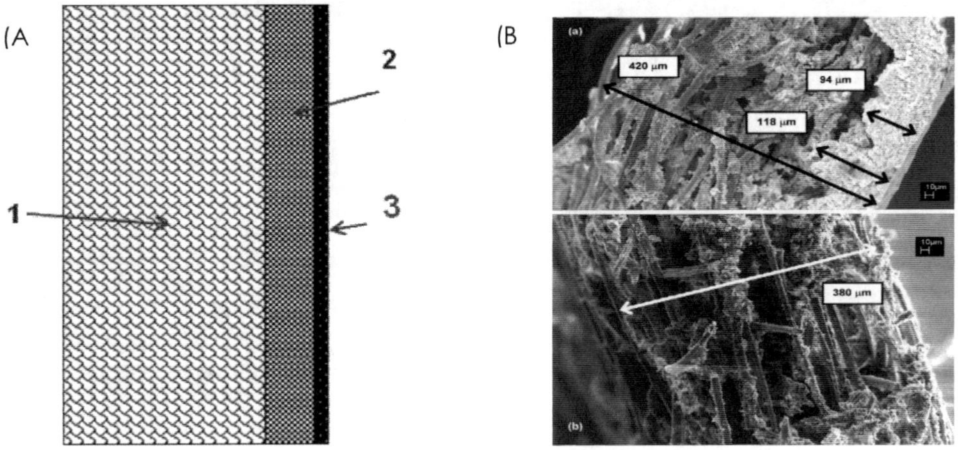

Figure 1.(A) - Gas diffusing electrode; 1 - gas diffusion layer (GDL); 2 - microporous layer (MPL, carbon-black); 3 –catalytic layer ((VPB) Pt black); (B)- Gas diffusing electrode Sigracet 10bb [14]

Catalyst for alkaline electrolysis based on VCB was formed on the front surface of porous nickel electrode (porosity 65%, thickness 1 mm) with biporous structure (large pores with diameter 20-30 μm and small pores with diameter 1-5 μm).

2.2. Structural changes

For research of porous structure of gas diffusing electrode, method of standard porosimetry was used (MSP) [13-15]. Its main advantage over method of mercury intrusion is ability to differentiate hydrophobic and hydrophilic pores [13]. Pore distribution by radius could be received by using of octane as wetting liquid. All pores are hydrophilic relatively octane. Using of water as wetting liquid allows defining hydrophilic pores. Research with MSP method was provided by automotive porosimeter «Porotech» that allows to receive porograms of dispersed samples with pores at range 2 nm - 30 μm [13-14]. Octane ACS (99.99%) and deionized water (>10 Mom) were used as working liquids. Also, scanning electron microscopy was used for receiving pictures of catalytic compositions (SEM TM-1000, Hitachi and SEM JSM-6390LA, Jeol).

2.3. Electrochemical measuring

For catalyst electrochemical research, potentiostat Solartron 1287 was applied. As auxiliary electrode platinized titan foil was used, $Hg/HgSO_4$ used as reference electrode. For measuring in alkaline 6M KOH solutions Hg/HgO was used as reference electrode. Potentials were reduced to reversible hydrogen electrode. Platinum catalytic black active surface was defined by adsorptive hydrogen oxidation currents. For anode activity definition «floating» gas diffusing electrode was used in electrochemical cell, represented on Fig.2.

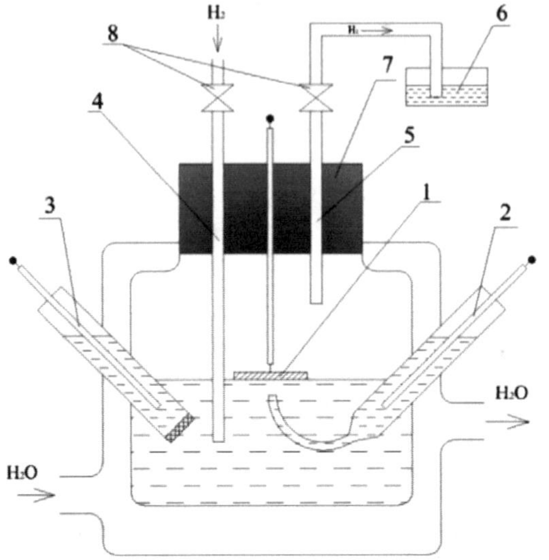

Figure 2. Scheme of electrochemical cell for GDE activity research by «floating electrode» method.: 1 – gas diffusing electrode; 2 – reference electrode; 3 – auxiliary electrode; 4 – hydrogen intake channel ; 5 – hydrogen output channel; 6 – hydroseal; 7 – leakproof cover; 8 –valves.

Electrode (Fig 2, position 1) look like 1 cm^2 round matrix with titan shunt, stuck with carbon-graphite glue from the back side of gas diffusion layer (GDL). At time of experiment electrode was touching electrolyte solution - 0.5 M H_2SO_4. For solution cleaning purposes hydrogen (99.99%) blowing was used. Hydrogen was feeding into cell under solution during measurement. Anode scanning speed (50 mV/s) was using to obtain polarization curves.

2.4. Elemental analysis

For research of catalyst elemental composition, energodispersed analyzer of electron scanning microscope Jeol JSM 6380LA was using.

3. RESEARCH RESULTS AND DISCUSSION

3.1.Structural measurement

Fig.3. shows scaled-up images of vacuum catalytic black based on smooth titan substrate. Visually, catalytic covering has black color. Structure of covering consists of fiber like brunches with thickness near 10 nm, length from 100 to 400 nm. With increasing amount of platinum on the surface of substrate, length of brunches also increase.

Fig.4. shows scaled-up SEM images of initial porous electrode based on carbonyl nickel particles (Fig.4.a) and the same electrode after platinum black deposition (Fig.4.b,c).

Apparently, smooth round (diameter 5-6 μm) carbonyl nickel particles covered with platinum dispersed catalytic coating with structure of thin branches with diameter 10-14 nm and length 200-300 nm. Such formations form the surface like bushes with size of a few hundred nm.

a) b) c)

Figure 3. Scaled-up images of vacuum catalytic black (VCB) based on smooth titan substrate: **a)** - m_s =0.594 mg/cm^2, f=336; **b)** - m_s =1.26 mg/cm^2, f=602; **c)** - m_s =2.0 mg/cm^2, f=1152

Figure 4. Scaled-up images of catalytic platinum black (a, c) on the surface of carbonyl nickel particles (0.36 mg/cm^2), (b) – initial sample.

Journal of Physics: Conference Series **291** (2011) 012003 doi:10.1088/1742-6596/291/1/012003

Fig.5. shows differential porograms of initial porous nickel electrode sample (curve 1) received by MSP method with octane as working liquid and after deposition of 0.36 mg/cm^2 PtIr(20%) black on front surface (curve 2). As seen, there is new nanosized structure with pore size of 10 nm, which forms between densely spaced platinum brunches with diameter 10-12 nm (Fig.5, curve 2)

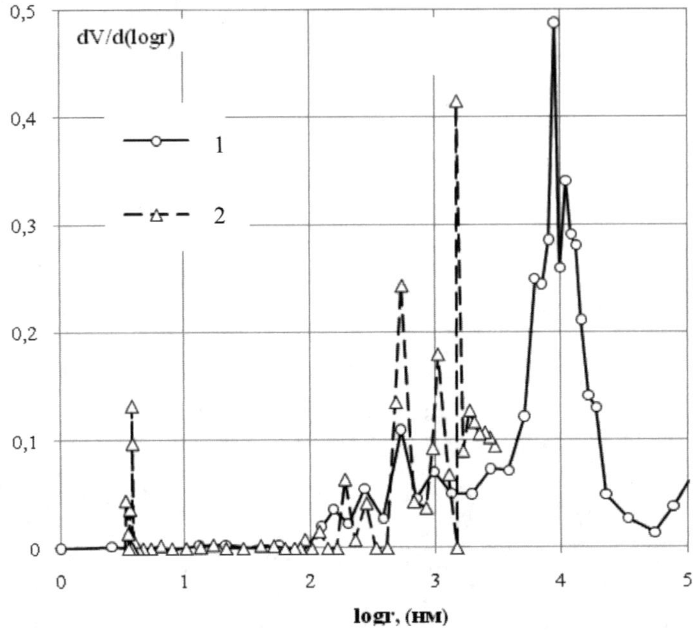

Figure 5. Biporous nickel electrode differential porograms before (1) and after (2) deposition of vacuum platinum black (VPB) on the front surface.

3.2. Electrochemical research.

Method of cyclic voltammetry (CV) was used in this research for definition of electrode active surface with vacuum catalytic black based on smooth titan substrate. Fig. 6. Shows CV-curves for dense thin covering on titan (1) and for platinum black-based electrodes with different amount of platinum (2-4). Apparently, CV-curves look like platinum curves.

Active catalyst surface based on platinum black significantly higher than smooth Pt/Ti electrode (1).

Definition of active surface (S_e), rough factor (f) and specific surface (m_s) based on CV-curves.

Platinum black active surface (S_e) calculated from electric charge value, spent for adsorbed hydrogen oxidation in the region of potentials between E_1 =0, E_2= -0.4 V relatively hydrogen electrode [15].

$$Q_e = \frac{1}{w}\int_{E_1}^{E_2} I \cdot dE \qquad (1)$$

where:

I - current instantaneous value, mA; w- electrode potential changing speed, mV/s

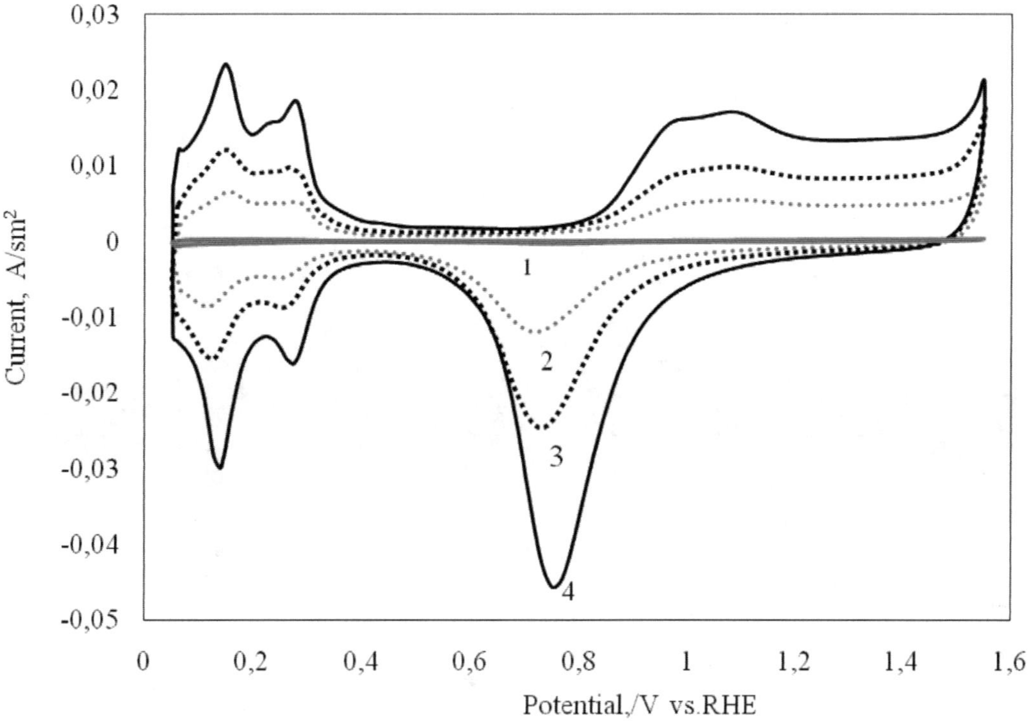

Figure 6. Cyclic I-E curves for Pt based on smooth titan substrate in 0.5 M H_2SO_4 solution, speed w=0.05 V/s and t=25°C: 1- Pt dense coating (m_s =0.12 mg/cm^2 , f=8.3); 2 - Pt black (m_s =0.594 mg/cm^2, f=336); 3- Pt black (m_s=1,26 mg/cm^2, f=602); 4- Pt black (m_s =2.0 mg/cm^2, f=1152)

Active electrode area evaluated with equation:

$$S_Э = \frac{Q_э}{Q_s} \qquad (2)$$

where Q_s– electricity amount during hydrogen monolayer oxidation on smooth platinum electrode with area 1 cm^2. Value Q_s = 210 μC cm^{-2} for platinum could be evaluated by equation (3) with assumption that any atom in crystal grid adsorbs one atom of hydrogen, which then oxidized with transfer of one electron:

$$Q_s = n_{Pt} \cdot e \qquad (3)$$

где e = 1.6 10 $^{-19}$ C - electron charge; n_{Pt} = 1,31·10^{15}·1.6 10 $^{-19}$ - amount of Pt atoms on 1 cm^2 smooth Pt area.

Value of catalyst active area defined as:

$$S_m = S_э / m_s \qquad (4)$$

Electrode rough factor was found by equation:

$$R_f = S_э / S \qquad (5)$$

where

Journal of Physics: Conference Series **291** (2011) 012003 doi:10.1088/1742-6596/291/1/012003

S - electrode geometrical surface

For dispersed vacuum catalytic black particles based on titan, dependence of R_f on time of catalyst deposition (or m_s value) is close to linear.

Table 1 shows parameters for dispersed sediments of vacuum catalytic Pt black on titan with different thickness.

Table 1 Parameters for dispersed sediments of vacuum catalytic Pt black on titan with different thickness.

m_s , mg/cm^2	0.594	1.26	2.0
Rf, sm^2/cm^2	336	602	1152
S_m , m^2/g	56.5	47.7	57.6

Comparison of vacuum catalytic black (on smooth titan substrate) with parameters of catalytic particles of platinized Pt [16] and Pt nanoparticles on carbon-black [17] is of big interest.

Besides S_e f and m_s there are other parameters for comparison: form of catalyst particles, relation h_2/h_1 which characterize texture of catalyst surface. Here:

h_1, h_2 - height of current desorption peak, solidly and weakly bound hydrogen respectively (Fig.7).

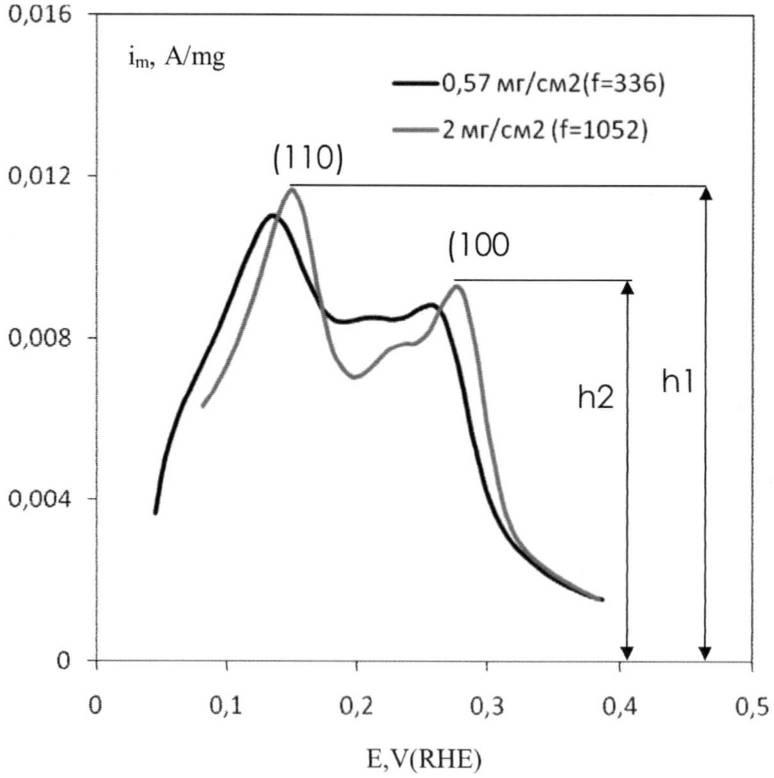

Figure 7. Hydrogen desorption peaks on cyclic volt-ampere characteristics of catalytic Pt black on titan substrate in 0.5 M H_2SO_4, V=0.05 V/s and T=25°C:

Journal of Physics: Conference Series **291** (2011) 012003 doi:10.1088/1742-6596/291/1/012003

As shown on Fig.7, specific hydrogen desorption currents i_m and its relation to potential in minimal degree depend on thickness of catalytic Pt black layer. It tells about equally accessible surface of Pt. Parameter h_2/h_l in this case equals 0.8 ± 0.02, and this is characteristic of samples with developed active surface.

It's known that higher values of h_2/h_l are indirect evidence of texture (111), and shorter textures (100) [16]

Table 2 shows basic characteristics of Pt catalyst.

Table 2

Sample	h_2/h_l	particle shape	D, nm	Rf, cm^2/cm^2	Sm, m^2/g
Pt sheet [16]	0.57	-	-	5-8	
Pt /Pt freshly prepared [16]	0.82	sphere	9-12	400-1400	20-35
Pt /Pt aged sample [16]	0.7	sphere	15-20	100-350	10-15
20 wt.% Pt/C Vulcan XC-72 [17]	0.9	sphere	2-3	-	65-73
Pt vacuum catalyst black (VCB)/Ti(sheet)	0.80	branch coral bush	d=12-14 L=100-300	300-1200	47-58

Petriy O.A. et al. explored structural characteristics of platinized Pt [16]. Particularly, dependence of Pt particles size on deposition potential in H_2PtCl_4 solution. Shown that linear dependence of rough factor from deposition time exists. Work shows that particles structure of Pt/Pt has normal distribution by size with maximum at 7 nm area for freshly made samples and near 10 nm area for aged samples. If rough factor R_f (defined by hydrogen desorption) for ultra thin freshly made particles was 200 and linearly raised to 1600 (near 10 times) depending on time, then for aged particles grows was only 3 times (from 100 to 300). Comparison of h2/h1 parameter is thing of interest (relation between current peek height and weekly bound hydrogen). For freshly made samples its value was 0.82 and for aged ones decreased to 0.70, that is became equal to h2/h1 for smooth Pt. In case of vacuum blacks on GDL (Pt/C) relation h_2/h_l was 0.79 for thick Pt films and 0.78 for thin (precisely calculated).

Pt nanoparticles, formed on carbon-black, have size near 2.6 nm, specific surface near 70 m^2/g and have high activity relatively hydrogen anode oxidation reaction [17].

1.2. Electrochemical activity research.

Anodic hydrogen oxidation

In contact of «floating» gas diffusing electrode (Sigracet 10bb) with deposited PtIr catalyst (0,36 mg/sm^2) with 1 M H_2SO_4 solution in three-electrode electrochemical cell (Fig.2.), three-electrode border appears on which hydrogen oxidation anodic reaction in acid electrolyte solution with hydrogen back feeding appears. This reaction speed defines by electrochemical reaction three-phase areas length, catalyst activity, temperature and transport limitations of reagent supply in reaction area.

Fig.8. shows polarization curves of hydrogen anode oxidation on «floating» gas diffusing electrode (Sigracet 10bb) without catalyst (1) and with PtIr catalyst (2) in different temperatures. Commercial GDL characteristics with 2.1 mgPt/cm^2 are presented for comparison purpose (4).

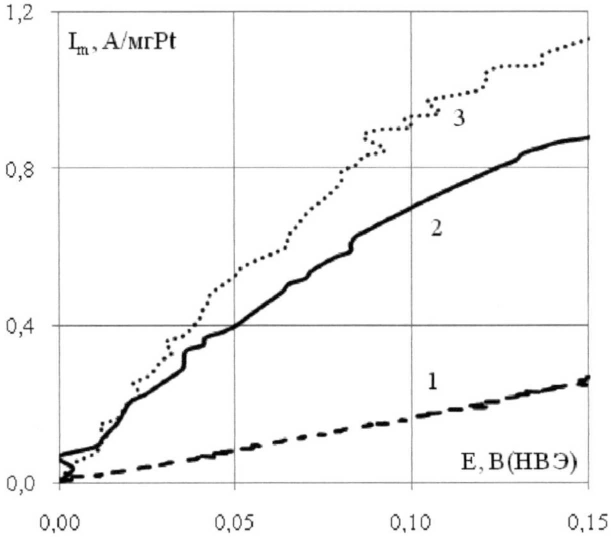

Figure 8. Anodic hydrogen oxidation on «floating» electrode curves: 1- GDL with commercial catalyst (Pt-2.1 mg/cm^2); 2- GDL (Sigracet 10bb) with Pt-Ir (10%) vacuum catalytic black (0.36 mg/cm^2) t=30°C; 3 - ГДС (Sigracet 10bb) with Pt-Ir (10%) vacuum catalytic black (0.36 mg/cm^2) t=70°C; Potential scanning speed is 50 mV/s; electrolyte 0.5 M H_2SO_4

Fig.8. shows that specific activity of researched catalyst reduced to Pt mass unit appeared higher than for commercial electrode. At polarization level 0.5 V it has value 1 A/mg Pt.

Hydrogen cathode recovery in alkaline media

High vacuum catalytic black activity on porous electrode found for reaction of hydrogen cathode recovery in alkaline solution. It's an important thing which lowering energy inputs in Fixed Alkaline Electrolyser (FAE) [18]. Fig.9. shows polarization characteristic for biporous nickel electrode with VPB catalyst (m_s=0.594 mg/cm^2).

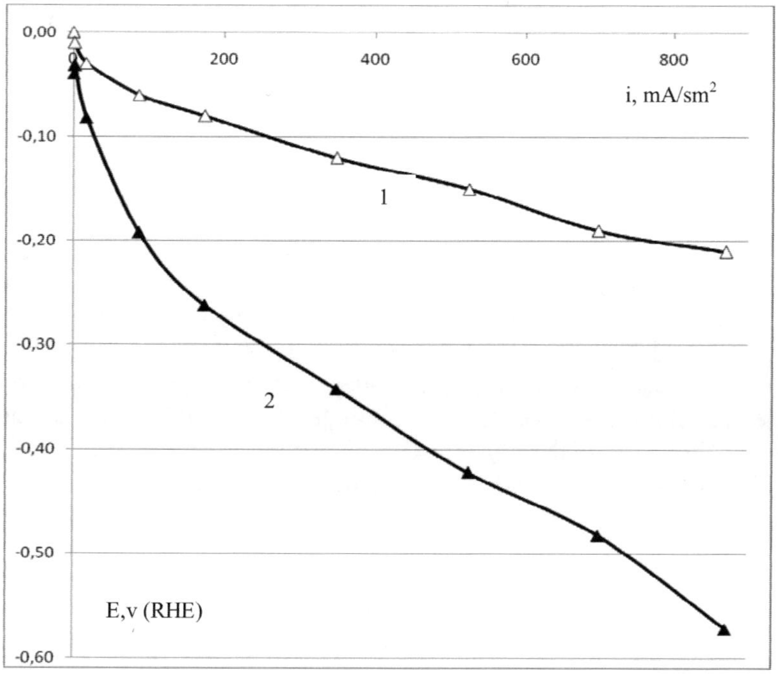

Figure 9. Polarization curves of cathode hydrogen recovery on VPB catalyst (0.36 mg/cm^2), deposited on front nickel electrode surface (1). Without catalyst (2) t=80 °C; 6M KOH electrolyte.

CONCLUSION

Thus, results of hydrogen-air fuel cell anode characteristic research showed that vacuum technologies allow receiving controlled by thickness catalytic compositions like noble metal blacks on GDL's microporous layer surface. Electrodes for low temperature electrochemical devices provide high activity. For example, in hydrogen oxidation anode reaction and cathode hydrogen recovery.

ACKNOWLEDGEMENTS

Financial support of this work has been conducted by Government contract with Russian Federal Agency of Science and Innovations from 18 June 2008. № 02.552.12.7046.

REFERENCES

[1] TIAX LLC Cost Analyses of Fuel Cell Stacks/System/ DE-FC 02-99EE50587/ Hydrogen and Fuel Cells Merit Review Meetig, Berkly 19-22 may 2003
[2] Saha, M.S., Gulla, A.F., Allen, R.J., Mukerjee, S., Electrochim. Acta 51 (2006) 4680–4692.
[3] Gulla, A.F., Saha, M.S., Allen, R.J., Mukerjee, S. Electrochem. Soc. **153** (2006) 366–371.
[4] Gulla, A.F., Saha, M.S., Allen, R.J., Mukerjee, S. Electrochem. Solid State Lett. **8** (2005)504–508.
[5] Ozturk, O., Ozdemir, O. I., Ulusoy, Ahsen, A.S., Slavcheva, E. Effect of Ti sublayer on the ORR catalytic efficiency of dc magnetron sputtered thin Pt films. Int. Journal of Hydrogen Energy **35**(2010) 4466-4473

[6] Hirano, S., Kim J., Srinivasan, S. High performance proton exchange membrane fuel cells with sputter-deposited Pt layer electrodes, Electrochim Acta **42** (1997) 1587–1593

[7] Radev, I., Slavcheva, E, Budevski, E. Investigation of nanostructured platinum based membrane electrode assemblies in "EasyTest" cell, Int J Hydrogen Energy **32** (2007) 872–877.

[8] O'Hayre, R., Lee, S.-J., Cha S.-W., Prinz, F.B. A sharp peak in the performance of sputtered platinum fuel cells at ultra-low platinum loading, J Power Sources **109** (2002) 483–493.

[9] Mukerjee, S., Srinivasan S., Appleby, A.J. Effect of sputtered film of platinum on low platinum loading electrodes on electrode-kinetics of oxygen reduction in proton-exchange membrane fuel-cells, Electrochim Acta **38** (1993) 1661–1669

[10] Ticianelli, E.A. Beery J.G., Srinivasan, S. Dependence of performance of solid polymer electrolyte fuel-cells with low platinum loading on morphological-characteristics of the electrodes, J Appl Electrochem **21** (1991) 597–605.

[11] Nefedkin S.I., Gladkikh, I.P., Mansurov, G.N., Korovin, N.V., Petrii, O.A. Electrochemical properties of thin film Ir / Ti electrodes. J. Electrochemistry **24 (**1988) 397-400.

[12] Labou, D., Slavcheva, E., Schnakenberg U., Neophytides, S. Performance of laboratory polymer electrolyte membrane hydrogen generator with sputtered iridium oxide anode, J. Power Sources **185** (2008) 1073-1078

[13] Volfkovich, Y.M., Bagotzky, V.S., Sosenkin, V.E., Blinov, I.A. The standard contact porosimetry, Colloids Surf. A 187-188 (2001) 349-365.

[14] Gostick, J. T., Fowler, M. W., Ioannidis, M. A., Pritzker, M. D., Volfkovich, Y.M., Sakars ,A. Capillary pressure and hydrophilic porosity in gas diffusion layers for polymer electrolyte fuel cells J. Power Sources **156** (2006) 375-387

[15] Nefedkin S.I. Physicochemical methods of research in hydrogen energy technologies. Moscow. MPEI, 2008, P.207.

[16] Petriy O.A., Tsirlina G.A., Pronkin S. N., Spiridonov F.M., Hruscheva M.L. Platinized platinum: Dependence of texture and particles size on preparation conditions. J. Electrochemistry **35** (1999) 12-22

[17] Esparbe, I., Brillas, E., Centellas, F., Garrido, J.A., Rodriguez, R.M., Arias, C., Cabot, P-L. Structure and electrocatalytic performance of carbon-supported platinum nanoparticles, J. of Power Sources **190** (2009) 201–209

[18] Vermeiren, Ph., Moreels, J.P., Claes, A., Beckers, H. Electrode diaphragm electrode assembly for alkaline water electrolysers. Int. Journal of Hydrogen Energy **34**(2009) 9305-9315

The sorption of influenza viruses and antibiotics on carbon nanotubes and polyaniline nanocomposites

V T Ivanova [1], G S Katrukha [2], A V Timofeeva[3], M V Ilyna [1], Y E Kurochkina [1], L A Baratova[3], I Yu Sapurina [4] and V F Ivanov [5]

[1] D.I. Ivanovsky Research Institute of Virology RAMS, Gamaleya st, 16, Moscow 123098, Russia; [2]G.F.Gause Institute of New Antibiotics RAMS, Moscow 119021, Russia [3]A.N. Belozersky Research Institute for Physico-Chemical Biology, M.V.Lomonosov Moscow State University, Moscow 119991, Russia; [4] Institute of Macromolecular Compounds RAS, 199004, St. Petersburgr. Bolshoy Pr.31, Russia.
[5]A.N. Frumkin Institute of Physical Chemistry and Electrochemistry, RAS, Leninsky prospect, 31, Moscow 119991, Russia.

Key words polyanilines, carbon nanotubes, influenza viruses, antibiotics-polypeptides
E-mail: **valivanova1946@mail.ru**

Abstract

The decontamination of the solutions from micropatogens and drug delivery are the important problems of modern life. It was shown that carbon nanotubes, polyaniline and their composites can interact with antibiotics-polypeptides and some viruses (pandemic strain of influenza viruses A(H1N1)v circulated in Russia in 2009-2010. During a short time drug and viruses can be absorbed by polyaniline and removed from aqueous solutions at the normal conditions. Polyaniline composites can be useful for the preparation of drug delivery and virus control filters and also in biotechnology for the improvement the methods of antibiotics purification.

Introduction

Drugs and especially antibiotics became the new type of environment pollution. The development drug delivery system is an actual problem for medical centers, hospitals, in cattle breeding and poultry farming. The modern drug delivery systems based on conducting polymers: polyaniline and polypyrrole [1]. These polymers were non-cytotoxicity, biocompatible and began to use in biomedical applications. Now polyaniline has been demonstrated to be the most important conducting polymer due to its good environmental stability low cost, ease of processibility, oxidation-protonation adjustable electrical and optical properties [2]. Due to electronic and electrochemical properties polyaniline is highly sensitive to molecular interactions and can provide signal transduction function for molecular detections. It can release the drug with the help of acid-base interaction or its reduction-oxidation activity. The possibility to use an electrical stimulation made this delivery system long-time functional and reversible. It was shown recently that polyaniline specifically interact with influenza viruses and surface HBs-antigen of hepatitis B [3, 4]. Such infection as poliomyelitis, Hepatitis A, influenza caused by pandemic strain A(H1N1)v, influenza bird viruses and others infections can be spread by the water way. The important task is to create the anti-viruses filters for prevention infections spreading.
This study continued the serial investigations devoted to the interactions between modern nanomaterials and viruses. The aim of this work manufacturing the antibiotic delivery system and virus control system based on polyaniline composite materials.

Experimental

Preparation of polyaniline and its composites

The chemical synthesis of polyaniline (Fig.1) was performed in the usual manner throw the oxidation of 0.1 M aniline by the 0.1 M ammonium persulphate in 0.5 M aqueous solution of HCl at room temperature. The formed polyaniline salt after precipitation was converted to the polyaniline base after the treating polyaniline salt by the 0.5 M aqueous solution of NH_3. After precipitation and washing this base was used as sorbent in our experiments. The polyaniline interpolymer complexes with polypropylamidopropanesulfonic acid (PAMPSA) was obtained in the same manner [5]. The sorbents – poyaniline base and composites were obtained in A.N. Frumkin Institute of Physical Chemistry and Electrochemistry, RAS.

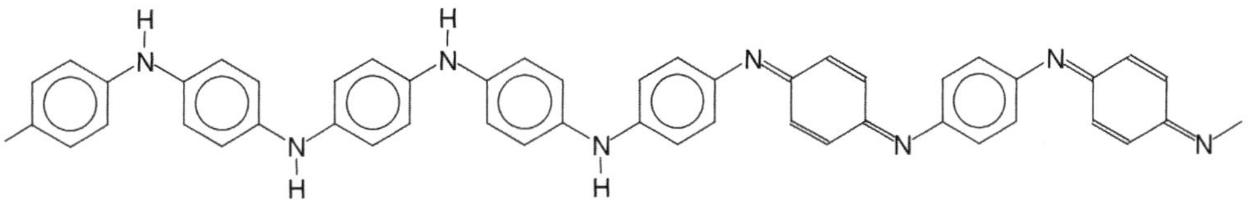

Figure 1. The structure of polyaniline

Preparation of Multiwalled carbon nanotubes
Multiwalled carbon nanotubes (MWCNT) were firstly described by Iijima [6].They manufactured by the catalytic chemical vapor deposition method (Nanotechcenter, Tambov, Russia). MWCNT consisted of particles 40-80 nm in diameter and up to 10 μm in length (Fig. 2). The specific conductivity of MWCNT is 40 S cm^{-1}. Specific surface area of the material is average 60-100 m^2/g, porosity more then 70%. The pore distribution is very narrow, more then 90% of pores belong to diapason 20-80 nm. Carbonaceous material was purified with mixture of concentrated acids and the residual catalyst content was equal to about 1% wt.

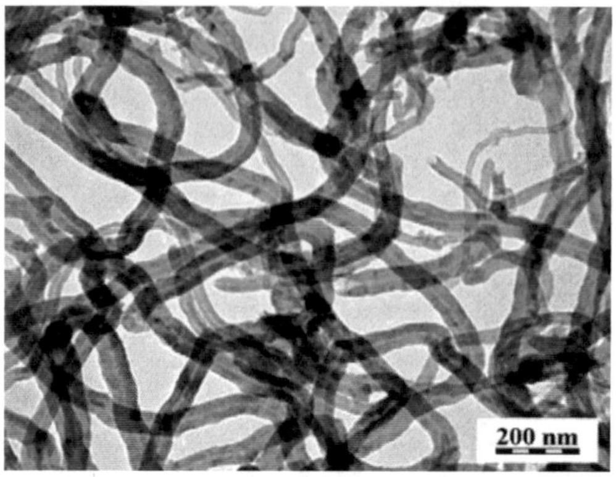

Figure 2. Multiwalled carbon nanotubes (MWCNT). Transmission electron microscopy.

MWCNT were coated with the polyaniline layer (Fig. 3) according [7]. This composite material was prepared by the method of *in-situ* aniline polymerization in the presence of MWCNT. Polyaniline content is 30% wt. The conductivity of composite material 10 S cm^{-1} independent of polyaniline doping level, specific surface area and porosity 50 m^2/g and 60% corresponding. The (MWCNT) covered polyaniline base were obtained in Institute of Macromolecular Compounds RAS.

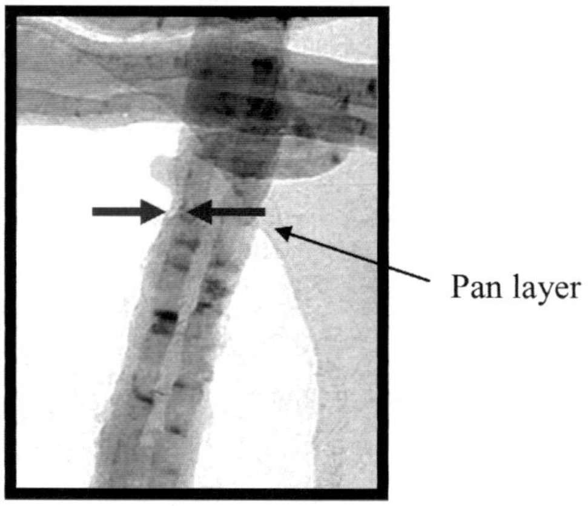

Figure 3. MWCNT covered with the polyaniline layer. (Transmission electron microscopy.)

Virus cultivation
The pandemic in Russia in 2009-2010 was caused by the active circulation of the variants of first influenza pandemic strain A/IIV-Moscow/01/2009 (H1N1)swl [8]. This virus was isolated in Institute of Virology RAMS. It was similar reference strain A/California/7/2009 (H1N1)v .The viruses were grown in 9-10 embrionated eggs strains were studied. The hemagglutination test with human erythrocytes was used to determined viruses in solutions.

Antibiotic cultivation and registration
Antibiotics-polypeptides were obtained in the Gauze Research Institute for the Search of Novel Antibiotics (Tab.1).

Table 1. Antibiotics-polypeptide used

Name	General chemical characteristics	UV spectrum, λmax, nm (solvent)	Molecular weight, D
Gramicidin S	Cyclopeptide	219 (MeOH)	1141
Teicoplanin A2	Cyclic lipoglycopeptide	280 (0,01 н. HCl)	1875
Bleomycetin	Heteropolypeptide	280 (H_2O)	1438
Polymyxin B	Peptide-cyclopeptide	214 (H_2O)	1202

RP HPLC analysis of the antibiotics in solution before and after sorption was performed on a Milichrom A-02 microcolumn liquid chromatograph (ZAO Econova, Novosibirsk, Russia) with stainless steel columns of the dimensions 2.0 x 75.0 mm packed with the Nucleosil 100-5C18 PAH sorbent from Macherey-Nagel (Germany). The columns were thermostated at 35°C. Antibiotic solutions in water with concentrations ranging between 0.2 mg/ml were used for the analysis. The volume of the probe injected was 15 µl. Two detection wavelengths λ 214 and 280 nm were used for the analysis. The analysis time ranged between 20 and 25 minutes for all the antibiotic preparations analyzed. Elution with linear acetonitrile gradient from 0 (eluent A, 0.1% solution of TFA in water) to 100% (eluent B, 0.1% solution of TFA in acetonitrile) at the flow rate of 100 µl/min was used.
The reagents were dissolved in deionized water purified by a Milli-Q® Plus device from Millipore (France); acetonitrile used for HPLC was from Sigma-Aldrich (Germany).

Antibiotics static sorption was studied according to the following procedure: 600 µl of the antibiotic solution with the concentration 0.2 mg/ ml was added to a portion of the sorbent (1.25, 2.5 or 5.0 mg). The resulting suspension was incubated for 15 min., 1.0, or 18 hours at 18–20°C. After sorption, the antibiotic solution was separated from the sorbent by centrifugation for 5–7 min at 5600–8850 g in a Beckman Coulter™ Microfuge® Centrifuge (United States).

Antibiotic amount determination in the starting solutions and the supernatants was performed using RP HPLC on a Milichrom A-02 liquid chromatograph as described above. Quantitative analysis of chromatographic data was performed using the program MultiChrom-SPEKTR for Windows 9. & NT, version 1.5x_E (Ampersend, Russia).

Interaction of viruses with sorbents
The method of investigation virus-sorbent interaction was developed by Ivanova et al [3]. It includes some stages: intensive contact virus + sorbent on shaker during different time diapazones (15min-1 h) in range of temperature (8-25°C), low speed centrifugation (2000 circle /min), the determination virus in solutions before and in supernatant solutions. In the case of influenza viruses was used hemagglutination test with human erythrocytes. The sediment (virus+ sorbent) after centrifugation represents the immunosorbent. The experiments were carried out with the follow solutions: H_20 after previously special treatment by heating to destroy other pathogens, 0.15 M NaCl in H_20, 0.1 M Tris-HCl buffer (pH 7.2), allantoic liquid of chicken embryons.

Results and discussion
The first biological object was the influenza virus–pandemic strain A/IIV Moscow/01/09, (H1N1)v antigenic related to influenza swine virus. Such viruses caused the pandemic in the world in 2009-2010 according the opining of World Health Organization experts and were widely spread in Russia in autumn 2009. These viruses had differences in protein structures compared with these properties of annual epidemic strains. It was established that adsorption activities of influenza strain A(H1N1)v varied and depended of the sorbent, The sorption of viruses in solutions were intensive with Pan sorbents, less with carbon nanotubes covered Pan and the least with nanotubes without Pan Fig.4. The hemagglutinin titers after sorption were in 64, 16, 8 и 4 times less then initial titers for these sorbents accordingly. All these experiments were carried out at T=25°C. So the result of the previous and current data directs

that influenza viruses A with hemagglutin A/H1,H3,H5,H7 isolated from human, birds and B viruses, can be immobilized at the sorbents. The sorption depends of sorbent structure. The most active sorption was observed with Pan sorbents.

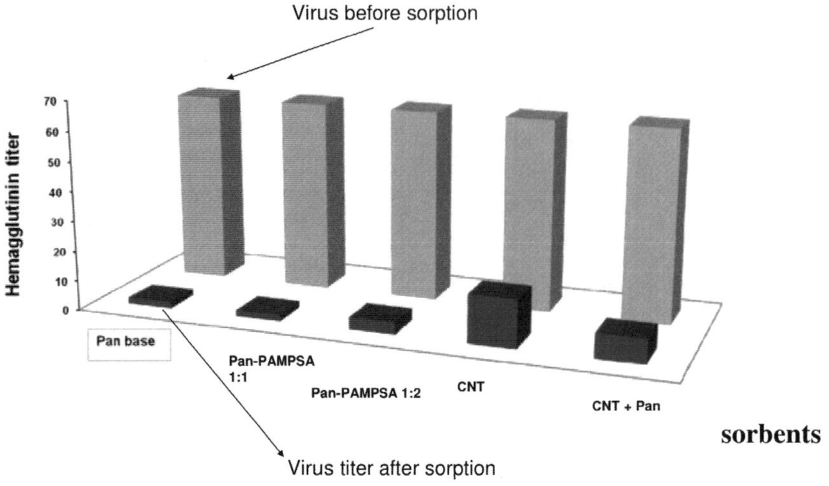

Figure 4. Sorption of influenza virus pandemic strain A/IIV Moscow/01/2009 (H1N1)swl on different sorbents from solutions.

The next objects for investigation were some practically important antibiotics-polypeptides: bleomycetin, gramicidin S, polymyxin B и teicoplanin A2. The influence of contact time (15 min, 1 h, 18 h) for the sorption on CNT was investigated in experiments at T=22°C (Fig.5).

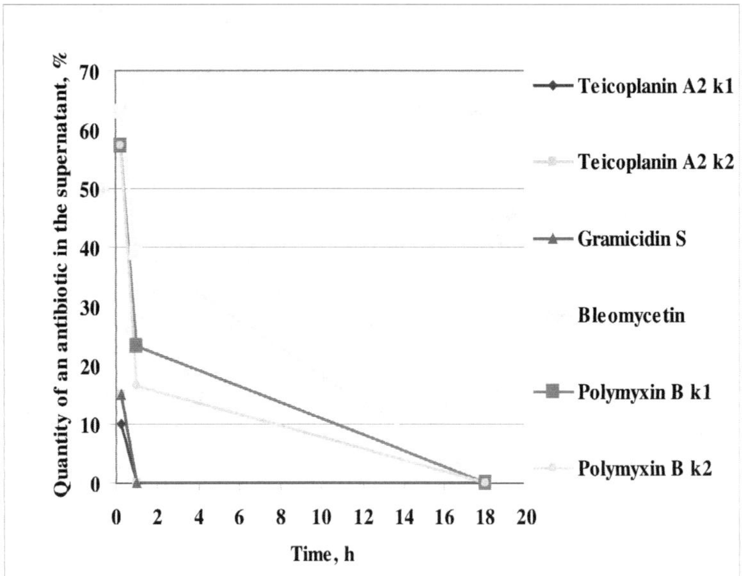

Figure 5. Sorption of antibiotics on CNT depending on contact time

It was shown that antibiotics (the concentration C=0,2 mg/ml) gramicidin S и teicoplanin A2 removed from water solutions during 1 hour, polymyxin B and и bleomycetin - 18 hours (Fig.5). The dependence of sorbent mass of CNT (1.25, 2.5, 5 mg) for the sorption was analyzed too at T=22°C, (Fig.6). It was established that the optimum sorbent mass is 5 mg, when all antibiotics (C= 0,2 mg/ml) adsorbed. The gramicidin S и teicoplanin A2 removed from water solutions during 1 hour, polymyxin B and bleomycetin - 18 hours.

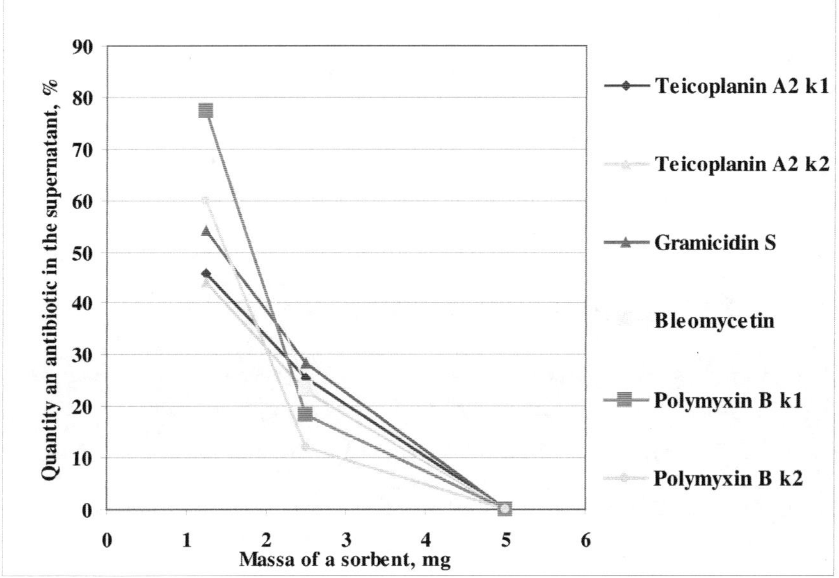

Figure 6. Sorption of antibiotics on CNT depending on sorbent mass

After the analysis of these data we make the conclusions that the relationship to the sorbent is determined by the structure of antibiotic. The hydrophobic antibiotics–- gramicitin S и teicoplanin A2 absorb more completely at the same condition, than hydrophilic positive charge antibiotics polymyxin B and bleomycetin.
Early we established that the antibiotics, gramicidin S и teicoplanin A2 removed from solution on sorbent CNT + Pan at 97-100% during 1h, in the case of Pan-base at 94-100% during 18 hours [9]. The data direct that affectivities of these antibiotics as an absorbent depends on hydrophobic properties of Pan-base, CNT+ Pan. The sorbent CNT+ Pan has more hydrophobic properties, then Pan-base, because when at the all parameters of the sorption hydrophobic antibiotics gramicidin S и teicoplanin A2 interacted with CNT+ Pan more intensive (1 hour). It is known that CNT can absorb gas (dioxin), liquid and dissolvable substances [10]. The interaction of CNT, Pan-base and composites prepared on their base with influenza viruses can be used for the creation of the effective filter for water decontamination from micro pathogens.

Conclusion

The analysis of the data obtained evidences that the elaborated sorbents (CNT, CNT+ Pan, Pan-base) are perspective for utilization as water filter for decontamination micropathogens from different solutions and in biotechnology for isolation and purification of antibiotics-polypeptides from culture liquids. It is established that antibiotics polypeptides – teicoplanin A2, gramicidin S, polymyxin B and bleomycetin are able to absorb by MWCNT most intensive.
It was shown that antibiotics sorption depends on as structure and properties of the antibiotics as on hydrophobic properties of sorbents.

Acknowledgments
The research was partially supported by ISTC grants 3070 and 3718.

References

[1] Ge D, Qi R., Mu J, Ru X, Hong S, Ji S, Linkov V, Shi W. 2010, *Electrochem. Commun.* 12 1087
[2] Liu S, Wang J, Zhang D, Zhang P, Ou J, Liu B, Yang S, 2010 *Appl Surf Sci* 256 3427
[3] Ivanova V T, Ivanov V F, Kurochkina Y E, Gribkova O L , Ilyina MV, Manykin A A 2009 *Problems of virology* 3 21
[4] Li X-N, Dai L, Liu Y,. Chen X-J, Yan W, Jiang L-P, Zhu J-J , 2009 *Adv Funct Mater.*19 3120
[5] Ivanov V F, Gribkova O L, Cheberjako KV, Nekrasov A A , Tverskoy VA, Vannikov A V 2004 *Rus. J. Electrochemistry*, 40 3. 339
[6 Iijima S 1991 *Nature. London* 354 56
[7] Sapurina I.Yu, Kompan M E , Zabrodskii A G, Stejskal J , Trchova M , 2007 *Rus J. Electrochemistry.* 43 5 528
[8] Lvov D K, Burtseva E I, Prilipov A G et al 2009 *Problems of virology* 5 10
[9] Ivanova V T , Sapurina I Yu, Ivanov V F , et al. 2009 *Rusnanoforum*, Moscow, 6-8 October 2009 , 274
[10] Dyakov P N 2006 *Carbon nanotubes. Structure, properties, application*. Moscow, 5 120.

Journal of Physics: Conference Series **291** (2011) 012005 doi:10.1088/1742-6596/291/1/012005

Modeling of self-organization of two-dimensional ordered structures

V V Egorov[1,2], Y P Garmay [1,2], A A Shaldzhyan[1], D V Lebedev[2], A V Vasin[1], N A Grudinina[3] and O I Kiselev[1]

[1] Research Institute of Influenza of the Ministry of Health and Social Development of the Russian Federation, Prof. Popova st. 15/17, St-Petersburg, Russia

[2] Department of Molecular and Radiation Biophysics Petersburg Nuclear Physics Institute of the Russian Academy of Science, Orlova Roscha, Gatchina, Leningrad Region, Russia

[3] Institute of Experimental Medicine, North-Western Branch of the Russian Academy of Medical Science , 12, Akademika Pavlova st., St-Petersburg,, Russia

E-mail: toizeg@gmail.com

Abstract. The problem of the search of biostructures capable to self-organization is quite urgent considering the prospects of application of nanostructured biomaterials as components of composite materials in transplantology and optics as well as "scaffolds" for the synthesis of nanostructured materials based on inorganic particles. The given study focuses on modeling of the growth of structures using the cellular automata with a set of states of the two values (0 and 1), with the value corresponding to the state is determined by the contribution of "the closest neighbor" (by the probability of induction of the state of the nextgeneration in the direction of the interaction) and the geometry of the field isdetermined by the vector of the direction of the particle and the direction of the interaction.

1. Introduction.

Materials structured "from bottom to top", from nanometer scale to millimeter one, are perspective to be used in different spheres of human activities – from creation of optically active media to development of extra strong materials. The idea of creation of the chain of machines capable of assembling materials from separate molecules ("billion tiny factories") which was put out in a lecture by Richard Feynman [1] has not been realized yet. Self-assembling processes are determined by a set of interactions between the elements of the system. To describe the process of self-organization the methods of molecular mechanics and molecular dynamics are commonly used [2]. Previously, we conducted experiments on modeling of self-organization of peptides using sequential molecular docking [3]. The applied methods, though they let us get the results consistent with the experimental data, are quite resource-intensive and do not always allow to quickly make the qualitative description of the system. For rapid qualitative assessment of the ability of the particles to self-organization modeling using cellular automata is possible. For example, work [4] describes the model of the dynamics of oligomerization of prion proteins basing on the theory of cellular automata. Modeling using cellular automata is less resource-intensive in comparison with molecular dynamic modeling. In contrast with the model based on molecular dynamics, this system can show only the equilibrium or quasi-equilibrium state of the system. The approximation of the game "Life" created by Conway and described by Martin Gardner [5] makes it possible to model only symmetric objects with a very limited range of valences determined by the geometry of a discrete cell of the game field. The probability of the presence of the object in a given cell can take only discrete values - 0 or 1 - and it is determined only by the nearest surrounding of the object. With the help of this model it is possible to describe various biological objects - for example, growth of colonies, however, this approximation is not enough to model physical interactions of the probabilistic nature. Another approach to the description of real interactions of objects is a model created by Ising, which has a probabilistic nature. Further development of the idea led to the creation of the theory of cellular automata, where the cellular automata are considered as a discrete model of real physical and biological processes [6] and the parameters of discrete objects can have different values, including those from continuous sets, depending on the statistically weighted contribution of the closest neighbors of each object in the modeling field.

The given study focuses on modeling of the growth of structures using the cellular automata with a set of states of the two values (0 and 1), with the value corresponding to the state is determined by the contribution of "the closest neighbor" (by the probability of induction of the state of the next generation in the direction of the interaction) and the geometry of the field is determined by the vector of the direction of the particle and the direction of the interaction.

The problem of the search of biostructures capable of self-organization is quite urgent considering the prospects of application of nanostructured biomaterials as components of composite materials in transplantology and optics as well as “scaffolds” for the synthesis of nanostructured materials based on inorganic particles. Modern biotechnological ecologically clean manufacturing technologies make it possible to get practically unlimited number of biomolecules at a low financial cost, and it also lets us talk about the perspective and the urgency of ongoing research.

2. Materials and methods.

Modeling the growth of the structures was performed with the help of the original program (Aggregator2d) which uses the data on the geometry of the interaction of the molecules and the probability of the generation corresponding to the effective energy of the interaction of the particles as the parameters. The result was obtained in the form of the table containing the coordinates of the centers and the vector of the directions of the objects, which was later converted into a graphic two-colored image.

3. Results and discussion.

A discrete model of cellular automata with probabilistic nature of filling of the cells (the growth of the structure) in the field determined by the direction and the nature of the connections between the cells was created. As the model of the growth the model of adherence of the particles with successive generations was chosen taking into consideration the probabilistic nature of the interaction between the active groups on the surface of the particles, and by specifying their location relating to the direction of the axis of the particle the peculiarities of the geometric arrangement of the connections were taken into account. The different nature of the interaction was taken into account by introducing a normalized probability of the formation of the connection.

Journal of Physics: Conference Series **291** (2011) 012005 doi:10.1088/1742-6596/291/1/012005

The algorithm was tested on the model of growth of the linear structures "Fibrils". Figure 1 shows the example of modeling of the growth of the structure from the geometric analogues of the peptide oligomers. Then with the help of the program the parameters of the interaction were varied. Several specific types of two-dimensional structures were revealed. Figure 2 shows a typical example of the structure of the "Net" type; Figure 3 shows an example of the structure of the same type, which is a typical example of structure remains regular fractals with a complete coverage of the space. Another revealed class of structures was the structures of the "Needle-shaped crystals" type (Figure 4), which are intermediate between the structure of "Fibrils" and "Net" types.

Figure 1. "Fibrils" type structure. example of modeling of the growth of the structure from the geometric analogues of the peptide oligomers.

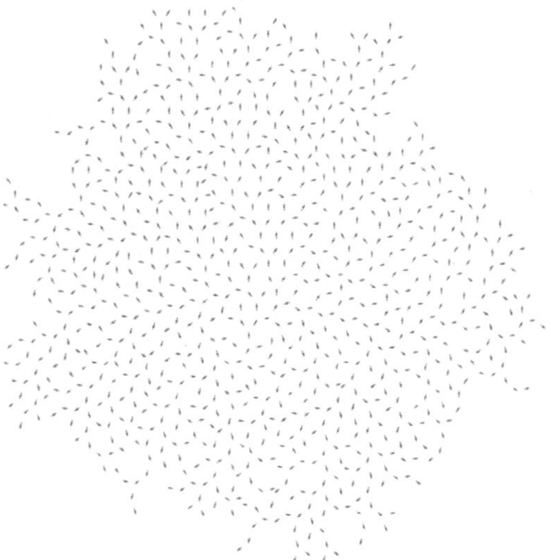

Figure 2. "Net" type structures.

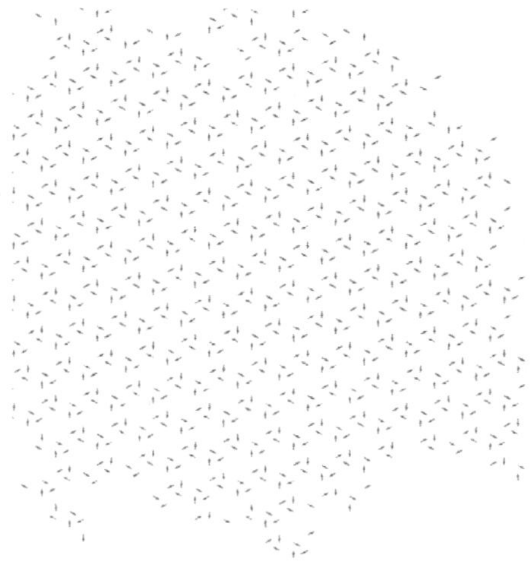

Figure 3."Net" type fractal-like structures.

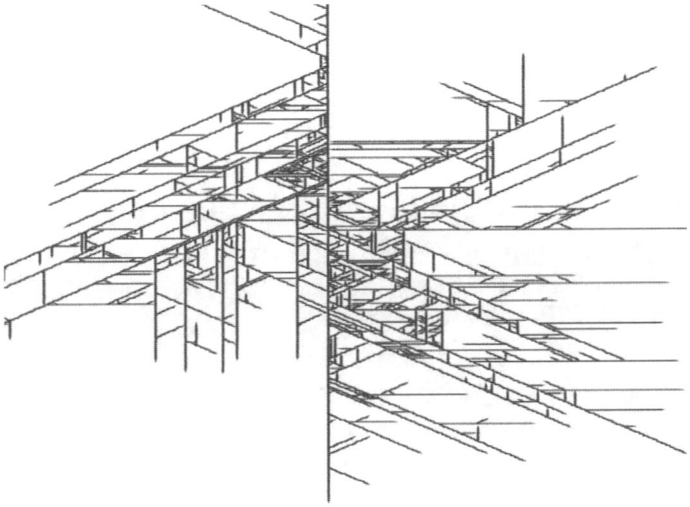

Figure 4. "Needle-shaped crystals" type structure.

The algorithm used in the research makes it possible to generate structures by changing the rules of the interaction for a single particle. The analysis of the rules of the interaction, which led to the formation of ordered structures, showed that their formation requires the sites of the interaction to be located symmetrically to the axis of the object, and besides, the probability of symmetric interactions were similar in meaning. At the introduction of a small asymmetry in the probability of the interactions the defects similar to defects of crystals appeared in the structures. No objects with an asymmetric interface of the interaction capable of forming ordered extended structures were found during the search. The acquired data are consistent with the data about the prevalence of symmetry in the amino acid sequence of interfaces of protein interactions capable of self-organization into an ordered quaternary structure, previously obtained by the author, and supplement them. The acquired model will be the basis for a three-dimensional model of formation of ordered structures.

4. Acknowledgments

The research was supported by the Grant of the Government of St Petersburg in 2009

5. References

[1] Feynman R 1959 Plenty of Room at the Bottom http://www.its.caltech.edu/~feynman/plenty.html

[2] Ritchie D Recent progress and future directions in protein-protein docking 2008 *Curr. Prot. Pep. Sci.* **9**(1): 1-15

[3] Egorov V, Dyukov M, Shaldzhyan,A, Vasin A, Nekrasov P, Sirotkin A, Plotnikova M, Garmay Y, Kiselev O The development of peptides capable of self-organization for creation of nano-structured materials 2008 *Innovations* **116**:84-87

[4] Mobley, Cox, Singh, Kulkarni, Slepoy Simulations of Oligomeric Intermediates in Prion Diseases 2003 *Biophysical Journal* **85**:2213–2223

[5] Gardner M MATHEMATICAL GAMES: The fantastic combinations of John Conway's new solitaire game "life" 1970 *Scientific American* **223**:120-123

[6] Packard, Wolfram Two-Dimensional Cellular Automata 1985 *Journal of Statistical Physics* **38**:901-946

Multisegment one-step RT-PCR fluorescent labeling of influenza A virus genome for use in diagnostic microarray applications

A V Vasin[1], N T Sandybaev[2], M A Plotnikova[1], S A Klotchenko[1], O V Chervyakova[2], V M Strochkov[2], E T Taylakova[2], E A Elpaeva[1], A B Komissarov[1], V V Egorov[1], J K Koshemetov[2], O I Kiselev[1] and S M Mamadaliev[2]
[1] Research Institute of Influenza of the Ministry of Health and Social Development of the Russian Federation, 15/17 Prof. Popova St., St. Petersburg, Russia
[2] Research Institute for Biological Safety Problems of the RK NBC/SC ME&S RK, Gvardeiskiy, Republic of Kazakhstan.

E-mail: vasin@influenza.spb.ru

Abstract. Microarray technology is one of the most challenging methods of influenza A virus subtyping, which is based on the antigenic properties of viral surface glycoproteins – hemagglutinin and neuraminidase. On the example of biochip for detection of influenza A/H5N1 virus we showed the possibility of using multisegment RTPCR method for amplification of fluorescently labeled cDNA of all possible influenza A virus subtypes with a single pair of primers in influenza diagnostic microarrays.

1. Introduction.

The subtyping of influenza A viruses (IAV) is based on the antigenic properties of two surface glycoproteins – hemagglutinin (HA) and neuraminidase (NA). IAV infect several animal species including humans, birds, pigs, horses and marine mammals. 16 subtypes of HA and 9 subtypes of NA have been described so far. In human population only three HA subtypes (H1, H2, H3) and two NA (N1, N2) subtypes predominate, but infections of humans with high pathogenic IAV strains, such as H5N1, H7N7 and H9N2, have been recently detected. Moreover, reassortment of viral segments can lead to the emergence of new virus variants, such as vH1N1 in 2009. For these reasons the development of new diagnostic methods for detection and subtyping of all existing and plausible new IAV strains is an extremely urgent problem of influenza surveillance [1].

Commonly used PCR methods are usually subtype specific and do not provide information about other subtypes. Recently the number of biochips for detection of 16 known subtypes of HA and 9 subtypes of NA have been developed [2,3], but one of the main problems of IAV microarray detection is associated with the choice of cDNA fluorescent labeling methods. In this article we propose the amplification of genomic RNA segments using universal primers to the conservative terminal regions of all IAV genomic segments for use in influenza subtyping microarrays. The application of these primers makes it possible to unify influenza microarray systems and to get rid of the complex set of multiplex primers specific for certain IAV subtypes.

2. Materials and methods.

2.1. RNA isolation

Total RNA was isolated from the allantoic fluid using TRIZol reagent (Invitrogen). RNA concentration and integrity were measured using NanoDrop ND-1000 spectrophotometer (NanoDrop Technologies, USA). Influenza virus A/H5N1 strains were 6:2 A/PR/8/34(H1N1) reassortants, obtained in St. Petersburg Influenza Institute using reverse genetics.

2.2. Oligonucleotide microarray design and printing

Oligonucleotides containing 5'-aminolinker were synthesized and HPLC purified by «DNA-synthesis» (Russia). The probes (300 ng/µl) in 3×SSC were immobilized on Vantage Aldehyde Slides «CEL Associates» using SpotBot 3 Personal Microarrayer («ArrayIt corporation», USA). After printing the slides were treated with UV (0.09 J/cm^2). Before hybridization the slides were boiled in miliQ H_2O for 1 min.

2.3. Fluorescent labeling of IAV genomic segments

Fluorescent labeling of cDNA was performed in the process of direct incorporation of Cy3-dCTP with M-RTPCR using SuperScript III One-Step Quantitative RT-PCR System (Invitrogen, USA) with universal MBTuni-12 (5'-ACGCGTGATCAGCAAAAGCAGG-3') и MBTuni-13 (5'-ACGCGTGATCAGTAGAAACAAGG-3') primers [4]. The reaction mixture additionally contained 20 µl 1mM Cy5-dCTP. Amplification conditions for Rotor-Gene 6000 thermocycler (Corbett Research, Australia) were the following: 1) 45°C for 60 min; 2) 94°C for 2 min; 3) 5 cycles at 94°C for 30 sec, 45°C for 30 sec and 68°C for 3 min; 4) 31 cycles at 94°C for 30 sec, 57°C for 30 sec and 68°C for 3 min.

2.4. Hybridization

30 µl of fluorescently labeled cDNA was incubated at 99°C for 2 min, chilled on ice for 2 min, afterwards 10 µl of 99% formamide and 10 µl of 5× hybridization buffer were added. Hybridization was performed in 50 µl using «Whatman» frames at 37°C for 2 hours with agitation at 250 rpm. After hybridization the slides were washed in the mixture of 1×SSC buffer and 0.1% SDS.

2.5. Microarray scanning

After hybridization the slides were scanned using ScanArray Express ("PerkinElmer", USA) with 5 µm resolution. The data analysis was made using ScanArray software.

3. Results and discussion.

For estimation of possibility of using fluorescently labeled viral cDNA in the process of M-RTPCR for IAV subtyping microarrays we developed the test biochip for human A/H5N1 detection. 50-mer oligonucleotide probes, complementary to HA and NA cDNAs of human A/H5N1 strains deposited in GenBank database, were designed using OligoWiz software [5]. Multiple alignment of these HA and NA sequences did not reveal any conservative 40-60-mer regions, which could be used as universal targets. For this reason HA sequences were divided into two and NA – into three groups, for which specific oligonucleotide probes denoted as H5-I^{H5N1}, H5-II^{H5N1}, N1-I^{H5N1}, N1-II^{H5N1} and N1-III^{H5N1} were designed (See figure 1). The biochip consisted of 16 identical subarrays containing 11 rows corresponding to 5 probes for A/H5N1, 4 probes for A/H3N2 and A/H1N1 reference strains detection (H3^{H3N2}, N2^{H3N2}, H1^{H1N1}, N1^{H1N1}) and 2 negative controls. The spot diameter was 200 nm, all the probes were immobilized in 10 repeats (See figure 2).

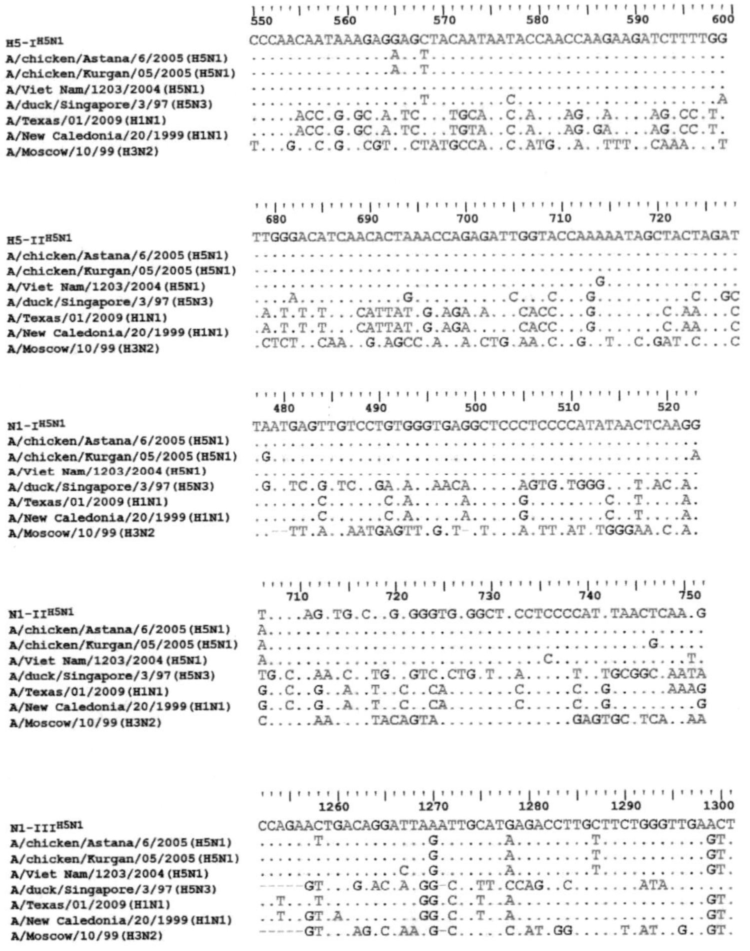

Figure 1. Multiple alignment of oligonucleotide probes, designed for influenza A/H5N1 subtyping, and HA and NA fragments of the analyzed influenza A virus H5N1, H5N3, H1N1 и H3N2 strains.

For viral cDNA fluorescent labeling one-step M-RTPCR method with universal primers for terminal regions of IAV gene segments was used. The incorporation of fluorescently labeled dNTP (Cy5-dCTP) into the reaction mixture changes the conditions of reaction dramatically. For this reason optimal Cy5-dCTP concentration (20 μM) and amplification conditions were evaluated on the basis of amplicons analysis. The results of hybridization of the fluorescently labeled cDNA samples of A/Kurgan/05/2005(H5N1), A/Moscow/10/99(H3N2) and A/New Caledonia/10/99(H1N1) (See figure 3) showed that they had specifically binded with the probes on the biochip. Significant levels of fluorescence were observed only in the probes specific to the corresponding IAV subtypes, for all other oligonucleotides fluorescence intensities were on the approximately equal level and comparable with standard error (See figure 4). The median values of the spot fluorescence intensities after deduction of fluorescence background were used as significant fluorescence values.

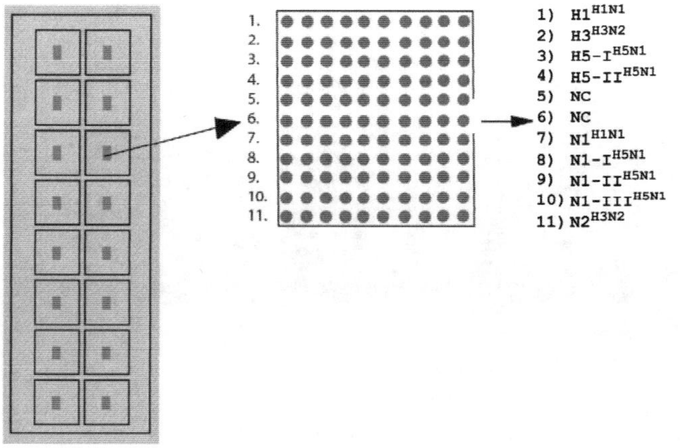

Figure 2. Layout of influenza A/H5N1 subtyping biochip. NC – negative control

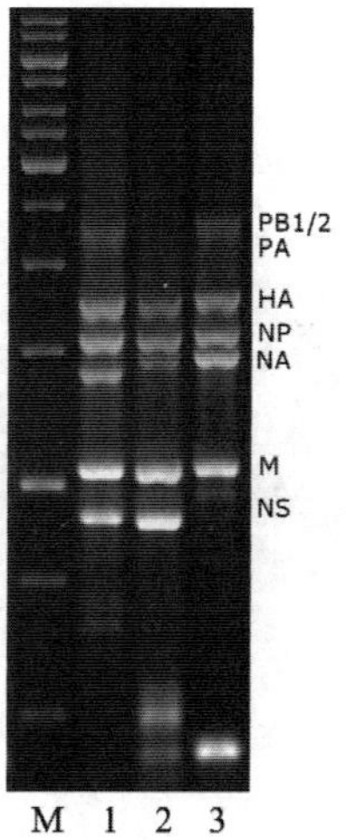

Figure 3. Agarose gel electrophoresis of M-RTPCR products, amplified using universal primers for terminal influenza A genome segments, from A/Kurgan/05/2005(H5N1) (1), A/Москва/10/99(H3N2) (2) and A/New Caledonia/10/99(H1N1) (3) virus samples. M – marker.

Moreover, for the analysis of specifity of the suggested approach we used A/chicken/Astana/6/2005(H5N1), A/duck/Singapore/3/1997(H5N3), A/VietNam/1203/2004 (H5N1) and A/Texas/01/2009(H1N1) strains. The data presented on figure 5 showed no nonspecific binding: only human and bird A/H5N1 virus strains were detected, A/duck/Singapore/3/1997 strain of A/H5N3 influenza virus subtype was not detected. Although «Swine» A/Texas/01/2009 (vH1N1) virus nonspecifically binds with N1-I^{H5N1} probe, the level of binding is comparatively low.

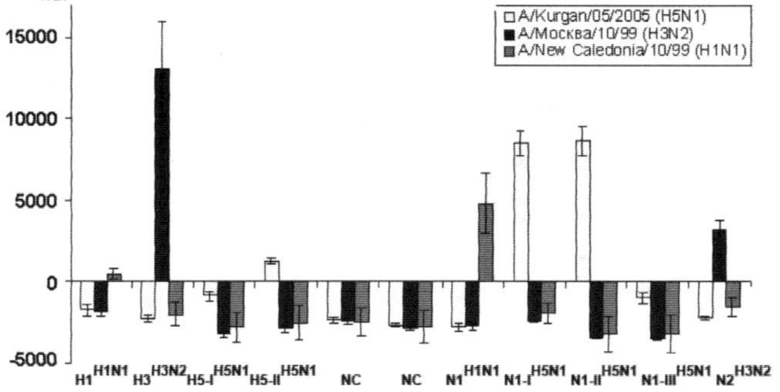

Figure 4. Relative values of the spot fluorescence intensities, derived after hybridization with A/Kurgan/05/2005(H5N1), A/Москва/10/99(H3N2) and A/New Caledonia/10/99 virus samples. Each column corresponds to the 10 spots mean intensity.

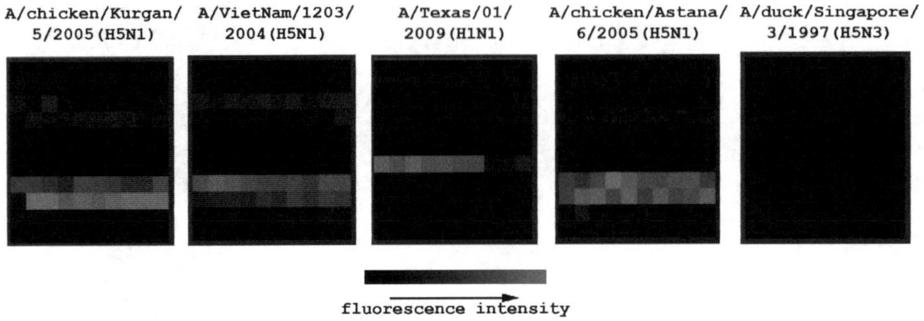

Figure 5. Graphical presentation of spot fluorescence intensities after hybridization with /chicken/Astana/6/2005(H5N1), A/duck/Singapore/3/1997(H5N3), A/VietNam/1203/2004 (H5N1) and A/Texas/01/2009(H1N1) virus samples

On the example of biochip for detection of influenza A/H5N1 virus we showed the possibility of using multisegment RTPCR method for amplification of fluorescently labeled cDNA of all possible influenza A virus subtypes with a single pair of primers in influenza diagnostic microarrays. Moreover, it allows molecular screening not only of HA and NA, but of all other IAV genes.

4. References

[1] Sengupta S, Onodera K, Lai A and Melcher U 2003 *J. Clin. Microbiol.* **41**(10) 4542-4550
[2] Fesenko E E, Kireyev D E, Gryadunov D A, Mikhailovich V M, Grebennikova T V, L'vov D K and Zasedatelev A S 2007 *Influenza Other Resp. Viruses.* **1**(3) 121-129
[3] Han X, Lin X, Liu B, Hou Y, Huang J, Wu S, Liu J, Mei L, Jia G and Zhu Q 2008 *J. Virol. Methods.* **152**(1-2): 117-121
[4] Wernersson R and Nielsen H B 2005 *Nuc. Acids Res.* **33** 611-615
[5] Zhou B, Donnelly M E, Scholes D T, St.George K, Hatta M, Kawaoka Y and Wentworth D E 2009 *J. Virol.* **83**(19) 10309-10313

Journal of Physics: Conference Series **291** (2011) 012007 doi:10.1088/1742-6596/291/1/012007

Nanostructure and new properties of hydrolyzed food globular proteins

I A Rogov[1], T N Danilchuk[1], J A Shushkevich[2], G V Semenov[1], O E Ovchinnikova[1]
[1]Moscow state university of applied biotechnology, 33, Talalihin street, Moscow, 109316, RU
[2] Company Base ELEMENT, 30, Rochdelskaya street, Moscow, 123022, RU
E-mail: iar@msaab.ru

Abstract. In the present work we obtained a hydrolysates of food proteins by enzyme hydrolysis, researched the comparative structural and the molecular – mass characteristics of proteins, and compared of hydrolysates particles structural characteristics on nanoscale with their biological properties.

Introduction

Now the requirement of manufacture of pure proteins commercially is rather great. Use of proteins concentrates allows improving and raising value of a foodstuff, improving their functional and medical and biologic properties. Such products are especially claimed at the strengthened physical activities when the norm of consumption of proteins for easy to acquire sharply increases. Among known sources of a protein, which use in food additives, a sports and children's food, the greatest interest is represented by soya proteins and proteins of dairy whey.

Soya protein is well balanced on amino acids. It is the unique protein possessing full of amino acid profile, unique, as a matter of fact, in a vegetative kingdom. After consumption of Soya proteins there is an accurate decrease in level of cholesterol in blood, therefore they are expedient for using in a diet of people with excess weight, and also people of dairy products suffering by intolerance. The main demerit of soya protein is presence inhibitor of digestive enzyme trypsin. For disposal from inhibitor additional processing of fiber with the help of enzyme hydrolysis is necessary [1].

Whey proteins of milk – a perspective and important class of protein products, nutrilites in food and medical appendices. The amino acid structure of milk whey proteins is closest to amino acid structure of a muscular fabric of the person, and under the maintenance of irreplaceable amino acids and amino acids with the branched out chain they surpass all other proteins of a vegetative and animal origin. Proteins of milk whey (α-lactalbumin, β-lactoglobulin, immunoglobulins) possess strongly pronounced antigene properties, have the highest speed of splitting among integral proteins and high assimilability.

Perspective use of hydrolysates whey proteins is represented. Last years by a number of researchers it is shown that as a result of proteolytic splitting whey milk proteins digestive enzymes form the peptides possessing biological activity. Physiological action of these peptides consists direct participation in regulation mechanisms in gastrointestinal tract or in their ability to an absorption through epithelium intestines, penetration into blood and transportation to peripheral bodies. There is a basis to believe that the spectrum of mechanisms of action of these peptides is wider, than at corresponding protein-predecessor. Depending on physical and chemical parameters of hydrolysis the received peptides can have both different structure, and different biological properties.

The purpose of the present work consisted in obtain of hydrolysis's of milk whey proteins by enzyme hydrolysis, research of comparative structural and the molecular -mass characteristics of proteins, and also in comparison of structural characteristics of particles of hydrolysis's on nanoscale with their biological properties.

Materials and methods

As objects of research have been chosen globe food proteins: the natural soya protein (Supro 760), taken directly from raw soybeans; the hydrolyzed soya protein (the experience sample); protein of whey of the cow milk an α-lactalbumin and its enzyme hydrolysis products.

The experience soya sample was received by innovative technology. Its key elements is treatment of initial soya protein by proteolytic enzymes, homogenization at a high pressure and then spray-type drying. For research of soya proteins properties modern methods of the physical and chemical analysis was used.

α-lactalbumin was isolation from bovine milk with use anion-exchange chromatography on DEAE-cellulose and the subsequent clearing by means of a gel-filtration [2]. The protein concentrarions were evaluated spectrophotometerically, using the extinction cjefficient $A_1{}^{1\%}{}_{cm,\ 280\ nm} = 18.2$ [3]. The received protein concentrate subjected to enzymatic hydrolysis with use of digestive enzymes: pork trypsin and pepsin (from Sigma). Hydrolysis of α-La was carried out in citrate buffer (pH 2.5) for pepsin and in Tris-HCl buffer (pH 8.0) for the trypsin. α -La (1 mg/mL) were incubated with the enzyme (1:20,enzyme-to-substrate ratio, w/w) for 24 h at 37°C. After incubation, enzymatic reactions were stopped by heating at 95°C for 15 min. Hydrolysates by pepsin were adjusted to pH 7.5 with 1M NaOH. The hydrolysates were centrifuged at 10000*g* for 30 min, the supernatants were removed, and an aliquot was frozen and kept at -20°C until use.

The form of particles and their sizes on nanoscale defined by method of the atomic force microscopy (AFM) with use scanning a microscope Solver NEXT. AFM images received in the air environment in a contact mode of scanning (cantilever CSG 01/10). Processing of images made with use of an information technology under program Image Analysis 3.0.

Results

In the process of hydrolysis the natural soya protein (Supro 760) by proteolytic enzymes occurs its structural updating which results are: truncation of molecules of separate polypeptides, formation new steady nanoscale globe structures, exhibiting of hydrophilic groups on a surface of protein substances rupture of cross-section communications, strengthening of superficial activity etc. Thus a protein product gets new properties: increase of solubility of a protein, increase of lathering ability, viscosity decrease, increase of thermo stability at heating (table 1). Besides, hydrolysis leads to occurrence of antioxidant activity and to increase of protein particles stability in the salt medium. Presence of the antioxidant activity in the hydrolyzed food soya proteins is one of key factors of its high biological value.

Table 1. The new properties attached for the product in the process of structural updating.

Properties	Samples	
	Supro 760	The experience sample
The maintenance of a soluble protein (method AOCS)	30	43
Lathering ability (on Wolf)	30,4 %	100 %
Dynamic viscosity (small speed of shift)	36.40 Pa·s-^{1}	0.48 Pa·s-^{1}
Dynamic viscosity (high speed of shift)	7.72 Pa·s-^{1}	0.09 Pa·s-^{1}
Part of an insoluble deposit at 50 °C	50 %	22 %

Results of the atomic force microscopy have shown that protein substances as natural, and the hydrolyzed protein of a soya have the globular form. The sizes of Supro-760 protein globules are in range from 100 to 400 nanometers, thus globules with a size 100 nanometers have tendency to agglomeration with a limit of 400 nanometers. During the analysis were able to deduce that agglomerates represent association of three globules (See the figure 1). On the contrary, at the experience sample the average size of visible granules makes 40-50 nanometers. (See the figure 2).

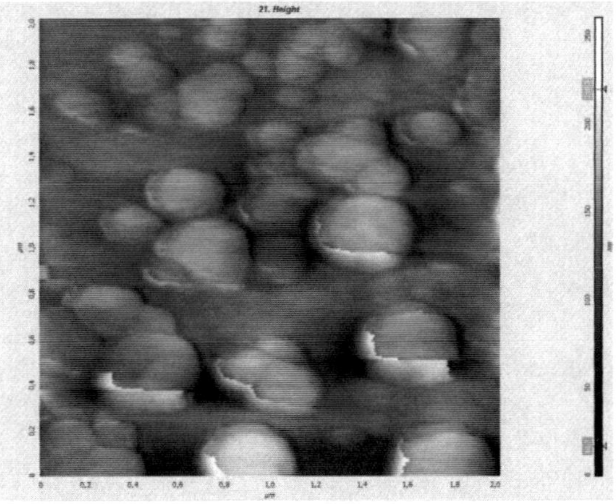

Figure 1. AMC image of the natural soya protein (Supro 760). The area of scanning is 2000 * 2000 nanometers.

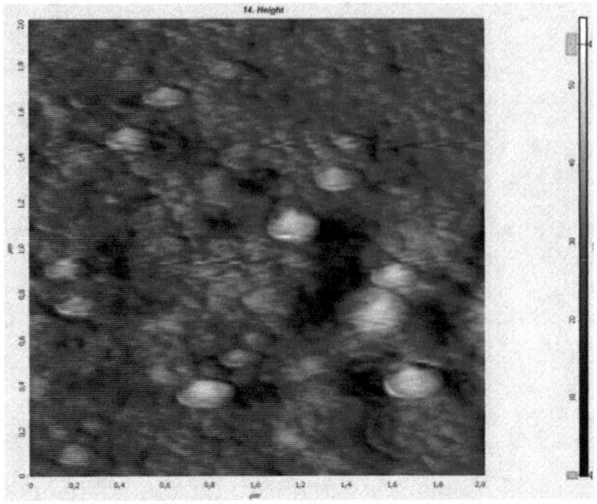

Figure 2. AMC image of globes the experience sample of soya protein after hydrolysis. The area of scanning is 2000 * 2000 nanometers.

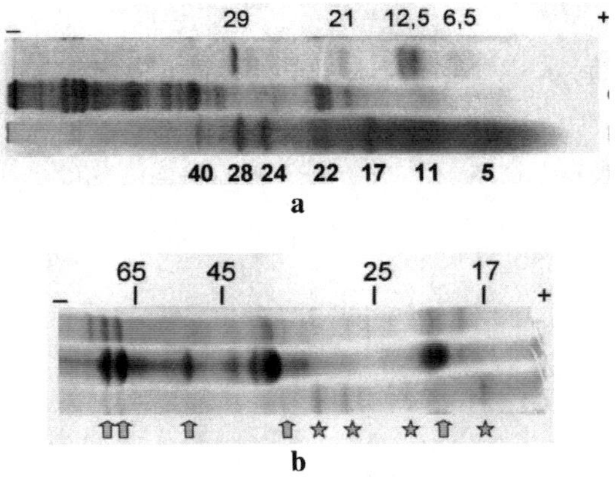

Figure 3. The results of electrophoresis of Soya proteins: a - Supro-760, b - the experience sample (hydrolysate). From above positions of markers of molecular weights in kilo Dalton (kDa) are designated. Below by arrows not hydrolyzed polypeptides are mark. By asterisks are mark hydrolysis products.

Research by method of electrophoresis has shown that in sample Supro-760 protein substances are presented, basically, by globules with molecular weights around 40 and 70 kDa. This in a linear scale corresponds to the sizes more than 100 nanometers (See the Fig. 3a). In the experience sample predominates polypeptides with the molecular weights in a range 5 and 11 kDa. The maintenance of these short polypeptides is estimated in 60-65 % from weight of fiber. Other products of hydrolysis (17, 22, 24, 28 and 40 kDa) make an order 1/3 from all quantity of protein (See the Fig. 3b). In a linear scale this distribution corresponds to fractions with the sizes from 4 to 10 nanometers (2/3 of general protein) and around 40-50 nanometers (1/3 of general protein). Globules of the second fraction are well identified by method AFM.

Proteins of the first fraction are not defined on AFM images because of the insufficient permission used cantilever, not allowing completely to scan objects with the sizes less, than 10 nanometers. Proteins the experience sample (hydrolysate) have globular form and are steady, not subject to agglomeration. Stability of experience sample globules can be connected with change of tertiary structure of protein as a result of rupture cross-section bisulphide S-S communications between cysteine molecules. Rupture cross-section S-S communications occurs in the process of protein homogenization under high (to 800 atmosphere) pressure which is a component of applied technology. Besides, stability to agglomeration influences the surface activity, obtained by protein structures in the biocatalysis process. Thanking nanosize of peptides and their stability the experience sample has additional properties of solubility, thermostability, stability in the salt medium and very low viscosity.

The great number of biologically active peptides, or directly present at foodstuffs, or isolated of proteins containing in these products by the use of enzymatic or chemical hydrolysis is revealed. In the course of natural processing of food proteins in a digestive system such peptides are liberated from structure of proteins and start to operate as independent regulation units possessing activity, similar to the hormonal. Now public recognition the leading part endogenous peptides in regulation of various systems of an organism. The direction of creation of parapharmaceutical preparations on a basis regulation peptides successfully develops [4].

The basic source of biologically active peptides of milk is its protein part, namely casein and whey proteins. α-lactalbumin (α-LA) is one of the basic components of protein fraction of milk whey (20-25 %), plays an important role in biosynthesis of lactose and it is capable to fixed ions of such metals calcium and zinc that causes a number of its biological functions. In the present work influence of conditions enzymatic hydrolysis on structural changes α-LA has been investigated.
Results of analysis AFM -images of particles an alpha-lactalbumina and its hydrolysates show that having the form of rigid balls globules an alpha-lactalbumin in the course of hydrolysis by proteolytic enzymes turn to the developed extended structures which length makes 1-2 microns (figure 1-4).

Figure 1. The AFM-images of the peptide fragments derived from α-LA digestion by trypsin

Figure 2. The AFM-images of the peptide fragments derived from α-LA digestion by pepsin.

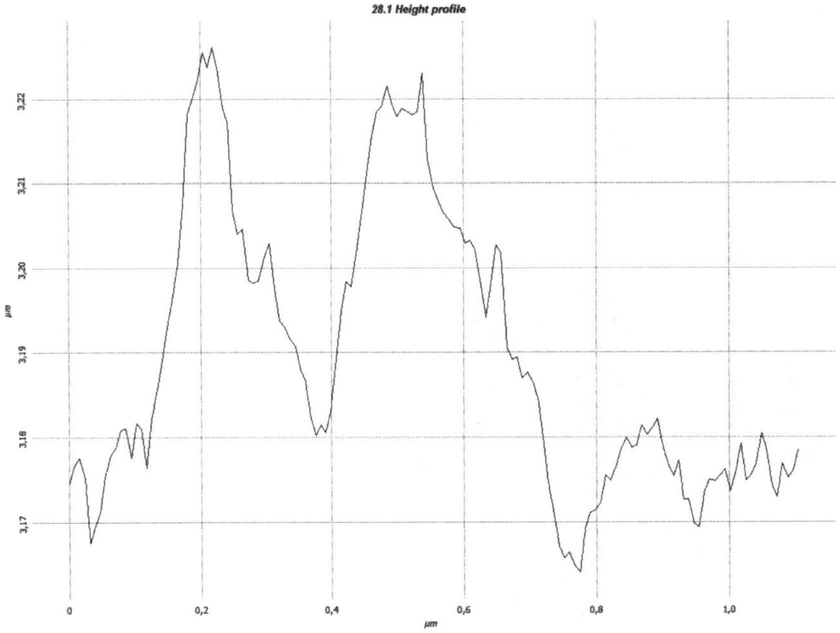

Figure 3. Profile peptide fragments derived from α-LA digestion by pepsin.

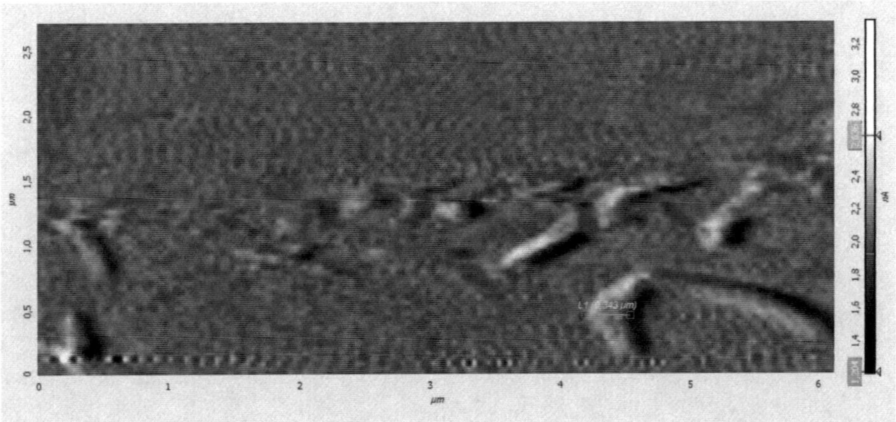

Figure 4. The AFM-images (2D) of the peptide fragments derived from α-LA digestion by pepsin.

Results electrophoresis researches of the received hydrolyzates, testify to presence of proteins particles since the sizes from 1,4 to 10 kDa. Except, the analysis of particles of hydrolysis's by AFM method shows presence of friable structures of the roundest form, having the nanoscale size, that can serve as the proof of that in the process of hydrolysis there is a formation disorder "fused globes ", which truncation leads to formation of the developed extended structures.
Thus there are the new properties shown, first of all, in increase of physiological activity of an protein preparation: antimicrobic action, anti-oxidative properties, ability to stimulate proliferation cages of an animal organism, etc.

Existence fused globes of α-lactalbumin was observed at studying of influence of denaturad (guanidinbichloride) on a protein solution [5]. The facts of formation of linear nanotubes are known at partial hydrolysis of α-LA by proteolytic enzymes after entering of calcium ions into solutions of hydrolysis's. The length of such tube makes thousand nanometers, external diameter is equal 20 nanometers, internal - 8 nanometers. These structures are stable enough, capable to transfer pasteurization, freeze- drying and are unique nanotubes, made from food protein. The internal cavity of such tubes can be used for capsulation various substances, for example, vitamins or enzymes, for the purpose of their masking or protection against destruction.

Thus, in the process of structural updating globe food proteins break up to particles, having the nanoscale size and forming a product possessing new unique properties, directly depending on their size and atomic structure of a surface. Use in the operated biocatalysts process of the technological influences, directly turned to macromolecular structure globe proteins, in aggregate with use of tools of nanotechnologies allows expanding possibilities nanobiotechnologies in creation of the preparations possessing preventive and medical effect.

References

[1] **www.medafarm.ru**
[2]Tirang R Neyestani, Mahmoud Djalali and Mohamad Pezeshki 2003 *Protein Expression and Purification* **29** 202-208
[3] Kronman M J, Andreotti R E 1964 *Biochemistry* **3** 1145-51
[4]Hartmann R and Meisel H 2007 *Current Opinion in Biotechnology* **18** 163-169
[5] Pfeil W, Bychkova V E and Ptitsyn O B 1986 *FEBS Letters* **198** 287-291.

Molecular-beam epitaxy growth and characterization of 5-µm quantum cascade laser

V.V. Mamutin, V.M. Ustinov, N.D. Ilyinskaya, M.V. Baydakova, B.Ya. Ber, D.Yu. Kasantsev

Ioffe Physical-Technical Institute of the Russian Academy of Sciences, St. Petersburg, 26 Polytechnicheskaya St, 194021 Russia

boris.ber@mail.ioffe.ru

Abstract. Molecular-beam epitaxy growth of 5 um emitting strain-compensated quantum semiconductor laser (QCL) is reported. The QCL structure is characterized by complementary techniques: high-resolution X-ray diffraction and dynamical secondary-ion mass-spectrometry, that reveal the high quality of QCL structure and in-depth distribution of chemical composition, respectively.

1. Introduction

The pioneering work on quantum cascade lasers was first done by Kazarinov and Suris in 1971 [1] describing what would later be called a 'diagonal' or 'photon-assisted' tunneling. Faist and Capasso demonstrated the first intersubband laser, designed to emit at 4.3 µm wavelength and grown by molecular beam epitaxy (MBE) in 1994 [2]. They named new laser as quantum cascade laser (QCL). Progress advanced quickly after that and is mostly due to the understanding of how bandstructure engineering, pioneered by Capasso, can be most successfully used to control the electron flow and thus increase population inversion and modal net gain, which ultimately controls the laser threshold [3]. After the first attempts the so-called 'three-well vertical' design was established in a wavelength range from 5 to 11 µm [4] and more, and the first far-infrared QCL with wavelength larger than 80 µm (87 and 130 µm) were demonstrated [5]. The wavelength is a function of an application for different QCL and one of the application is using atmospheric windows at 3 - 5 µm and 8 – 13 µm for very high speed communication, with satellites also, without turbulence and absorption as in near infrared and visible spectrum ranges [3] (see Figure 1).

In this paper we report the formation and characterization of QCLs for 5 µm range lasing based on AlInAs/GaInAs lattice matched superlattices grown by MBE on the InP substrate with 'three-well' design of its active zone. Such design allows to achieve both the shortest ground state life time (~0.4 ps) and high injection efficiency [6].

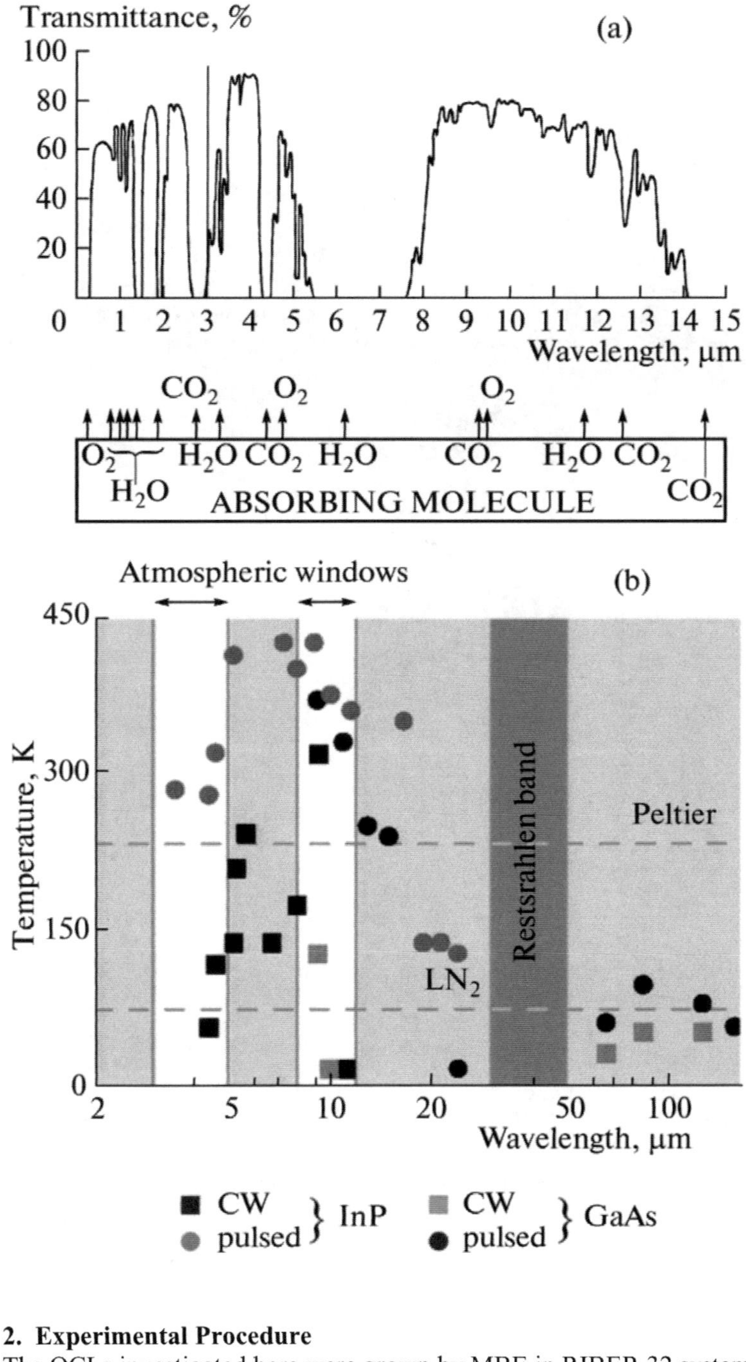

Figure 1. Spectral transmittance of atmosphere, molecule's absorbing and overlapping of wavelength of QCL on InP and GaAs substrates.

2. Experimental Procedure

The QCLs investigated here were grown by MBE in RIBER 32 system. The growth temperatures were 500°C, and slightly As-stabilized conditions were applied for all samples. The whole layer structure was designed following [6] and was deposited onto n-InP:Sn substrate which was doped only at a level of 2×10^{17} cm^{-3}. Active regions of AlInAs/GaInAs heterostructures were grown in a lattice matched mode. The active region of the QCL structure consisted of 25 periods. One period (Figure 2) constitutes a sequence of 18 alternating AlInAs/GaInAs layers in growth direction and is composed as follows (thicknesses in Å): 30/36/28/45/65/10/27/**13**/23/**17**/20/**16**/19/**21**/21/21/21/21, starting with the AlInAs exit barrier where the underlined numbers represent well layers and the **bold** numbers indicate n-doped layers, that were doped with Si at a level of $4\times10^{17}cm^{-3}$.

The whole structure consists, from substrate to top, of the digitally graded AlGaInAs (Si, 2×10^{17} cm^{-3}, 30 nm thickness) layer, the GaInAs (Si, 1×10^{17} cm^{-3}, 300 nm thickness) waveguide, the active region (thickness 1135 nm) consisting of 25 periods, the GaInAs (Si, 1×10^{17} cm^{-3}, 300 nm thickness) waveguide, the AlGaInAs (Si, 2×10^{17} cm^{-3}, 30 nm thickness) digitally graded layer, the AlInAs (Si, 2×10^{17}, 3×10^{17}, and 7×10^{18} cm^{-3}, 2500 nm thickness) cladding layer, the AlGaInAs (Si, 7×10^{18} cm^{-3}, 30 nm thickness) digitally graded layer, and the GaInAs (Si, 2×10^{19} cm^{-3}, 10 nm thickness) top contact layer. Each period comprises of a three quantum-well (QW) active region and a graded-gap superlattice injection region (Figure 2). The whole structure was grown following [7] for achieving 5 μm laser generation. Finally, QC lasers was fabricated emitting around 5 μm.

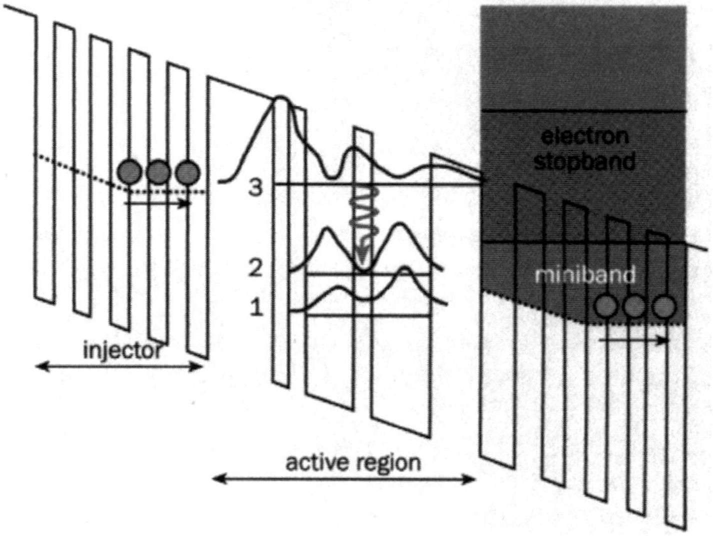

Figure 2. Schematic diagram of the active zone conduction band under bias.

The QCL structure was characterized by high-resolution X-ray diffraction (HRXRD) and dynamical secondary-ion mass-spectrometry (D-SIMS) to reveal the crystalline quality and the chemical composition, respectively.

The HRXRD was studied using a D8 Discover X-ray diffractometer (Bruker AXS, Germany) in a double-crystal arrangement with a channel-cut analyzer. The source of X-ray radiation was a tube with a rotating copper anode of power of 6 kW. The HRXRD curves were obtained by $2\theta/\theta$ scanning in the region of the reflection angle from the InP (004).

D-SIMS was carried out using the CAMECA IMS-7f secondary ion microprobe. In-depth profiling of matrix elements was done using Cs^+ primaries with an impact energy of 3 keV and CsX^+ (X=Al, Ga, In, and As) secondary ions. High depth resolution profiling of the QCL active region was done using O_2^+ primaries with an impact energy of 2 keV impinging at an angle of 45°. The depth distribution of silicon was measured using Cs^+ primaries while following the Si^- secondaries under high-mass resolution mode $M/\Delta M = 3000$ to avoid mass interference of the $^{28}Al^1H^-$ and $^{28}Si^-$ analytical signals. The depth of the sputtering crater was measured by mechanical stilus profilometer AMBIOS XP-1.

3. Results and discussion

Figure 3 shows the SEM image of the mesa surface of the QCL structure under post-treatment preparation. One can see distinctly 25 periods of the active zone with waveguides around it.

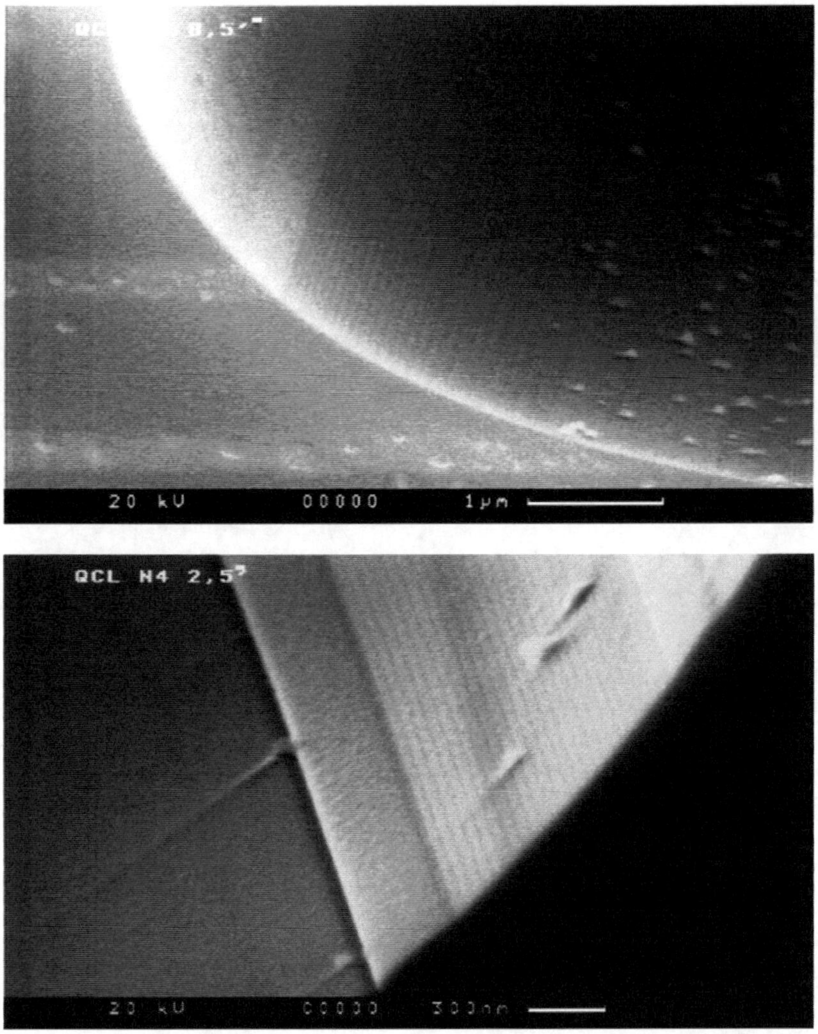

Figure 3. SEM image of the QCL mesa stripe under processing.

The multiple mode lasing spectra of the QCL above the threshold are shown on Figure 3 for 78 K and 300 K operating temperatures. The QCL emits at a wavelength of 5.05 μm and 5.24 μm at 78 K and at 300 K, respectively. The lasing characteristics of the QCL been described in detail elsewhere [7].

The diffraction curve for QCL sample is shown in Figure 4. On the curve an intense central peak is observed. The location of this peak corresponds to the Bragg diffraction from the InP substrate. In addition to this peak, the developed pattern caused by the structure of the active region Sl is observed in curve. The period in the spatial arrangement of active region Sl , which was calculated from the distances between interference fringes, is equal to 470±20 Å. This result is in good agreement with the SIMS profiling data.

As seen from insert on Figure 4 the central peak of the interference pattern Sl0 as well as peaks corresponded AlInAs and GaInAs layers are overlapped with the peak of InP. It means that these layers show acceptably low lattice-mismatch Δc/c. The Δc/c is equal to 1.2×10^{-3}, 4.0×10^{-4}, and -8.7×10^{-4} for AlInAs, GaInAs and Sl0 (or GaAlInAs), respectively.

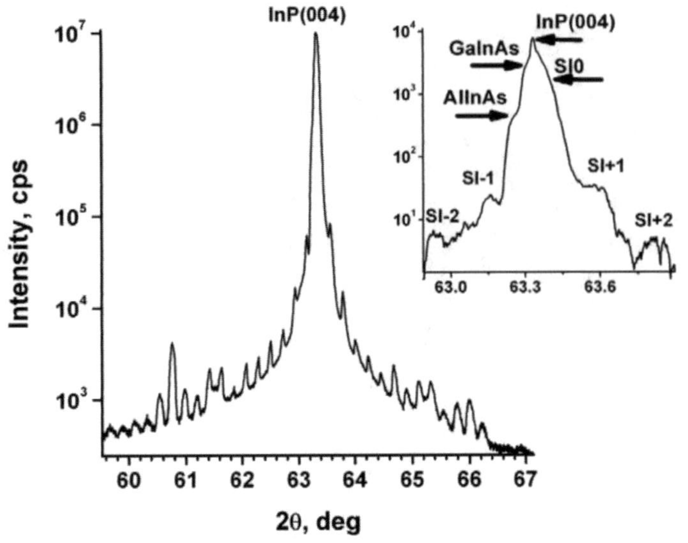

Figure 4. X-ray diffraction curves for the QCL sample recorded near substrate (004) InP maximum. A well developed interference patterns indicating a high planarity of the interfaces in the active region SL. The fragment in the inset ($2\theta \approx 63.3^o$) shows the curve recorded in a high-resolution mode to separate substrate and Sl0 peaks.

From the values of $\Delta c/c$ we can adjust composition of the upper layers as well as average one in the Sl. The initial compound was taken from the technological data. The molar concentration of InAs in AlInAs and GaInAs layers are equal to 0.528 and 0.524, respectively. The molar concentration of GaAs and AlAs, in the Sl GaAlInAs layer forming the active region is equal to 0.21 and 0.27, respectively. It should be noted that the concentration obtained from the analysis of X-ray diffraction is in good agree with the values obtained from SIMS profiling.

A well developed interference patterns indicates a high planarity of the interfaces in the active region of the QCL structure. The inset ($2\theta \approx 63.3^o$) shows the curve recorded in high-resolution mode to separate substrate and Sl0 peaks.

Al, Ga, In, As and Si depth profiles are shown in Figure 5. The results of high depth resolution SIMS profiling of the active region are presented in Figure 6. The quantification SIMS procedure for chemical composition was done using the set of single-layer epitaxial structures as external standards. Quantification SIMS procedure for in-depth distribution of the silicon dopant was done using implanted standards according to the well defined SIMS protocol [8].

The in-depth distributions of the chemical composition and distribution of donor atoms revealed by D-SIMS profiling are in good agreement with the structure of QCL.

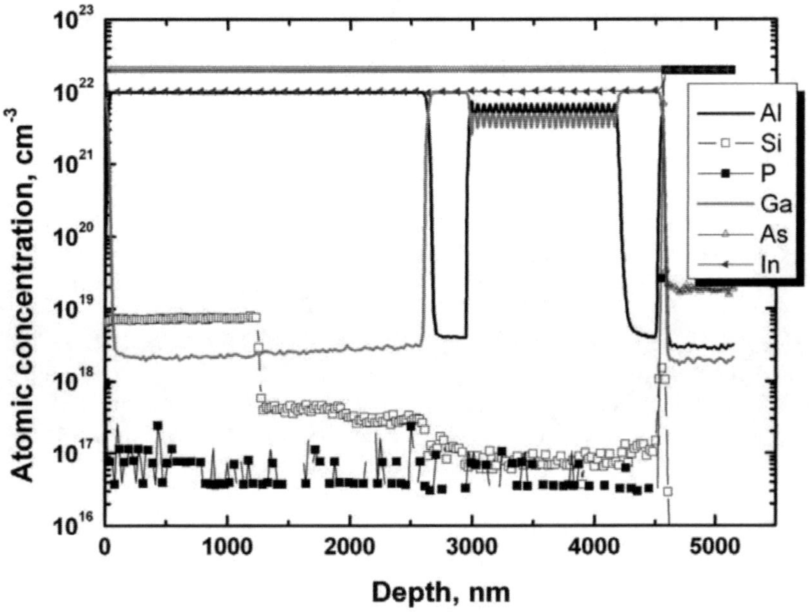

Figure 5. SIMS profiles of epitaxial QCL device structure of matrix elements and the donor impurity.

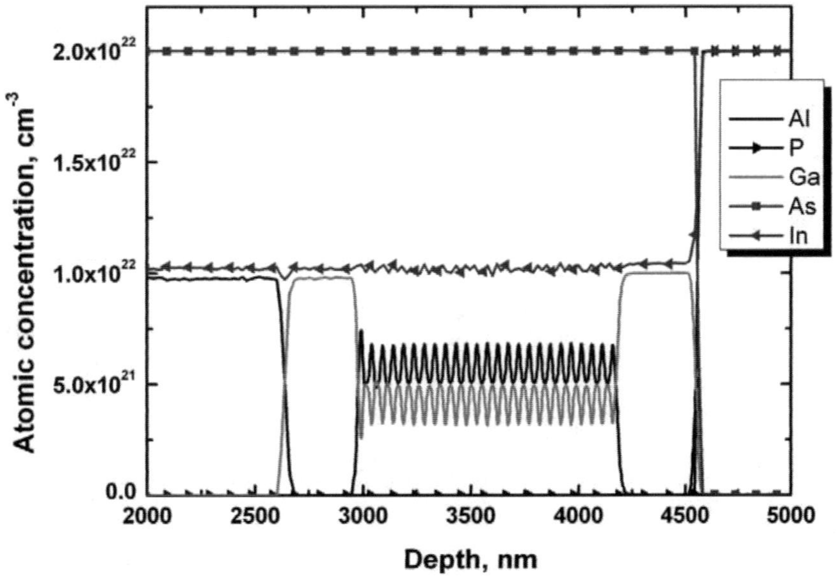

Figure 6. High-depth resolution SIMS profiles of matrix elements in the active region of QCL

4. Conclusions

Molecular-beam epitaxy was used to fabricate strain-compensated quantum cascade lasers emitting at a wavelength of 5 μm. The quality of the QCL structure was characterized by high-resolution X-ray diffraction and dynamical SIMS in-depth profiling.

5. Acknowledgements

This work was supported partly by the Russian Foundation of Basic Research under the Grant No. 11-02-00856-a. The materials characterization was done by using the instruments of the Center of Shared

Journal of Physics: Conference Series **291** (2011) 012008 doi:10.1088/1742-6596/291/1/012008

Facilities "Material Science and Characterization for Advanced Technologies" supported in part by the Ministry of Education and Science of the Russian Federation under the contract No. 02.552.11.7054.

References

[1] Kazarinov R F and Suris R A 1971 *Sov. Phys. Semiconductors* **5** 797
[2] Faist J, Capasso F, Sivco D L, Sirtori C, Hutchinson A L and Cho A Y 1994 *Science* **264** 553
[3] Gmachl C, Capasso F, Sivco D L and Cho A Y 2001 *Rep. Prog. Phys.* **64** 1533
[4] Tahraoui A, Matlis A, Slivken S, Diaz J, Razeghi M 2001 *Appl. Phys. Lett.* **78** 416
[5] Ajili L, Scalari G, Faist J, Beer H, Linfield E, Ritchie D and Davies G. 2004 *Appl. Phys. Lett.* **85** 3986
[6] Faist J, Capasso F, Sirtori C, Sivco D L, Baillargeon J N, Hutchinson A L, Chu S-N G and Cho A Y 1996 *Appl. Phys. Lett.* **68** 3680
[7] Mamutin V V, Ustinov V M, Boetthcher J and Kuenzel H 2010 *Semiconductors* **44** 962
[8] R.G. Wilson R G, Stevie F A and Magee C W 1989 *Secondary Ion Mass Spectrometry. A Practical Handbook for Depth Profiling and Bulk Impurity Analysis* (New York: Wiley & Sons)

Journal of Physics: Conference Series **291** (2011) 012009 doi:10.1088/1742-6596/291/1/012009

Magnetic nanoparticles. Metrological aspects.

V N Nikiforov[1,4], B L Oxengendler[2], N N Turaeva[2], A V Nikiforov[1], V G Sredin[3].
[1.] ***M.V.Lomonosov Moscow state university, Moscow.***
[2.] ***Institute of physics and chemistry of polymers AN Uz, Tashkent***
[3.] ***Peter the Great Military academy, Moscow***

E-mail: pppnvn@yandex.ru

Abstract. The experiments on influence of the iron oxide cluster size on the specific magnetic moment are performed. Both free and covered clusters are investigated. The experiments are interpreted on the base of core-shell model by analogy to Weizsäcker formula in the nuclear physics. Metrological parameters for the cluster size investigation are obtained.

1. Introduction

By now, it has been firmly established by various studies that nanostructures (including nanoclusters) demonstrate a significant difference of many physical and physical–chemical properties compared to bulk materials [1,2]. For example, nanoclusters can melt at temperatures both above and below their bulk analogues [3-5]. The ionization energy, the electron affinity [1,6], the photoluminescence lifetime of semiconductor nanoclusters [1,7], etc. can not only differ from those of bulk objects but also depend strongly on the nanocluster size [2-8]. In the radiation physics field the investigations of nanoclusters have been fulfilled too [9-11].

As regards the works in the field of nanocluster magnetic physics, the situation here is considerably rich [12]; some models have been constructed and many experiments have been performed. However, the systematical analysis of experimental situation was yet not performed, but now with the large generality it is possible to believe, that transition to study of nanoparticles leads to necessity of the account of two new positions [1,2]:

1. Quantum confinement of elementary excitations, caused by the small sizes of nanoparticles.
2. The total number of surface states in nanoparticle is comparable with the number of the bulk ones

Using of these two principles concerning the magnetism in nanophysics can be realized by means of “core-shell” idea. In present work the approach is used on the base of experiments and model construction.

Methods of magnetic nanoparticles metrological control are being developed for the usage in magnetic separation, magnetic resonance imaging, magnetic hyperthermia and other applications [13]. Our research deals with iron oxide nanoparticles synthesized by different methods. The proposed model, based on the analogy with nuclear physical Weizsäcker model [14], allows to separate contribution of surface and bulking states into the nanoparticle total magnetic moment.

2. Materials and methods of experiments

The proposed method is applied to iron oxide nanoparticles 4-22 nm synthesized by co-precipitation [15]. The samples were analyzed by X-ray diffractometer DRON2 with CoKα ($\lambda = 0.1789$ nm).Crystal structure was depicted and composition was controlled to avoid multiphase samples. Test results confirm the presence of one phase of the spinel structure, as well as the average particle size of 4-22 nm. Magnetic methods of diagnostics were used: SQUID and ESR [16]. Magnetic moment dependence on temperature and magnetic field was measured and ESR spectra for nanoparticles were obtained at temperatures 4.2- 380K.

---

[4] To whom any correspondence should be addressed

3. Results of experiments

Experimentally we obtained distribution of magnetic nanoparticles in size and determine the average size. The results are in good agreement with X-ray data and the results of transmission electron microscopy. The experimental results can be summarized as follows:

1. Specific magnetic moment of nanoclusters changes monotonically with the increase of a cluster size μ (N) (figure 1).

2. This monotonic dependence of μ (N) can be either increasing or falling for the different samples.

3. At dependencies M (H) absence of hysteresis was observed for all the magnetic nanoparticles, that indicates the absence of the coercive force and, consequently, the superparamagnetic state of nanoparticles.

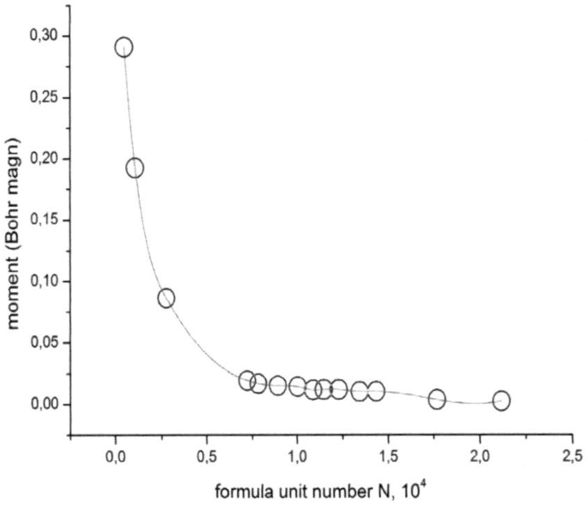

Figure 1. Magnetic moment deviation from formula unit one in iron oxide nanocluster.

According to Langeven expression $M(H)=N\mu_{eff}(cth(\mu_{eff}H/kT)-1/(\mu_{eff}H/kT))$ the average values of effective magnetic moment of one formula unit, the effective magnetic moment of one particle, the average number of formula units in one particle were estimated as in [15-17].The dependence of the average magnetic moment' change per number of formula units in nanoparticle is obtained. In our case, it was falling.

In all presented here results samples were in the form of nanopowder. Moreover investigations of nanoparticles in liquid were carried out. Effective magnetic moment of nanoparticles didn't change sufficiently (values were estimated according to Langeven equation). Magnetic characteristics of nanoparticles on magnetite basis in the form of magnetic liquid, polymer matrix, quarts matrix were also under investigation.

The influence of nanoparticles environment and synthesis conditions on the blocking temperature for nanoparticles Tg were found. The dependence of blocking temperature of the synthesis conditions is revealed.

4. Phenomenological model

Presented experimental data can be interpreted on the basis of several different models: Heisenberg model, Ising model, Weitzecker model and quantum-chemical model. Despite the difference of these models, the basic position they are united is coexistence of different magnetic moments of surface and bulk states: μ_S^0 and μ_V^0. The simplest phenomenological model is a modification of Weitzecker model, used for analysis of specific characteristics of complex nucleus [14]. Generalizing this model on specific magnetic moment of nanoparticles, we can write:

$$\mu(N) = (1/N)*\{\sum(a_j * N^{\chi_j})\}, \text{ where } j = 1, 2, ... \quad (1)$$

In this expression N is the number of atoms in the cluster, a_j is the phenomenological parameters (in our case μ_S^0, μ_V^0, etc.), χ_j are the rational numbers.

In the simplest case the number of surface atoms is $N^{2/3}$, and the number of volume atoms is N - $N^{2/3}$, so we obtain:

$$\mu(N)= \mu_V^0+(\mu_S^0-\mu_V^0)*N^{1/3}. \quad (2)$$

It is obvious that formula (2) is consistent with the qualitative conclusions of the experiments. Indeed, the theoretical dependence of μ(N) is monotone. It can be either increasing or decreasing depending on which is more: μ_S^0 or μ_V^0. In the estimations of other authors (for example [18]) analysis of isolated cobalt clusters by quantum chemical methods showed μ(N) decrease with increasing cluster size. This refers to our case when $\mu_S^0>\mu_V^0$. Another interesting conclusion from formula (1) is that at the transition from a free native cluster to the similar one covered with passivating substance the sign of derivative d(μ(N))/dN can be changed. This is due to, for example, valence saturation in Tamm surface states and results in the following condition: $\mu_S^0>\mu_V^0$. Thus, in this simple model, based on a modification of the Weizsäcker formula, μ_S^0 and μ_V^0 are the metrological parameters, and calculating them by measuring $\mu_{exp}(N)$, can get the size of the cluster.

$$N = [(\mu_S^0-\mu_V^0)/(\mu_{exp}-\mu_V^0)]^3 \quad (3)$$

When the sum expression (1) is used in modified Weizsäcker formula, taking into account the third and subsequent terms of the expansion, of course complicate the expression (3). The developing model is based on assumptions, that surface states are located only in one monolayer of external cluster's atoms, but it can be easily generalized to multilayer location of "magnets" on nanoparticle surface. In this case a third phenomenological parameter appears: a/R, where a is the thickness of the surface layer, and R is the radius of the nanoparticle.

5. Conclusions

The combination of informative methods of magnetic diagnostics with phenomenological model of nanoparticles magnetic properties allows us to obtain the dependence of specific magnetic moment of nanoparticles depending on the number of its constituent magnetic formula units, as well as highlight the contributions to the magnetic moment of bulk and surface states. This approach allows us to formulate the task of metrology as finding constants in each order of Weitzecker sum expression. The work has the prospect, namely, the addition of developed phenomenology to quantum-chemical calculations. Concerning the applications these results may be applied in medicine, materials science and criminological applications [19]. Co-usage of new magnetic methods with well known methods of nanodiagnostics showed its effectiveness.

References

[I] Suzdalev P 2006 *Physical chemistry of nanoclusters, nanostructures and nanomaterials* (Moscow: Com.Kniga; *in Russian*)
[2] Pool Ch and Owens F 2007 *Introduction to nanotechnology* (New York: Wiley–Interscience)
[3] Kellermann G and Graievich A 2002 *Phys. Rev. B: Condens. Matter* **65** 134204
[4] Kellermann G and Graievich A 2008 *Phys. Rev. B: Condens. Matter* **78** 054106
[5] Pakarinen J, Backman M, Djurabekova F, and Nordlund K 2009 *Phys. Rev. B: Condens. Matter* **79** 085426
[6] Melnikov D and Chelikowsky J 2004 *Phys. Rev. B: Condens. Matter* **69** 113305
[7] Garcia C, Garrido B, Pellegrino P, Ferre R, Moreno J A, Morante J R, Pavesi L and Cazzanelli M 2003 *Appl. Phys. Lett.* **82** 1595
[8] Kashtanov P V, Smirnov B M 2010 *Usp. Phys. Nauk* **48** №6 886 (*in Russian*)

[9] Backman M, Djurabekova F, Pakarinen O H, Nordlund K, Araujo L L, and Ridgway M C 2009 *Phys. Rev. B: Condens. Matter* **80**, 144109
[10] Krasheninnikov A V and Nordlund K 2010 *J. Appl. Phys.* **107** 1
[11] Oksengendler B L, Turaeva N N, Maksimov S E and Djurabekova F C 2010 *Journal of Experimental and Theoretical Physics* **111** №3 415
[12] Gubin S P, Koksharov J A, Homutov G B and Jurkov G J 2005 *Uspekhi Khimii* **74** 539 (*in Russian*)
[13] Nikiforov V N and Filinova E Yu 2009 *Biomedicical application of magnetic nanoparticles* (Wiley-VCH Verlag GmbH) pp 393-455
[14] Blatt J M and Weisskopf V F 1952 *Theoretical Nuclear Physics* (New York: Wiley)
[15] Brusentsova T N, Brusentsov N A, Kuznetsov V D and Nikiforov V N 2005 *JMMM* **293** 298
[16] Nikiforov V N, Kuznetsov V D, Nechipurenko Yu D et al. 2005 *JETP Letters* **81** 264.
[17] Nikiforov V N, Sredin V G, Nikiforov A V et al. 2009 *2 Nanotechnology International Forum.6-8 Oct.* 254. (*in Russian*)
[18] Hernando A, Crespo P, Garcia M A 2006 *Phys Rev Lett .* **96(5)** 057206.
[19] Lobanov N N, Nikiforov V N, Gudoshnikov S A et al. 2009 *Doklady Chemistry* **426** №1 96.

Journal of Physics: Conference Series **291** (2011) 012010 doi:10.1088/1742-6596/291/1/012010

Effect of abnormal high solubility of impurity in nanocrystalls and its metrological aspects

B L Oksengendler[1,2,5], V N Nikiforov[3], V G Sredin[4], N N Turaeva[1] and S E Maksimov[2]

[1]Institute of chemistry and physics of polymers, Uzbek Academy of Sciences, Tashkent, Uzbekistan;
[2]Arifov Institute of electronics, Uzbek Academy of Sciences, Tashkent, Uzbekistan.
[3]M.V.Lomonosov State University, Moscow, Russia;
[4]Great Peter Military Academy, Moscow, Russia;

E-mail: oksengendlerbl@yandex.ru

Abstract.
The problem of doping impurity intrusion into nanoparticles is analyzed. Thermodynamical model of the effect is developed, taking into account specific properties of nanoparticles: quantum confinement of elementary excitations and the impact of nanoparticle surface states compared with bulk ones. We obtained the size dependence of impact equilibrium solubility in nanoparticles. Magnetometric experiments, carried out on nanoparticle ensemble of iron oxide doped with gadolinium and other impurities, show lack of contradictions and good qualitative accordance between the model and the experiment.

1. Introduction

The large investigations carried out in present time in the field of nanomaterial science, have revealed a considerable quantity of the various effects [1,2] which main feature is dependence of properties of nanoscale objects from their characteristic size. In turn, one of leading places in the physical chemistry of objects of nanomaterial science [3] is the problem of the doping of nanoparticles by various impurities. This phenomenon is extremely interesting as for the understanding of features of penetration of particles in nanoobject, and it is exclusively important from the applied point of view [4]. The problem of the doping of nanoparticles in some degree have been studied from the point of view of features of penetration of impurities in nanoobjects depending on its size [5,6,7] in the experimental aspect. The theoretical analysis of penetration of impurities and defects in nanoobjects is poor [8-12]. Moreover, theoretical works have basically the modelling character and are not connected with the fundamental properties of nanoparticles. This fact does a problematics of influence of the size of nanoparticles on their alloying properties (in particular, solubility of impurities) the topical problem.

If we summarise experimental results on doping of nanoparticles then it is possible to present them in a following kind:

1. The effect of increase of impurities solubility in nanocrystalls in comparison with that is observed in volume materials has been found;
2. The effect increases with the decrease of nanoparticle radius;
3. The effect that more than is less solubility in a volume material.
4. The effect of increase of impurities solubility in nanoparticle decreases with the temperature growth in comparison with the bulk material.

Considering the fundamental nature of the found effect and that circumstance, that these properties appear the general for various materials, we will analyse it from the basic thermodynamic positions.

[5] To whom any correspondence should be addressed

2. Model

With the large generality it is possible to believe, that transition to study of nanoparticles leads to necessity of the account of two new positions:

1. Quantum confinement of elementary excitations, caused by the small sizes of nanoparticles [13].
2. The total number of surface states in nanoparticle is comparable with the number of the bulk ones [1].

In the problem of nanoparticles doping as it will be shown below, both these features of nanoobjects are presented. So, really, the confinement of electron and phonon excitations leads to that the quasicontinuous spectra turn to the discrete spectra of elementary excitations. It immediately has an effect on, first, the statistical sums Z describing the elementary excitation (and, also, on their thermodynamic parametres [14]). Secondly, the equilibrium constants of reactions between electrons and the charged impurity and defects will depend on the differences of energy levels of the lowest unoccupied orbital and the highest occupied one [15], i.e. from the size of nanoparticles. Furthermore, the comparable numbers of surface and bulk conditions lead to that in expressions for free energies the contributions of these states should be taken into account as comparable values. We will construct, proceeding from these positions, the simple thermodynamic model of nanoparticles doping.

We consider the following system: very large volume of a phase A (a material of the future doping impurity) with an impregnation of a nanocrystall of phase B in it having the volume $V=4/3\pi R^3$ and the surface $S=4\pi R^2$. Let the density of the phase B is N, the surface tension between A and B is σ, the cohesion energy of phase B is U_B, and the temperature of the A+B system is T. The free energy of A+B system before the penetration of atoms A in nanoparticle B is:

$$F1=F_{1A}+F_{1B}+NV\{U_B+3kT\ln Z_B\} \tag{1}$$

Here Z is the statistical sum of phonons of each atom in nanoparticle. We will calculate Z in Einstein's model, considering, however, unlike last, *finiteness* of the potential well, i.e. the cohesion energy, which, that is the characteristic for nanoparticle and depends on the size of nanoparticle [5] and on the concentration of the penetrated dopants [17], that is:

$$U_B = U_B(R, C) = Uo(C)[1 - f(aS/V)], \tag{2}$$

where a is the interatomic distance in nanoparticle. Indeed, as it has been shown theoretically [16] and experimentally [8], the cohesion energy is the drop-down function of f(aS/V) under decrease of the nanocluster radius. In turn, the value of the cohesion energy of the alloy is the function of concentration of its components, that is C, that has been shown by various methods, for example, in a ε-dielectric formalism [17]. As a first approximation, the cohesion energy is equal $U=U_0 + CdU/dC \approx U_0 - \gamma C$.

In approach *of a finite* potential well, for the individual Einstein's oscillator we have:

$$\ln Z_B = \ln\Sigma\exp[-(n+1/2)\hbar\omega/kT]=\{(1-\delta^m)/(1-\delta)\},\text{ where } \delta=\exp\{-(\hbar\omega_B/kT)\}. \tag{3}$$

Here summation is made on n, changing from n=0 up to $m=U_B/\hbar\omega_B-½\approx U_B/\hbar\omega_B$, where m is the maximum number of phonons located in the potential well, which is equal to the cohesion energy, and ω_B is the frequency of Einstein's phonon. Then $\ln Z_B \approx -\hbar\omega_B/kT - \ln m$.

Now we write the free energy of A+B system after the end of the doping process, assuming, that a part of A atoms has replaced a part of B atoms, and taking into account, that the changes in the large system A are negligibly small ($F_{2A}\approx F_{1A}$).

$$F_2\approx F_{1A}+NV(1-C)\{U_B+3kT\ln Z_B\}+NVC(U_{A/B}+3kT\ln Z_{A/B})+NVCkT\ln C+(\sigma+Cd\sigma/dC)S \tag{4}$$

Here A/B indexes mean the A atoms, penetrated in B, the derivative $d\sigma/dC = -\beta$ means the "speed"

Journal of Physics: Conference Series **291** (2011) 012010 doi:10.1088/1742-6596/291/1/012010

of change of surface tension under transition of A atoms in B (see [9]). The change of free energy under doping is equal:

$$F_2 - F_1 = NVC(U_{A/B} - U_B) + 3NVC\,\{\ln(Z_B/Z_{A/B})\} + C(d\sigma/dC)S + NVkTC\ln C. \quad (5)$$

Equilibrium value C corresponds to the solution of equation:

$$d\{F_2 - F_1\}/dC = 0 \quad (6)$$

If $\Delta U(C,S/V)=U_{A/B}-U_B=\Delta U_0\{1-f(aS/V)\}(1-\gamma C)$, then the solution of the equation (6) leads to the transcendental equation relative unknown concentration C:

$$C = A_0 \exp\{2\gamma C/kT\}; \quad (7)$$

Here $A_0=[U^0_B/U^0_{A/B}]^3[\omega_{A/B}/\omega_B]^3\exp\{-[\Delta U_0/kT(1-f(aS/V)]-1]+\beta S/NkTV\}$; U^0_B, $U^0_{A/B}$ are the cohesion energies of atoms B and A in the phase B (that is A/B), in a hypothetical case of a bulk phase of C.

The equation (7) is convenient for solving graphically (figure 1). However, for small γ the simple analytical solution takes place:

$$C=(A_0)/(1-A_0 2\gamma/kT). \quad (8)$$

It is evident, that the concentration C sharply depends on the nanoparticle radius, equal $3(S/V)^{-1}$, and super exponentially increases at decrease of the nanoparticle size. This dependence is much stronger, than found in [9]. It is obvious, that at $\gamma>0$ and $\beta>0$ all properties 1-4of Introductions are satisfied, that speaks about a rationality of the suggested model. Note, that at $\gamma=0$ our results (8) also are not reduced to [9] since in our model the finiteness of cohesion energy and its dependence on the nanoparticle size are considered.

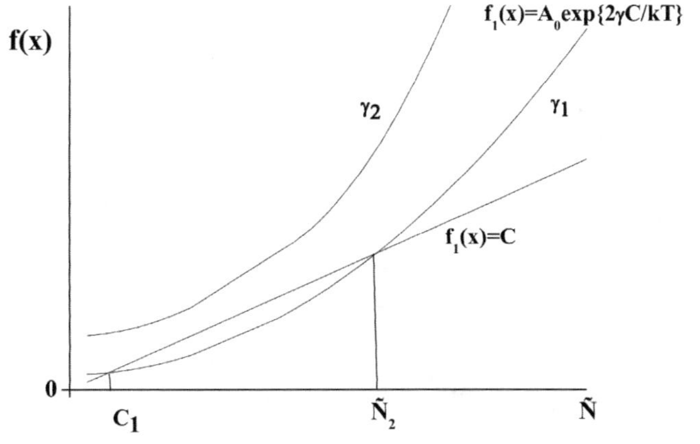

Figure 1. The graphic solution of the equation (7); $\gamma_2 > \gamma_1$.

3. Discussion

The experimental check of effect of increase of solubility of a doping impurity in nanoparticle at reduction of its size represents a not trivial problem. Apparently, the magnetic nanometrology [18] methods can play a special role here. In this connection we performed the experimental research of magnetic parametres of the iron oxide nanoparticles Fe_3O_4 doped as by gadolinium, as by other impurities. The magnetic properties were measured on SQUID magnetometer and on the vibratory magnetometer. Were investigated the electron paramagnetic resonance (EPR) spectra of these nanoparticles. The standard technology of chemical co-deposition at pH 10 - 12 has been used for synthesis of the gadolinium-replaced ferrite. The size of Fe $[Gd_xFe_{2-x}]$ O_4 nanoparticles was≈10 nm and has been supervised by TEM. The Curie temperature of these nanoparticles is enough high, and the specific magnetisation decreases with the increase of gadolinium content (table 1, figure 2). It is established, that the Gd doping influences on the Curie temperature weakly. It would seem, for the increase of nanoparticles magnetisation it is necessary to partially replace Fe^{3+} by ions with the big magnetic moment, for example, by Gd^{3+}. However, the result was opposite (see table 1). The growth of gadolinium percentage reduced considerable (by two order) the magnetic moment of nanoparticles (figure 2).

Table 1. Magnetic parameters of $Fe[Gd_xFe_{2-x}]O_4$ nanoparticles

x	T_{Curie}, K	M_S, emu/g	$N=M_S/\mu_{Fe3O4}$
0.1	738	29,1	$7{,}657*10^{20}$
0.16	873	27,1	$7{,}124*10^{20}$
0.71	893	19,0	$4{,}998*10^{20}$
0.97	891	15,1	$3{,}211*10^{20}$
1	724	14,68	$3{,}862*10^{20}$
1.12	833	12,94	$3{,}404*10^{20}$
1.9	626	1,36	$3{,}590*10^{19}$

Note, that in view of the big difference of the ionic radiuses of trivalent gadolinium ions and of iron ions (for example, for Gd^{3+} ionic radius in spinel is 1.15 Å, and for Fe^{3+} it is 0.67 Å) gadolinium in a volume variant cannot penetrate into a spinel matrix [12]. It follows from the data of the roentgen-phase analysis, that in system $(1\text{-}x)Fe_2O_3 + xGd_2O_3$ the hard solutions [19] are not formed. At the same time, under doping of nanoparticles as it has been shown preliminary by us [5], the situation is opposite. Indeed, on the base of measurement of magnetic characteristics of nanoparticles system of "iron oxide with the impurity of gadolinium" it is possible to show, that gadolinium replaces the ions of iron and it transits in Gd^{3+} state [5,20]. It is obvious for the system of $Fe3O_4$ with gadolinium, that $\gamma < 0$ and $\beta < 0$. Thus, it is evident, that gadolinium became a very good test for model, satisfying to points 1 and 3 (see Introduction).

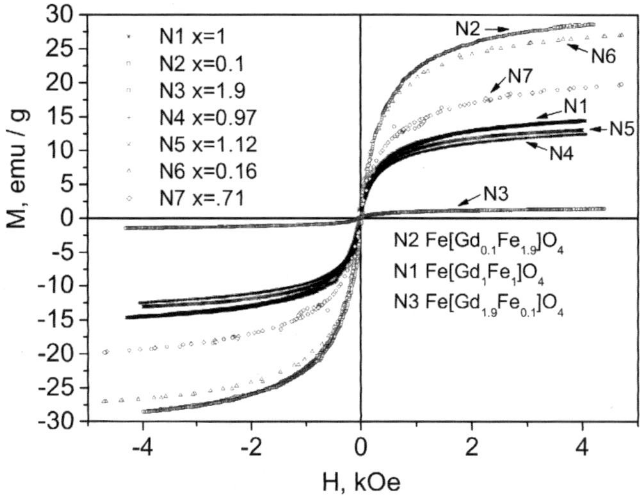

Figure 2. The magnetic field dependence of the specific magnetic moment for iron oxide nanoparticles doped by gadolinium. X is the Gd content.

It is possible to explain the paradoxical result presented in a figure 2 as follows. The basic contribution to formation of the total magnetic moment of the nanoparticles ensemble give one-domain small nanoparticles. The curves of magnetisation presented in figure 2 have superparamagnetic character, that is the contribution to formation of the magnetic moment of the nanoparticles ensemble give small one-domain particles with radius, smaller than R~5 nanometer. This fact, that the gadolinium solution "worsens" the magnetic properties, i.e. reduces the full magnetic moment of ensemble, confirms, that the basic adolinium solutionoccurs in small nanoparticles. As to the microscopic mechanism of "damage" of one-domain small particles, that, probably, here can play a role the "facilitated" mechanism of formation of dislocations, vacancies and other defects in small and ultrasmall nanoparticles under reduction of the characteristic sizes of nanoobjects, found in [8, 21, 22].

As to the possible mechanism of the "facilitated" formation of dislocations in small particles it is possible to note the following. The large difference of radiuses of Fe^{3+} ions, replaced by Gd^{3+}, causes the critical tension [20]. In the case of large difference of the sizes of the doping impurity and the replaced nanocrystall atoms, the realization of other mechanism of the reduction of free energy of nanocrystall+impurity system is hypothetically possible in principle, i.e. the spontaneous formation of misfit dislocation with the core of an impurity atom. The similar phenomenon is known in the heteroepitaxy physics, namely formation of the van der Merve dislocation [22]. The dislocations in small particles of radius R ~ 5 nanometers possess considerably smaller elastic energy according to the formula [23]: $U_{elast} \sim Gb^2\ln(R/a^*)$, where G is the elastic module of shift, b is the Burgers vector of the misfit dislocation, a* is the size of the impurity atom. As the linear size R of nanocrystall is much less than the linear size L of bulk samples, the energy of formation of the misfit dislocation in nanocrystall is much more small, than in a bulk crystal. Hence, the process of dopant penetration in the nanocrystall becomes possible, whereas in the bulk material it is impossible. The same mechanism, apparently, is more significant and for the impregnation of the doping impurity in nanocrystall in the form of ultrasmall cluster. Note, that the presented microscopic mechanism of abnormal of the impurity penetration in nanocrystall has been not discussed till now. In case of penetration of magnetic, fluorescent, plasmon etc. labels in the nanocrystall, their concentration and measured amount refer to one nanoparticle are directly connected according to the formula (1) with the size R of the nanoparticle and therefore can serve as a reference point for determination of the metrological parameters of nanoparticle.

4. Conclusions

In present work is constructed the model of dependence of the effect of increase of solubility of the doping impurity in nanocrystall from the nanoparticle size, i.e. from the amount of atoms penetrating into it. The comparison of experiment with the theory allows to receive the parametres essential for the diagnostics of nanocrystalls: "speeds" of decrease of cohesion energy from the nanocrystall size and "speeds" of reduction of surface tension coefficient at the change of concentration of the penetrating impurity. These two characteristics are essentially important in a problem of metrology of the nanocrystalls doping process. The given approach is actual, because multipurpose nanoparticle are used in nanodiagnostics, in medicine, in criminalistics and many other things areas.

Acknowledgements

We are grateful to Prof. V.D. Kuznetsov, and Prof. N.A.Brusentsov for their support and assistance.

References

[I] Suzdalev P 2006 *Physical chemistry of nanoclusters, nanostructures and nanomaterials* (Moscow: Com.Kniga; *in Russian*)
[2] Gubin S P, Koksharov J A, Homutov G B and Jurkov G J 2005 *Uspekhi Khimii* **74** 539
[3] Pool Ch and Owens F 2007 *Introduction to nanotechnology* (New York: Wiley–Interscience)
[4] Nikiforov V N and Filinova E Yu 2009 *Biomedicical application of magnetic nanoparticles* (Wiley-VCH Verlag GmbH) pp 393-455
[5] Brusentsova T N, Brusentsov N A, Kuznetsov V D and Nikiforov V N 2005 *Journal of*

Magnetism and Magnetic Materials **293** №1 298
[6] Zubkov A I, Ilinsky A I, Podgornaja O A, Sevruk V I and Sokol-Prussky J G 1990 *Fizika metallov i metallovedenije (Physics of metals and metallurgical science)* **№10** 197 (*in Russian*)
[7] Gleiter H, Weissmuller J, Wollersheim O and Wurschum R 2001 *Acta Mater.* **49** №4 737
[8] Qi W.H and Wang M.P 2003 *Physica B: Condensed Matter* 432
[9] Koshkin V M and Slezov V V 2004 *Pis'ma v ZhTF* **30** №9 38 *(in Russian; translated to: Technical Physics Letters)*
[10] Narayan J 2006 *J. Appl. Phys.* **100** 034309
[11] Wesselinowa J M and Apostolova I 2007 *J. Phys.: Condens. Matter* **19** 406235
[12] Malavosklish Bikram . *Patent USA.* Ultrasmall superparamagnetic iron oxide nanoparticles and uses. IPC8 Class: AA61K4912FI ,USPC Class: 424 9322
[13] Cen C, Thiel S, Mannhart J. and Jeremy Levy February 2009 *Science 20* **323** №5917 1026
[14] Landau L D and Lifshits E M 1976 *The theoretical physics in ten volumes. V.5. The statistical physics. Part.1.* (Moscow: Fizmatlit; *in Russian*)
[15] Vinetsky V L and Holodar G A 1969 *Statistical interaction of electrons and defects in semiconductors* (Kiew: Naukowa dumka; *in Russian*)
[16] Tereshchuk P L, Khakimov Z M, Umarova F T and Swihart M T. 2007 *Phys. Rev. B* **76** №12 125418
[17] Animalu A O 1977 *Intermediate Quantum Theory of Crystalline Solids* (New Jersey: Prentice-Hall, Englewood Cliffs) 516
[18] Nikiforov V N, Sredin V G, Nikiforov A V etc. 2009 *Second International forum on nanotechnologies*. 254
[19] Music S, Ilakovac V, Ristic M and Popovic S 1992 *J. Mater. Sci.* **27** 1011
[20] Obaidat I M, Mohite V, Issa B, Tit N and Haik Y 2009 *Cryst. Res. Technol.* **44** №5 489
[21] Qin S, Liu D, Liu H, Zuo Z 2008 *J. Phys. Chem. C* **112 (44)** 1717
[22] Jain Harker A H and Cowley R A 1997 *Philosophical Magazine A* **75** №6 1461
[23] Kosevich A M 1981 *Physical mechanic of real crystals* (Kiew: Naukowa dumka; *in Russian*)

Journal of Physics: Conference Series **291** (2011) 012011 doi:10.1088/1742-6596/291/1/012011

Moving liquid surfactant as a way of assessing the properties of surfactant, liquids and surfaces.

A O Titov, O P Titov, M O Titov, A N Karbainov

RUSSIA. GOU VPO East Siberian State Technological University.

E-mail - fibrilla45@mail.ru

Abstract.
In the study of surface phenomena of the main and only instrumentally-defined parameters are surface tension and wetting angle, including in the field of nanotechnology. These indicators were introduced more than 200 years ago, and any new inventions in this field was no more. The university developed a new method and device for determining the surface activity. The basis of the method and device is the use of video cameras to record the droplet size and changes on the surface of the liquid layer of known thickness from the impact of drops of surfactant (surfactant). Committed changes are then processed using computer software and calculated parameters, which can be characterized by a surfactant and surface properties, which is fluid and very liquid. Determine the surface tension or contact angle is not necessary. Measures of surface activity using the method and device are:

1. The amount of fluid that can move one kilogram of surfactant. The value of this index varies from tens of nanometers to hundreds of thousands of units. The indicator can be converted to energy units, joules.
2. The amount of fluid confined by a surface per unit time is calculated based on the first indicator, complements the characterization of surfactant and may be an indicator of surface characteristics and fluid.
3. propagation speed of the capillary and microwaves. This indicator complements the first two.

1. Introduction

Known methods for determining the surface activity using two instrumental parameters to be determined the surface tension and wetting angle. To determine them, developed a lot of ways. According to these parameters is judged on the ability of emulsifying agents is suspended, or form a foam [1].In addition, the solutions themselves surface-active substances are composed of self-assembled particles - micelles, which have dimensions of nanoparticles and the aptly Victor Balobanova are the miracle of nanotechnology [2].

And the contact angle and surface tension in certain conditions does not provide a sufficiently complete characterization of solutions on the surface activity of the impossibility of defining or low variability index determined by various conditions, such as substance concentration, temperature, use of substances inactive or do not alter the surface activity, etc. And improve the accuracy of identification, such as surface tension requires a high purity of the material. In addition to defining the work of adsorption, wetting or spreading requires knowledge of other parameters, the definition of which is associated with significant complications or impossible to determine because of the lack of reliable methods to determine [1].

The introduction of surface-active substances (SAS) in the liquid alters the surface tension is not more than ten units, and the wetting angle from 0 to 180, therefore, share or identify surfactants on these indicators is practically impossible.

Particularly relevant is the separation and identification of substances with surface activity in the development and evaluation nanomotors using surface activity as the driving force, as well as to evaluate the surface properties obtained with the use of nanotechnology

2. Experimental part

Determination of surface activity is carried out at the facility (see Figure 1). Detailed scheme of the setup is given in [3].

On the table with adjustable horizontal surface of a stack plate made of material 2, the properties of the surface which is necessary to investigate (in the diagram in Fig. 1 denotes an object - the drug). For retention on the test surface of a liquid layer, such as 0,1-1 mm thick, the material applied to the circumference of a hydrophobic substance, if liquid is polar, or hydrophilic substance, liquid or solutions of various substances, whose influence is necessary to investigate, not polar. Then set the camera 4 or a movie camera so that the restrictive line and the center of bounding shapes are clearly visible in the viewfinder and possibly cover the entire area frame (adjustment of image sharpness). After setting the image sharpness set line with a scale of 1 mm. And fix the camera for subsequent scaling measurements. Ruler set perpendicular to the optical axis of the lens fixing process chamber 4, exactly the diameter of the circle. After that line removed.

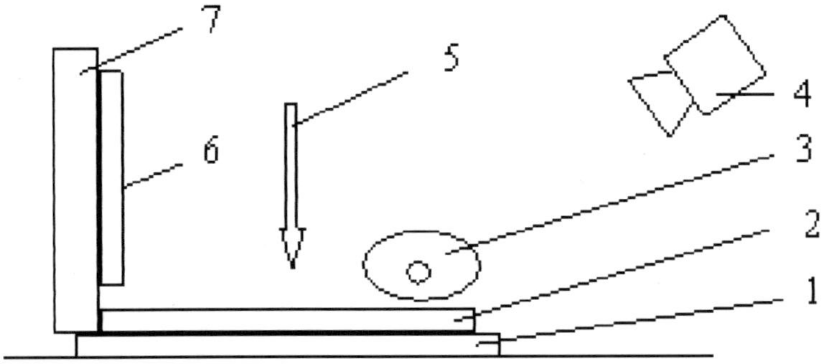

Scheme 1. The general scheme of the installation.
1. Desk with adjustable horizontal; 2. Object - a drug (investigated surface); 3. Camcorder to record the droplet size; 4. Camcorder to record changes on the surface of the object - a drug; 5. Pipette; 6. Indicator grid; 7. Illuminator.

In limited hydrophilic or hydrophobic substance circle made by the investigated liquid in an amount necessary to create a liquid layer thickness selected researcher.

Exactly over the center of bounding shapes, such as a circle, set the calibrated mass of the drop and capillary diameter tip pipette 5 so that a drop of it fell as far as possible the exact center of

Journal of Physics: Conference Series **291** (2011) 012011 doi:10.1088/1742-6596/291/1/012011

the figure. Edge of the tip of the pipette set at a height of 4-30 mm. Illuminator 7 with flat grid 6, set so that reflected from the surface of the liquid image grid in the fixing chamber 4 was clearly visible.

The camera includes the fixation image, both to determine the drop volume at the time of separation from the capillary pipettes include a chamber 3, retaining a larger scale drop, and a drop of surfactant solution or the liquid is made into the center of the circle. Film frames, fixing the process of moving the liquid successively studied by determining the distance from the center of the fall of the drop to the ground «wave of displacement» and in accordance with the scale of the transfer of units of length. Similarly, define the diameter of the drop at the time of separation from a capillary pipette.

If you want to define or compare the properties of surfactants, we can use the «standard» surface, as which can be used hydrophobic heat-resistant film or writing paper, or paper with a modified surface, such as gelatin.

When working with paper on it cause the circle with the necessary internal diameter of the hydrophobic dye. Linewidth bounding figures 5 - 6 mm. Paper coated with a bounding shape (an object - a drug) are soaked in a solvent such as water for some time, say 10 minutes and put on the table, or stacked on his plane-parallel plate (thick glass). In this paper, straighten, and from beneath it removes the air, extruded through a glass tube with rounded ends, for example, pipette diameter of 10 - 15 mm, or other devices, such as roller glossing photos or cushion for straightening and pasting wallpaper. To the square of paper, limited deposited lines (circle, square), cause under investigation in the amount of liquid required to give a layer thickness is determined by the experimental conditions. In the center sets a pipette tip, include fixing the camera and make the center of the bounding shape of the object - a drug drop of the test surfactant. For more details see [4]

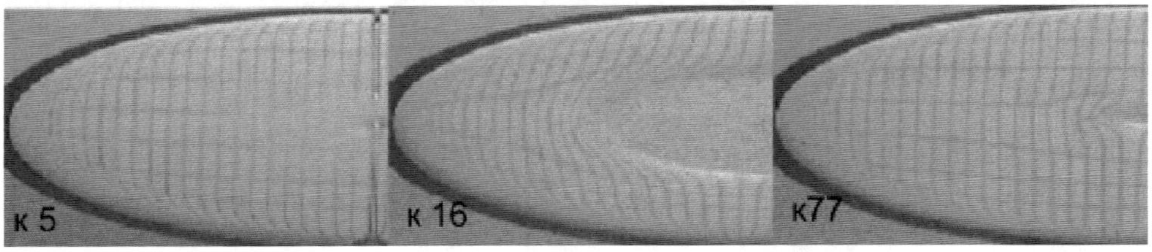

Fig. 1. View of the surface after exposure to drops of surfactant «Equalizer», a concentration of 5 kg / m^3. From left to right frame k5 - The fall of the drop, shot k16 - the largest radius of the displaced layer frame k77 - the closure of the funnel.

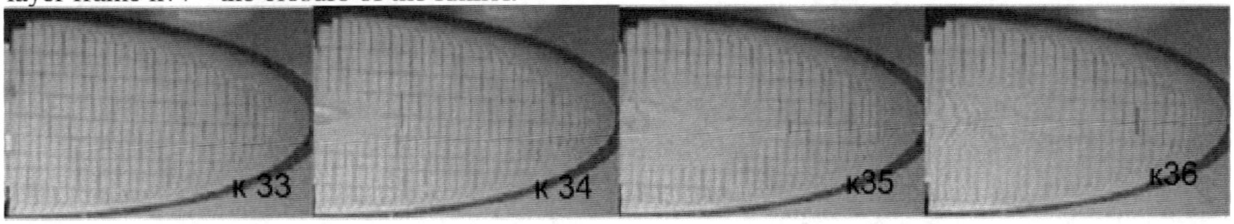

Fig. 2. View of the surface after exposure to a drop of clean water. From left to right frame of k33 - a drop in flight, shot k34 - a drop fell formation of capillary waves, the frame k35 - scattering of waves, the frame k36 - the extinction of the wave.

Method illustrated by the following images in the figures. After touching a drop of surfactant solution (see Fig. 1 frame k5) on the surface of the liquid layer, a small «crater». From it in different directions begins to propagate the «wave of displacement» of the liquid, which after a time reaches the maximum displacement (see frame k16 in Fig 1). Then release part of the surface closes the liquid (see frame k77 in Fig. 1). On these images, Fig. 1 shows the process of moving the liquid surface-active substances called «Equalizer», a concentration of 5 kg / m^3. Accordingly, frame k5, k16, k77 movie. The inner diameter of the bounding circle of 122 mm. Water depth - 0,5 mm. For comparison, Figure 2 shows the change in the water from falling a drop of clean water.

3. Determining the amount of fluid moved surfactant.

To determine the amount of fluid displaced by time-lapse unfolding find a frame corresponding to the maximum radius (diameter) of the displaced fluid layer and using standard computer programs, measure it. Figure 3 shows an example of the measurement.

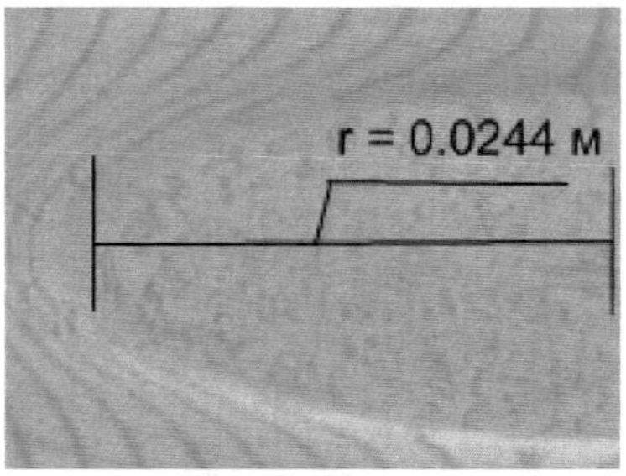

Fig. 3. Example of measuring the radius of the displaced fluid layer to frame 16 in Fig. 1

After finding the radius is determined by the volume of displaced fluid, given its density. In this example, the density of water was equal to 1000 kg/m3. The volume of displaced fluid was found to be 9.4 × 10-7 m3. Water depth in the experiment was 5.0 x 4.10 m. In terms of mass of the displaced fluid was 0.00094 kg. When the surfactant concentration «EQUALIZER» 5 kg / m^3 and 0.00229 m diameter drops of surfactant, that you move is equal to 3.15 x 8.10 kg. Then a specific amount of fluid displaced will be 29688.55 kg. One kg of surfactant can move 29.7 tons of water to a distance of 0.0244 m. This corresponds to work on a 7106.37 joules.

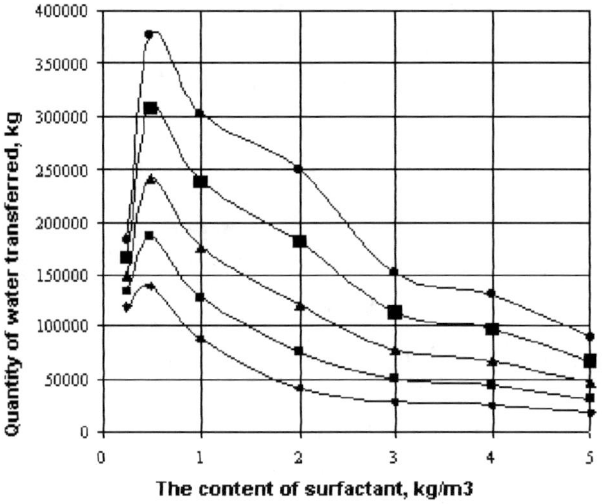

Fig. 4. Changing the specific amount of water transferred, surfactants «Equalizer», depending on the content of the surfactant and the diameter of the bounding circle (the thickness of the liquid layer 0.0005 m). The radii of the bounding circle (curves from bottom to top) 0.04, 0.05, 0.06, 0.07, 0.08 m.

The amount of fluid displacement affects the size of the bounding shape and concentration of surfactant in the solution. Fig. 4 shows the results of a study on the influence of these parameters on the amount of fluid displacement surfactants.

4. Determination of surface properties

To determine the properties of the surface on which the liquid was, found time to reach maximum radius of travel by the number of personnel from the moment of touching a drop of surfactant liquid

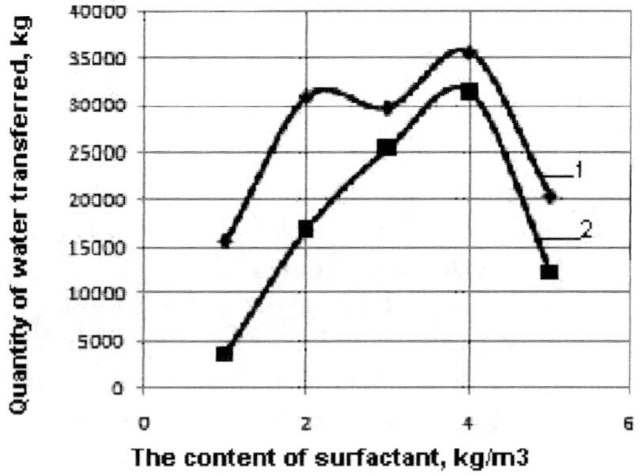

Fig. 5. Changing the parameters depending on the surfactant concentration:
1. Specific amount of fluid displacement (kg liquid / kg of surfactant
2. The specific amount of fluid displacement per unit time ((kg liquid / kg surfactant) * second)

to a frame with the largest radius of movement. Knowing the frame rate, found the time between shots and multiplied by the number of frames to achieve the greatest range of fluid movement. Then multiplying by the time required to achieve the greatest movement of fluid, the number of displaced one kilogram of surfactant liquid, were found in surface. The results are shown in Fig. 5. The results presented show that the number of displaced fluid is changed, passing through a maximum which is reached at a concentration of 4 kg / m^3. Perhaps this is due to the properties of surfactant solutions, which vary significantly with increasing concentration and the formation of micellar nanostructures. [2]

5. Determining the properties of liquids

To determine the range distribution of the microwaves used by reflection indicator grid, which on the surface of the liquid is clearly seen in the movie. Excitation of the surface of the liquid was carried out with the falling water drops height of 0,02 m at the surface of the water

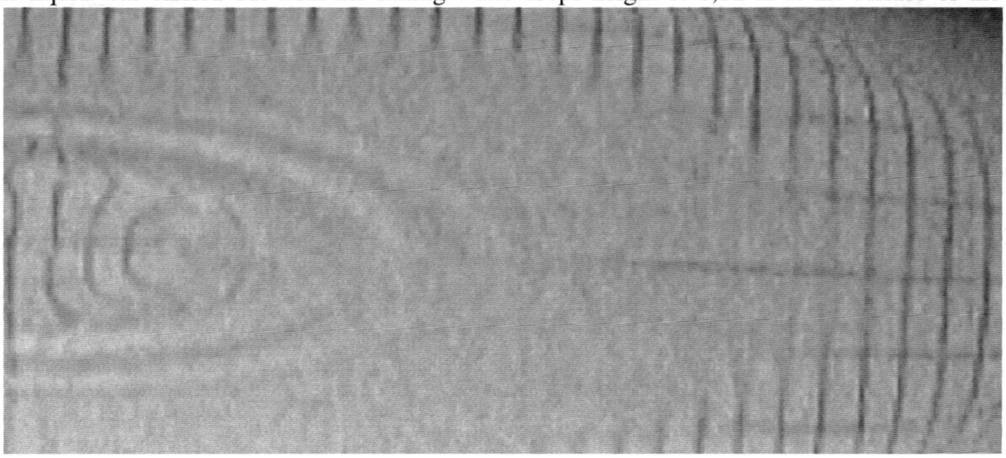

Figure 6. Propagation of waves on the surface of the liquid third frame from falling drops of water.

poured into the restrictive circle on the paper surface. Different amounts deposited in the bounding circle of the liquid changes the thickness of the layer from 0.0002 m to 0,001 m on the surface of the water-cut paper. Changes occurring in the interaction of water droplets with surface layer of water was

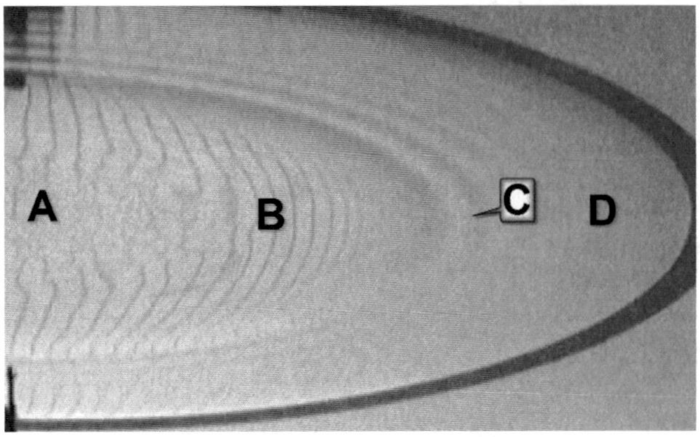

Figure 7. Form a layer of water after exposure to a drop of surfactant solution.

fixed with a video camera. Then on the frame scan selected shots for the measurements of these observed changes. (See appendix) for this use standard computer programs WINDOWS. Similarly, there were at studying the effect of surfactant concentration on the velocity of the waves. Studies have shown that water droplets in contact with the surface water initially formed microwaves (see Fig. 6.) that move across the water with almost constant velocity (70 - 75) x10-2 m / sec. (See Figure 4) Completion of the microwave image of the indicator grid disappears. Formed and then great waves move over the surface of the liquid at a lower rate, on average, 32h10-2 m / sec. (See Figure 6) C increase in water depth range propagation of microwaves increases. Given the study of American astronauts, one can assume that microwaves are generated by vibrations of the surface water film whose thickness is 0.00015 m [5]. And their height (amplitude) may not exceed twice the thickness of the film.

The interaction of drops of surfactant with the surface water also causes the appearance of these two types of waves, however, larger wave, starting from the center of the fall of the drop, then move to the edges of the bounding circle is under the influence of «wave moving fluid» (see Figure 7. Mark with). She seemed to be pushed in front waves and the trailing edge of the wave becomes steeper. This wave slows its movement when approaching the edge of the bounding circle, and the beginning of their movement reveals a surface on which the liquid (see Fig. 7. A tag) and «boundary layer» of fluid (see Fig. 7. In the tag, the line indicator grid are clearly visible), under the influence of the force field of this surface. Moreover, at low concentrations of surfactants and strong interaction with the fluid lying below the surface «wave moving fluid» rolled over the surface «boundary layer». He is brilliant at all times and determine the zone A free fluid is observed. The introduction of surfactant lowers the speed of propagation of large capillary waves. (Without the surfactant average speed of capillary waves, 32 x 10-2 m / s and with surfactant 20.5 x10-2 m / s) And microwaves are observed on the surface of large waves, because of what the indicator grid on the surface of the liquid is not visible (Fig. 6 and 7.) Large Wave «movement of the fluid» slow down when approaching the edge of the bounding circle to the (13-15) x10-2 m / sec. The speed of microwaves remains almost constant between 65 - 75 x10-2 m / sec. At this speed can be detected speed only twice during the passage of microwaves on the investigated surface. Microwaves quickly reach the restrictive circle.

The effect of opening the boundary layer of fluid wave displacement can be used to estimate the thickness of boundary layers and their visual observations. A property of microwaves to increase the range of distribution with increasing thickness of the layer of fluid can be used to estimate the thickness of the layer of water associated with the surface on which the liquid is, and this estimate of interaction energy of the liquid to the surface. Method for moving liquid surface-active agents are encouraged to use to characterize the properties of surfaces and identification of surfactant, including nanomaterials and in the metrology of nanomaterials.

References

[1] Surface phenomena and surface-active substances: A Handbook, A. Abramson, LE Bobroff, LP Zaichenko and others, ed. A. Abramson and ED Shchukin, - L., Chemistry, 1984, s.392, illus., P.165.

[2] V. Balabanov Nanotechnology. Future Science Series: Discoveries that Shook the World Publisher: Penguin Books, 2009, 256 pages http://www.nanonewsnet.ru/blog/nikst/nanotekhnologii-v-meditsine-0

[3] AO Titov Titov, OP, Titov, MO U.S. Patent 2362979 Device for determining distance propagation of microwaves on the surface of the liquid layer.
[4] AO Titov Titov, OP, Titov, MO Patent 2362141 Method for determining the amount of fluid moved surface - active substances.
[5] Experiments with soap film without soap. Magazine "Science and Life» № 1, 2004.

Journal of Physics: Conference Series **291** (2011) 012012 doi:10.1088/1742-6596/291/1/012012

A complex of complementary pulsed neutron sources, neutron and radiographic nano-diagnostic instruments at the Institute for Nuclear Research RAS

E.A. Koptelov[1], Yu.V. Ryabov[1], A.A. Alekseev[1], S.F. Sidorkin[1], M.I. Grachev[1], V.A. Fedchenko[1], R.A. Sadykov[1,2], Yu.B. Lebed[1], V.S. Litvin[1], S.P. Kuznetsov[3], E.S. Clementyev[1,4], P.A. Alekseev[4], V.A, Trunov[5], A.P. Bulkin[5], V.A. Ulianov[5] and S.N. Axenov[1]

[1] Institute for Nuclear Research RAS, pr. 60-letiya Oktyabrya 7a, 117312 Moscow, Russia

[2] Institute for High Pressure Physics RAS, 142190 Troitsk, Moscow Region, Russia

[3] Lebedev Physical Institute RAS, 119991 GSP-1 Moscow, Russia

[4] RRC "Kurchatov Institute", 123182 Moscow, Russia

[5] Petersburg Nuclear Physics Institute, 188300 Gatchina, Russia

E-mail: koptelov@inr.ru

Abstract. A new large-scale research center for nano diagnostics and neutron and X-Ray studies of nanomaterials is briefly described. This center at the Institute for Nuclear Research of the Russian Academy of Sciences (INR RAS) is based on three specially designed spallation neutron sources driven by primary proton beams. Similar research centers on the basis of the high flux spallation neutron sources are created in the USA, Japan, Great Britain, Switzerland and will be build in China and the EU (Sweden). We discuss neutron and X-Ray instrumentation at the INR RAS and the corresponding domains of research of various materials including nano-systems.

1. Introduction

Neutron scattering is an indispensable tool for the structural and dynamical investigations of new materials, including nanomaterials [1]. As a consequence, a lot of new research centers for nano-diagnostics and for the detailed studies of nanomaterials are created on the basis of the spallation neutron facilities in the USA, Japan, Great Britain, Switzerland and a few will be build in China and Sweden (the European Spallation Neutron Source) [2]. The intense pulsed neutron beams at such neutron research centers are generated by the primary proton beams, which are provided by large accelerating facilities. A similar research center is in operation in the Russian Federation at the experimental area of the large-scale neutron complex of the tree accelerator-driven spallation neutron sources at the Institute for Nuclear Research of the Russian Academy of Sciences (INR RAS). The pulsed neutron sources provide different neutron energy distributions so one can choose an appropriate one depending on the scientific task and the required energy (or neutron wavelength) range. Due to a variety of tasks and objects to study the neutron instruments are designed for various branches of the neutron experimental technique: neutron diffraction, inelastic neutron scattering, neutron reflectometry, small-angle neutron scattering, neutron radiography [3]. All these branches are presented at INR RAS and the corresponding neutron instruments are either already in operation or under construction.

X-Ray experimental technique (both synchrotron and laboratory X-Ray sources) is a valuable complementary tool for neutron scattering, that is why new synchrotrons are built very close to the neutron scattering research facilities in France, Great Britain, Germany, etc. A few state of-the art laboratory X-ray instruments are available at the INR RAS for nano-diagnostics.

The research center under discussion is located in Troitsk (a home for many academic and non-academic research institutes) close to Moscow. Taking into account the unprecedented concentration of scientific institutes and universities in Moscow and its region, the research center at INR RAS provides an access to neutron and X-ray instruments to a vast scientific community specialized in condensed matter physics, Earth physics, chemistry, biology, nuclear energy, materials science and nano-technology.

A very important aspect is an ecologically-friendly regime of the neutron generation at the INR RAS since no fissile materials are involved and no radioactive waste is produced at the neutron facility. This complex is the only ecologically-friendly neutron research center in Russia since all other intense neutron sources in the country (stationary or pulsed reactors) rely on the nuclear fission process.

2. Spallation neutron sources at the INR RAS

The Neutron Complex of the INR RAS includes three spallation neutron sources: the pulsed thermal neutron source IN-06, the high luminosity 100-ton spectrometer for neutron slowing down in lead LSDS-100 and the neutron source RADEX with a time-of-flight neutron spectrometer based on a beam-stop irradiation facility (RADEX-n-TOF spectrometer). The operation of the INR RAS Neutron complex is governed by the high-current proton Linac. The total length of the Complex including the Linac exceeds 800 m. The Complex is located in Troitsk, a multi-purpose scientific center of the RAS in the vicinity of Moscow.

The Linac consists of a 400-keV proton injector under operation and a H^- injector under commissioning, lowenergy beam transport lines, a 750-keV RFQ booster cavity, five Drift Tube Tanks of the DTL initial Linac part with the energy 100 MeV, and 27 modules (four sections each) of the Disk and Washer accelerating structure of the Coupled Cavity Linac (CCL) with energy 600 MeV. A DTL RF system consists of six 5-MW pulsed-power 198.2-MHz RF stations. The CCL system has 31 RF stations based on 3.75-MW 991-MHz klystrons. A low-level RF system allows keeping the RF setting of the accelerating field with stability at the level 0.1% for amplitude and 0.1^o for phase. The details on the proton driver, the neutron sources and experimental areas are provided in [4]. The general layout of the neutron sources and the proton driver is depicted in figure 1. See Table 1 for the proton driver parameters.

Table 1: The INR RAS Linac parameters

Parameter	Designed	Achieved	November 2010	Nearest future
Energy, MeV	600 H^+,H^-	502 H^+	209 H^+ (160 H^-)	502 H^+,H^-
Pulse current, mA	50	22	12	16
Repetition rate, Hz	1÷100	1÷50	1÷50	1÷100
Pulse length, μs	100	0.5÷200	0.25÷170	0.25÷200
Average current, μA	500	150	120 (160 MeV) 40 (209 MeV, n-TOF)	300

The former name of the whole facility was the Moscow Meson Factory. Nowadays it is converted into the Neutron Complex of the INR RAS due to the shift of the focus of the main interest from the physics of mesons and nuclear physics to the studies of condensed matter including nano-systems. However the branches of nuclear energy investigations and fundamental physics of neutron interactions are still active at the INR and a few instruments are designed for these domains in neutron science.

The pulsed neutron source IN-06 was designed primarily for the neutron studies of structure and dynamics of condensed matter by thermal neutron scattering method on different instruments and a variety of experimental techniques. Measurements at the IN-06 may cover different fields of solid matter and soft matter investigations in science and technology. Two other spallation neutron sources (LSDS-100 and RADEX-n-TOF spectrometer) are used mainly for neutron data collection for nuclear energy technologies and nuclear physics. The RADEX-n-TOF spectrometer has a capability in the field of the neutron studies of condensed matter by epithermal and thermal neutrons.

The integrated neutron flux depends on the Linac proton-beam pulse duration (0.25 - 160 μs) and it could be up to 1.2×10^{15} n/s in the full solid angle (4π Sr). In the current time the Linac provides the primary beam with the proton energy 209 MeV and the current up to 150 μA. The project parameters are the following: the proton energy 600 MeV and the current up to 500 μA, which will be achieved in the near future. The intensity of the secondary neutrons and the neutron pulse length are controlled by the accelerator output and its parameters (power and pulse length).

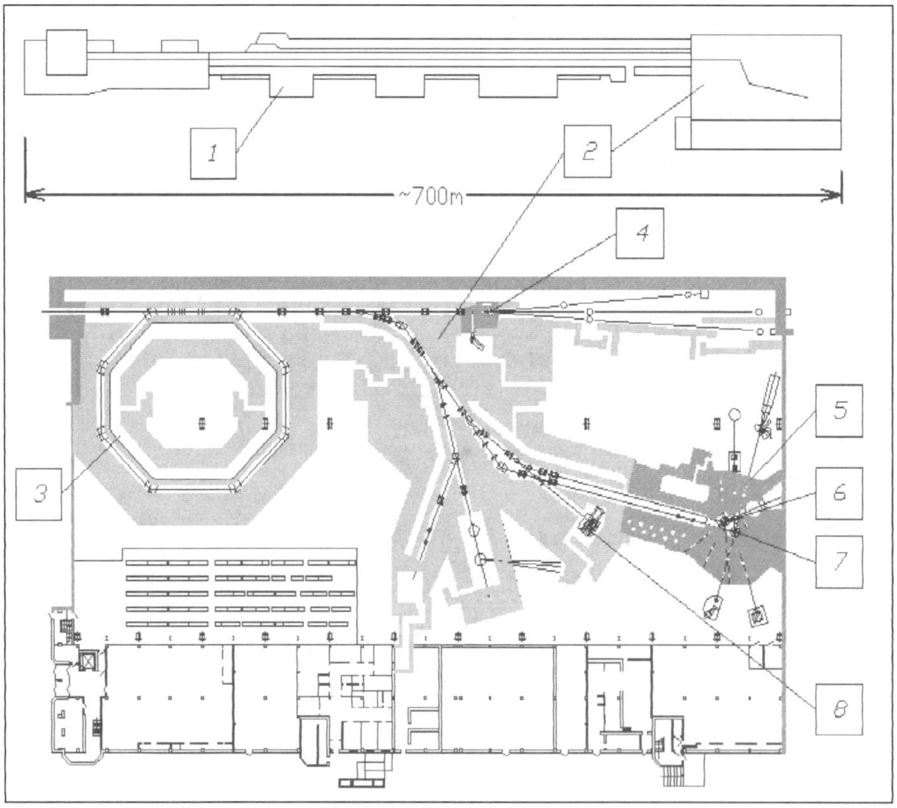

Figure 1. The Neutron Complex of INR RAS. 1- Linac; 2- Experimental Complex; 3- Proton storage ring (not completed); 4- RADEX neutron soutce with neutron guides for TOF spectroscopy; 5- spallation neutron source IN-06 with neutron beamlines for the condensed matter studies; 6- target box the or accelerator driven systems studies; 7- tungsten target; 8- LSDS-100 neutron source.

3. Neutron and X-Ray instrumentation at the INR RAS

The first stage of the Neutron Complex includes of a number of neutron beamlines and instruments at IN-06. The plan of the experimental hall with the neutron instruments of the 1st stage at this source is shown in figure 2. Another three instruments (the 2nd stage) will be located in the second experimental hall (design project is under way) on the neutron beamlines in the forward direction with respect to the incoming proton beam.

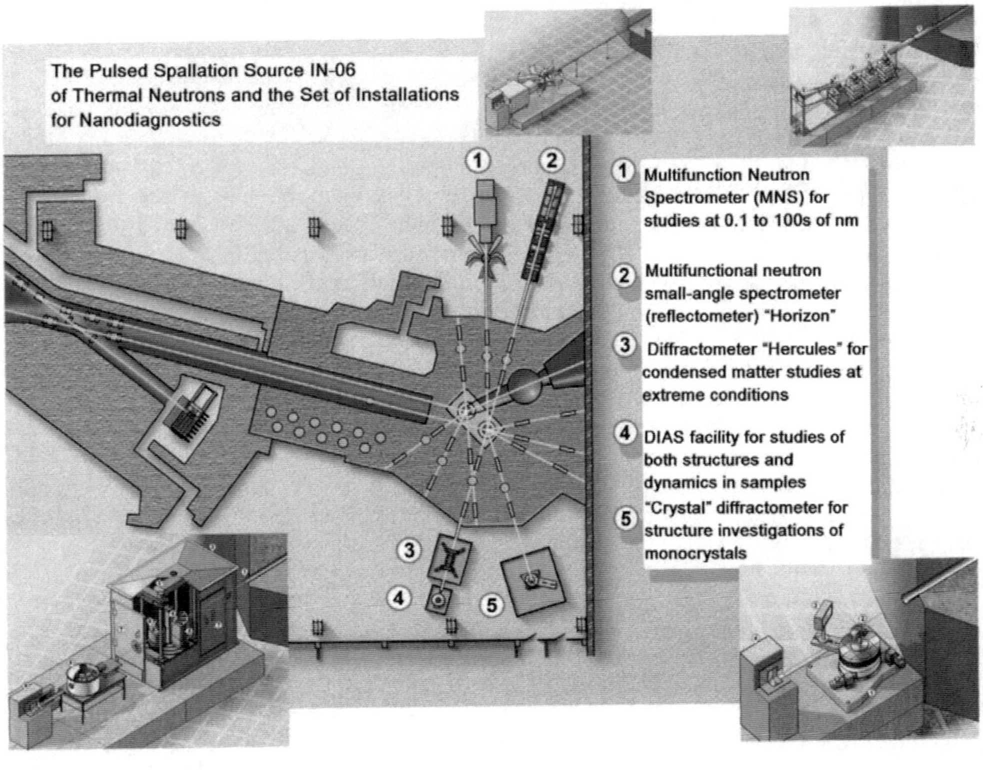

Figure 2. The set of the neutron scattering instrument on the spallation neutron source IN-06 (1st stage). The names of the instruments and the corresponding research branches are indicated on the figure.

The IN-06 instruments for condensed matter studies and nanoscience were designed and constructed at the INR RAS in collaboration with several research institutes by research groups involved in the collaboration. Measurements at high pressures and "*in situ*" experiments of hydrogen containing materials are the specific features of the neutron scattering center at the INR RAS.

The following partner research institutes have made substantial contributions to the development of the neutron scattering instruments at the INR RAS: the Russian Research Centre "Kurchatov Institute", the Institute for High Pressure Physics RAS, the Petersburg Nuclear Physics Institute RAS, the Lebedev Physical Institute of the RAS.

A multifunctional neutron small-angle spectrometer (and neutron reflectometer) "Horizon is ready for operation at one of the channels of the pulsed neutron source IN-06. This instrument was designed and assembled by the researchers from the INR RAS and the Petersburg Nuclear Physics Institute RAS. "Horizon" is intended for studying both liquid and solid samples by the methods of neutron reflectometry and small-angle neutron scattering. Reflectometric studies of liquid samples assume that the surface of these objects is aligned in the horizontal plane by definition. So the scattering plane must be vertical in such a case. The primary neutron beam of the spectrometer (reflectometer) "Horizon" is formed by the NiMo/Ti supermirror neutron guide, which is seven meter long and bent in the vertical plane. The cross section of the beam just after the neutron guide is equal to (10×70) mm^2 (10 mm size is in the vertical plane, and 70 mm -- in the horizontal one). The characteristic wavelength of this neutron guide is about 1.7Å. The small angle scattering option of "Horizon" provides an opportunity to study samples with the characteristic length scales of nanometers. A closed cycle refrigerator for neutron scattering is available on this instrument. "Horizon" has been successfully tested in November 2010.

A neutron diffractometer "Hercules" is intended for the condensed matter studies at extreme conditions at the sample position, including high pressure, high and low temperatures. This instrument was designed in collaboration with the Russian Research Centre "Kurchatov Institute" and the Institute for High Pressure Physics RAS. A special press (force up to 300 ton) and a set of high pressure cells give us a possibility to study the structural properties "*in situ*". The high pressure cells (HPC) has capillaries for filling samples (clatrates, for example) by a gas (D_2, Ar, etc). The HPC operates in a special cryostat. This instrument is well-suited for the studies of hydrogen-containing materials. The sample volume at high pressure can be as large as a few cm^3, which places "Hercules" to the top of the list of the neutron instruments designed to study big samples at high pressure (from a few GPa up to 10 GPa). The maximum of the incoming neutron spectrum on the beamline is about 1.2Å. A special neutron concentrator (TiNi supermirror-based technology) was designed, build and placed into the beamline to enhance the neutron flux density at the longer wavelength side of the spectrum. The first experiments on "Hercules" have been performed in November 2010. These measurements demonstrated a good counting rate and a reasonable resolution of this neutron diffractometer.

DIAS (under construction) is a powder diffractometer, a high-resolution diffractometer of reverse geometry with temporal focusing, and an inelastic scattering diffractometer of inversed geometry (fixed final neutron energy). This multi-functional instrument was designed in collaboration with the Russian Research Centre "Kurchatov Institute". DIAS is intended for studies of both structures and dynamics in samples of different nature and at various conditions including low and high temperatures, high pressure and magnetic field.

A first stage of the Multifunctional Neutron Spectrometer (MNS) has been tested successfully in November 2010. The instrument was designed by the physicist from the Lebedev Physical Institute RAS. The full scale MNS will consist of four modules for the structural studies with different characteristic length scales (up to several hundred nanometers). In order to wider the range of investigated neutron energies and to reduce their losses in air, the supermirror neutron guide is planned. This instrument will be well-suited for the studies of the quasielastic spectral response, lattice dynamics, diffuse scattering.

"Crystal" is a neutron diffractometer for investigations of single crystals at different temperatures and pressures is also ready for measurements. The instrument was designed by the INR RAS. "Crystal" (its first stage) has been successfully tested in November 2010.

Further development of the neutron studies at the INR RAS will include an upgrade of the IN-06 neutron channels by supermirror neutron guides. This work will be implemented in cooperation with the Petersburg Nuclear Physics Institute. This second stage of development will include also a substantial enhancement of the accelerator performance. A project of a new experimental hall with a few neutron spectrometers is under development now. The instruments of the 2nd stage will provide us a possibility to study quasiparticles in condensed matter, including magnetic excitations and lattice excitations in nanosystems in the energy transfer range up to a few hundred meV or even at the eV range. Such a high energy transfer range is unique in Russia since the INR RAS Neutron Complex is the only spallation neutron facility in the country and the spectrum is very rich in hot and epithermal neutrons compared to the reactor-based neutron sources. Thanks to the relatively high incoming neutron energy, neutron diffraction on highly absorbing systems (samples containing Cd, Gd, Sm, Eu, B) is possible at INR RAS. One small-angle neutron diffractometer for highly absorbing matter is under development now.

A set of complementary installations for material studies by neutrons of thermal and epithermal energies may be arranged at the experimental area of the RADEX-n-TOF neutron source.

In addition to the neutron facilities a set of modern X-ray instruments is available for the structural research at ambient and low temperatures, ambient and high pressure. These instruments (German-made by Stoe and Huber) include X-Ray instruments for powder samples (Stadi MP diffractometer) and single-crystals (4-circle diffractometer Huber and image plate diffractometer Stoe IPDS II). X-Ray measurements are possible using MoKalpha1 and CuKalpha1 radiation thanks to monochromators installed on the instruments.

Detailed information about the experimental facility at the INR RAS including the linac, the neutron sources and instruments descriptions, schedule of operation, etc., is presented at www.inr.ru.

4. Conclusion

To sum up, the Neutron Complex of the INR RAS outlined above is available now for the studies and diagnostics of different materials including solid and liquid samples, soft matter, functional materials and nanosystems. A set of modern X-Ray instruments for the structural studies and nano-diagnostics is available as a complementary experimental tool. These facilities will provide an easy access to various neutron scattering techniques for a vast scientific community from Moscow and its region, numerous research institutes and universities.

References

[1] Neutron Applications in Earth, Energy and Environmental Sciences, Springer Series: Neutron Applications and Techniques, Liang L, Rinaldi R, Schober H (Eds.), Vol. **XVIII**, 2009.

[2] Carpenter J M and Yelon W.B, *Neutron Sources*, in *Neutron Scattering*, Skold K and Price D L (Eds.) *Methods of Experimental Physics* **23** A, Academic Press, New York 1986.

[3] Shober H, *Neutron Scattering Instrumentation*, Springer 2008.

[4] Koptelov E A and Kravchuk L V, *Proc. Int. Conf.* ICANS-XVIII, Dongguang, 2007.

Journal of Physics: Conference Series **291** (2011) 012013 doi:10.1088/1742-6596/291/1/012013

New opportunities provided by modernized small-angle neutron scattering two-detector system instrument (YuMO)

A.I.Kuklin[1, 2*], **D.V.Soloviov**[1, 3], **A.V.Rogachev**[1, 4], **P.K.Utrobin, Yu.S. Kovalev**[1], **M.Balasoiu**[1, 5], **O.I.Ivankov**[1], **A.P. Sirotin**[1], **T.N.Murugova**[1], **T.B.Petukhova**[1], **Yu.E.Gorshkova**[1], **R.V.Erhan**[1, 5], **S.A.Kutuzov**[1], **A.G.Soloviev**[1], **V.I. Gordeliy**[1, 2, 6, 7]

[1] Joint Institute for Nuclear Research, Dubna, Russia.
[2] Bionanocenter, Moscow Institute of Physics and Technology, Dolgoprudny, Russia.
[3] Kyiv National Taras Shevchenko University, Kyiv, Ukraine.
[4] Lomonosov Moscow State University Skobeltsyn Institute of Nuclear Physics (MSU SINP), Moscow, Russia.
[5] Horia Hulubei National Institute for R&D in Physics and Nuclear Engineering, Bucharest-Magurele, Romania.
[6] ICS-5, Forschungszentrum Juelich, Germany.
[7] Institute of Structural Biology J. P. Ebel, Grenoble, France.

* E-mail: kuklin@nf.jinr.ru

Abstract. Main features of the modernized small-angle neutron scattering spectrometer (YuMO) at IBR-2M pulsed reactor are described. New installations for sample environment of the spectrometer are highlighted. The modernized SANS instrument (YuMO) is equipped with a new type of position sensitive detector as well as two detector system which provide a unique dynamic range (Q_{max}/Q_{min} ratio is about 90).Sample environment is extended with a magnetic system (magnetic field about 2.5 Tesla), automated high pressure setup which allows simultaneous SANS and volumetric high pressure studies and light illumination system. In particular, these developments led to considerable improvements of resolution of the instrument (about 1%) and opened the possibility to study anisotropic materials and perform efficient high pressure studies.

1. Introduction

In 2011 the IBR-2M pulsed reactor (JINR, Dubna, Moscow region) will face new challenges. One of the most efficient instruments at the IBR-2M is small angle neutron scattering spectrometer YuMO which allows one to obtain I(q) curves in absolute scale. The spectrometer is characterized by relatively short data acquisition time which depends on the type of the studied material and varies from minutes to hours. The YuMO spectrometer is used for wide range of scientific and technical applications.

Main features of the YuMO spectrometer are as follows: high flux, absolute scale, two-detector system, high (relative to typical SANS instruments at the steady neutron sources) resolution and circle geometry. Due to direct view of the surface of the moderator of the reactor and time-of-flight method for wavelength determination the instrument has higher flux [1, 2].

Most of the studied materials via small-angle neutron scattering (SANS) require data collection in a wide range of momentum transfer (Q-range). Larger Q-range means more complete and reliable determination of structural parameters of the investigated materials. The dynamic range (a ratio of

Q_{max} to Q_{min} measured simultaneously) of SANS instruments is normally determined by the sizes of the detector, which are limited mainly by technical reasons, and by the wavelength range of available thermal neutrons in the neutron beam. It was shown [3,4] that the time-of-flight method provides necessary conditions for a considerable increase of the dynamic range of a SANS instrument by using two or more detectors of scattered neutrons with central holes placed at different distances from the sample. Excellent resolution of YuMO spectrometer was recently further improved by using new type of position sensitive detector PSD and is about 1% of module scattering vector [5-6].

In addition, the sample environment of the SANS instrument was also improved providing new opportunities for the researches working in different fields of science. We describe here recently installed at the YuMO magnet (2.8 T), high pressure setup (4 kbar) and a system for light illumination of the samples.

2. High-pressure setup

High penetration of neutrons into the materials and the possibility of contrast variation by using different sample preparation techniques make neutron scattering a valuable tool for high pressure studies. First high pressure setup at the small angle neutron scattering facility was realised using the Ti/Zr "null matrix" alloy [7]. The experiments with the surfactants C14MDAO and TDMAO provided new important information [7, 8].

Recently we used developed high pressure setup (HPS) to perform the studies of main (chain melting) phase transition in lipid membranes. Membrane fluidity predetermines many of the membrane properties. Pressure influences membrane fluidity, structure and dynamics of membranes. Some bacteria (for example algae) live under extreme conditions. Therefore, high pressure studies of the membranes are of significance for better understanding of their biological and biophysical properties [9].

However high pressure setup (HPS) with Ti/Zr alloy chamber provides satisfactory results only in q-range up to 0.03 $Å^{-1}$. Second step of the development of HPS was based on a new approach, namely: separation of the liquid used to create a pressure from the studied samples. One of the parts of the device is the SITEC hand pump. The HPS was automated by homemade mechanical and electronic devices which allowed us to considerably increase the efficiency of the corresponding studies. New scientific results with surfactant systems have been obtained with the mentioned above high pressure setup [10, 11].

Next step of modernization of hydrostatic high-pressure setup was adaptation of existing device to the volumetric measurements. For this purpose the rotation angle sensor was mounted at the hand pressure pump. It provides control of the change of the volume with accuracy up to $\pm 2*10^{-6}$ ml. In addition, pressure pump was equipped with a motor which gives the possibility to change volume of the system with a rate about $2*10^{-3}$ ml/s. Electronic pressure sensor allows one to control pressure with accuracy of about ± 0,1 bar. The temperature in the pressure cell is controlled by the LAUDA thermostat with an error ± 0,05°C. First P-V-T experiments with the water-dimyristoylphosphatidylcholine (DMPC) system have been done in [8]. Pressure dependence of isothermal compressibility of lipid membranes was studied. Dependence of the pressure corresponding to phase transition from gel to liquid crystal phase in the studied system was obtained in the temperature range (24 ÷ 35)° C and a P-V-T phase diagram was determined in the pressure range (1 ÷ 800) bar.

Figure 1. Volumetric high-pressure setup for the YuMO spectrometer, adapted for SANS measurements.

The typical experimental curve of compressibility via pressure near main phase transition point is shown in figure 2.

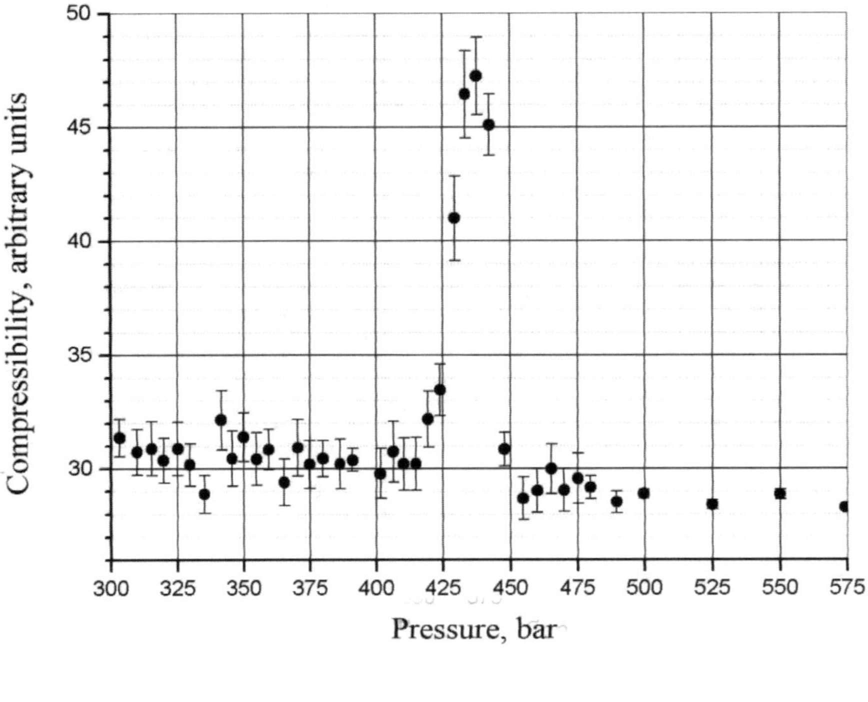

Figure 2. Compressibility of DMPC at 33.2°C as a function of pressure.

3. Magnetic field system

The magnetic field, like temperature or pressure, is one of the important parameters influencing the state of many materials. The applied magnetic field is tangible not only for magnetic materials but also in nonmagnetic ones such as paramagnetic and diamagnetic materials.

It is very important to control the microstructure of magnetic materials because textured or anisotropic magnetic materials usually show better performance in practical applications. Magnetic field processing is a well-proven technique to impose the desired texture and magnetic properties of magnetic materials. Macroscopic materials comprising a large number of nanostructures in crystalline alignment with one another would constitute highly anisotropic materials with wide potential applications. It has been also found that magnetic field can make inclusions migrate in a melt and align grains in fabrication processes for the magnetic and non-magnetic materials.

The "control of materials" for materials science by means of a magnetic field, during the materials synthesis, chemical reactions or crystal growth in high magnetic fields is useful method for creating and investigating new types of materials [13].

A new dimension in advanced materials (such as magnetic materials, oxide superconductors and organics) processing, and in the microstructure research of nanometer-scaled materials, spintronics, polymer, bio-chemistry and so on, through the possibility of SANS investigation in situ of the phenomena induced by applied magnetic fields has been discovered.

Other aspect of using magnetic field in small angle neutron scattering experiment refers to the possibility of determination of the nuclear and magnetic elastic scattering contributions to the scattering. When investigating magnetic materials in a SANS with nonpolarized neutrons experiment, measuring the scattering intensity without magnetic field and with a saturation magnetic field oriented perpendicular to the scattering momentum transfer, the nuclear and magnetic contributions can be differentiated [14-16]. Previously, this differentiation was possible to be obtained at SANS instrument only in the case of magnetic liquids by means of the contrast variation method [17-19].

For common goals the magnitude of magnetic field inducing changes in samples, (namely: anisotropy and orientation of magnetic domains in magnetic materials, as well as in the liquid crystal polymers [20]) is about few Tesla.

A new magnetic system for small angle neutron scattering at YUMO instrument was put into operation. The system includes about 3T (factory setup) electromagnet fixed on a two-axes goniometric table, power supply, cooling system and PC based control equipment. Main features of magnetic system are: big changeable gap for the samples (up to 130 mm size), computer controlled horizontal and vertical rotation and sufficiently big space for the sample holders [21, 22].

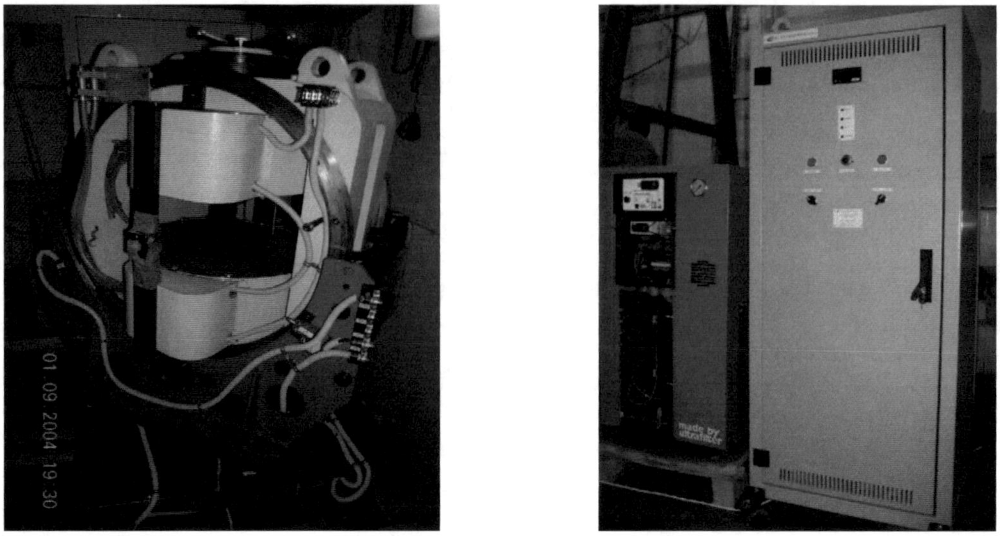

Figure 3. Electromagnet setup on a two-axes goniometric table (left); power supply and chiller system (right).

First experimental results of SANS in ferrofluids, magnetic elastomers, ferromagnetic powder have been obtained at YUMO spectrometer equipped with the new magnetic system.

4. Illumination of the samples

A sample in a SANS experiment is located in the air gap between neutron guide and detector tube. Thus there is a possibility to illuminate a sample with a laser beam or a special light beam. The YuMO system comprises the cold source of continuous high-intensity visible light (Schott KL 1500 electronic (SCHOTT Glas)). Light is guided to the object using flexible movable light guides. Taking an advantage of this light source system, functional structural changes of a membrane protein - bacteriothodopsin (BR) have been investigated [23].

BR absorbing a photon uses its energy to transport protons H^+ through the membrane in which BR is located. It results in electro-chemical gradient on the membrane. This first step of accumulation of energy (creation of electro-chemical gradient by proton pumping) is an universal step in energy transformation in bacteria, plants and animals. BR is widely used for investigation of mechanism of proton pumping. A major question to be answered is whether protein conformational changes are involved in proton pumping and what is their amplitude. The SANS experiments revealed a substantial difference between the thickness of purple membrane (a part of bacterial membrane containing solely BR) dark-adapted (T=50.3 Å) and exposed to continuous illumination (T=44.7 Å). The reduction of the membrane thickness triggered by light illumination most probably is connected to structural changes of BR molecule. Difference of the scattering intensities of light illuminated and nonilluminated purple membranes BR is presented in figure 4.

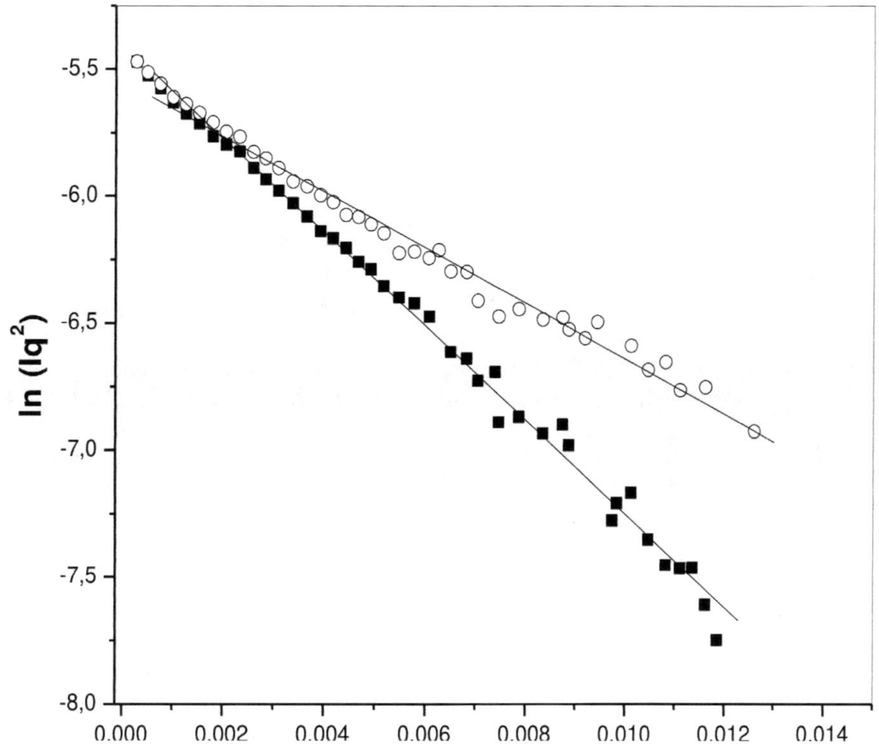

Figure 4. Dependence of the scattering intensity of PM in D_2O at T=16.8 °C on the scattering vector in $Ln(I(q)q^2)$ vs. q^2 plot: (■) dark-adapted state; (O) upon illumination.

5. Conclusions

The modernized SANS instrument (YuMO) was equipped with a new type of position sensitive detector as well as two detector system which provide a unique dynamic range (Q_{max}/Q_{min} ratio is about 90).

Sample environment was extended with a magnetic system (magnetic field about 3 Tesla), automated high pressure setup which allows simultaneous SANS and volumetric high pressure studies and light illumination system.

In particular, these developments led to considerable improvements of resolution of the instrument (about 1%), opened possibility to study anisotropic materials and perform efficient high pressure studies.

Acknowledgements

The work was supported by BMBF grant №03DU03G2, the grants of Romanian, Slovak Republic, Czech Republic, Poland Plenipotentiary in JINR, grant of Ministerium of Education and Science № P1160 of of the Federal Target Program ''Scientific and academic research staff of innovative Russia''for2009–2013 and Seventh Framework Programme [FP7/2007-2013 №211800]. The authors are grateful to Prof. S.A. Sergeenkov for attentive reading of the manuscript and valuable observations and remarks.

References

[1] Kuklin A I, Bobarikina G N, Bogdzel A A, Gordeliy V I, Islamov A Kh, Konovalov V Yu, Rogov A D, Florek M 2002 Measurements and Simulation of Neutron Beam Spectrum by Monte-Carlo Method. Beam Parameters of Small-Angle Spectrometer YuMO at the Fourth Channel of IBR–2 *Preprint* P13–2004–77.

[2] Kuklin A I, Islamov A Kh, Rogov A D, Gorshkova Yu E, Utrobin P K, Kovalev Yu S, Rogachev A V, Ivankov O I, Kutuzov S A, Soloviev D V and Gordeliy V I 2009 The Results of Tests of Methane Cryogenic Moderators *JINR Communication* P3–2009–114.

[3] Kuklin A I, Islamov A Kh and Gordeliy V I 2005 *Neutron News* **16**(3) 16–18.

[4] Kuklin A I, Islamov A Kh, Kovalev Yu S, Utrobin P K, Gordeliy V I 2006 *Poverkhnost* **6** 74–83 (in russian).

[5] Kuklin A I, Kutuzov S A, Gabriel A, Eckold G, Utrobin P K, Smirnov A A, Ivankov A, Islamov A Kh, Kovalev Yu S, Feoktistov A, Rogachev A V, Kirilov A S, Gordeliy V I 2007 Measurements with New Type of Position Sensitive Detector in Backscattering and Small-angle Setup *Annual report FLNP* (Dubna: JINR Publishing Department) pp 129–132.

[6] Ostanevich Y M 1988 *Makromol.Chem., Macromol.Symp.* 15 91–103.

[7] Gorski N I, Ivanov A N, Kuklin A I, Smirnov L S 1995 Small angle neutrons scattering setup for high pressure measurements at IBR-2 *J.of High Pressure Research* **14** 215–220.

[8] Gorski N, Kalus J, Kuklin A I and Smirnov L S 1997 A Small-Angle Neutron Scattering Investigation of the TDMAO Micelle System at High Hydrostatic Pressure *J. Appl. Cryst* **30** 739–743.

[9] MacDonald A G 1987 The role of membrane fluidity in complex processes under high pressure *Current Perspectives in High Pressure Biology* (London: Academic Press).

[10] Islamov A, Haramagatti C R, Gibhardt H, Kuklin A, Eckold G 2006 Pressure-induced phase transitions in micellar solutions *Physica* B **385–386** 791–794.

[11] Haramagatti C R, Islamov A, Gibhardt H, Gorski N, Kuklin A and Eckold G 2006 Pressure induced phase transitions of TTAB-micellar solutions studied by SANS and Raman spectroscopy *Phys.Chem.Phys.* **8** 994–1000.

[12] Solovjov D V, Kuklin A I, Utrobin P K, Ozerin A N, Kurkin T S, Ivankov O I, Bulavin L A, Gordeliy V I 2011 X-ray scattering and volumetric P-V-T investigation of dimyristoylphosphatidylcholine-water system *Journal of Surface Investigations, X-ray, Synchrotron and Neutron Techniques* In print.

[13] Ma Y, Xiao L, Yan L 2006 *Chinese Science Bulletin* **51(**24) 2944–2950.

[14] Bacon G 1962 *Neutron Diffraction* (Oxford: University Press).

[15] Kohlbrecher J, Wiedenmann A, Hollenberger H 1997 *Z. Phys. B* **104** 1–2.

[16] Loffler J F, Braun H B, Wagner W, Kostorz G, Wiedenmann A 2005 *Phys. Rev.* B **71** 134410–15.

[17] Grabcev B, Balasoiu M, Bica D, Kuklin A I 1994 *Magnetohydrodynamics* **10(**2) 156–162.

[18] Balasoiu M 1998 *PhD Thesis* (Bucharest: Institute for Atomic Physics).

[19] Grabcev B, Balasoiu M, Tarziu A, Kuklin A I and Bica D *J. Mag. Mag. Mater* **201** 140–143.

[20] Kuklin A I, Cherny A Yu, Cherezov V G, Kostromin S, Rogachev A V, Shibaev V P 2008 X-ray investigation of comb-like liquid crystal polymers, oriented by a strong magnetic field for SANS measurements *Annual report FLNP* (Dubna: JINR Publishing Department) pp 43–44.

[21] Balasoiu M, Kappel W, Cios M, Stancu N, Cios A, Kirilov A S, Kutuzov S A, Smirnov A A, Gordeliy V I, Erhan R, Ivan'kov A E, Rogacev A V, Kovalev Yu S, Kuklin A I 2007 *Book of abstracts 4th European Conf. on Neutron Scattering (Lund, Sweden25–29 June 2007)* p 484.

[22] Balasoiu M, Kirilov A S, Kutuzov S A, Smirnov A A, Kappel W, Cios M, Cios A, Kuklin A I 2008 Magnetic System For Small-Angle Neutron Scattering Investigations at YuMO Instrument of Nanomaterials *JINR Communication* E14–2008–193.

[23] Gordeliy V I, Dencher N, Hauss T, Kuklin A, Tougan-Baranovskaya A, Teixeira J, Yaguzhinskiy L S and Bueldt G 1999 Light-Induced Long-Living Changes of Bactreirhodopsin Structure *Annual report FLNP* (Dubna: JINR Publishing Department) pp 91–92.

Problem of safety in nanoparticle mass production and its solution by means of advanced scanning electron microscopy

S.K. Maksimov; K.S. Maksimov

Moscow Institute of Electronic Technology, Russia

E-mail: maksimov_sk@comtv.ru

Abstract. It is suggested possible approaches to the comprehensive solution of the problem of structural-morphological control in mass production of nanoparticles only by the use of advanced methods of scanning electron microscopy.

1. Introduction

Outgoing production control is the integral part of any technology. It must secure conformity of a controlled portion of production with its technical regulations and thereby lock in its quality and compatibility with the biosphere. Risks connected with nanodimensional particles for the life are associated with their capacity to control reactions in ambient environment [1], i.e., with their catalytic activity. Catalytic activity of solid objects is controlled by their structure and crystallographic faceting or a habit [2]. Size, structure and habit are thermodynamically interconnected in the nanoworld, and decrease of particle sizes leads to change of its equilibrium structure and habit [3, 4]. However thermodynamics determines only a direction of possible phase transitions, but their realisation is controlled by factors controlling the kinetics of nucleation and growth of particles. Technologies, which allow obtaining particles of only one single size, are impossible. Therefore particles of the same size and obtained in the same process can be of different structural-morphological characteristics and, naturally, of different catalytic properties. Environmental safety in mass production of particles can be provided only by means of revealing of fractions of particles with peculiar structural-morphological characteristics [5, 6]. The urgent need of the factional control is emphasized by such problems as the property of nanodimensional anatase particles to modify DNA [7] (the anatase is present at medicines comprising TiO_2, e.g., at the Tamsulosin, Ofloxacin, etc) or the possibility of use of nanocrystallization for increasing solubility of pharmaceutical substances [8].

A spatial resolution that makes it possible to solve problems of structural-morphological characterization of nanoparticles is provided by scanning probe microscopy (SPM), transmission electron microscopy (TEM) and scanning electron microscopy (SEM). SPM and TEM allow determining a habit of a sole object with the highest accuracy and precision [2, 9]. However massifs numbering billions of particles have to arise in the large-scale manufacturing, whereas investigations of influence of nanoparticles on the metabolism have been carried out only for subjects with small duration of a life cycle [10], and it is no way of telling: "how much a number of particles can result in irreversible consequences throughout human lifetime". The environmental safety requires identification of fractions, a portion of which can be equal, at least, to $10^{-4} – 10^{-6}$ of total number of nanoobject of a massif and, accordingly, controlled massifs must exceed $10^6 – 10^8$ particles [2, 11].

Productivity of SPM and TEM are not sufficient for the control of massifs numbering $\geq 10^6$ objects [2, 9].

SEM techniques allow, in principle, getting images of massifs of nanoobjects practically infinite by numbers. SEM images depend only from an object habit and its tilting about an electron beam direction, if conditions of an image obtaining (electron energy, beam convergence, focusing distance, etc) as well an object composition are constant [12]. Images of objects can be dividing into groups in compliance with such image characteristics as a size, shape and intensity distribution (Fig. 1).

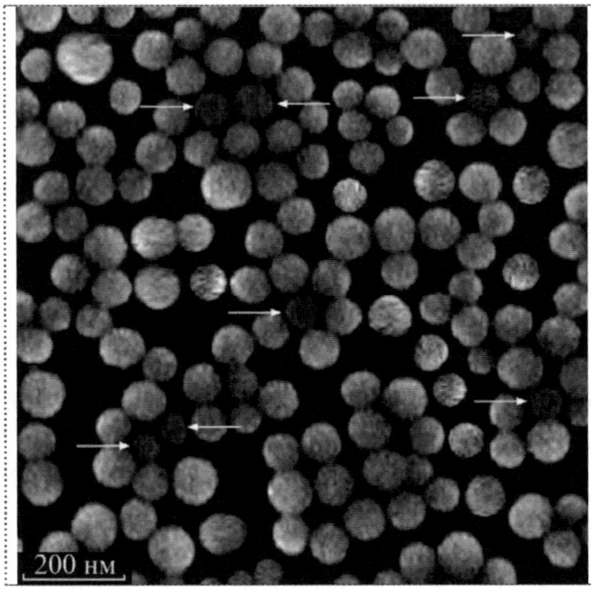

Figure 1. SEM images of SiO_x particles. Particle images differ in average intensity and contrast range. Images with anomalously lowered intensity are shown by arrows.

Images with less average intensity and contrast in Fig.1 can correspond to particles of another nature than images with more typical characteristics, and revelation of SEM images with different characteristics could be the basis for control of massifs with any quantity of nanoparticles. How-ever only a direct determination of structural-morphological characteristics particles with unusual images can assure a valid conclusion concerning their nature [13].

2. Approach to the methods of the habit determination.

SEM images arise in the course of electron beam scanning over the surface of the object and represent the sum of sequential responses generated by the object in response to inelastic scattering of electrons of an illuminating beam. The intensity profiles have the shape of curves on which points corresponding to the edges of the object cannot be determined; i.e., the sizes of the object cannot be directly measured [12]. Methods of the stereo microscopy, basing on images corresponding to different projections of object, are not valid with decreasing sizes for this reason. Technique allowing studying their habit is suggested in this paper. It can be realized with the help of secondary electrons with energy of 20 – 50 eV (this energy region provides the minimal free path of recoil electrons [13]).

When an illuminating electron beam scans the surface of a relief object in SEM, the probe becomes alternately farther and closer to the focusing plane and its sizes change; i.e., a movement provokes defocusings. Modern SEM techniques are based on illuminating electron beams with convergence angles which are equal to $\geq 3\times10^{-3}$ rad, and therefore a depth of the focus of SEM at maximal magnifications is equal to ≈ 0.5 μm [13]; i.e., in the majority of cases, it exceeds a height of an object, and an image is insensible to these defocusings. But if the angle of convergence reaches $\approx 10^{-1}$ rad, the depth of the focus becomes equal to or less than 10 nm, that gives rise to noticeable defocusings even for objects with a height of ≤ 10 nm [11] (Fig. 2).

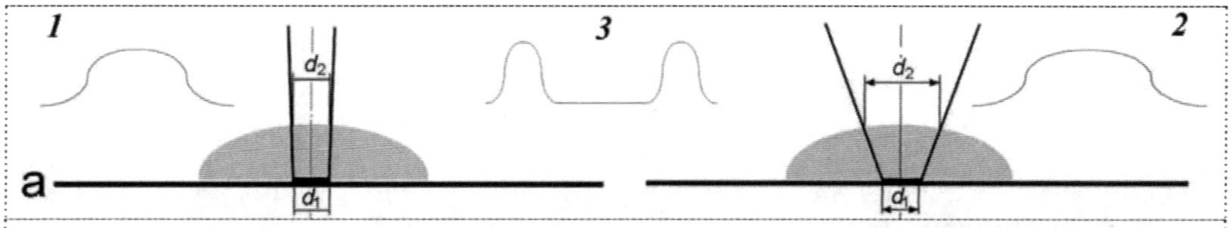

Figure 2. Diagrammatic representation of influence of the habit defocusing on BE images. ***1*** – the illuminating beam with a small angle of convergence and typical intensity profile in this case, ***2*** – the illuminating beam with a large angle of convergence and typical intensity profile in this case; ***3*** – a differential curve; "*a*" is a substrate plane; "d_1" is a size of a beam cross-sections by a substrate (focusing plane, datum plane); "d_2" is a size of a probe.

Hereinafter, 2 variants of convergences are subsequently referred to as the 1st or 2nd convergences in accordance with Fig. 1, the term "probe" applies to a cross-section of an illuminating beam by an object surface, and the term "beam-section" means the a cross-section of an illuminating beam by a datum plane or focusing plane. These defocusings (in contrast to the instrumental defocusings associated with specific excitation of an objective lens) can be named as the habit defocusings. These defocusings are considered in this paper in the frameworks of the 2-D approximation with the help of intensity profiles, but results obtained and recommendations elaborated here are true for intensity distributions i.e., for 3-D situations.

Action of a probe on an object in the presence of habit defocusings is described by an action curve which is characterized by the expression [13, 14]:

$$I_{\Sigma} = \int_X \frac{J}{\sqrt{\pi}\sigma} \exp\left(-\frac{(x - x_c)^2}{\sigma^2} \right) \mathrm{d}x_c \tag{1}$$

Here, in the linear approximation,

$$\sigma = k\left(|z - f| \cdot \frac{R - r_f}{f} + r_f \right) \tag{2}$$

I_{Σ} is the total action of the probe at scanning, J is the total intensity of the probe, x_c is the coordinate of the centre of the probe, z is the distance from the point on the surface of the object to the output diaphragm of the objective lens, f is the focal length, $|z - f|$ is the distance from the point at the surface of the object to the focusing plane, therefore σ depends on an object height, r_f is the effective radius of the beam-section at the focusing (datum) plane, R is the radius of the exit aperture of the objective lens, r_f is the effective cutoff radius of the beam cross-section in the focusing plane, k is the coefficient depends on the effective cutoff radius, the region of integration X is determined by a lateral size of the object along the specified direction. In these formulas, $\sigma = \sigma_G / \sqrt{2}$ where σ_G is the root-mean-square deviation (the characteristic of the Gauss distribution) and the proportionality of σ to σ_G is the important property of the approach, since it allows using the software instruments, already created for modelling SEM images, also for processing of images obtained with the help of convergence beams.

The intensity distributions for images obtained in conditions of a 2nd convergence can be described as the sum of two components. The first component is a distribution that is independent of a convergence of the illuminating electron beam and, hence, does not reflect the shape of the object. The second component is determined by the changes in the sizes of the probe during its motion over the surface of the object, i.e., by the habit defocusing. Therefore, in order to determine the component dependent on the habit defocusing, the component independent of the convergence of the illuminating electron beam should be eliminated from the total distribution. A curve arising in the result of the subtraction of action curves corresponding to probes with different convergences is referred subsequently as the differential curve. The differential curve is described by the expression:

$$I_- = \int_X \frac{J}{\sqrt{\pi}\sigma_2} \exp\left(-\frac{(x-x_c)^2}{\sigma_2^2}\right) \mathrm{d}x_c - \int_X \frac{J}{\sqrt{\pi}\sigma_1} \exp\left(-\frac{(x-x_c)^2}{\sigma_1^2}\right) \mathrm{d}x_c \quad (3)$$

where I_- is the total extent (intensity) of action due to the habit defocusing, and σ_1 and σ_2 correspond to 1st and 2nd convergences.

3. Experimental techniques to the habit determination.

3.1. Technique based on reconstruction a three-dimensional image in the real space.
If cross-sections of the illuminating beams with different convergences by certain plane are of identical in sizes so the Exp. (3) for a point of a differential curves at this plane converts to [13]:

$$\int_X \frac{J}{\sqrt{\pi}\sigma_2} \exp\left(-\frac{(x-x_c)^2}{\sigma_2^2}\right) \mathrm{d}x_c - \int_X \frac{J}{\sqrt{\pi}\sigma_1} \exp\left(-\frac{(x-x_c)^2}{\sigma_1^2}\right) dx_c = 0 \quad (4)$$

Therefore effect of these beams on all points of this plane is also identical.

It is evidently on physical grounds, if two impacts are identical, so responses to these impacts are also identical, i.e., in line with Exp. (4), yields of recoil electrons are independent of convergences of the illuminating beams in all points of a focusing plane. If the 1st convergence is used, all points of an object are concurrently in-focus. Therefore if an intensity profile corresponding to the 1st convergence to lay over an intensity profile corresponding to the 2nd convergence, they intersect, to a first approximation, at a point where the focusing plane intersects the intensity profile conforming to the 2nd convergence. Consequently, if to focus the probe initially on the datum plane and thereupon to shift the focusing plane for known distance to a certain object cross-section so this procedure provides determining a distance between points of an intensity profile, lying in this cross-section, and the datum plane. Reiteration of this procedure allows to determine distances for many points of profile (including a height of an object summit) and to transform an intensity profile in the intensity space into a three-dimensional image in the real space (Fig.3). Overlapping of intensity profiles can be changed by profile subtraction and in this case, an extreme of the differential curve (maximum or minimum depending on what profile is the deduction) or a sign inversion of the differential curve can be observed.

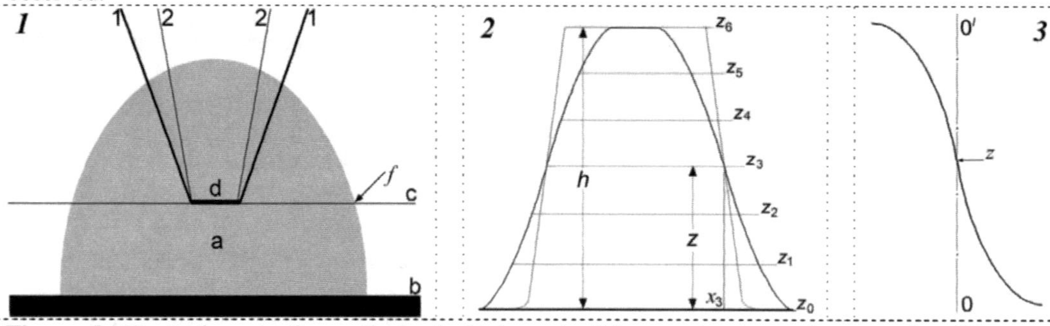

Figure 3. Illustrations to the method of obtainment of images in the real space. *1* – Diagram illustrating conditions of obtaining of contention profiles. 1, 2 are generatrixes of illuminating beams, "a" is an object, "b" is a substrate surface, "c" is the focusing plane, "d" is a cross-section of illuminating beams by the focusing plane, "*f*" is the point at an object surface, wherein impacts of illuminating beams are equal. *2* – Illustration to superposition of intensity profiles corresponding

Journal of Physics: Conference Series **291** (2011) 012014 doi:10.1088/1742-6596/291/1/012014

different convergences for an object with a trapezoidal shape; a thin line profile corresponds to the 1st convergence; a heavy line profile conforms to the 2nd convergence. z_0 is the datum plane (a substrate surface); z_i are traces of focusing planes; z_3 is the focusing plane for a beam with 2nd convergence; z is a distance of the focusing plane from the datum plane; h is an object height; x_3 is an x-coordinate of the point z_3. **3** – Differential profile corresponding to Fig. 2; 00′ is the zero line; z is a point of a sign reversal of the profile.

3.2. Technique based on direct reconstruction of a habit.

Approach based on focusings upon object cross-sections is applicable only to objects which are of height exceeding 10 nm and, accordingly, provide sizeable habit defocusings. Problems of morphological identification of particles with sizes ≥10 nm can be solved on the base of another approach that can be named as the method of the direct habit reconstruction.

Two terms of Exp. 3 reflect identical dependences on the same variable and in the same interval; therefore Exp. 3 can be transformed as:

$$I_{-}=\frac{J}{\sqrt{\pi}}\int_{-\infty}^{\infty}\left[\frac{1}{\sigma_1}\exp\left(-\frac{(x-x_c)^2}{\sigma_1^2}\right)-\frac{1}{\sigma_2}\exp\left(-\frac{(x-x_c)^2}{\sigma_2^2}\right)\right]dx_c \tag{5}$$

Let $|\sigma_2-\sigma_1|\ll\sigma$, that allows realizing the linearization of the Exp (5) by means of the replacement of the difference of functions by the difference of arguments multiplied by the derivative. If σ_i to represent as $\sigma_i = \sigma_0 + \tau$ (where σ_0 is a size of the beam-section by a focusing plane and $\tau = k\{|z-f|\times(R-r_f)/f\}$ reflects σ changes due to the habit defocusing) and to take into account, that σ is equal to σ_0 over an extent of a profile for the 1st convergence, so the Exp. (5) can be put into the form:

$$I_{-}=\frac{J}{\sqrt{\pi}}\int_{-\infty}^{\infty}\frac{\tau}{\sigma_0^2}\left[1-2\frac{(x-x_c)^2}{\sigma_0^2}\right]\exp\left(-\frac{(x-x_c)^2}{\sigma_0^2}\right)dx_c \tag{6}$$

Exp. (6) emphasizes one-to-one correspondence between habit defocusings and differential curves.

Differential curves differ fundamentally from intensity profiles, since they carry identified information (which misses for profiles) concerning all three dimensions. On the strength of Exp. (2) an electron distribution in a probe conforming to a beam with the 2nd convergence and its size differ from those of a beam with the 1st convergence and these differences increase with distance from the focusing plane (linearly for sizes and inversely quadratically for intensity distributions). Therefore calculations on the base of differential curves allow reconstructing a habit. In some cases, when $\sigma_0 \gg \tau$, habits can be built in accordance with the Eq. (6) on the base of differential curves by means of the reciprocal integral transformation or matrix inversion, but in the majority instances, the condition $\sigma_0 \gg \tau$ is difficult implemented and reconstruction must realized on the base of these curves and software resulting from Exp. (5).

Besides, software must take into account that electrons registered in SEM correspond not to electrons of illuminating beams but to recoil electrons, and, accordingly, obtained curves are not the differential curves but curves of differences of intensity profiles. Therefore soft must include operations reflecting links of regularities of the yield of recoil electrons and, correspondingly, allowing transforming an experimental profile differences into the differential curves. These operations must take into consideration, that: firstly, intensity of recoil electrons depends on a tilting of a facet about a microscope axis; secondly, if a probe conforming to the 2nd convergence is located wholly on a plane facet, so the yield of recoil electrons is equal to their yield for the 1st convergence, i.e., this portion of profile difference must be of a zero level; thirdly, if a probe comes nearer to a line of an intersection of facets at a distance less then the free path of electron of the given energy, so recoil electrons can be emitted through two surfaces and an intensity maximum arises, fourthly, if a probe is distributed between surfaces with different yields of recoil electrons, so their yield is a sum of yields emitted by each facet; fifthly, the antecedent rule is true also for probes distributed between an object and substrate. Listed rules cannot restrict the applicability of the described approach since each change of conditions of the yield of recoil electrons reflects on intensity profiles and a curves corresponding to their differences and these notches must be included in the software. At that, the information associated with differences of notch positions at intensity profiles, corresponding to different convergences, and a distance between notches at a curve, reflecting their difference, is the major factor allowing correct reconstruction of the differential curve.

Identification of minute particles is hampered since very small habit defocusings are associated with these objects. This impediment can be surmounted with 2 variants of a special method of obtaining a differential curve. In the framework of this method, 3 planes (central and 2 lateral) are used in SEM (Fig. 4). In the first variant, two defocused images are formed by means of alternate coincidences of the focusing planes with two planes that are equidistant with respect to the plane of the object localization (Fig. 4-a). In the second variant, the focusing plane is brought into coincidence with the central plane and the object is coincided with two planes that are equidistant with respect to the central plane (Fig.4-b). It should be shown that the equidistance is optical one; i.e., the cross sections of the illuminating beam by two profile planes are identical to each other upon focusing onto the central plane, but, upon focusing onto the lateral planes, there arise equal cross sections in the central plane irrespective of the plane onto which the beam is focused. Since these two images correspond to the same datum plane, the subtraction of one of the relevant intensity distributions from the other distribution also gives rise to a differential profile that depends uniquely only on the habit of the object.

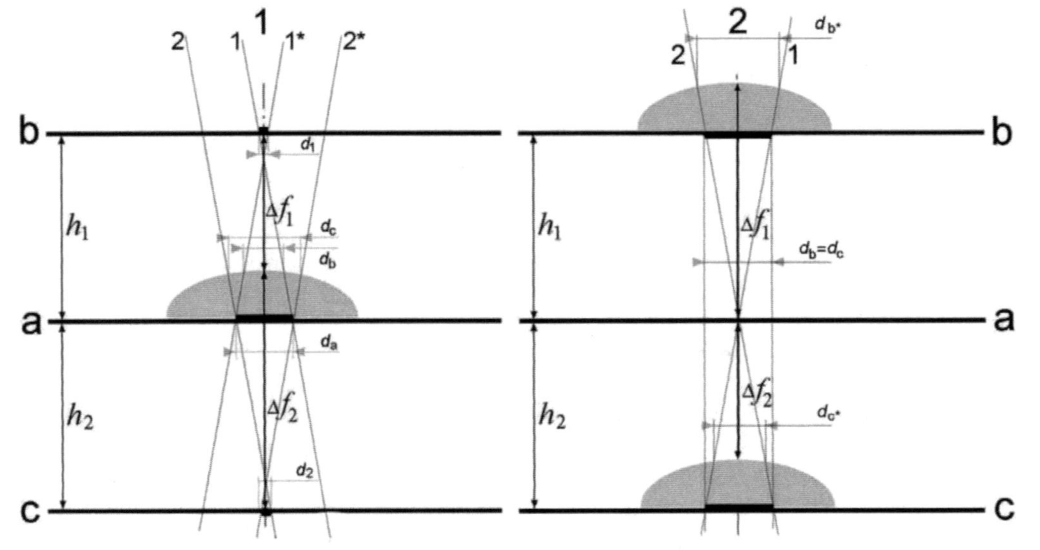

Figure 6. Schemes illustrating the specific features of the implementation of the technique: (**1**) the scheme reflecting the specific features of the image formation in the case where the beam is focused onto two equidistant planes (**b** and **c**) and the datum plane is brought into coincidence with the central plane (**a**); (**2**) the scheme reflecting the specific features of the image formation in the case where the datum plane lies in equidistant planes (**b** and **c**) and the beam is focused onto the central plane (**a**). In both schemes, h_1 and h_2 are identical distances due to changes in the sizes of the cross sections of the illuminating beam. Designations in scheme (**1**): 1 and 1* are the generatrices of the cone of rays in focusing onto the **b** plane; 2 and 2* are the generatrices of the cone of rays in focusing onto the **c**

plane; d_1 is the size of the cross section of the beam by the **b** plane and d_2 is the size of the cross section of the beam by the **c** plane ($d_1 = d_2$); d_a is the size of the cross section of the beam by the **a** plane, which is identical for both defocusings; Δf_1 is the distance from the top point of the object to the **b** plane and Δf_2 is the distance from the top point of the object to the **c** plane ($\Delta f_1 < \Delta f_2$); d_b is the probe size in focusing onto the **b** plane and d_c is the probe size in focusing onto the **c** plane ($d_b < d_c$). Designations in scheme (**2**): 1 and 2 are the generatrices of the cone of rays in focusing onto the **a** plane; Δf_1 is the distance between the top point of the object with the datum plane in the **b** plane and the focusing plane **a;** Δf_2 is the distance between the top point of the object with the datum plane in the **c** plane and the focusing plane **a** ($\Delta f_1 > \Delta f_2$); d_b and d_c are the sizes of the cross sections of the beam by the **b** and **c** planes, ($d_b = d_c$); and d_{b*} and d_{c*} are the probe sizes in the case where the zero-height planes of the object lie in the **b** and **c** planes, respectively ($d_{b*} > d_{c*}$).

In the both variants of the technique, when images corresponding to different defocusings for each point on the surface, it can be write $|\tau_1| = |\tau_2|$, but the quantities τ_1 and τ_2 have different signs depending on the defocusing sign. As a result, expression (3) can be represented in the form:

$$I_- = \frac{J}{\sqrt{\pi}} \int_{-\infty}^{\infty} \frac{2\tau}{\sigma_0^2} \left[1 - 2\frac{(x - x_c)^2}{\sigma_0^2} \right] \exp\left(-\frac{(x - x_c)^2}{\sigma_0^2} \right) dx_c \tag{6}$$

Consequently, the technique provides obtaining a differential curve which, at that, displays twofold increasing sensitivity to the habit defocusings. The latest circumstance allows, firstly, to use this technique for investigations of the minute particles which give rise to negligible habit defocusings, and, secondly, to minimize a necessary value of convergence that simplifies the problem of the correction of the spherical aberrations and provides more precise execution of the condition $\tau \to 0$; that is especially important when the minute particles are studied.

4. Conclusion.

Enlargement of a convergence of illuminating beam enlarges dramatically spreading a probe in the result of the spherical aberration. Possibilities of limitation of the spherical aberration effect (including aberration associated with large angles of the beam convergence) have been solved in [15]. For the implementation of the method, it is necessary to vary the convergence of the illuminating electron beam at precisely specified values. The problem of variation in the angles of convergence is solved by using a two-lens condenser and sets of condenser and objective diaphragms in [16].

The key problem governing the implementation of the proposed approach is to bring the datum plane into coincidence with the focusing plane. For nanodimensional objects, these planes must be brought into coincidence with angstrom accuracy. The coincidence of the planes with this accuracy (especially the planes with the a priori unknown relief) cannot be achieved in frameworks of using SEM techniques (even with images arising in conditions of large convergences of illuminating beams). Therefore microscope must have the "limit stop" which breaks off movement of a microscopic stage when the datum plane coincides with the focusing plane. Controllers of the type used in SPM [17] provide solution of this problem.

Both the suggested techniques (the three-dimensional image and the direct habit reconstruction) cannot provide obtaining an exact habit on the strength of specific peculiarities of SEM images if even all shown above technical problems are solved. However their results reflect sizes and a shape of an object and, consequently, can be used as virtual prototypes of an object, and, at that, a further correction of object parameters can be carried out with the help of the familiar procedure involving receipt of the prototype image in conditions corresponding to the conditions of obtaining the original object image and subsequent fitting of this prototype image to the experimental object image by means of variations of prototype shape and sizes. Different software products intended for the computation of virtual images (e.g. Joy's PC Monte Carlo programs, Casino Monte Carlo Program, SEMLP, etc) can be used for this aim. Joint use of the suggested methods and developed computer methods of SEM image processing provides the true habit determination.

It must be emphasized that although approaches grounded on the transformation of intensity distributions into three-dimensional images or obtaining a differential profiles are very time consuming, they can be recommended also for outgoing control in nanotechnologies, since the shown shortcoming becomes unobtrusive in this case. Habit identification is obligatory process only on the stage of designing a technique for a new objects or new technologies, but current control can be carried out only by scanning of massifs, obtaining their images, and the classification of these images on the base of found dependences between object shapes and image characteristics. Therefore the suggested methods can be suggested also as the constituent part of the outgoing inspection.

Although the process of determination of dependences between characteristics of SEM images and structural-morphological characteristics of particles is carries out only during the stage of development of technology of a given type of produce, it remains the key phase of the suggested approach to all outgoing control, since only this phase allows ascertaining potential risky kinds of particles. Therefore this phase demands a thorough inspection of even single images if the slightest doubts are cast upon their nature. The phase of obtaining of images of large massifs of particles is principal in the frameworks of the given approach, and not only because that it is the single inspection process of a production stage but also since it is process allowing exposing objects demanding further control. Therefore this process demands rigorous identity of conditions of image formations of a massif as a whole.

Acknowledgments. The authors would like to thank the staff of the Abdominal Department of the Moscow P.A.Herzen Oncology Research Institute whose efforts have made possible the appearance of this publication.

References

[1] Nanotoxicology. Characterization, Dosing and Health Effects. Eds. Montairo-Riviere N., Tran L.C. // Informa Healthcare USA Inc. 2007. 450 P.

[2] S. Zuin, G. Pojano, A. Marcomini. Effect-Oriented Physicochemical Characterization of nanomaterials.19 – 57. // Nanotoxicology. Characterization, Dosing and Health Effects. Eds. Montairo-Riviere N., Tran L.C. // Informa Healthcare USA Inc. 2007. PP. 19-58.

[3] A.S. Barnard A special issue on theory and simulation of nanomorphology. // J. of Computational and Theoretical Nanoscience. 2007. V. 4. No 2. PP. i-ii

[4] G. Guisbiers, G. Abulimu, F. Clement, M. Wautelet. Effects of Shape on the Phase Stability of Nanoparticles. // J. of Computational and Theoretical Nanoscience. 2007. V. 4, No 2, PP. 309-315.

[5] S.K. Maksimov; K.S. Maksimov. Controlling the surface functionality of nanomaterials. Nanotechnologies in Russia, 2009. V. 4, No 3–4, PP.188–200.

[6] S.K. Maksimov, K.S. Maksimov.. Principles of nanomaterial control for development of safety standards as illustrated by establishing of Laws of Nanostructuring in $Ca_yLa_{1-y}F_{3-y}$ and $La_xCa_{1-x}F_{2+x}$ solid solution. Technical Physics Letters, 2009, V. 35, No 3. PP. 224 – 227.

[7] Linglan Ma, Jinfang Zhao, Jue Wang, et al. The Acute Liver Injury in Mice Caused by Nano-Anatase TiO_2. // Nanoscale Res Lett. 2009, No 4, PP. 1275–1285.

[8] R. Ravichandran. Physico-chemical evaluation of gymnemic acids nanocrystals. Int. J. Nanoparticles, 2010, No 3, PP. 280-296.

[9] Nanocharacterization. Eds. Hutchison J., Kirkland A. // RSC Publishing, 2007, xii + 304 P.

[10] Fan Yang, Fashui Hong, Wenjuan You, et al. Influences of Nano-anatase TiO_2 on the Nitrogen Metabolism of Growing Spinach. // Biological Trace Element Research, 2006, V. 110, No 2, PP. 179–190.

[11] S.K. Maksimov, K.S. Maksimov. // Concept of safety standard and outgoing inspection in

technologies of nano-particles. // Nanotechnics, 2009, No 18, PP. 5 – 12. (in Russian).

[12] D.E. Newbury, D.C.Joy, P. Echin, C.E. Fiory, J.I.Goldstein. Advanced Scanning Electron Microscopy and X-Ray Microanalysis. // Plenum Press. N.Y., L. 2003, 634 P.

[13] S.K. Maksimov, K.S. Maksimov. New approach in metrology for nanotechnologies. // Technical Physics Letters, 2010, Vol. 35, No. 10, PP. 938–941.

[14] K. S. Maksimov. Systematic Features of Defocused Images in Scanning Electron Microscopy and Nanoscale Size Measurements. // Semiconductors, 2009, Vol. 43, No. 13, PP. 1725–1727.

[15] Kawasaki, T. Yoshida, Y. Ose, H. Todokoro. // Electron beam apparatus with aberration corrector. // USPat. 7,199,365. 2007.

[16] H. Kitsuki; K. Aoki; M. Sato. Scanning electron microscope. // USPat. 7,442,929. 2008.

[17] B. Bhushan, O. Marti. Scanning Probe Microscopy – Principle of Operation. Instrumentation and Probes. (2007). Springer Handbook of Nanotechnology, 2nd ed. (Ed. by Bharat Bhushan), PP. 239–278.

Journal of Physics: Conference Series **291** (2011) 012015 doi:10.1088/1742-6596/291/1/012015

The experimental setup for studying the molecular composition of nanoscale films and coatings

Turiev A.M., Butkhuzi T.G. ,Ramonova A.G., Magkoev T.T., Tsidaeva N.I.

The state University of North Ossetia,Vladikavkaz,46 Vatutina Street, Russia

ra6jt@mail.ru

Abstract. It is offered the method of measurement and design of the experimental setup, allowing to control the flow of particles from the surface of organic films during annealing by pulsed laser radiation. The method is based on the TOF(Time Of Flight) principle of detecting particles, desorbed from the surface by laser pulses, used for annealing. The principle of registration and the structure (block- scheme) of the experimental setup and its constituent parts are detailed in the work The setup consists of the analytical part, the system of laser irradiation and computer measurement system. The basis of the analytical part of the installation is a TOF(Time Of Flight) mass spectrometer of original construction.

1. Introduction

For the synthesis of nanoscale organic films, along with the spin-coating and Langmuir - Blodgett, widely used method of thermal spraying in vacuum. Quality of films deposited in vacuum, is controlled by spectral analysis techniques AES, XPS, SIMS and TOF SIMS [1]. The last of these methods of learning the best indicators for the study of nanoscale organic films, because , besides recording the elemental composition of the films, it enables to study molecules and their fragments, containing large groups of atoms [2]. Synthesis of thin organic films by mentioned methods usually implies their subsequent annealing, which can be carried out at a certain temperature. Along with thermal annealing , the laser annealing is used. Especially promising is annealing, carried out by a pulsed laser, allowing a delicate influence on the surface

2.Experimental Part

In this paper we propose an experimental setup , that allows to control the flow of particles from the surface of organic films during this process. The method is based on TOF principle of registration of particles , desorbed from the surface by laser pulses ,used for annealing. The main advantage of TOF mass spectrometer is that the registration of products of desorption is carried out in the area of direct flight from the sample to the IS(ion source), i.e. the particles, reflected from the walls and from other elements do not get in the mass spectrum. In order to study the fragmentation of the films of organic materials by laser radiation, we have developed a technique, that allows to record the energy distribution of all sorts of particles, desorbed from the surface by a single laser pulse. This is achieved by existence of two span areas of desorbed particles: in the first area OA(fig.1) the particles in the neutral state divided by the initial velocity, in the second area CD the division by masses of ionized in the IS particles takes place

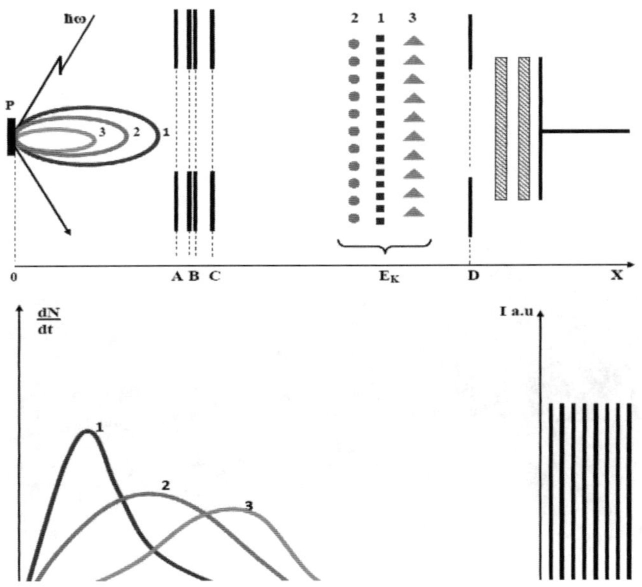

Figure1.The principle of registration of mass spectra and energy distributions of desorption flux of particles.

Frequency mode of a mass spectrometer allows to measure distribution of the kinetic energy of particles, and for all particle of the desorption flux simultaneously. It should be noted that other dynamic mass spectrometers (monopole and quadrupole) can monitor only at one specific mass during the registration process dynamics.

We studied the mass spectra and energy distributions of particles desorbed from the surface of organic semiconductors films by a pulsed (t = 10ns) of the neodymium YAG laser: Nd3 + (9) with wavelengths λ1 = 1064 nm and λ2= 532 nm. A laser beam is directed at an angle of 45 ° to the sample surface (8), focused to a spot size of about 10-3 cm2. Area of irradiation on the surface of the film is selected with an optical device (13). Changing the laser intensity is carried out discretely by a set of neutral filters (10).

In addition, the use of the mass spectrometer TOF-type in the study of desorption processes [2,3] allows us to investigate not only the spectral composition, but also to measure kinetic energy of particles leaving the surface.

This is achieved by introduction of two areas of flight of desorbed particles: in the first area the particles in the neutral state divided by the initial velocity, in the second area the division by masses of particles, ionized in the ion source, takes place

Particles that left the sample surface under the action of a single laser pulse, move along normal to the surface and successively pass two drift space. In the first drift space (1) particles in a neutral state are separated according to the initial kinetic energy and fall into the ion source (3) TOF mass spectrometer. In the ion source are formed the packages of charged particles of different masses and then enter the acceleration region, where they additionally obtain the same energy E = 1200 eV, far surpassing their initial kinetic energy Ek In the second drift space (2) charged particles divided according to the masses and fall to the detector SEM-2A (4). The signal from the detector increases by means high-frequency amplifier (5) and digitized in ADC La n1USB (6). The final result in the form of mass spectrum is recorded and processed by a computer measuring complex (7).

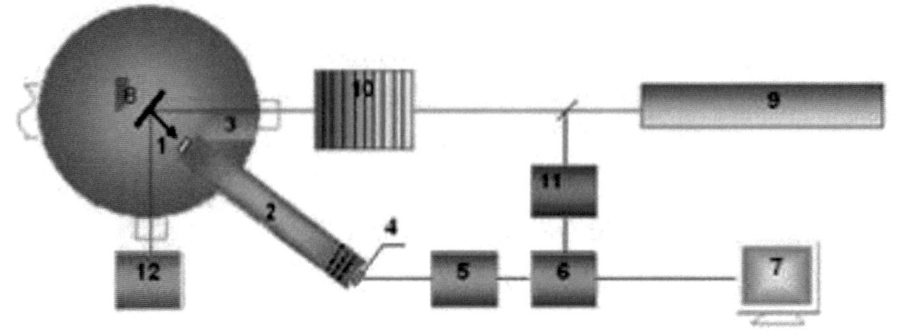

Figure 2.Block scheme of setup for recording the mass spectra of particles, desorbed from the surface by laser irradiation
1-The first space flight(OA in the fig.1);2-The second space flight(CD in the fig.1);3-Ion source (IS); 4-detector (SEM), 5-high-frequency preamplifier; 6-ADC and the buffer memory (La n1USB); 7- Computer measuring system (DL-1700E); 8-Sample (O);9- Neodymium laser (...); 10- a set of neutral filters;11-Photocell (FEC-19A); 12-Monitoring device (optical microscope)

The Synchronization of the recording system and the laser radiation is carried out with the help of a coaxial photocell FEK-19A (11) The pulse-frequency registration mode of mass spectra of desorbed particles allows us to observe the development of the desorption process within tens of milliseconds Functionally, the installation can be represented as consisting of: the analytical part, the sample excitation system and computer-measuring system. The basis of analytical part of the installation - TOF- mass spectrometer of original construction . All other blocks are now commercially produced:

ASK-4106, a combined device company AKTAKOM (generates packets of pulses of a given configuration);

- La n1USB, analog-digital converter (ADC) company ZAO Rudnev-Shilyaev "(used to convert the analog signal to digital);
- DL 1700E and the PC company YOKOGAWA (applied for registration of mass spectra and processing of experimental results).

3. Conclusions

On the example of the films of NTCDA and copper phthalocyanine peripherally substituted by F atoms , are shown the mass spectra and time distribution of particles of the desorption flux.

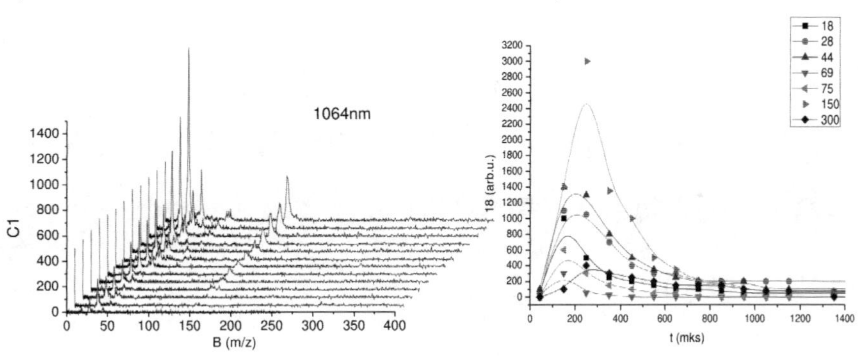

Figure 3. a)TOF distribution of the particles, desorbed from the NTCDA film surface (on the GaAs substrate) by laser irradiation with a wavelength λ =1064 нм,

b) The development of desorption from the NTCDA film surface(on the GaAs substrate) by laser radiation with a wavelength =532 нм

Work is executed at support of Federal agency by training within the limits of the analytical departmental target program «Development of scientific potential of the higher school» (the project № 2) and the Russian fund of basic researches within the limits of the joint program of cooperation with the Japanese fund of advancement of a science-JSPS (the project № 09-02-92109-ЯФ-а)

References

[1] Zaikin V, Varlamov A , Mikaya A and Prostakov N.2001 Fundamentals of mass spectrometry of organic compounds., Moscow, Nauka.Interperiodika, pp. 286.

[2] Lazneva E, Turiev A and Komolov SA. 2009 Laser-induced fragmentation and desorption from the surface of organic films: 1-perylene derivatives.Technical Physics Letters, T. 35, № 16, S. 88-94.

[3] Turiev A 2010.Modification of films of organic semiconductors by laser radiation / / Proceedings of the7-th International Conference on Amorphous and Microcrystalline Semiconductors' .- St. Petersburg, S. 175-176.

Journal of Physics: Conference Series **291** (2011) 012016 doi:10.1088/1742-6596/291/1/012016

Design of numerical algorithms for the problem of charge transport in a 2D silicon MOSFET transistor with a silicon oxide nanochannel[1]

Alexander Blokhin
Sobolev Institute of Mathematics
and Novosibirsk State University,
Novosibirsk, 630090, Russia
e-mail: blokhin@math.nsc.ru

Boris Semisalov
Novosibirsk State University,
Novosibirsk, 630090, Russia
e-mail: vibis@ngs.ru

Abstract. We are concerned with the problem of charge transport in a 2D silicon MOSFET transistor occupying a domain Ω with a silicon oxide nanochannel occupying a domain Ω_G.After proposing an additional boundary condition for the electric potential on the common boundary of the domains Ω and Ω_G we design two numerical algorithms for funding approximate solutions of this problem. The first algorithm is a new one and uses interpolation polynomials of spline-collocation and the sweep method. The second algorithm is based on the well-known longitudinal-transverse sweep (l.t.s.) method. By using these algorithms we obtain graphs of stationary solutions of our problem. We also compare the workability and efficiency of the proposed algorithms for various values of parameters.

Introduction

By now there are plenty of mathematical models describing physical phenomena in semiconductor devices with one or another degree of reliability. For finding approximate solutions of problems of semiconductor physics we should design numerical algorithms for these models. Undoubtedly, the design of such algorithms is urgent because now semiconductor devices are essential parts of many electron appliances. The effective power and reliability of modern computers depend on features of these devices. And the main goal of mathematical modelling is to calculate all the features of a semiconductor with an accuracy requirement.

In this paper we consider a hydrodynamical model proposed recently in [1, 2]. This so-called MEP (maximum entropy principle) model is a quasilinear nonstationary system of conservation

[1] This work is supported by RFBR grant No. 10-01-00320-a, the interdisciplinary integration project No. 91 of SB RAS, the programme "Development of scientific potential of the Russian Higher School (2009–2010)" (grant No. 2.1.1/4591) and the programme "Research and educational manpower of innovation Russia (2009–2013)" (grant No.1180/27.08.09)

laws obtained from the moment system of the Boltzmann kinetic equation. It is called MEP model because for closing the moment system the maximum entropy principle (MEP) was used. The system of MEP model for the 2D case and in a dimensionless form reads

$$\left.\begin{aligned} &R_t + \operatorname{div}(\mathbf{J}) = 0, \\ &\mathbf{J}_t + \frac{2}{3}\nabla\sigma = R\mathbf{Q} + c_{11}\mathbf{J} + c_{12}\mathbf{I}, \\ &\sigma_t + \operatorname{div}(\mathbf{I}) = (\mathbf{J}, \mathbf{Q}) + cP, \\ &\mathbf{I}_t + \nabla(æR) = \frac{5}{3}\sigma\mathbf{Q} + c_{21}\mathbf{J} + c_{22}\mathbf{I}. \end{aligned}\right\} \tag{1}$$

Here R is the electron density, E is the electron energy, $\mathbf{J} = R\mathbf{u}$, $\mathbf{I} = R\mathbf{q}$; $\mathbf{u} = (u^{(x)}, u^{(y)})$ is the electron velocity in the Cartesian coordinate system (x, y); $\mathbf{q} = (q^{(x)}, q^{(y)})$ is the energy flux, $P = R\Big(\frac{2}{3}E - 1\Big)$, $\sigma = RE$, $æ = \frac{10}{9}E^2$, $\mathbf{Q} = \nabla\varphi = (\varphi_x, \varphi_y)$; $\varphi = \varphi(t, x, y)$ is the electric potential satisfying the Poisson equation

$$\triangle_{x,y}\varphi = \triangle\varphi = \varphi_{xx} + \varphi_{yy} = \beta(R - \rho), \tag{2}$$

and $\rho = \rho(x, y)$ is the doping density (a given function in the domain Ω). The coefficients c_{11}, ..., c_{22}, c of system (1) are smooth functions of the energy E detailed in [3, 4], $\beta > 0$ is a constant (see [5] where the reader can also find the detailed reduction of system (1) and the Poisson equation (2) to dimensionless forms). We note that in [6]–[10] various numerical algorithms were proposed for funding approximate solutions of the mathematical model (1), (2).

The present work is devoted to the construction and realization of numerical algorithms for finding stationary solutions of the problem on charge transport in a 2D MOSFET (Metal Oxide Semiconductor Field Effect Transistor). The detailed description of this semiconductor device with electron conductivity is given in [7]. Its characteristic is the presence of a silicon oxide nanochannel. The MOSFET transistor is sketched on Fig. 1 (in terms of dimensionless variables).

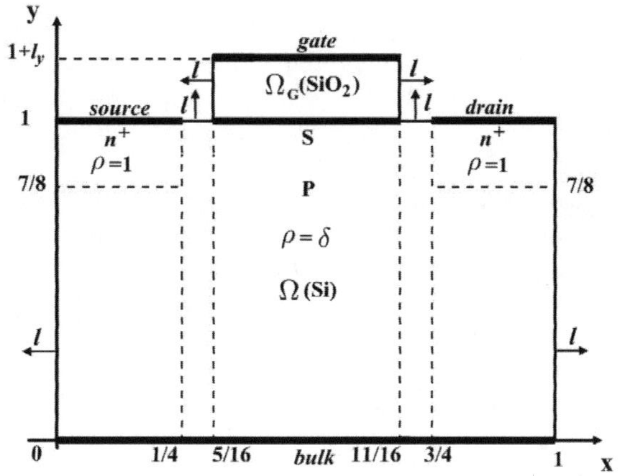

Fig. 1. Schematic sketch of the 2D silicon MOSFET transistor

Remark 0.1. Since there is no charge transport in the nanochannel Ω_G (see Fig. 1), in the domain Ω_G the electric potential $\Phi(t, x, y)$ satisfies the Laplace equation

$$\triangle_{x,y}\Phi = \Phi_{xx} + \Phi_{yy} = 0. \tag{3}$$

The statement of the problem on charge transport in MOSFET for the equations of the MEP model is given in Section 1. Then, according to results of [8, 9], relations (1), (2) in the stationary case are rewritten as a system of three Poisson equations. For finding solutions of these equations we propose two efficient algorithms based on a nonstationary regularization, the stabilization method [11]) and the method of lines (see [12, 13]). The first of these algorithms is detailed [8, 9]. Its main idea is to approximate in the regularized equations the time derivative by a difference relation and the x derivative by an interpolation polynomial with interpolation nodes at zeros of the Chebyshev polynomial. As the result, the original problem for the equations of the MEP model is reduced to a boundary value problem for a second-order ODE system. In [9] the solution of this problem is written in the form of a cubic interpolation $\mathbb{C}^2$ spline. Then, we get the three point scheme

$$\Big\{I_N - \frac{h_y^2}{6}\mathcal{B}\Big\}\widehat{\mathbf{Y}}_{k-1} - 2\Big\{I_N + \frac{h_y^2}{3}\mathcal{B}\Big\}\widehat{\mathbf{Y}}_k + \Big\{I_N - \frac{h_y^2}{6}\mathcal{B}\Big\}\widehat{\mathbf{Y}}_{k+1} =$$

$$= \frac{h_y^2}{6}\{\boldsymbol{\mathcal{F}}_{k-1} + 4\boldsymbol{\mathcal{F}}_k + \boldsymbol{\mathcal{F}}\},\ k = \overline{1, K-1} \tag{4}$$

with the boundary conditions

$$\widehat{\mathbf{Y}}_1 = A_0\widehat{\mathbf{Y}}_0 + \mathbf{B}_0,\ \widehat{\mathbf{Y}}_K = A_K\widehat{\mathbf{Y}}_{K-1} + \mathbf{B}_K. \tag{5}$$

Here $\mathbf{Y}$ is the vector of values of the unknown in the nodes of the interpolation function, N is the number of these nodes, I_n is the unit matrix of order N, h_y is the mesh width of spline-interpolation; the elements of the matrices A_0 and A_K and the components of the vectors $\mathbf{B_0}$ and $\mathbf{B_K}$ are determined from the boundary conditions at $y = 0$ and $y = 1$, and concrete expressions for the elements of the matrix $\mathcal{B}$ and the components of the vector $\boldsymbol{\mathcal{F}}$ are written down in [9].

Thus, the solution of the regularized equation at each time layer can be found from (4) and (5) by the matrix sweep method. Then, using the idea of the stabilization method, we pass from the previous to the next layer until the stationary solution is found. In this paper, we omit calculations towards the construction and justification of the described numerical scheme and pass at once to its realization.

There is only one serious trouble in the application of the proposed algorithm for finding solutions of the problem on charge transport in MOSFET. This is the presence of nanochannel Ω_G. However, in Section 2 under certain assumptions we give an *additional mixed boundary condition* for the potential φ on the set S, where the nanochannel adjoins to the rest part of the transistor. As the result, we will be able to use the proposed numerical algorithm without any principal difficulties.

For the verification of the described methods in Section 3 we construct a second algorithm for finding numerical solutions of the boundary value problem for the Poisson equations obtained in Section 1. Here, together with the stabilization method and the nonstationary parabolic regularization we use the well-known longitudinal-transverse sweep method (this method is described in full details, e.g., in [14, 15]). Section 4 is devoted to the realization of constructed numerical algorithms. There we also detail how did we manage to get the convergence of the stabilization method, and we present graphs of obtained numerical solutions. It should be noted that we have managed to obtain solutions for any desired set of parameters of the problem only by the l.t.s. method. But, if the algorithm based on ideas from [8, 9] works for some set of parameters, then it turns out to be much more efficient than the sweep method.

1. Statement of the problem

Following [7, 16], for the mathematical model (1)–(3) we set boundary conditions (for a reason which will become clear below we restrict ourself by the statement of boundary conditions only for the potentials φ and Φ):

$$\varphi = \begin{cases} \ln\left(\dfrac{N^+}{n_i}\right), \text{ for } y=1,\ 0 \le x \le \dfrac{1}{4} \textbf{ (source)}, \\ D \text{ for } y=1,\ \dfrac{3}{4} \le x \le 1 \textbf{ (drain)}, \\ B \text{ for } y=0,\ 0 \le x \le 1 \textbf{ (bulk)}; \end{cases} \quad (6)$$

$$\Phi = G \text{ for } y = 1 + l_y, \frac{5}{16} \le x \le \frac{11}{16} \textbf{ (gate)}, \quad (7)$$

$$(\mathbf{l}, \nabla\varphi) = 0 \text{ on } \Gamma_l, \quad (8a)$$

$$(\mathbf{l}, \nabla\Phi) = 0 \text{ on } \Gamma_l^{(G)}, \quad (8b)$$

and the matching conditions (see [16]) on S:

$$\frac{1}{3}\Phi_y = \varphi_y,\ \Phi = \varphi, \quad (9)$$

where (see Fig. 1): $\Omega = \{(x,y) : 0 < x, y < 1\}$,
the nanochannel $\Omega_G = \left\{(x,y) :\ \frac{5}{16} < x < \frac{11}{16},\ 1 < y < 1 + l_y\right\}$, $\beta = \frac{e^2 L^2 N^+}{\zeta K_B T_0}$ (see [5]);

$$\rho = \rho(x,y) = \begin{cases} 1,\ (x,y) \in \overline{\Omega}_+, \\ \delta\left(= \dfrac{P}{N^+} = -10^{-3}\right),\ (x,y) \in \overline{\Omega} \setminus \overline{\Omega}_+, \end{cases} \quad (10)$$

$\Omega_+ = \left\{(x,y) :\ \frac{7}{8} < y < 1,\ \left(0 < x < \frac{1}{4}\right) \cup \left(\frac{3}{4} < x < 1\right)\right\}$,
$\Gamma_l = \left\{(x,y) : x = 0 \cup x = 1,\ 0 < y < 1;\ y = 1, \left(\frac{1}{4} < x \le \frac{5}{16}\right) \cup \left(\frac{11}{16} \le x < \frac{3}{4}\right)\right\}$,
$\Gamma_l^{(G)} = \left\{(x,y) :\ x = \frac{5}{16} \cup x = \frac{11}{16},\ 1, y < 1 + l_y\right\}$,
$\mathbf{l}$ is the unit outward normal vector,
$S = \left\{(x,y) :\ y = 1,\ \frac{5}{16} < x < \frac{11}{16}\right\}$,
$D = \frac{e}{K_B T_0} V_D + \ln\left(\frac{N^+}{n_i}\right)$, $G = \frac{e}{K_B T_0} V_G - 10 + \ln\left(\frac{N^+}{n_i}\right)$, $B = -\ln\left(\frac{P_-}{n_i}\right)$; V_D and V_G are the bias voltages (measured in volt). The dimensional parameter e, L, N^+, ζ, K_B, T_0, P_-, and n_i are given in Table 1. From Fig. 1 (see also (10)) we see that the doping density ρ is a piecewise constant function (in numerical computations we use some of its smooth approximations).

Table 1. Values of the physical parameters

Parameter	Description	Value
e	Electron charge	$1,6 \times 10^{-19}$C
L	Characteristic linear dimension	4×10^{-7} m
N^+	Doping density in the n^+ zone	$10^{23} \frac{1}{m^3}$
T_0	Lattice temperature	300K
K_B	Boltzman constant	$1,38 \times 10^{-23} \frac{J}{K}$
$\zeta(Si)$	Dielectric constant (Si)	$1,03545 \times 10^{-10} \frac{C}{Vm}$
$\zeta(SiO_2)$	Dielectric constant (SiO_2)	$\frac{1}{3}\zeta(Si)$
P_-	Doping density in the P zone	$10^{20} \frac{1}{m^3}$
n_i	Intrinsic electron concentration	$10^{16} \frac{1}{m^3}$

Following [9], we recall that the mathematical model (1), (2) in the stationary case can be reduced to the following system of quasilinear elliptic equations for the three unknowns ϑ, R and φ:

$$\begin{aligned}\triangle\vartheta = \mathcal{F}^{(\vartheta)}(\nabla\vartheta, \mathbf{X}, \mathbf{Q}, \vartheta) = a_1|\nabla\vartheta|^2 + a_2(\nabla\vartheta, \mathbf{X})+ \\ + a_3(\nabla\vartheta, \mathbf{Q}) + a_4(\mathbf{X}, \mathbf{Q}) + a_5|\mathbf{Q}|^2 + bc\vartheta,\end{aligned} \tag{11}$$

$$\begin{aligned}\triangle R = \mathcal{F}^{(R)}(\nabla\vartheta, \nabla R, \mathbf{Q}, \vartheta, R, \rho) = -b_1 R|\nabla\vartheta|^2 + b_2(\nabla\vartheta, \nabla R)+ \\ + b_3 R(\nabla\vartheta, \mathbf{Q}) + b_4(\nabla R, \mathbf{Q}) + b_5 R|\mathbf{Q}|^2 + \frac{\beta}{1+\vartheta}(e^{\chi} - \rho)R + ncR\vartheta.\end{aligned} \tag{12}$$

$$\triangle\varphi = \mathcal{F}^{(\varphi)}(\chi, \rho) = \beta(e^{\chi} - \rho), \tag{13}$$

where $\vartheta = \frac{2}{3}E - 1$, $\mathbf{X} = \frac{\nabla R}{R}$. The components of the vectors of electron velocity $\mathbf{u}$ and the energy flux $\mathbf{q}$ are found from the relations

$$\left.\begin{aligned}\mathbf{u} = F(E)\{\mathbf{Q} - (1+\vartheta)\mathbf{X} - F_0(E)\nabla\vartheta\}, \\ \mathbf{q} = G(E)\{-\mathbf{Q} + (1+\vartheta)\mathbf{X} + G_0(E)\nabla\vartheta\},\end{aligned}\right\}$$

where $|\nabla\vartheta|^2 = \vartheta_x^2 + \vartheta_y^2$, etc.,

$$\begin{aligned}&a_1 = -a' F(E)F_0(E) + b' G(E)G_0(E), \quad a_2 = -1 + (1+\vartheta)\{b' G(E) - a' F(E)\},\\
&a_3 = a' F(E) - b' G(E) - bF(E)F_0(E), \quad a_4 = -b(1+\vartheta)F(E), \; a_5 = bF(E),\\
&b_1 = -m' F(E)F_0(E) + n' G(E)G_0(E), \quad b_2 = (1+\vartheta)\{n' G(E) - m' F(E)\},\\
&b_3 = -\frac{1}{(1+\vartheta)^2} + m' F(E) - n' G(E) - nF(E)F_0(E), \quad b_4 = \frac{1}{1+\vartheta} + bF(E)F_0(E),\end{aligned}$$

$$b_5 = nF(E), \quad a' = \frac{da}{d\vartheta} = \frac{3}{2}\frac{da}{dE}, \quad b' = \frac{db}{d\vartheta} = \frac{3}{2}\frac{db}{dE},$$

$$\begin{aligned}a = a(E) = \frac{2}{5}\frac{c_{21}}{1+\vartheta} - c_{11}, \quad b = b(E) = \frac{2}{5}\frac{c_{22}}{1+\vartheta} - c_{12},\\
m = m(E) = \frac{c_{11} - a}{1+\vartheta}, \quad n = n(E) = \frac{c_{12} - b}{1+\vartheta},\end{aligned}$$

$$F(E) = -\frac{c_{22} - (5/3)Ec_{12}}{det}, \quad G(E) = -\frac{c_{21} - (5/3)Ec_{11}}{det},$$

$$F_0(E) = 1 - \frac{(5/3)Ec_{12}}{c_{22} - (5/3)Ec_{12}}, \quad G_0(E) = 1 - \frac{(5/3)Ec_{11}}{c_{21} - (5/3)Ec_{11}},$$

$$det = c_{11}c_{22} - c_{21}c_{12}.$$

Thus, in the stationary case the original mathematical model (1), (2) in the domain Ω can be reduced to the system of quasilinear elliptic equations (11)–(13). To complete the statement of the problem we formulate boundary conditions for ϑ and R (the boundary conditions for the potential φ were given above):

$$\left.\begin{array}{l} R = 1, \text{ for } y = 1, \ \left(0 \le x \le \dfrac{1}{4}\right) \cup \left(\dfrac{3}{4} \le x \le 1\right); \\ R_y = 0 \text{ for } y = 0, \ 0 \le x \le 1; \\ \vartheta = 0, \text{ for } y = 1, \ \left(0 \le x \le \dfrac{1}{4}\right) \cup \left(\dfrac{3}{4} \le x \le 1\right), \\ \vartheta_y = 0 \text{ for } y = 0, \ 0 \le x \le 1; \\ (\mathbf{1}, \nabla\vartheta) = (\mathbf{1}, \nabla R) = 0 \text{ on } \Gamma_l, \\ \vartheta_y = 0, \ R_y = \dfrac{R}{1+\vartheta}\varphi_y \text{ on } S. \end{array}\right\} \quad (14)$$

The boundary conditions (14) are set in the accordance of recommendations from [7, 8, 9].

There is only one trouble towards the application of the numerical method designed in [8, 9] for finding approximate solutions of the boundary value problem for ϑ, R and φ in the domain Ω. This is the matching conditions (9) for the potentials φ and Φ. In the next section we show that under certain conditions we can redefine the boundary value problem for the potential φ in the domain Ω. Namely, one can formulate on the set S an additional boundary condition for the function φ. A concrete form of this condition is determined as the result of simplification of the procedure of funding the potential Φ in the nanochannel Ω_G.

2. Additional boundary condition for φ on the set S

As we can see on Fig. 1, a MOSFET transistor consists of the two parts: the domain Ω and the nanochannel Ω_G which adjoins to Ω along the boundary S. In the next sections we apply the numerical model from [8, 9] for funding stationary solutions of the problem of charge transport in MOSFET. But, if we try to design a numerical algorithm like that in [9] directly to equations (11)–(13) with the boundary conditions (6)–(9), (14), we have essential difficulties. The point is that in this case we have to introduce grids on the sets Ω and Ω_G and approximate the unknowns along the x axis by interpolation polynomials and along the y axis by a cubic $\mathcal{C}^2$ spline. Then, there appears a problem with the realization of the matching conditions (9) for the spline-function as well as a number of other difficulties connected with the conformance of grid steps and sweep methods in the domains Ω and Ω_G. In this section we propose an idea which enables one to be saved from the mentioned difficulties and perform computations only in the domain Ω. This idea is based on the smallness of the width of the nanochannel compared to its length. Starting from such an assumption we obtain an additional boundary condition on the set S for computing zero and first approximations of the potentials φ and Φ in their expansions into the series in the small parameter ε_M which is the relation of the width of the nanochannel to its length. By numerical simulations using the l.t.s. method it was shown in [20] that such approximations are accurate enough.

We demonstrate the way of constructing the additional boundary condition on the example of the model problem (2), (3), (6)–(9) assuming that the function R appearing in (2) is a *known function* $R(x, y)$ in the domain Ω.

Remark 2.1. We can simplify somewhat the model problem if instead of φ and Φ we introduce the functions $\tilde{\varphi} = \varphi - \ln\left(\frac{N^+}{n_i}\right)$, $\tilde{\Phi} = \Phi - \ln\left(\frac{N^+}{n_i}\right)$ (below we drop tildes). Then we get the boundary condition $\varphi = 0$ on the drain for $0 \le x \le \frac{1}{4}$.

We now describe the procedure of finding an approximate solution of equation (3). Let us make the change of independent variables

$$\xi = \frac{x - 5/16}{l_x}, \ \eta = \frac{y-1}{l_y}, \ 0 \le \xi, \eta \le 1, \tag{15}$$

where l_x and l_y are the length and the width of the nanochannel respectively.

For the MOSFET transistor sketched on Fig. 1 the length $l_x = \frac{3}{8}$. In terms of the new variables ξ and η equation (3) becomes

$$\varepsilon_M^2 \Phi_{\xi\xi} + \Phi_{\eta\eta} = 0, \tag{16}$$

where the parameter $\varepsilon_M = \frac{l_y}{l_x}$. Assuming ε_M to be small enough and dropping the first term in (16) after some transformations (see [20]) we get

$$\varphi(x,1) + 3l_y\varphi_y(x,1) = G, \ (x,1) \in S. \tag{17}$$

Then, (17) is the desired additional boundary condition for the potential φ. The potential $\Phi(x, y)$ is found from

$$\Phi(x,y) = 3l_y\varphi_y(x,1)(\eta - 1) + G = [G - \varphi(x,1)]\eta + \varphi(x,1). \tag{18}$$

Thus, the boundary value problem (2), (3), (6)–(9) is reduced to the problem for the potential φ in the domain Ω with the boundary conditions (6), (8a), (17). For finding an approximate solution to this problem we use the numerical algorithm designed in [8, 9] which was successfully applied for the computation of concrete semiconductor devices.

The rest of the paper is organized as follows. In the next section, basing on the well-known l.t.s. method, we construct a numerical scheme for finding approximate solutions to equations (11)–(13) and the boundary conditions (6)–(8b), (17), (14). In the last section we compare results obtained by the scheme based on the longitudinal-transverse sweep method and the algorithm which uses the numerical model from [9].

3. Longitudinal-transverse sweep method (l.t.s.)

The l.t.s. method is often used for finding numerical solutions of various nonstationary boundary value problem of mathematical physics (see, for example, [14, 15]). In this connection, it is interesting to compare the efficiency and performance of this method and the numerical scheme from [9]. We will use the l.t.s. method, which is based on difference relations approximating derivatives of unknown functions, together with the method of lines, a regularization and the stabilization method. We introduce space and time grids. The passage from the previous time layer to the next one is performed in two steps (the diagram on Fig. 2):

a) In the *longitudinal sweep*, going from the left boundary of Ω to the right one along the lines $y = y_k$, $k = 0, ..., K$ of the grid, we calculate the values of sweep coefficients by recurrence formulas. Then, we resolve the right boundary condition and in the return step (from the right to the left) we compute the unknown functions by using the found sweep coefficients.

b) In the *transverse sweep*, going from the lower boundary of the domain Ω to the upper one along the lines $x = x_j, j = 1, ..., N$ of the grid, we find the values of sweep coefficients. Using them and taking into account the right boundary condition, we get the values of unknown functions.

According the idea of the stabilization method (see [11]), we will perform these operations until the solution is stabilized. As the result we find a stationary solution of the problem of charge transport in MOSFET.

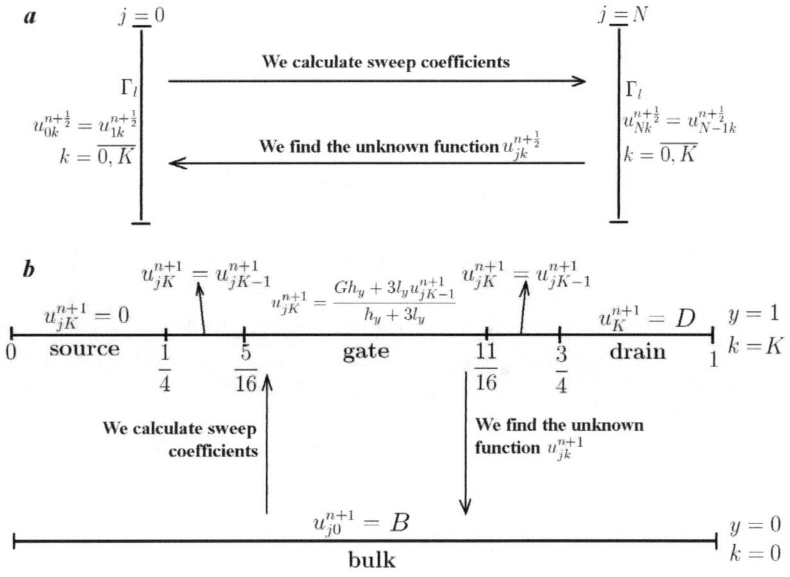

Fig. 2. Diagram of computations by the l.t.s. method (parts a and b are for longitudinal and transverse sweeps respectively)

We now pass to the construction of the numerical model for our problem based on the l.t.s. method. This model for problem (11)–(13), (6)–(8b), (17) for a MOSFET transistor is convenient to be constructed on the example of the *model problem* for the Poisson equation

$$\triangle_{x,y}\psi = \psi_{xx} + \psi_{yy} = f(x,y),\ (x,y) \in \Omega \tag{19}$$

with the mixed boundary conditions

$$\left.\begin{array}{l} \psi = \begin{cases} 0 \text{ for } y=1,\ 0 \le x \le \frac{1}{4}, \\ D \text{ for } y=1,\ \frac{3}{4} \le x \le 1\ , \\ B \text{ for } y=0,\ 0 \le x \le 1, \end{cases} \\ (\mathbf{l}, \nabla\psi) = 0 \text{ on } \Gamma_l. \\ \psi + 3l_y\psi_y = G \text{ for } y=0,\ \frac{5}{16} \le x \le \frac{11}{16}, \end{array}\right\} \tag{20}$$

where ψ is the unknown function and $f(x,y)$ is a sufficiently smooth right-hand side.

Performing a parabolic regularization of the Poisson equation (19) and denoting the solution of the regularized problem by $u(t,x,y)$, we obtain the relation

$$u_t = \triangle_{x,y}u - f(x,y),\ u = u(t,x,y),\ t > 0,\ (x,y) \in \Omega. \tag{21}$$

Remark 3.1 By deriving a priori estimates for solutions of the original and regularized problems, it was shown in [9] that $u(t,x,y) \to \psi(x,y)$ as $t \to \infty$. Moreover, it was also proved there that the solution ψ of the model problem (19), (20) is asymptotically stable (by Lyapunov). This, in particular, justifies on the differential level the applicability of the stabilization method [11].

We now make time and space digitizations in equation (21). We introduce on Ω a uniform grid with the mesh points (x_j, y_k) and the steps h_x and h_y ($j = \overline{0,N}$, $k = \overline{0,K}$, $x_j = h_x j$, $y_k = h_y k$). We consider the time grid with the mesh points $n\Delta$ and the step Δ ($n = 0, 1, 2, ...$).

Let $u = u^n_{jk} = u(n\Delta, jh_x, kh_y)$ be the mesh function

$$\Lambda = \frac{\Psi_x - 2 + \Psi_x^{-1}}{h_x^2} + \frac{\Psi_y - 2 + \Psi_y^{-1}}{h_y^2},$$

$\tau = \chi - 1$, χ, $\Psi_x^{\pm 1}$, $\Psi_y^{\pm 1}$ the shift operators $(\Psi^{+1}_{x,y} = \Psi_{x,y})$:

$$\chi u^n_{jk} = u^{n+\frac{1}{2}}_{jk} = \hat{u},\ \Psi_x^{\pm 1} u^n_{jk} = u^n_{j\pm 1,k},\ \Psi_y^{\pm 1} u^n_{jk} = u^n_{j,k\pm 1},\ \mathcal{F} = f_{jk} = f(x_j, y_k).$$

Then, approximating in (21) the derivatives of u by difference relations, we find

$$\tau u - \Delta \cdot \Lambda \hat{u} = -\Delta \mathcal{F}$$

or

$$\hat{u}_{jk} - a(\hat{u}_{j-1,k} + \hat{u}_{j+1,k}) - b(\hat{u}_{j,k-1} + \hat{u}_{j,k+1}) = f^n_{jk}, \tag{22}$$

where

$$f^n_{jk} = \frac{u - \Delta \mathcal{F}}{d},\ a = \frac{a_x}{d},\ b = \frac{b_y}{d},\ a_x = \frac{\Delta}{h_x^2},\ b_y = \frac{\Delta}{h_y^2},\ d = 1 + 2a_x + 2b_y.$$

We will find a solution of difference relations (22) on each time layer $n = 0, 1, 2, ...$ by the l.t.s. method (see [14, 15] and fig. 2).

As the result, starting from the values u^n_j, we found the solution u^{n+1}_j ($j = 0, ..., N$) on the $(n+1)$th time layer. Below in Section 4 we detail the numerical scheme which gives us the stationary solution of the problem of charge transport in a MOSFET transistor by passing from a previous time layer to the next one and using the idea of the stabilization method.

4. Realization of numerical algorithms

For funding stationary solutions of the problem of charge transport in a MOSFET transistor we propose two numerical models. The first one uses the ideas from [8, 9] (interpolation polynomials of spline-collocation and the sweep method). The second one exploits the l.t.s. method and is described in the previous section. These models are based on principally different ideas, but both of them use time regularization and the stabilization method for funding stationary solutions. In this section we describe numerical schemes and details of the realizations of the algorithms constructed on the basis of the two proposed models. We also compare the efficiency of these algorithms and corresponding numerical results.

At each time layer in the process of stabilization (for both the l.t.s. method and the algorithm based on numerical model from [8, 9]) we should step by step solve three boundary value problems for the regularized Poisson equations (11)–(13). Thus, under the construction of our algorithms we first perform a regularization of equations (11)–(13). If we use the l.t.s. method, it is the parabolic regularization (21). For the technology proposed in [9] we can apply one of the two nonstationary regularizations: the parabolic or Sobolev's one.

Remark 4.1. For example, after the application of Sobolev's regularization to equation (19) of the model problem we get the relation

$$u_t - \triangle u_t = \triangle u - f(x, y),\ u = u(t, x, y),\ t > 0,\ (x, y) \in \Omega.$$

Arguments justifying the stabilization method in the case of this regularization can be found in [9].

For numerical calculations we should define initial data $u_0 = u(0, x, y)$.

Remark 4.2. To get an approximate solution for a desired set of parameters of the problem we set initial data in different ways and "pull values of parameters". That is, we first set original initial data, for instance, such as

$$\vartheta(t, x, y)|_{t=0} \equiv 0, \;\; R(t, x, y)|_{t=0} \equiv 1, \;\; \varphi(t, x, y)|_{t=0} \equiv 0,$$

and then we perform calculations for such a set of parameters that the stabilization method converges. After that as the initial data we take the obtained solution and set the values of parameters to be close to the desired ones (but so that the stabilization method converges). Then we perform calculations and again as the initial data we take the stabilized solution. Continuing such a procedure, we can finally get the stationary solution for the desired set of parameters. It should be noted that in spite of the high efficiency of the procedure of "pulling parameters" its application does not guarantee that we can find a solution for any desired range of parameters. Therefore, in numerical simulations we use this procedure together with a number of other methods (we will talk about them below).

After setting the initial data we start iterations of the stabilization method in which using the variables computed at the previous and present time layers we calculate the right-hand sides $\mathcal{F}^{(\vartheta)}$, $\mathcal{F}^{(R)}$, $\mathcal{F}^{(\varphi)}$ and solve the equations for ϑ, R and φ respectively. These iterations work until the solution is stabilized, i.e., until the norm of the difference between the solutions at the next and previous time layers is close to zero. Such a numerical algorithm is described on Fig. 3.

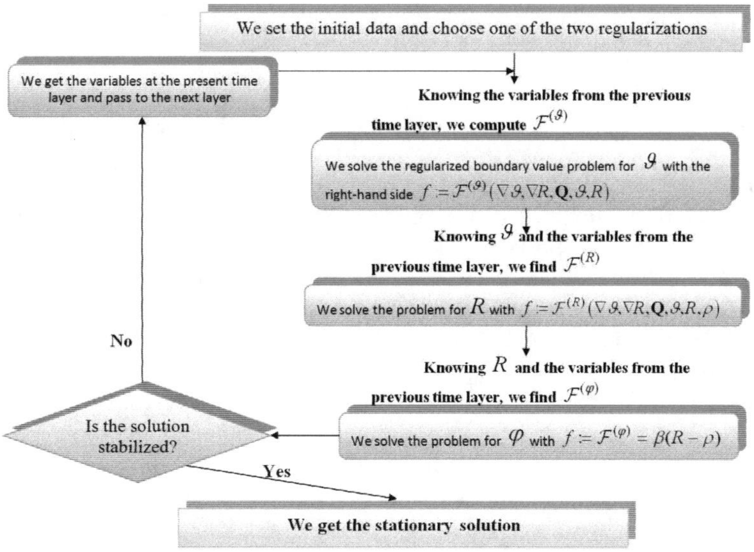

Fig. 3. Diagram of numerical algorithm

The proposed numerical algorithm was realized by Delphi 6 (Object Pascal) and Java. As input parameters the computer program took values of physical and numerical parameters of the problem (the description of some of them is given in Table 1 above). In Table 2 we describe the set of those parameters which varied in numerical simulations.

Table 2. Values of physical and other parameters

Parameter	Description	Value
V_G	Gate voltage	0,36 – 1 V
V_D	Drain voltage	0,36 – 1 V
B	Dimensionless bulk voltage	-25,328 – 0
δ	Dimensionless doping density in the domain $\Omega \setminus \Omega_+$	-0,001 – 0,8
l_y	Width of the nanochannel	1 –20 nm
N	Number of mesh points along the x axis	20 – 40
K	Number of mesh points along the y axis	20 – 40
Δ	Time step of the grid	0,0001 – 0,1
N_{it}	Number of nonlinear iterations (see (23))	1 – 10
n_{sgla}	Nonlinear smoothing is used after each n_{sgla} steps (see (24))	2 – 10
θ	Parameter of nonlinear smoothing (see (24))	$0,1$
λ	Parameter of nonlinear smoothing (see (24))	1
ε_1	Accuracy of stabilization	$10^{-4} - 10^{-8}$

The algorithm stops if the necessary accuracy ε_1 is achieved (see [11]):

$$\sum_{j=0}^{N}\sum_{k=0}^{K}(|R_{jk}^{n+1} - R_{jk}^{n}| + |\varphi_{jk}^{n+1} - \varphi_{jk}^{n}| + |\vartheta_{jk}^{n+1} - \vartheta_{jk}^{n}|) \leq \varepsilon_1.$$

Remark 4.3. We note that in numerical simulations we aimed to obtain the stationary solution of the problem of charge transport in a MOSFET transistor for the following values of parameters (below we call these values the *desired set of parameters*):

$$V_D = 1V,\ V_G = 1V,\ B = -25,328,\ \delta = -0,001,\ l_y = 20nm,\ \varepsilon_1 = 10^{-5}.$$

This set of values is a standard test (see, e.g., [7]) which is often used in real physical and numerical experiments. Finding the stationary solution for the desired set of parameters demanded some efforts.

In computations we met some difficulties. Under the usage of the stabilization method there appeared a jump growth of the unknowns caused by nonlinearity of the problem. Namely, the norm of solution became very big that led to the buffer overflow and the program stop until stabilization. To overcome this difficulty we used *nonlinear iterations*.

The main idea of the algorithm based on nonlinear iterations is the calculation of parameters and variables of the problem by formulas assigned for a next time layer whereas we stay at the present time layer. To clarify this we use nonlinear iterations for a modification of the scheme (4), (5). In this case the scheme (4), (5) should be reduced to the form

$$\begin{aligned}\left\{I_N - \frac{h_y^2}{6}\mathcal{B}_k^{[l-1]}\right\}\widehat{\mathbf{Y}}_{k-1}^{[l]} - 2\left\{I_N + \frac{h_y^2}{3}\mathcal{B}_k^{[l-1]}\right\}\widehat{\mathbf{Y}}_{k}^{[l]} + \left\{I_N - \frac{h_y^2}{6}\mathcal{B}_k^{[l-1]}\right\}\widehat{\mathbf{Y}}_{k+1}^{[l]} = \\ = \frac{h_y^2}{6}\{\boldsymbol{\mathcal{F}}_{k-1}^{[l-1]} + 4\boldsymbol{\mathcal{F}}_{k}^{[l-1]} + \boldsymbol{\mathcal{F}}_{k+1}^{[l-1]}\},\ k = \overline{1, K-1},\ l = \overline{1, N_{it}},\end{aligned} \quad (23)$$

Journal of Physics: Conference Series **291** (2011) 012016 doi:10.1088/1742-6596/291/1/012016

where the components of the vector $\mathcal{F}_k^{[l-1]}$ and the elements of the matrix $\mathcal{B}_k^{[l-1]}$ are calculated at the $(l-1)$th nonlinear iteration; N_{it} is the number of nonlinear iterations at each time layer (see Table 2). The elements of the matrix $\mathcal{B}_k^{[0]}$ and the components of the vector $\mathcal{F}_k^{[0]}$ are taken from the previous time layer. At the lth nonlinear iteration we compute the values of the components of the vector $\widehat{\mathbf{Y}}_k^{[l]}$, $k = 1, ..., K-1$ according to (23). Then, using these values we calculate the elements of the matrices $\mathcal{B}_k^{[l]}$ and the components of the vector $\mathcal{F}_k^{[l]}$. After that the program passes to the $(l+1)$th nonlinear iteration. For $l = N_{it}$ we pass to the next time layer.

In numerical calculations based on the proposed algorithm there also appear short-wave oscillations of the unknowns of the problem. These oscillations precluding the convergence of the scheme with a desired accuracy have no physical meaning and are only a numerical effect. For removing these oscillations we use nonlinear smoothing. Calculations were carried out on a mesh with the points (x_j, y_k) and the steps $h_x = \frac{1}{N}$, $h_y = \frac{1}{K}$. Let g_j be the value of one of the unknowns R, ϑ or φ at the point (x_j, y^*), where y^* is one of the horizontal lines $y = y_k$, $k = 0, ..., K$. Then the filter of nonlinear smoothing looks as follows (see [21]-[23]):

$$(g_j)_{fil} = g_j - \xi\, \overline{M}_{j+\frac{1}{2}} - \xi\, \overline{\overline{M}}_{j+\frac{1}{2}}, \tag{24}$$

where

$$\overline{M}_{j+\frac{1}{2}} = \widetilde{\theta} g_j + \overline{N}, \quad \overline{\overline{M}}_{j+\frac{1}{2}} = -\widetilde{\theta} g_{j+1} + \overline{\overline{N}},$$

$$\overline{N} = \widetilde{\vartheta}_j \mathrm{P}(\frac{\widetilde{\theta}}{2}\widetilde{\vartheta}_j \xi g_j,\ \widetilde{\theta}|\xi g_j|,\ \widetilde{\theta}\lambda\widetilde{\vartheta}_j \overline{\xi} g_j), \quad \overline{\overline{N}} = \widetilde{\vartheta}_j \mathrm{P}(\frac{\widetilde{\theta}}{2}\widetilde{\vartheta}_j \xi g_j,\ \widetilde{\theta}|\xi g_j|,\ \widetilde{\theta}\lambda\widetilde{\vartheta}_j \xi g_{j+1}).$$

Here $0 \le \widetilde{\theta} \le \frac{1}{2}$, $\lambda > \frac{1}{2}$ are constants (see Table 2), $\widetilde{\vartheta}_j = sign(\xi g_j)$, $\mathrm{P}(a_1, ..., a_m) =$ $= max\{0, min(a_1, ..., a_m)\}$, $(g_j)_{fil}$ are discrete values of the component g_j after the nonlinear smoothing (24). The values $\widetilde{\theta}$ and λ are connected by the inequality $2\widetilde{\theta}(1+\lambda) \le 1$. Moreover, $(g_j)_{fil}$ and g_j have the same growth direction.

The calculations using (24) were performed along each line $y^* = y_k$, $k = 0, ..., K$ (longitudinal smoothing). A filter like (24) was also used for the nonlinear smoothing of the unknown function along the y axis (transverse smoothing). It should be noted that in order to absolutely remove short-wave oscillations we have to apply the filter of nonlinear smoothing after each 2–10 time steps.

In numerical calculations using the described technique, we have managed to get stationary solutions for the following sets of parameters:

a) $V_D = 0,36V$, $V_G = 0,36V$, $B = 0$, $\delta = 0,8$, $l_y = 20nm$, $\varepsilon_1 = 10^{-5}$ for the algorithm based on the numerical model proposed in [9]. The graphs for the electron energy E and the electric potential φ calculated in this case are given on Fig. 4;

b) $V_D = 1V$, $V_G = 1V$, $B = -1$, $\delta = -0,001$ $l_y = 20nm$, $\varepsilon_1 = 10^{-5}$ for the algorithm based on the l.t.s. method. The graphs for the electron energy E, the electron density R and the electric potential φ calculated in this case are given on Fig. 5.

However, we could not find a solution for the desired set of parameters (see Remark 4.3) because the stabilization method did not converge in this case. This trouble caused by a fast growth of the variables, the buffer overflow and the program stop prompted us to transform equations (11)–(13). To this end we introduce the auxiliary variables $\varphi^*(x,y)$, $\widetilde{\varphi}(x,y)$, $R^*(y)$, $\Psi(x,y)$ such that $\varphi = \varphi^* + \widetilde{\varphi}$,

$$\Delta\varphi^* = \beta(R^* - \rho), \tag{25}$$

$$\Delta\widetilde{\varphi} = \beta(R - R^*). \tag{26}$$

Journal of Physics: Conference Series **291** (2011) 012016 doi:10.1088/1742-6596/291/1/012016

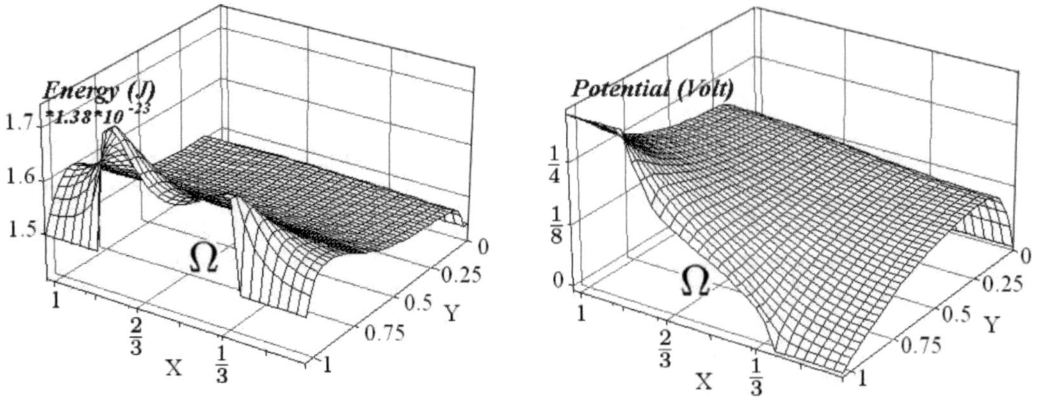

Fig. 4. Numerical results obtained by the algorithm from [9] with the parameters
$V_D = 0,36\ V,\ V_G = 0,36\ V,\ B = 0,$
$\delta = 0,8,\ l_y = 20\ nm,\ N = 40,\ K = 30,\ \varepsilon_1 = 10^{-5}$

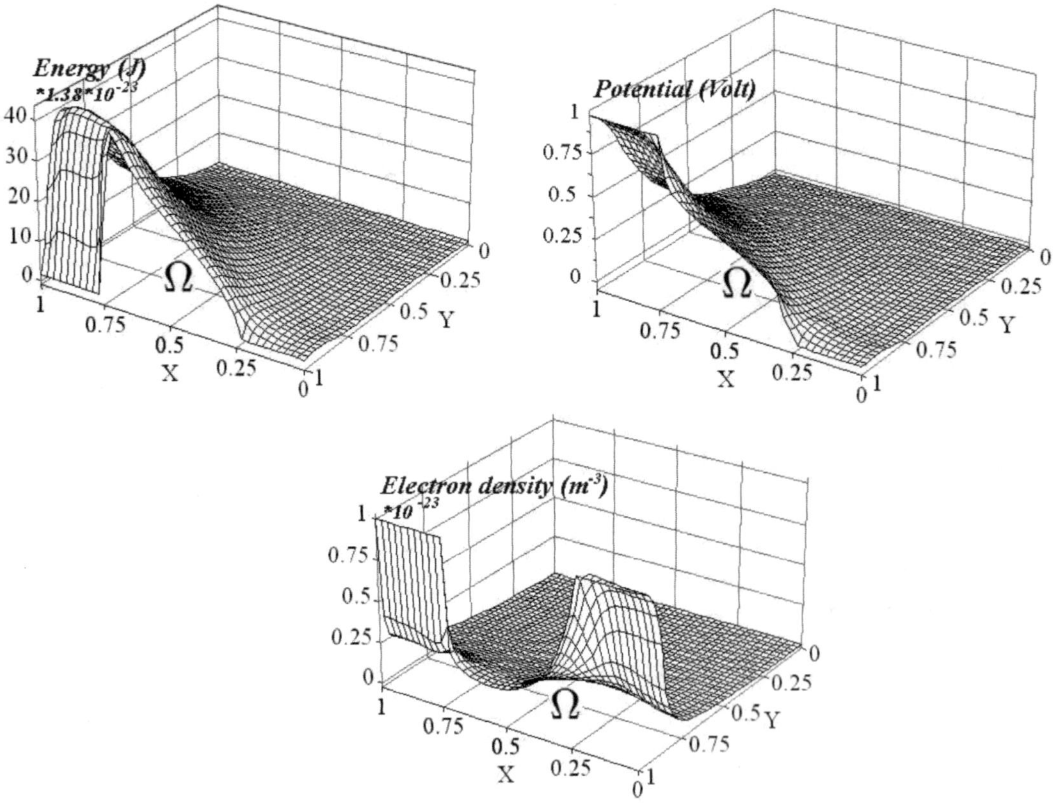

Fig. 5. Numerical results obtained by the algorithm based on the l.t.s. method with the parameters
$V_D = 1\ V,\ V_G = 1\ V,\ B = -1,\ \delta = -0,001,\ l_y = 20\ nm,\ N = 40,\ K = 40,\ \varepsilon_1 = 10^{-5}$

The boundary conditions for φ^*, $\widetilde{\varphi}$ read

$$\begin{cases} \varphi^* = ln\Big(\dfrac{N^+}{n_i}\Big),\ \widetilde{\varphi} = 0,\ \text{for } y = 1,\ 0 \le x \le \dfrac{1}{4}\ \textbf{(source)}, \\ \varphi^* = D,\ \widetilde{\varphi} = 0,\ \text{for } y = 1,\ \dfrac{3}{4} \le x \le 1\ \textbf{(drain)}, \\ \varphi^* = B,\ \widetilde{\varphi} = 0,\ \text{for } y = 0,\ 0 \le x \le 1\ \textbf{(bulk)}; \end{cases} \tag{27}$$

$$(\mathbf{l}, \nabla\varphi^*) = 0,\ (\mathbf{l}, \nabla\widetilde{\varphi}) = 0 \text{ on } \Gamma_l, \tag{28}$$

$$\varphi^* + 3l_y\varphi^* = G,\ \widetilde{\varphi} + 3l_y\widetilde{\varphi} = 0 \text{ on } S. \tag{29}$$

Below we drop tildes by φ.

In numerical calculations we used two forms of the auxiliary function $R^*(y)$ (we tried to define it to be "close", in some sense, to the functions obtained by cutting the graph of the given doping density $\rho(x, y)$ by the lines x=const for $0 \le x \le \frac{1}{4}$, $\frac{3}{4} \le x \le 1$):

$$1)\ R^* = R^*(y) = 3y^2 - 2y^3. \tag{30}$$

Moreover, $R^*_y = 6y(1-y)$, $R^*_{yy} = 6(1-2y)$. While deducing relation (30) we used the conditions $R^*(1) = 1$, $R^*(0) = 0$, $R^*_y(0) = 0$, $R^*_y(1) = 0$.

$$2)\ R^* = R^*(y) = \begin{cases} 0,\ 0 \le y \le \kappa, \\ -3\Big(\dfrac{y-\kappa}{1-\kappa}\Big)^4 + 4\Big(\dfrac{y-\kappa}{1-\kappa}\Big)^3,\ \kappa < y \le 1. \end{cases} \tag{31}$$

Here $\kappa = \frac{7}{8} - \varepsilon_{small}$, where $\varepsilon_{small} << 1$. In the deduction of (31) we started from the condition $R^*(1) = 1$, $R^*(\kappa) = 0$, $R^*_y(\kappa) = 0$, $R^*_y(1) = 0$. Applying (31), it is easy to calculate

$$R^*_y = -12\Big(\frac{y-\kappa}{1-\kappa}\Big)^3 \frac{1}{1-\kappa} + 12\Big(\frac{y-\kappa}{1-\kappa}\Big)^2 \frac{1}{1-\kappa},$$

$$R^*_{yy} = -36\Big(\frac{y-\kappa}{1-\kappa}\Big)^2 \frac{1}{(1-\kappa)^2} + 24\frac{y-\kappa}{1-\kappa}\frac{1}{(1-\kappa)^2}.$$

Remark 4.4. The cut of the doping density $\rho(x, y)$ (see (10)) by the lines x=const for $0 \le x \le \frac{1}{4}$, $\frac{3}{4} \le x \le 1$ has the form of a step function with the discontinuity at the point $y = \frac{7}{8}$. This fact was crucially used for determining the auxiliary function $R^*(y)$ by setting the parameter κ (see (31)). Below we will see that the usage of such kind of function $R^*(y)$ has proved to be very perspective for finding stationary solutions of the problem for MOSFET with the desired set of parameters.

We define the auxiliary function $\Psi(x, y)$ by the relation

$$\Psi = \triangle\varphi = \beta(R - R^*). \tag{32}$$

Then

$$\triangle\Psi = \mathcal{F}^{(\Psi)} = \beta\triangle R - \triangle R^*. \tag{33}$$

In the case of determining $R^*(y)$ in the first way (see (30)) the equation for $\Psi(x, y)$ reads

$$\triangle\Psi = \mathcal{F}^{(\Psi)} = \beta\mathcal{F}^{(R)} - 6\beta(1 - 2y).$$

If we define $R^*(y)$ in the second way (see (31)) for the unknown function $\Psi(x,y)$ from (33) we obtain the relation

$$\triangle\Psi = \mathcal{F}^{(\Psi)} = \beta\mathcal{F}^{(R)} + 36\left(\frac{y-\kappa}{1-\kappa}\right)^2 \frac{1}{(1-\kappa)^2} + \frac{24(y-\kappa)}{(1-\kappa)^3}.$$

The boundary conditions for Ψ follow from the conditions for R (see (14)):

$$\begin{cases} \Psi = 0, \text{ for } y = 1, \ \left(0 \le x \le \dfrac{1}{4}\right) \cup \left(\dfrac{3}{4} \le x \le 1\right); \\ \Psi = 0, \text{ for } y = 0; \\ \Psi_y = 0 \text{ for } y = 0, \ 0 \le x \le 1; \\ (\mathbf{1}, \nabla R) = 0 \text{ for } \Gamma_l; \\ (\Psi + \beta R^*)_y = \dfrac{\Psi + \beta R^*}{1+\vartheta}(\varphi^* + \varphi)_y \text{ for } S. \end{cases} \tag{34}$$

After above manipulations the numerical calculations are described as follows (see Fig. 6):

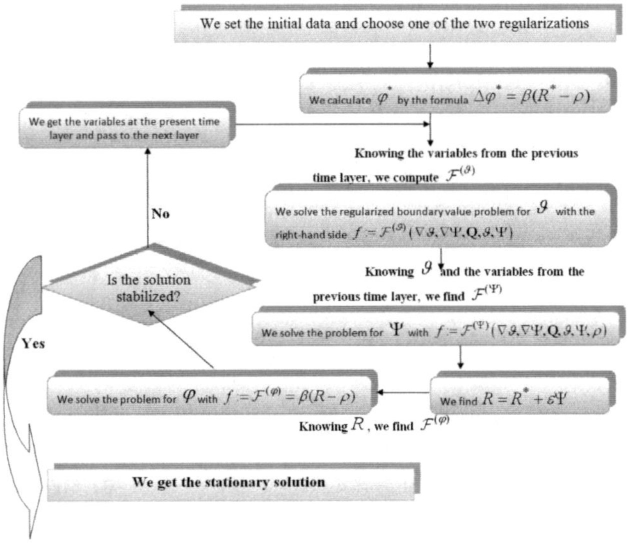

Fig. 6. Diagram of the numerical algorithm using auxiliary functions

1.Before starting iterations of the stabilization method we should state initial data for the unknowns (see arguments in Remark 4.2) and compute the values of φ^* by (25) with a given right-hand side.

2. In the stabilization process while solving the boundary value problems for the Poisson equation for the unknowns Ψ, φ, ϑ we apply one of the numerical models described above: either the model based on the l.t.s. method and described in Section 3 or the numerical model from [9]. At each time layer we step by step calculate the following functions:

$\vartheta(t,x,y)$ by using the same relation (11) as above,

$\Psi(t,x,y)$ by (33) (by choosing one of the two ways of determining the auxiliary function $R^*(y)$),

$R(t,x,y) = R^*(y) + \frac{1}{\beta}\Psi$,

$\varphi(t, x, y)$ by the formula $\triangle\varphi = \Psi$ (see (32)).

3. We stop these operations when the norm of the difference between the solutions at the next and previous time layers becomes less then ε_1.

Performing calculations according to the above scheme, using the l.t.s. method, nonlinear iterations, the filter of nonlinear smoothing, the "pulling parameters" technique described in Remark 4.2, and defining the auxiliary function $R^*(y)$ in the second way (see (31)), we have managed to get the stationary solution for the desired set of parameters (see Remark 4.3). The graphs for the electron density, the electron energy and the electric potential calculated in this case are given on Fig. 7.

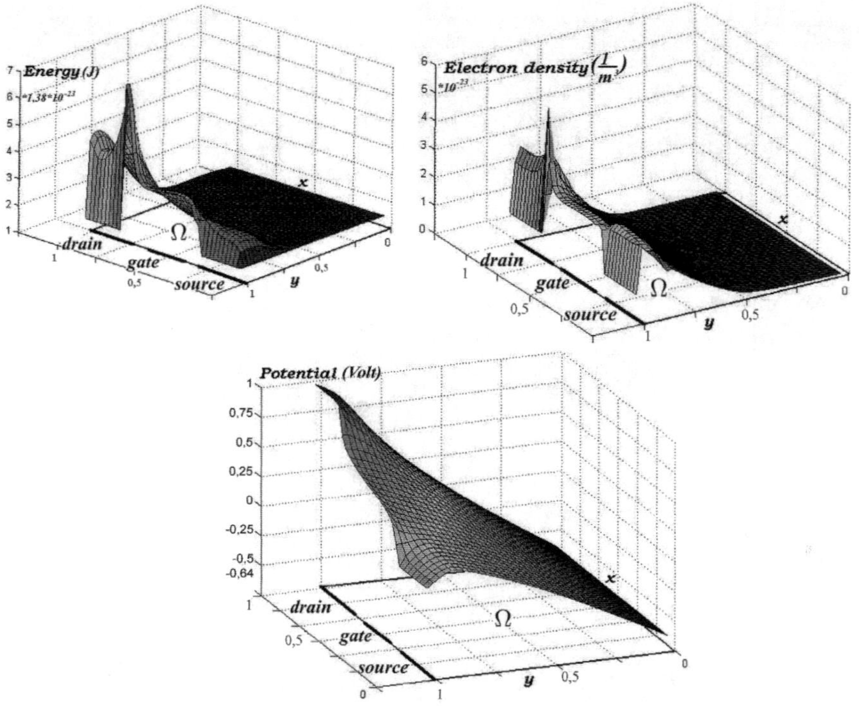

Fig. 7. Numerical results obtained by the algorithm based on the l.t.s. method with the parameters $V_D = 1\ V$, $V_G = 1\ V$, $B = -25,328$, $\delta = -0,001$, $l_y = 20\ nm$, $N = 40$, $K = 40$, $\varepsilon_1 = 10^{-5}$

However, for the algorithm based on the numerical model from [8, 9] we are not able find a solution for such values of parameters. But, in the case when we take other values of parameters (e.g., $V_D = 0,36\ V$, $V_G = 0,36\ V$, $B = 0$, $\delta = 0,8$, $l_y = 20\ nm$, $\varepsilon_1 = 10^{-5}$, see the numerical results for this set of parameters on Fig. 4), the stabilization method together with the technique described in [8, 9] converges much faster than the numerical algorithm based on the l.t.s. method.

Conclusions

In this paper for finding solutions of the problem on charge transport in MOSFET we proposed two efficient numerical algorithms: the algorithm using interpolation polynomials, spline approximations and the matrix sweep method, and the algorithm based on the approximation of derivatives by difference relations and the application of the longitudinal-transverse sweep method.

We hope that the obtained results stimulate a further development and improvement of numerical algorithms for problems of physics of semiconductors (this will be useful for the construction of real devices) and the proposed methods will be adopted for finding numerical solutions of various applied problems outside semiconductor subjects.

References

[1] *Anile A.M., Romano V.* Non parabolic band transport in semiconductors: closure of the moment equations // Cont. Mech. Thermodyn. 1999. Vol. 11. P. 307–325.

[2] *Romano V.* Non parabolic band transport in semiconductors: closure of the production terms in the moment equations // Cont. Mech. Thermodyn. 2000. Vol. 12. P. 31–51.

[3] *Blokhin A.M., Bushmanov R.S., Romano V.* Asymptotic stability of the equilibrium state for the hydrodynamical model of charge transport in semiconductors based on the maximum entropy principle // Int. J. Engineering Science. 2004. Vol. 42(8–9). P. 915–934.

[4] *Blokhin A.M., Bushmanov R.S., Romano V.* Nonlinear asymptotic stability of the equilibrium state for the MEP model of charge transport in semiconductors // Nonlinear Analysis. 2006. Vol. 65. P. 2169–2191.

[5] *Blokhin A. M., Bushmanov R. S., Rudometova A. S., Romano V.*, Linear asymptotic stability of the equilibrium state for the 2D MEP hydrodynamical model of charge transport in semiconductors, Nonlinear Analysis, 65 (2006), pp. 1018–1038.

[6] *Romano V.* 2D simulation of a silicon MESFET with a non-parabolic hydrodynamical model based on the maximum entropy principle, J. Comp. Phys., 176 (2002) pp. 70–92.

[7] *Romano V.* 2D Numerical Simulation of the MEP Energy-Transport Model with a Finite Difference Scheme // J. Comp. Fhys, v. 221, p. 439–468 (2007).

[8] *Blokhin A.M., Ibragimova A.S.* Numerical method for 2D Simulation of a Silicon MESFET with a Hydrodynamical Model Based on the Maximum Entropy Principle // SIAM J.Sci. Comput. 2009. Vol. 31, Issue 3. pp. 2015–2046.

[9] *Blokhin A.M., Ibragimova A.S., Semisalov B.V.* Design of a numerical algorithm for the system of moment equations of the charge transport in semiconductors. Math. Modelling. 2009. V. 21. P.15–34 (in Russian).

[10] *Blokhin A.M., Boyarsky S.A., Semisalov B.V.* On an approach to the construction of difference schemes for the moment equations of charge transport in semiconductors // Le Matematiche. 2009. Vol. LXIV. Fasc. I. P. 77-91.

[11] *Babenko K.I.* Fundamentals of numerical analysis. Moscow–Izhevsk, Regular and chaotic dynamics, 2002 (in Russian).

[12] Berezin I.S., Zhidkov N.P. Computing methods. Vol. II. Oxford–London–Edinburgh–New York–Paris–Frankfurt, Pergamon Press, 1965.

[13] Krylov V.I., Bobkov V.V., Monastyrnyj P.I. Numerical methods of higher mathematics. Vol. II. Minsk, Higher School, 1975 (in Russian).

[14] *Yanenko N.N.* The method of fractional steps (The solution of problems of mathematical physics in several variables) // Springer-Verlag, Berlin etc., 1971.

[15] *Godunov S.K. and Ryabenki V. S.* Difference Schemes: an introduction to the underlying theory // North Holland, Amsterdam, 1987.

[16] *Lab C. and Caussignac P.* An energy-transport model for semiconductor heterostructure devices: application to AlGaAs/GaAs MODFETs // COMPEL. 1999. Vol. 18, 1. pp. 61–76.

[17] *Zav'yalov Yu.S., Kvasov B.I., Miroshnichenko V.L.* Methods of spline functions. Moscow, Nauka, 1980 (in Russian).

[18] *Blokhin A.M., Semisalov B.V., Semenko R.E.* Numerical study of parametric instability in layered structures // Math. Modelling, in press.

[19] *Blokhin A.M., Alaev R.D.* Energy integrals and their applications to the analysis of stability of difference schemes. Novosibirsk, Novosibirsk State University, 1993 (in Russian).

[20] *Blokhin A.M., Ibragimova A.S.* On calculation of the electric potential for 2D silicon transistor with a silicon oxide nanochannel // Math. Modelling. 2010. V. 22. N. 9, pp. 79-94 (in Russian).

[21] *Blokhin A.M., Iordanidi A.A., Merazhov I.Z.* Numerical analysis of a hydrodynamical model of charge transport in semiconductors. Novosibirsk, 1996, Preprint No. 33 of the Sobolev Institute of mathematics (in Russian).

[22] *Pinchukov V.I.* Adaptive operators of smoothness of arbitrary order //Comput. Tech.: Proc. ICT SD RAS. 1993. Vol. 2. No. 6. P.232–245.

[23] *Pinchukov V.I.* Algorithms monotonization of schemes of advanced exactness for equations of type $\frac{\partial f}{\partial t} + \mu \frac{\partial^k f}{\partial x^k} = 0,\ k \geq 0$ // Simulation in Mechanics: Proc. IATM SD RAS. 1993. Vol. 7(20). No. 2. P. 150–159.

Net-like structured materials for gas sensors

I.E. Gracheva, V.A. Moshnikov, S.S. Karpova and E.V. Maraeva

Saint-Petersburg state electrotechnical university "LETI", 5 Prof. Popova str., 197367, Saint-Petersburg, Russia

E-mail: iegrachova@mail.ru, jenvmar@mail.ru

Abstract. Silicon dioxide-based fractal aggregates as an example of self-assembly in sol-gel processes were prepared. Main evolution stages of tin dioxide-silicon dioxide fractal systems were demonstrated by atomic force microscopy (AFM): diffusion-limited aggregation, cluster-cluster aggregation, formation of percolation net and 3D-net nanostructures. The formation possibilities of porous nanomaterials based on different metal oxides, including those based on tin dioxide, iron oxide and zinc oxide, were shown. New chemical etching method to obtain microreactors was developed. Specific surface area of nanostructures was investigated by thermal desorption of nitrogen and gas sensitive properties of tin dioxide nanocomposites were also studied.

1. Introduction

Developing methods for synthesis of materials with hierarchical architecture is one of the most advancing directions among nanotechnologies of functional materials. Such materials consist of elements of different scales; they are organized so that the elements of smaller scales are built into the elements of larger scales. New unique properties of hierarchical materials arise from composition changing, size changing and (or) from construction of substructures (clusters, blocks) on one or several different levels [1]. One of the methods to obtain such hierarchical nanomaterials is self-assembly [2, 3]. Hierarchical self-assembly processes can be realized not only due to ionic or covalent interactions or by metallic, hydrogen or coordination bonds, in this case weaker forces, such as Van der Waals forces, Coulomb interaction, Casimir forces, forces caused by special features of medium and outer fields characteristics (colloidal, capillary, π-bonds, convective, magnetic, electrical, optical, gravity forces and hydrophilic-hydrophobic interactions), generally take place [4].

The main point of hierarchical self-assembly is synthesis of initial constructive "blocks" with different sizes and shapes (and size-dependent properties) and their further integration. Hierarchical self-assembly can be a multilevel one, when the integrated blocks are basic elements for larger consolidations (with lager interaction radius).

The products of hierarchical self-assembly are materials with hierarchical porous structure. One of the simplest idealized examples of such self-assembly is Julien fractal (Figure 1) [5]. It is an appropriate example for understanding of the principal possibility to obtain the materials with pores of calibrated sizes. That is very important for development of new-generation devices, because the pores with different sizes have different functions (nanoreactors with different capillary phenomena, canals for reaction products delivery and withdrawal, adsorption centers etc.).

The sol-gel processes are based on hierarchical self-assembly [6]. In sol-gel processes the initial fractal aggregates are created throughout the volume of colloid as a result of diffusion-limited

Published under licence by IOP Publishing Ltd

aggregation. The subsequent processes are caused by growing role of cluster-cluster interactions [7]. The processes of formation, growth and evolution of fractal aggregates in sol-gel synthesis can be controlled. Sol-gel methods are ones of the molecular design technologies [8]. Fractal aggregates in the process of their growth can be transformed into percolation composites with pores of controlled and reproducible sizes from nanometer to micrometer (Figure 2), or formation of porous structure can take place because of spinodal decomposition and phase separation. Sol-gel techniques are often supplemented with the template synthesis methods [9, 10, 11], such as insertion of dendrites, fullerenes or carbon nanotubes, which are burnt later. Large surface area (Figure 2) and high modulation of current canal sections during gas adsorption on the surface can be effectively used in new generation of semiconductor gas sensors, including multisensor "electronic-nose" systems [12].

Figure 1. Projection of 3-D Julien fractal.

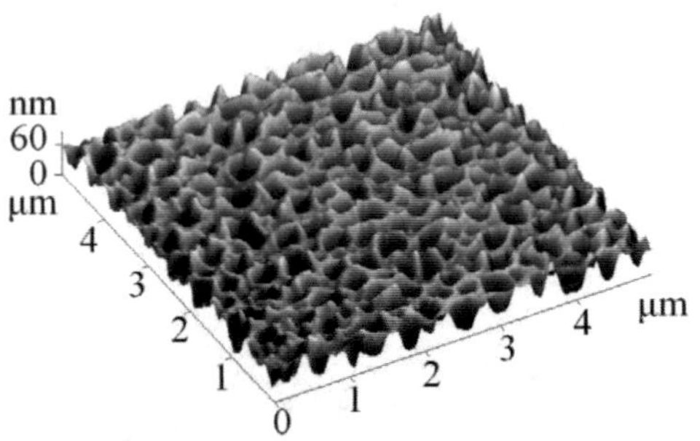

Figure 2. AFM image of fractal SiO_2-SnO_2 nanocomposite synthesized by sol-gel method.

The systems based on n-type semiconductor metal oxide nanocomposites (such as SnO_2, ZnO, Fe_2O_3, NiO, V_2O_3, WO_3, Co_3O_4) are very sensitive to surface states at high temperatures, so they are suitable materials for detection and identification of minute quantity of explosive, narcotic and dangerous substances.

2. Experimental details

AFM was used to measure the surface morphology of the thin-film nanostructures synthesized from sol solutions in the precursor medium based on tetraethoxysilicate (TEOS). Tapping mode AFM experiments were performed using NTEGRA-Therma nanolaboratory (NT-MDT, Zelenograd, Russia). Commercial etched silicon tips NSG 01 (NT-MDT, Zelenograd, Russia) with typical resonance frequency of 150 kHz were used as AFM probes.

Specific surface area measurements were made using Sorbi № 4.1 (CJSC «META», Novosibirsk, Russia) that realizes physical adsorption of noble gas by the sample to be studied. Operating principle of the instrument is based on the comparison of the volume of gas-adsorbate which is absorbed by the studied sample with the volume of gas absorbed by the standard sample with known value of specific surface area. In current research standard sample pressed in globe-shaped granules (diameter is 2 mm, weight is 0.01 g) and having specific surface area 67 m^2/g was used for calibration of Sorbi. Nitrogen was used as gas-adsorbate. For the measurement of absorbed gas volume the thermal desorption method was applied. A thermal conductivity sensor was a detector of gas mixture composition. The signal of thermal conductivity sensor (TCS signal) is a desorption curve (Figure 3). The measurements were made under variation of the relative partial pressure (P/P_0) of gas-adsorbate (Figure 4, where V is absorbed gas volume normalized to 1 gram of the sample). Specific surface area was calculated according to the model of Brunauer, Emmett, Teller.

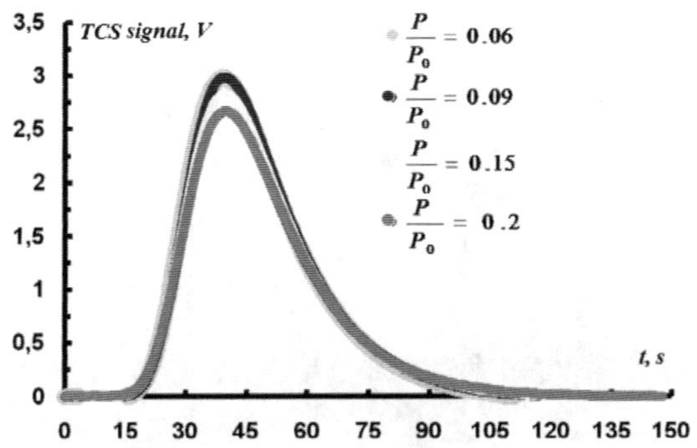

Figure 3. Typical desorption curves for porous nanocomposites based on the tin dioxide (the relative partial pressure of nitrogen are 0.06, 0.09, 0.15 and 0.20).

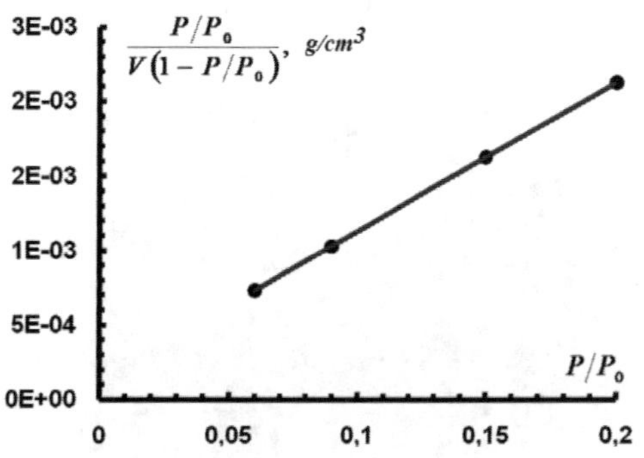

Figure 4. Typical plot $\frac{P/P_0}{V(1-P/P_0)} = f(P/P_0)$ for porous nanostructures.

Preliminary thermal action influence on the structures was studied with the use of SorbiPrep (the station for preliminary treatment of the samples). Operating principle of the station is based on the degassing of disperse and porous materials by their heating in the stream of noble gas (helium). The station works in various operation modes: temperature and time can be changed as 50-400 °C, 0-99 minutes. In this work the SiO_2-SnO_2 nanomaterials with various SiO_2 content (0 – 100 %) obtained by sol-gel method under other equal conditions (pH level, temperature, annealing time) were chosen for studying.

Gas-sensitivity measurements were done to acetone and ethanol vapors. Value of sensitivity was calculated as $S = (Rair - Rgas)/Rgas$, where *Rair* is the electrical resistance in dry air, *Rgas* is the electrical resistance in the presence of gas to be tested. Automatic cycle of sensitivity measurement consists of several stages: heating of sensor film composite to required temperature in air atmosphere; measurement of the sensor resistance while the temperature is maintained; gas-reagent supply (the concentration of gas-reagent is 1000 ppm) until the sensor resistance in the presence of gas vapors is stabilized.

3. Results and discussion

Figure 5 illustrates the self-assembly in sol-gel processes and shows SEM (in secondary electron mode) microphotographs of SiO_2 fractal aggregates synthesized by sol-gel method. The series of images (Figure 5 a – d) was analysed and the film structure was found to consist of rather large clusters of spherical particles. The size of these clusters ranges from several to tens of micrometers. In sol-gel processes growth of fractals according to the model of diffusion-limited aggregation occurs simultaneously with the random cluster-cluster aggregation. In that case occasional contacts among fractal aggregates and particles, cross-linking reactions among the branched structures and also the formation of large percolation clusters take place.

Sol-gel processes occurring in unstable states lead to the spontaneous formation of materials with different morphology as a result of spinodal decomposition [14]. In unstable states any fluctuation gains energy so that the fluctuations spontaneously develop without activation energy [15].

Generally the results of AFM studies show that evolution of fractal aggregates based on metal oxides can be controlled by varying the thermodynamic and kinetic conditions of sol-gel nanotechnology (composition of the materials, precursors, amount of solvent, polymer sol solutions aging time, annealing temperature). Several stages of evolution for fractal-aggregated systems based on tin dioxide and silicon dioxide were determined: spherical structures formation (nucleophilic growth) (Figure 6 a, the scanning area size is 25 μm x 25μm); co-continuous structures formation (spinodal decomposition) (Figure 6 b, the scanning area size is 10 μm x 10 μm); percolation net-like structures formation (Figure 6 c, the scanning area size is 5 μm x 5 μm).

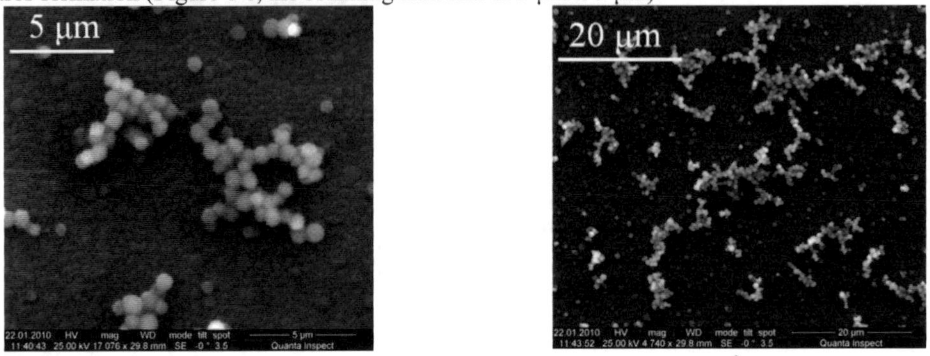

a b

Figure 5. SEM images of SiO_2 fractal structures synthesized by sol-gel method.

The formation possibilities of porous nanomaterials based on different metal oxides, including those based on tin dioxide (Figure 2, Figure 6), iron oxide (Figure 7), zinc oxide (Figure 8), were shown in this work.

The shape and size of obtained particles depend on the synthesis conditions [16, 17]. The specific advantages of zinc oxide micro- and nanocrystlas are related to the chemical properties but also strongly dependent on the size and shape of the crystals. Thus control over the size and morphology of nanometer- and micrometer-sized semiconductor materials is a great challenge to realize new functional devices.

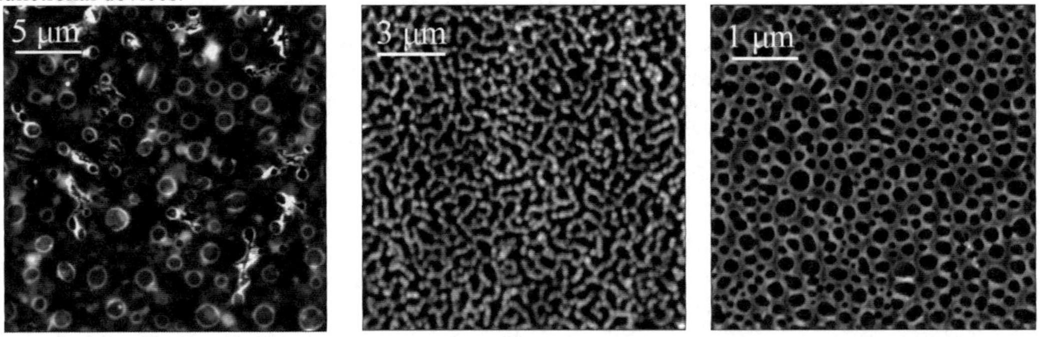

a b c

Figure 6. Stages of evolution for fractal-aggregated systems based on tin dioxide and silicon dioxide.

It was found for ZnO-SiO_2 nanocomposites that particles are formed inside the pores at low annealing temperatures (300°C) and the shape and size of these particles depend on amount of solvent used for preparation of sol solution. When the amount of solvent is small crystalline structures with a size of about 1 micron grow perpendicular to the substrate surface inside the pores (Figure 8 a, the scanning area size is 20 μm x 20 μm). When the amount of solvent increases the nanoparticles smaller than 100 nm are observed inside the pores (Figure 8 b, the scanning area size is 2 μm x 2 μm). It was suggested that zinc oxide phase crystallizes in the pores as a result of spinodal decomposition and phase separation processes. However, the proof of this fact requires further experiments. It is found that $ZnO-SiO_2$ samples annealed at higher temperatures do not contain crystalline structures inside the pores.

Figure 7. AFM image of iron oxide – silicon dioxide nanocomposite.

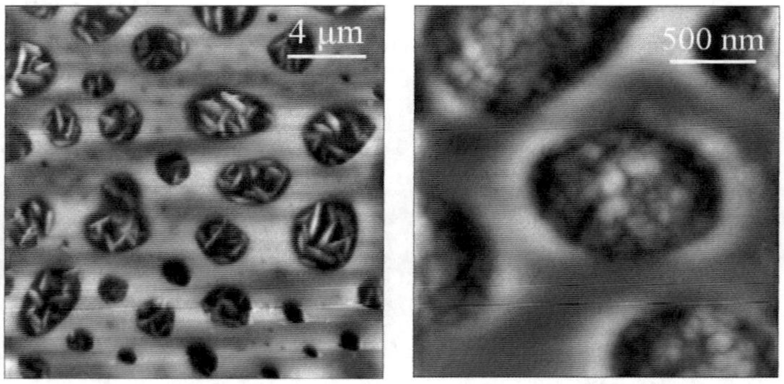

a b

Figure 8. AFM images of zinc oxide-silicon dioxide nanocomposites annealed at 300 °C.

It should be noted that sol-gel derived porous silica (Figure 9, the scanning area size is 80 µm x 80 µm) has been used as a matrix for synthesis of different nanoparticles [18, 19]. Synthesis of such functional materials is usually requires several stages: first the porous matrix is produced by sol-gel method and then nanoparticles are precipitated into the pores. The advantage of the proposed method for synthesis of $ZnO-SiO_2$ nanocomposites is the ability to synthesize nanomaterials and porous matrix in a single process.

There is no doubt that studies on the compatibility of sol-gel synthesis with other methods at nanoscale are promising. The compatibility of sol-gel processes with spray pyrolysis is of particular interest. It is associated with the fact that the formation of new phases occurs in the vapor phase in spray pyrolysis and in the liquid phase in sol-gel processes [20]. The combination of two technologies similar in physical and chemical laws along with the easy introduction of catalyst particles by spray pyrolysis is promising for development of multisensor systems. Formation of intermediate porous layers having a special form is often required to create the multi-layer structures. The effective method for formation of microreactors was developed. It is etching (in hydrofluoric acid, HF) of net-like percolation nanocomposites based on tin dioxide (Figure 10 a, the scanning area size is 5 µm x 5 µm). Decreasing conductive branches (Figure 11) and formation of hexagonal pores are occurred (Figure 10 b, the scanning area size is 20 µm x 20 µm) during etching processes.

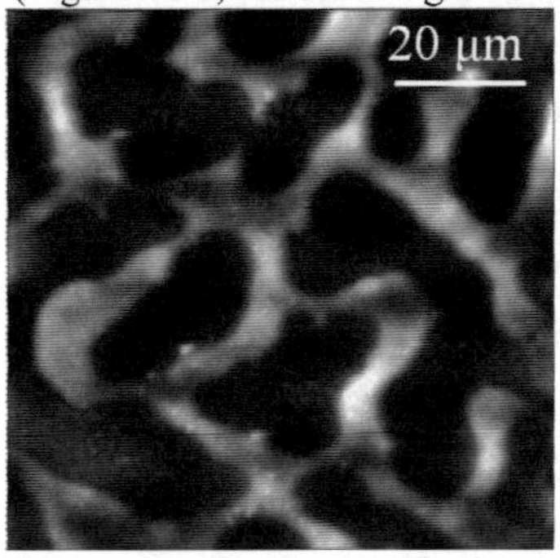

Figure 9. AFM image of sol-gel derived porous silica.

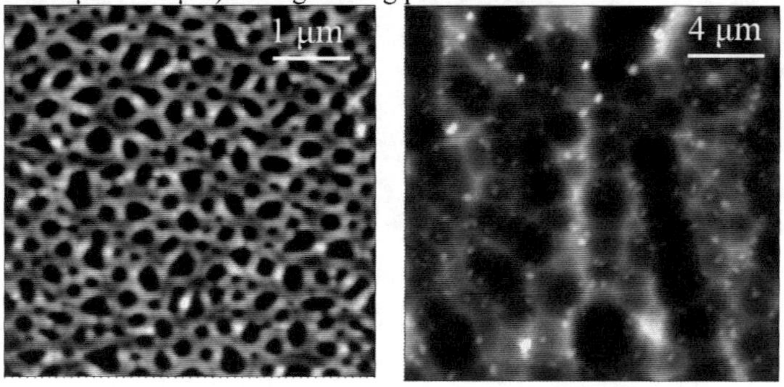

a b

Figure 10. AFM images of tin dioxide net-like nanocomposite before (a) and after (b) etching.

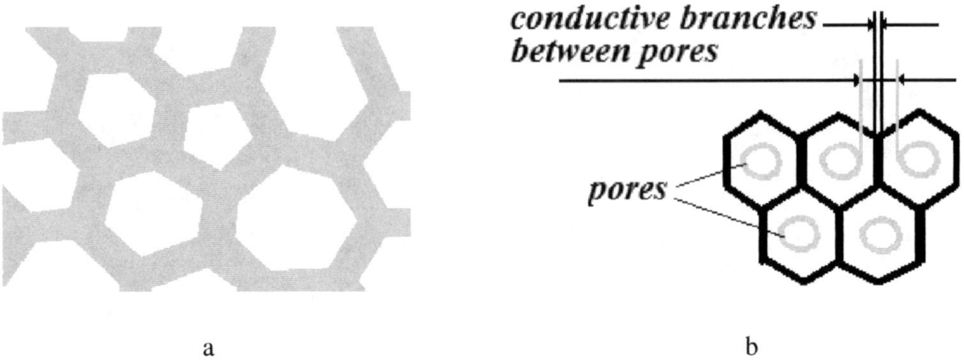

Figure 11. Schemes of microreactors obtained by etching of net-like nanocomposites (a), of decreasing conductive branches and the formation hexagonal pores (b).

High specific surface area and the porous structure with different sizes of pores are very important factors which are necessary for gas-sensitive structure to work efficiently. That is why studying the specific surface is one of the crucial methods for analysis of gas sensor characteristics.

In Figure 12 the dependence between specific surface area (SSA) of sol-gel SiO_2-SnO_2 nanocompisites and SnO_2 content is shown. The character of concentration dependence was repeated while several xerogel series were studied, so it can be considered that the specific surface area dependence reflects a common regularity for sol-gel system SiO_2-SnO_2.

It can be seen from Figure 12 that the plot can be divided in two parts having different characters of concentration dependence of specific surface area. In each part the experimental data can be approximated by a linear function. In the first area where tin content is low (0 – 10 % SnO_2) prompt growth of specific surface area is observed with increasing tin dioxide content. It is supposed that at low SnO_2 content monomeric branches of tin dioxide are incorporated into polymeric net of silicon dioxide, i.e. a monophase polymer solution in the binary system SiO_2-SnO_2 is created. The second area (10 – 100 % SnO_2) is characterized by gradual recession of the specific surface area with increasing tin dioxide content. An increase in tin oxide content leads to the formation of tin dioxide monophase aggregates, aggregate's size increases in the process of gel annealing resulting in decreasing porosity and descending the specific surface area of the samples.

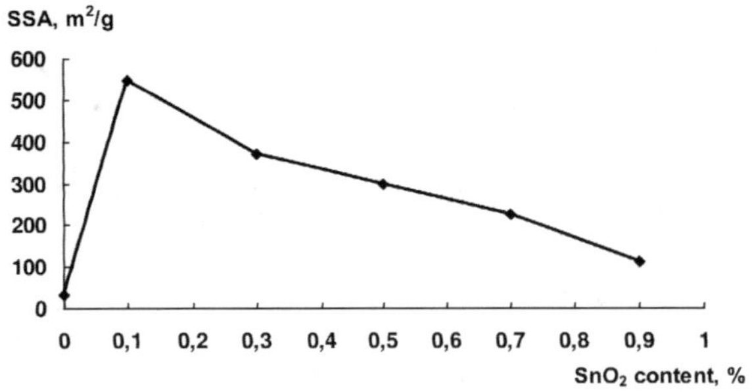

Figure 12. The dependence between specific surface area of sol-gel SiO_2-SnO_2 nanocompisites and SnO_2 content.

Preliminary thermal action influence on the structures was studied under various temperatures and times: T = 150 °C, 200 °C, 250 °C; t = 60, 90 minutes. Optimal conditions for preliminary treatment of xerogels with high specific surface area were determined (T=200 °C, t=60 minutes). Necessity of using stricter conditions of preliminary degassing of powders in SorbiPrep can be connected with changing the character of porosity when SnO_2 content is increasing (more than 10%). The micropores arise along with macro- and mesopores, i.e. under those conditions the xerogels with multilevel system of pores are formed. Removal of residual gases adsorbed in micro- or ultramicropores takes stricter conditions in comparison with larger pores.

Gas sensitivity is determined by the processes in the near-surface layer, namely, by adsorption of gas molecules and their reactions with chemisorbed oxygen ions on the surface. The electrical resistance of sensor changes reversibly under exposing to reducing gases (in proportion to the gas concentration).

Typical time dependence of the netting nanocomposite resistance in the presence of gas-reagent is shown in Figure 13. The main results of the investigation of tin dioxide gas-sensitive nanostructures are following:

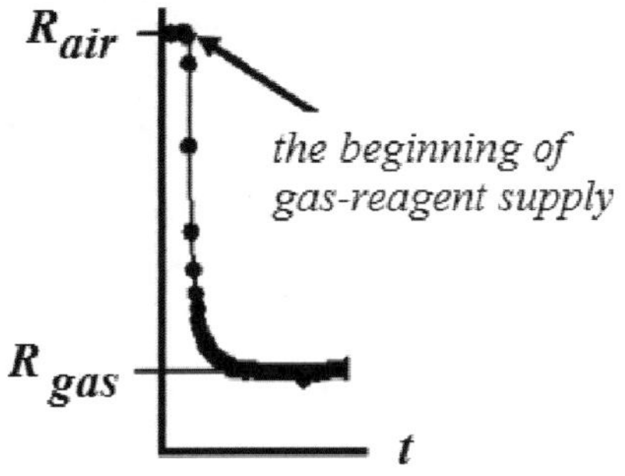

Figure 13. Typical time dependence of the netting nanocomposite resistance when gas-reagent is supplied and stopped.

- the nanostructures formed during nucleophilic growth (Figure 6 a) and co-continuous structures caused by spinodal decomposition (Figure 6 b) have very low sensitivity to acetone and ethanol vapors ($S \leq 0.5$);
- value of sensitivity to gas-reagents is $S > 20$ for netting percolation nanostructures (Figure 6 c).

4. Conclusions

The experimental results show that handling of self-assembly during sol-gel processes allows one to control the structure of fractal systems and their gas-sensitive properties. Main evolution stages of fractal systems based on tin dioxide were demonstrated by atomic force microscopy: diffusion-limited aggregation, cluster-cluster aggregation, formation of percolation net. Gas sensitive properties of nanocomposites under exposing to reducing acetone and ethanol vapors were investigated. The chemical etching method resulting in decreasing conductive branches of nanostructures and generation of hexagonal pores was developed for effective formation of microreactors. It was found that at low annealing temperatures pores of zinc oxide nanomaterials contain particles. The shape (sheet-like or spherical) and size (10-1000 nm) of these particles depend on amount of solvent used for preparation of sol solution. Study of the material properties by nitrogen thermal desorption after preliminary thermal action showed that the tin dioxide xerogels with multilevel porous system can be formed in sol-gel processes.

Acknowledgements
The results of this research were used in works under state contracts № P1249 (07.06.2010), № P399 (30.07.2009), № P2279 (13.11.09).

References

[1] Alfimov M V 2010 *Russian Nanotechnologies* **5** 2
[2] Bufon C, Gonzalez J, Thurmer D, Grimm D, Bauer M and Schmidt O 2010 *Nano Lett.* **10** 2506
[3] Cui H, Pashuck E, Velichko Y, Weigand S, Cheetham A, Newcomb C and Stupp S 2010 *Science* **327** 555
[4] Ozin G A Hou K, Lotsch B V, Cademartiri L, Puzzo D P, Scotognella F, Ghadimi A and Thomson J 2009 *Materials Today* **12** 12
[5] Julien R 1989 *Uspehi fizicheskih nauk* **157** 339
[6] Maximov A I, Moshnikov V A, Tairov Yu M and Shilova O A 2008 *Basics of sol-gel technology of nanocomposites* (SPb: Elmor)
[7] Moshnikov V A, Gracheva I E, Kuznezov V V, Maximov A I, Karpova S S and Ponomareva A A 2010 *J. Non-Cryst. Solids* **356** 2020
[8] Gracheva I E, Maksimov A I and Moshnikov V A 2009 *Journal of Surface Investigation. X-ray, Synchrotron and Neutron Techniques* **3** 761
[9] Bystrova A V, Parshina E V and Tatarinova E A 2007 *Russian Nanotechnologies* **2** 83
[10] Marsuda A, Matsuno Y, Tatsumisago M and Minami T 1998 *J. Am. Ceram. Soc.* **81** 2849
[11] Haruvy Y, Ryabov Y and Arkhipov V 2002 *J. Non-Cryst. Solids* **305** 226
[12] Gracheva I E and Moshnikov V A 2009 *Izvestiya of Russian State Pedagogical University* **11** 100
[13] Gracheva I E, Maximov A I, Moshnikov V A and Plekh M E 2008 *Instruments and Experimental Techniques* **51** 462
[14] Moshnikov V A and Gracheva I E 2009 *Supplement to journal "Vestnik RGRTU"* **4** 92
[15] Nakanishi K 2006 *Bull. Chem. Soc. Jpn.* **79** 673
[16] W Peng W, Qu S, Cong G and Wang Z 2006 *Crystal Growth and Design* **6** 1518
[17] Polsongkram D, Chamninok P, Pukird S, Chow L, Lupan O, Chai G, Khallaf H, Park S and Shulte A 2008 *Physica B* **403** 3713
[18] Huang X and Chen Z 2006 *Chin. Sci. Bull.* **51** 2529
[19] Buso D, Guglielmi M, Martucci A, Cantalini C, Post M L and Hache A 2006 *J. Sol-Gel Sci. Technol.* **40** 299
[20] Gracheva I E and Moshnikov V A 2008 *Izvestiya of Saint-Petersburg Electrotechnical University* **6** 19

Journal of Physics: Conference Series **291** (2011) 012018 doi:10.1088/1742-6596/291/1/012018

Near-field radiative heat transfer and van der Waals friction between closely spaced graphene and amorphous SiO_2

A.I.Volokitin[1,2]* and B.N.J.Persson[1]
[1] *Institut für Festkörperforschung, Forschungszentrum Jülich, D-52425, Germany and*
[2] *Samara State Technical University, 443100 Samara, Russia*

We study the radiative heat transfer and the van der Waals friction between graphene and an amorphous SiO_2 substrate. We study the surface phonon-polaritons contribution to the low-field mobility as a function of temperature and of carrier density. We find that the electric current saturate at a high electric field, in agreement with experiment. The saturation current depends weakly on the temperature, which we attribute to the "quantum" friction between the graphene carriers and the substrate optical phonons. We calculate the frictional drag between two graphene sheets caused by van der Waals friction, and find that this drag can induce a high enough voltage which can be easily measured experimentally. We find that for nonsuspended graphene the near-field radiative heat transfer, and the heat transfer due to direct phononic coupling, are of the same order of magnitude at low electric field. The phononic contribution to the heat transfer dominates at high field. For large separation between graphene and the substrate the heat transfer is dominated by the near-field radiative heat transfer.

PACS: 47.61.-k, 44.40.+a, 68.35.Af

Introduction. Graphene, the recently isolated single-layer carbon sheet, consist of carbon atoms closely packed in a flat two-dimensional crystal lattice. The unique electronic and mechanical properties of graphene[1,2] is being actively explored both theoretically and experimentally because of its importance for fundamental physics, and for possible technological applications[3]. In particular, a great deal of attention has been devoted to the applications of graphene for electronics and sensoring[1,3].

Graphene, as for all media, is surrounded by a fluctuating electromagnetic field due to the thermal and quantum fluctuations of the current density. Outside the bodies this fluctuating electromagnetic field exists partly in the form of propagating electromagnetic waves and partly in the form of evanescent waves. The theory of the fluctuating electromagnetic field was developed by Rytov[4–6]. A great variety of phenomena such as Casimir-Lifshitz forces[7], near-field radiative heat transfer[8], non-contact friction[9–12], and the frictional drag in low-dimensional systems[13,14], can be described using this theory.

In the present paper we apply the theories of the near-field radiative heat transfer and the van der Waals friction to study the electric transport properties, and the heat generation and dissipation in graphene, mediated by the fluctuating electromagnetic field. The charge carriers in graphene absorbed on, or located close to, a substrate experience a friction due to interaction with substrate. This friction affects the graphene carrier mobility and leads to the heat generation. The existing microscopic theories[15,16] involve several fitting parameters, while our theory is macroscopic. In this theory the electromagnetic interaction between graphene and a substrate is described by the dielectric functions of the materials which can be accurately determined from experiment.

The development of more powerful microelectronic devices results in increasing dissipated power (per unit volume), and developing structures with smaller heat production and better cooling strategies, is one of the "grand challenge" of modern electronics[17]. The heat generated in microelectronic devices leads to elevated device operation temperatures, performance reduction, and ultimately to hardware failures. The three classical mechanisms of heat transfer are convection, conduction, and radiation. The former is not important in existing circuit architecture having no fluid or gas flowing inside.

Graphene interact very weakly with most substrates mainly via van der Waals forces. According to theoretical calculations[18] the thermal contact conductance due to the direct phononic coupling for the interface between graphene and a perfectly smooth (amorphous) SiO_2 substrate is $\alpha_{\rm ph} \approx 3 \times 10^8 \rm{Wm^{-2}K^{-1}}$, and according to experiment[19] (at room temperature) the thermal contact conductance ranges from 8×10^7 to $1.7 \times 10^8 \rm{Wm^{-2}K^{-1}}$ (however, these values are probably influenced by the substrate surface roughness).

At large separation $d \gg \lambda_T = k_B T/\hbar$ the radiative heat transfer is determined by the Stefan-Boltzman law, according to which $\alpha = 4\sigma T^3$. In this limiting case the heat transfer between two bodies is determined by the propagating electromagnetic waves radiated by the bodies, and does not depend on the separation d. At $T = 300$K this law predicts the (very small) thermal contact conductance, $\alpha \approx 6 \ \rm{Wm^{-2}K^{-1}}$. However, as was first predicted

* Corresponding author. *E-mail address*:alevolokitin@yandex.ru

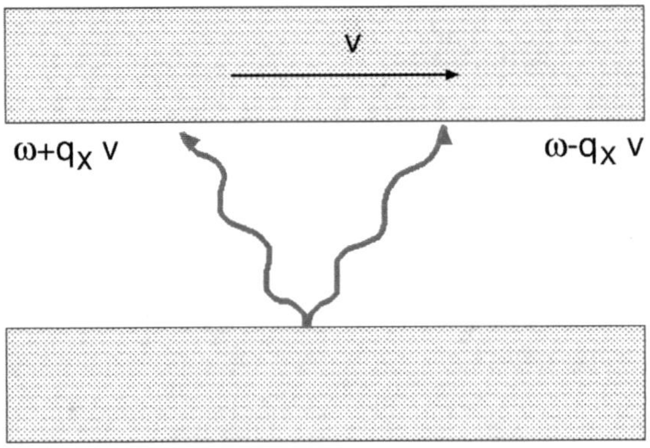

FIG. 1: Two bodies moving relative to each other will experience van der Waals friction due to Doppler shift of the electromagnetic waves emitted by them.

theoretically[8], and recently confirmed experimentally[20,21], at short separation $d \ll \lambda_T$, the heat transfer may increase by many orders of magnitude due to the evanescent electromagnetic waves; this is often referred to as photon tunneling. Particularly strong enhancement occurs if the surfaces of the bodies can support localized surface modes such as surface plasmon-polaritons, surface phonon-polaritons, or adsorbate vibrational modes[10,22]. In two recent experiments[20,21] the thermal contact conductance between a silica plate and a silica sphere was measured from 30 nm separation out to few micrometers. Heat transfer across the plate-sphere gap causes the cantilever to bent very slightly, and this was measured by optical fiber interferometry.

The theory of the radiative heat transfer developed in Ref.[8] is only valid for bodies at rest. A more general theory of the radiative heat transfer between moving bodies, with arbitrary relative velocities, was developed in Ref.[12]. This theory can be applied to calculate the radiative heat transfer between carriers (moving with the drift velocity v) in graphene and the substrate. According to theory[10], for the graphene-SiO_2 interface the maximum of the contribution of the radiative heat transfer to the thermal contact conductivity is $\alpha_{max} \sim (k_BT)^2n/\hbar$, where n is the concentration of carriers in graphene. At room temperature and $n \approx 10^{13}\mathrm{cm}^{-2}$ this gives $\alpha_{max} \sim 10^7\mathrm{Wm}^{-2}\mathrm{K}^{-1}$. However this estimation does not take into account that the charge carriers in graphene are moving relative to the substrate. Relative motion of charge carriers in graphene corresponds to an effective increase in the temperature difference between graphene and substrate, what leads to an increase in thermal contact conductance. Thus, for nonsuspended graphene the thermal contact conductivity due to near-field radiative heat transfer, and due to direct phononic mechanism, can be of the same order. However for suspended graphene, the main contribution results from radiative heat transfer since this process exhibit a much weaker distance dependence than for the phononic contribution.

The radiative heat transfer is closely connected with the van der Waals friction. This friction is due to the Doppler effect. Assume that a graphene sheet is separated from the substrate by a sufficiently wide insulator gap, which prevents particles from tunneling across it. If the charge carriers inside graphene move with velocity v relative to the substrate, a frictional stress will act on them. This frictional stress is related to an asymmetry of the reflection amplitude along the direction of motion; see Fig. 1. If the substrate emits radiation, then in the rest reference frame of charge carriers in graphene these waves are Doppler shifted which will result in different reflection amplitudes. The same is true for radiation emitted by moving charge carriers in graphene. The exchange of "Doppler-shifted-photons" will result in momentum transfer between graphene and substrate, which is the origin of the van der Waals friction.

Theory. Let us consider graphene and a substrate with flat parallel surfaces at separation $d \ll \lambda_T = c\hbar/k_BT$. Assume that the free charge carriers in graphene move with the velocity $v \ll c$ (c is the light velocity) relative to other medium. According to Ref.[9–12] the frictional stress F_x acting on charge carriers in graphene, and the radiative heat flux S_z across the surface of substrate, both mediated by a fluctuating electromagnetic field, are determined by

$$F_x = \frac{\hbar}{\pi^3}\int_0^\infty dq_y \int_0^\infty dq_x q_x e^{-2qd}\left\{\int_0^\infty d\omega\left(\frac{\mathrm{Im}R_d(\omega)\mathrm{Im}R_g(\omega^+)}{\mid 1-e^{-2qd}R_d(\omega)R_g(\omega^+)\mid^2}\times\right.\right.$$

$$\left.[n_d(\omega)-n_g(\omega^+)] + \frac{\mathrm{Im}R_d(\omega^+)\mathrm{Im}R_g(\omega)}{\mid 1-e^{-2qd}R_d(\omega^+)R_g(\omega)\mid^2}[n_g(\omega)-n_d(\omega^+)]\right)+$$

$$\int_0^{q_xv} d\omega \frac{\mathrm{Im}R_d(\omega)\mathrm{Im}R_g(\omega^-)}{\mid 1-e^{-2qd}R_d(\omega)R_g(\omega^-)\mid^2}[n_g(\omega^-)-n_d(\omega)]\Bigg\}, \quad (1)$$

$$S_z = \frac{\hbar}{\pi^3}\int_0^\infty dq_y \int_0^\infty dq_x e^{-2qd}\Bigg\{\int_0^\infty d\omega\Bigg(-\frac{\omega\mathrm{Im}R_d(\omega)\mathrm{Im}R_g(\omega^+)}{\mid 1-e^{-2qd}R_d(\omega)R_g(\omega^+)\mid^2}\times$$

$$[n_d(\omega)-n_g(\omega^+)] + \frac{\omega^+\mathrm{Im}R_d(\omega^+)\mathrm{Im}R_g(\omega)}{\mid 1-e^{-2qd}R_d(\omega^+)R_g(\omega)\mid^2}[n_g(\omega)-n_d(\omega^+)]\Bigg)+$$

$$\int_0^{q_xv} d\omega \frac{\omega\mathrm{Im}R_d(\omega)\mathrm{Im}R_g(\omega^-)}{\mid 1-e^{-2qd}R_d(\omega)R_g(\omega^-)\mid^2}[n_g(\omega^-)-n_d(\omega)]\Bigg\}, \quad (2)$$

where $n_i(\omega) = [\exp(\hbar\omega/k_BT_i - 1]^{-1}$ $(i = g, d)$, $T_{g(d)}$ is the temperature of graphene (substrate), R_i is the reflection amplitude for surface i for p -polarized electromagnetic waves, and $\omega^+ = \omega \pm q_xv$. The reflection amplitude for graphene (substrate) is determined by[13]

$$R_{g(d)} = \frac{\epsilon_{g(d)}-1}{\epsilon_{g(d)}+1}, \quad (3)$$

where $\epsilon_{g(d)}$ is the dielectric function for graphene (substrate).

In the study below we used the dielectric function of graphene, which was calculated recently within the random-phase approximation (RPA)[23,24]. The small (and constant) value of the graphene Wigner-Seitz radius r_s indicates that it is a weakly interacting system for all carries densities, making the RPA an excellent approximation for graphene (RPA is asymptotically exact in the $r_s \ll 1$ limit). The dielectric function is an analytical function in the upper half-space of the complex ω-plane:

$$\epsilon_g(\omega,q) = 1+\frac{4k_Fe^2}{\hbar v_Fq} - \frac{e^2q}{2\hbar\sqrt{\omega^2-v_F^2q^2}}\Bigg\{G\Bigg(\frac{\omega+2v_Fk_F}{v_Fq}\Bigg) - G\Bigg(\frac{\omega-2v_Fk_F}{v_Fq}\Bigg) - i\pi\Bigg\}, \quad (4)$$

where

$$G(x) = x\sqrt{1-x^2} - \ln(x+\sqrt{1-x^2}), \quad (5)$$

where the Fermi wave vector $k_F = (\pi n)^{1/2}$, n is the concentration of charge carriers, the Fermi energy $\epsilon_F = \gamma k_F = \hbar v_F k_F$, $\gamma = \hbar v_F \approx 6.5$ eVÅ, and v_F is the Fermi velocity.

The dielectric function of amorphous SiO_2 can be described using an oscillator model[25]

$$\epsilon(\omega) = \epsilon_\infty + \sum_{j=1}^{2}\frac{\sigma_j}{\omega_{0,j}^2-\omega^2-i\omega\gamma_j}, \quad (6)$$

where parameters $\omega_{0,j}$, γ_j and σ_j were obtained by fitting the actual ϵ for SiO_2 to the above equation, and are given by $\epsilon_\infty = 2.0014$, $\sigma_1 = 4.4767\times10^{27}\mathrm{s}^{-2}$, $\omega_{0,1} = 8.6732\times10^{13}\mathrm{s}^{-1}$, $\gamma_1 = 3.3026\times10^{12}\mathrm{s}^{-1}$, $\sigma_2 = 2.3584\times10^{28}\mathrm{s}^{-2}$, $\omega_{0,2} = 2.0219\times10^{14}\mathrm{s}^{-1}$, and $\gamma_1 = 8.3983\times10^{12}\mathrm{s}^{-1}$.

The equilibrium or steady state temperature can be obtained from the condition that the heat power generated by friction must be equal to the heat transfer across the substrate surface

$$F_x(T_d,T_g)v = S_z(T_d,T_g) + \alpha_{ph}(T_g-T_d), \quad (7)$$

where the second term in Eq. (7) takes into account the heat transfer through direct phononic coupling; α_{ph} is the thermal contact conductance due to phononic coupling.

Results. At small velocities ($v \ll v_F$) the friction force depends linearly on v: $F_x = \Gamma v$, where the friction coefficient Γ is determined by

$$\Gamma = \frac{\hbar}{2\pi^2}\int_0^\infty d\omega\Bigg(-\frac{\partial n_d(\omega)}{\partial\omega}\Bigg)\int_0^\infty dq q^3 e^{-2qd}\frac{\mathrm{Im}R_d(\omega)\mathrm{Im}R_g(\omega)}{\mid 1-e^{-2qd}R_d(\omega)R_g(\omega)\mid^2} \quad (8)$$

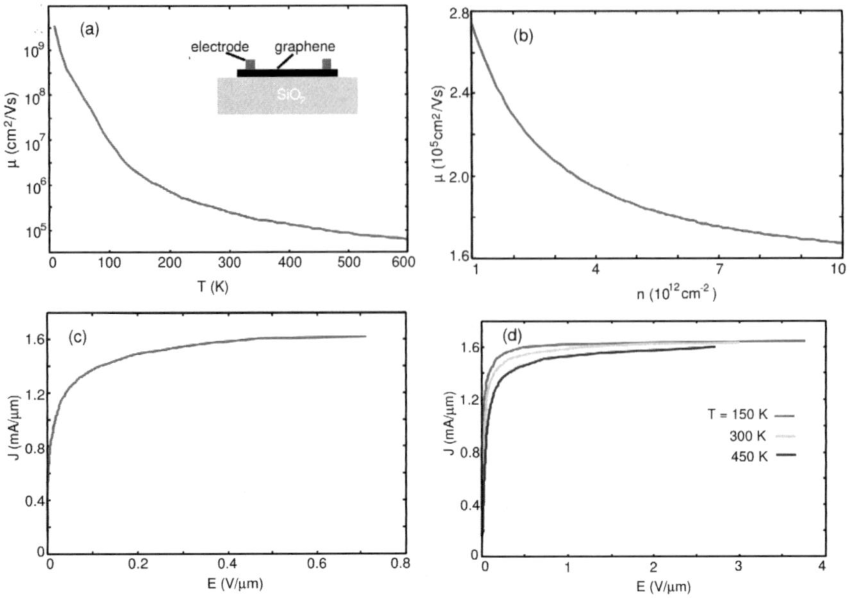

FIG. 2: The role of the interaction between phonon polaritons in SiO_2 and free carriers in graphene for graphene field-effect transistor transport. The separation between graphene and SiO_2 is $d = 3.5$Å. (a) Dependence of the low-field mobility on the temperature. Concentration of charge carriers $n = 10^{12}\text{cm}^{-12}$. (b) Dependence of the low-field mobility on the concentration of charge carriers n for $T_d = T_g = 300$ K. (c) Current density-electric field dependence at $T = 0$ K, $n = 10^{12}\text{cm}^{-12}$. (d) The same as in (c) but for different temperatures.

The low-field mobility can be described using Matthiessen's rule

$$\mu^{-1} = \mu_{int}^{-1} + \mu_{ext}^{-1}, \tag{9}$$

where μ_{int} is the intrinsic mobility due to scattering of the graphene free carriers against the acoustic and optical phonons in graphene, and μ_{ext} is the extrinsic mobility due to scattering against the optical phonons in the substrate. According to the theory of the van der Waals friction $\mu_{ext} = ne/\Gamma$. Scattering of the graphene carriers by the acoustic phonons of graphene places an intrinsic limits on the low-field room temperature mobility given by $\mu = 200000\text{cm}^2/\text{Vs}$ at the carriers density 10^{12}cm^{-2} [26]. However, for graphene on SiO_2 the mobility of the graphene carriers is reduced to $\mu = 40000\text{cm}^2/\text{Vs}$. This reduction must result partly from interface imperfections, and partly from scattering of the graphene carriers from the optical phonons of the SiO_2 substrate.

Fig. 2a shows the calculated dependence of the low-field mobility on the temperature, due to scattering from the optical phonons in SiO_2. The calculations were performed using the theory of the van der Waals friction. The mobility is characterized by a strong temperature dependence. At room temperature the calculated mobility $\mu \approx 3 \times 10^5$ cm^2/Vs, which is approximately one order of magnitude larger than the mobility calculated in Ref.[16]. This indicates that for graphene adsorbed on SiO_2 the main contribution to the low-field mobility comes not from scattering from the optical phonons in SiO_2, but from scattering from point defects on the interface.

Fig. 2b shows the dependence of the mobility at $T_d = 300$ K on the charge density n. The mobility monotonically decreases when the concentration of charge carriers increases. Figs. 2c and 2d show the dependence of the current density on the electric field at $n = 10^{12}\text{cm}^{-2}$, and for different temperatures. In obtaining these curves we have used that $J = nev$ and $neE = F_x$, where J and E are current density and electric field, respectively. Note that the current density saturate at $E \sim 0.5 - 2.0\text{V}/\ \mu\text{m}$, which is in agreement with experiment[27]. The saturation velocity can be extracted from the $I - E$ characteristics using $J_{sat} = nev_{sat}$, where 1.6 mA/μm is the saturated current density, and with the charge density concentration $n = 10^{12}\text{cm}^{-2}$: $v_{sat} \approx 10^6\text{m/s}$. Fig. 2c was calculated at $T_d = 0$ K. At zero temperature the van der Waals friction is due to quantum fluctuations of charge density, and is determined by the second term in Eq. (2)[9–12,28]

$$F_x(T_d = T_g = 0) = -\frac{\hbar}{\pi^3}\int_0^\infty dq_y \int_0^\infty dq_x \int_0^{q_x v} d\omega q_x e^{-2qd} \frac{\text{Im}R_d(\omega)\text{Im}R_g(\omega^-)}{\mid 1 - e^{-2qd}R_d(\omega)R_g(\omega^-)\mid^2} \tag{10}$$

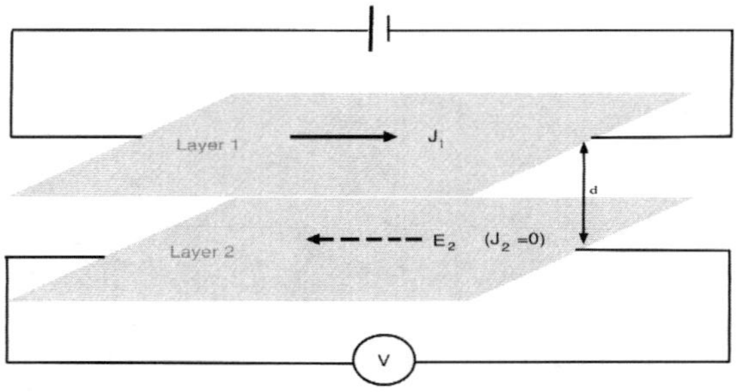

FIG. 3: A schematic diagram of a drag experiment. A current $\mathbf{J_1}$ is passed through layer **1**, and a voltmeter is attached to layer **2** to measure the induced electric field $\mathbf{E_2}$ due to the inter-layer interaction.

The existence of "quantum" friction is still debated in the literature[29–31]. The theory[9–12,28] predicts that a solid moving relative to another experiences a force due to quantum fluctuations, that is opposite to the direction of motion. The van der Waals friction can be studied in non-contact experiments, and in frictional drag experiments[10,11]. In both these experiments the solids are separated by a potential barrier thick enough to prevent electrons or other particle with a finite rest mass from tunneling across it, but allowing the interaction via the long-range electromagnetic field, which is always present in the gap between bodies. In non-contact friction experiments the damping of cantilever vibrations is typically measured, while in frictional drag experiments a current density is induced in one medium. The friction between the moving charge carriers and nearby medium gives rise to a change of $I - E$ characteristics, which can be measured.

The friction force acting on the charge carriers in graphene for high electric field is determined by the interaction with the optical phonons of the graphene, and with the optical phonons of the substrate. The frequency of optical phonons in graphene are a factor 4 larger than for the optical phonon in SiO_2. Thus, one can expect that for graphene on SiO_2 the high-field $I - E$ characteristics will be determined by excitations of optical phonons in SiO_2. According to the theory of the van der Waals friction[10,11], the "quantum" friction, which exist even at zero temperature, is determined by the creation of excitations in each of the interacting medium, the frequencies of which are connected by $vq_x = \omega_1 + \omega_2$. The relevant excitations in graphene are the electron-hole pairs which frequencies begining from zero, while for SiO_2 the frequency of surface phonon polaritons $\omega_{ph} \approx 60$meV. The characteristic wave vector of graphene is determined by Fermi wave vector k_F. Thus the friction force is strongly enhanced when $v > v_{sat} = \omega_{ph}/k_F \sim 10^6$m/s, in the accordance with numerical calculations. Thus the measurements of the current density-electric field relation of graphene adsorbed on SiO_2 give the possibility to detect "quantum" friction, what has fundamental significance for physics.

An alternative method of studying of the van der Waals friction consists in driving an electric current in one metallic layer and studying of the effect of the frictional drag on the electrons in a second (parallel) metallic layer (Fig. 3). Such experiments were first suggested by Pogrebinskii[32] and Price[33], and were performed for 2D-quantum wells[34,35]. In these experiments, two quantum wells are separated by a dielectric layer thick enough to prevent electrons from tunneling across it, but allowing inter-layer interaction between them. A current density $J_1 = n_1 ev$ is driven through layer **1** (where n_1 is the carrier concentration per unit area in the first layer), see Fig. 3. Due to the inter-layer interactions a frictional stress $\sigma = \Gamma v$ will act on the electrons in layer **2** from layer **1**, which will induce a current in layer **2**. If layer **2** is an open circuit, an electric field E_2 will develop in the layer whose influence cancels the frictional stress σ between the layers. Experiments[34] show that, at least for small separations, the frictional drag can be explained by the interaction between the electrons in the different layers via the fluctuating Coulomb field. However, for large inter-layer separation the frictional drag is dominated by phonon exchange[36].

Similar to 2D-quantum wells in semiconductors, frictional drag experiments can be performed (even more easy) between graphene sheets. Such experiments can be performed in vacuum where the contribution from phonon exchange can be excluded. To exclude noise (due to presence of dielectric) the frictional drag experiments between quantum wells were performed at very low temperature ($T \approx 3$ K)[34]. For suspended graphene sheets there are no such problem and frictional drag experiment can be performed at room temperature. In addition, 2D-quantum wells in semiconductors have very low Fermi energy $\epsilon_F \approx 4.8 \times 10^{-3}$eV[34]. Thus electrons in these quantum wells are degenerate only for very low temperatures $T < T_F = 57$ K. For graphene the Fermi energy $\epsilon_F = 0.11$eV at $n = 10^{12}$cm^{-2}, and the electron gas remains degenerate for $T < 1335$ K.

At small velocities the electric field induced by frictional drag depends linear on the velocity, $E = (\Gamma/ne)v = \mu^{-1}v$, where μ is the low-field mobility. For $\hbar\omega \ll \epsilon_F$ and $q \ll k_F$ the reflection amplitude for graphene is given by the same

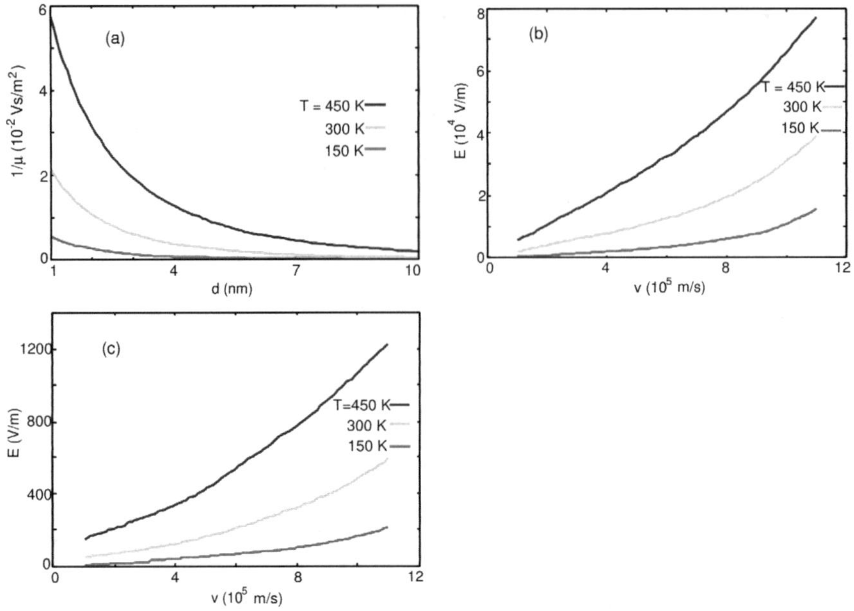

FIG. 4: Frictional drag between two graphene sheets at the carrier concentration $n = 10^{12}\mathrm{cm}^{-2}$. (a) Dependence of friction coefficient per unit charge, $\mu^{-1} = \Gamma/ne$, on the separation between graphene sheets d. (b) Dependence of electric field induced in graphene on drift velocity of charge carriers in other graphene sheet at the layer separation $d = 1$ nm. (c) The same as in (b) but at $d = 10$nm.

expression as for a 2D-quantum well[10]:

$$R_g = 1 + \frac{\hbar\omega}{2k_F e^2} \tag{11}$$

Substitution of Eq. (11) into Eq. (8) gives

$$\Gamma = 0.01878\frac{\hbar}{d^4}\left(\frac{k_B T}{k_F e^2}\right)^2 \tag{12}$$

Fig. 4a shows the dependence of the friction coefficient (per unit charge) μ^{-1} on the separation d between the sheets. For example, $E = 5\times10^{-4}v$ for $T = 300$ K and $d = 10$ nm. For a graphene sheet of length 1 μm, and with $v = 100$m/s this electric field will induce the voltage $V = 10$ nV. Figs. 5b and 5c shows the induced electric field-velocity relation for high velocity, with $d = 1$nm (b) and $d = 10$nm (c).

As discussed above, for graphene on SiO_2 the excess heat generated by the current is transferred to the substrate through the near-field radiative heat transfer, and via the direct phononic coupling (for which the thermal contact conductance $\alpha \approx 10^8\mathrm{Wm^{-2}K^{-1}}$). At small temperature difference ($\Delta T = T_g - T_d \ll T_d$), from Eq. (7) we get

$$\Delta T = \frac{F_{x0}v - S_{z0}}{\alpha_{ph} + S'_{z0} - F'_{x0}v} \tag{13}$$

where $F_{x0} = F_x(T_d, T_g = T_d)$, $S_{z0} = S_z(T_d, T_g = T_d)$,

$$F'_{x0} = \left.\frac{dF_x(T_d, T_g)}{dT_g}\right|_{T_g=T_d}, \quad S'_{z0} = \left.\frac{dS_z(T_d, T_g)}{dT_g}\right|_{T_g=T_d}$$

The thermal contact conductance is given by

$$\alpha = \frac{S_z(T_d, T_g) + \alpha_{ph}\Delta T}{\Delta T} \approx \frac{(\alpha_{ph} + S'_{z0})F_{x0}v - S_{z0}F'_{x0}v}{F_{x0}v - S_{z0}} \tag{14}$$

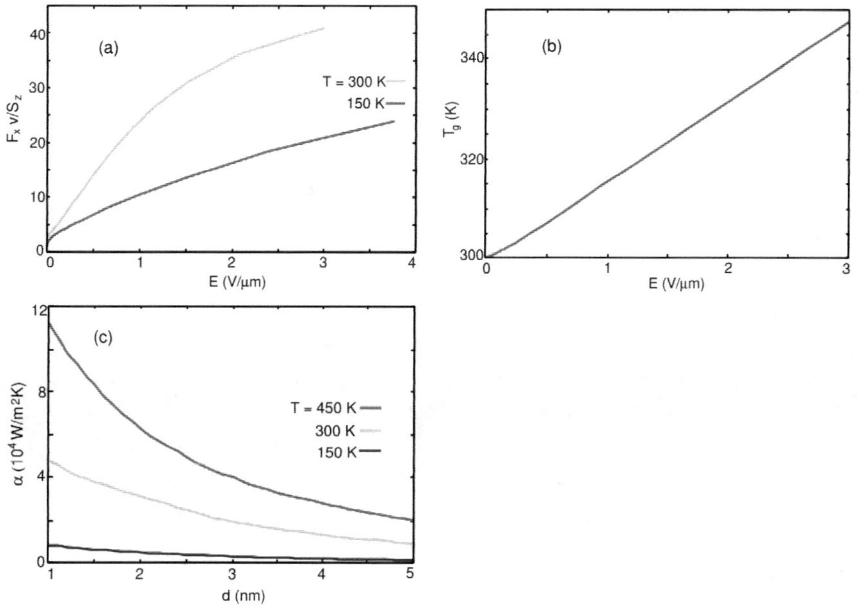

FIG. 5: Radiative heat transfer between graphene and SiO_2. (a) The dependence of the ratio between the total heat generated by current and the radiative heat flux, on electric field. $n = 10^{12}\text{cm}^{-2}$, $d = 3.5$Å, $\alpha_{ph} = 1.0 \times 10^8\text{Wm}^{-2}\text{K}^{-1}$ (b) Dependence of graphene temperature T_g on the electric field. The substrate temperature $T_d = 300$ K. Parameters are the same as in (a). (c) Dependence of the thermal contact conductance on the separation d for low electric field, $\alpha_{ph} = 0$.

From Eq. (14) it follows that the thermal contact conductance can be strongly enhanced when $F_{x0}v \approx S_{z0}$. Fig. 5a shows the ratio of the total heat generated by the current to the radiative heat flux. For low field this ratio ~ 2 so that in this case both radiative heat transfer and phononic heat transfer give contributions of the same order. The phononic mechanism of heat transfer dominates for high electric field. For high electric field the equilibrium graphene temperature due to self-heating depends linear on the electric field (see Fig. 5b) in contrast to the low-field case where this dependence is quadratic. This can be explains by the fact that the saturation velocity depends weakly on electric field; therefore heat, generated by friction, will linearly depend on the electric field. Fig. 5c shows the dependence of the thermal contact conductance on the separation d for low electric field. At $d \sim 5$ nm and $T = 300$ K the thermal contact conductance, due to the near-field radiative heat transfer, is $\sim 10^4\text{W}\cdot\text{m}^{-2}\cdot\text{K}^{-1}$, which is ~ 3 orders of magnitude larger than that of the black-body radiation. In comparison, the near-field radiative contact conductance in SiO_2-SiO_2 for the plate-plate configuration, when extracted from experimental data[20] for the plate-sphere configuration, is $\sim 2230\text{W}\cdot\text{m}^{-2}\cdot\text{K}^{-1}$ at a $\sim$30 nm gap. For this system the thermal contact conductance depends on separation as $1/d^2$; thus $\alpha \sim 10^5\text{W}\cdot\text{m}^{-2}\cdot\text{K}^{-1}$ at $d \sim 5$ nm what is one order of magnitude larger than for the graphene-SiO_2 system in the same configuration. However, the sphere has a characteristic roughness of ~ 40 nm, and the experiments[20,21] were restricted to separation wider than 30 nm (at smaller separation the imperfections affect the measured heat transfer). Thus the extreme near-field-separation, with d less than approximately 10 nm, may not be accessible using a plate-sphere geometry. On the other hand suspended graphene sheet has a roughness $\sim$1 nm[37], and measurements of the thermal conductance can be performed from separation larger than ~ 1 nm. At such separation one would expect the emergency of nonlocal and nonlinear effects. This range is of great interest for the design of nanoscale devices, as modern nanostructures are considerably smaller than 10 nm and are separated in some cases by only a few Angstroms. Another advantage of using graphene for studying the radiative heat transfer result from the fact that under steady-state condition the heat flow is equal to the heat generated in the graphene by the current density: $S_z = EJ$. This quantity can be accurately obtained from $I - V$ characteristics. The temperature of graphene can be measured accurately using Raman scattering spectroscopy[37].

Conclusion. We have used theories of the near-field radiative heat transfer and the van der Waals friction to study transport, heat generation and dissipation in graphene due to the interaction with phonon-polaritons in the (amorphous) SiO_2 substrate. In contrast with existing theories, based on the semiclassical Boltzmann transport equation, our approach is macroscopic. In the latter approach all microscopic properties are included in dielectric functions of material. Explicit formulas were obtained for the low-field mobility, the high-field transport and the near-field radiative heat transfer between graphene and the substrate. The low-field mobility exhibit a strong temperature dependence, which (according to the theory of the van der Waals friction) is associated with the thermal fluctuation inside the media. High-field transport exhibit a weak temperature dependence, which can be considered as manifestation of quantum fluctuations. Thus the study of transport properties in graphene gives the possibility to detect "quantum" friction, the existence of which is still debated in literature. We have calculated the frictional drag between graphene sheets mediated by the van der Waals friction, and found that it can induce large enough voltage to be easily measured experimentally. This effect can be used in electronics and for sensoring. We have shown that for the low-field heat transfer between graphene and the substrate, both radiative heat transfer and phononic heat transfer give contributions of the same order. High-field heat transfer is determined by the phononic mechanism. We have pointed out that graphene can be used to study near-field radiative heat transfer in the plate-plate configuration, and for shorter separations than it is possible now in the plate-sphere configuration.

Acknowledgment
A.I.V acknowledges financial support from the Russian Foundation for Basic Research (Grant N 08-02-00141-a) and ESF within activity "New Trends and Applications of the Casimir Effect".

---

[1] Novoselov, K.S.; Geim, A.K.; Morosov, S.V.; Jiang, D.; Zhang, Y.; Dubonos, S.V.; Grigorieva, I.V.; Firsov, A.A. *Science* **2004**, *306*, 666-669.
[2] Novoselov, K.S.; Geim, A.K.; Morosov, S.V.; Katsnelson, M.I.; Grigorieva, I.V.; Dubonos, S.V.; Firsov, A.A. *Nature (London)* **2005**, *197*, 197-200.
[3] Geim, A.K.; Novoselov, K.S. *Nat. Mater.* **2007**, *6*, 183-191.
[4] Rytov, S. M. *Theory of Electrical Fluctuation and Thermal Radiation* (Academy of Science of USSR Publishing, Moscow, 1953)
[5] Levin, M. L.; Rytov, S. M. *Theory of eqilibrium thermal fluctuations in electrodynamics* (Science Publishing, Moscow, 1967)
[6] Rytov, S. M.; Kravtsov, Yu. A.; Tatarskii, V. I. *Principles of Statistical Radiophyics*(Springer, New York.1989), Vol.3
[7] E. M. Lifshitz *Zh. Eksp. Teor. Fiz.* **1955**,*29* , 94-110 [Sov. Phys.-*JETP* **1956**, *2*, 73-83]
[8] Polder, D.; Van Hove, M. *Phys. Rev. B* **1971**, *4*, 3303-3314
[9] Volokitin, A.I.; Persson,B.N.J. *J.Phys.: Condens. Matter* **11**, 345 (1999);*Phys.Low-Dim.Struct.***7/8**,17 (1998)
[10] Volokitin, A. I.; Persson, B. N. J. *Rev. Mod. Phys.* **2007**, *79*, 1291-1329
[11] Volokitin, A. I.; Persson, B. N. J. *Usp. Fiz. Nauk* **2007**, *177*, 921-951 [*Phys. Usp.* **2007**, **50**, 879-906
[12] Volokitin, A.I.; Persson, B.N.J. *Phys. Rev. B* **2008**, *78*, 155437.
[13] Volokitin, A.I.; Persson, B.N.J. *J.Phys.: Condens. Matter* **2001**, *13*, 859-873
[14] Volokitin, A.I.; Persson, B.N.J. *Phys. Rev. B* **2008**,*77*, 033413
[15] Rotkin, S.V.; Perebeinos, V.; Petrov, A.G.; Avouris, P. *Nano Lett.* **2009**, *9*, 1850-1855
[16] Perebeinos, V.; Avouris, P. *Phys. Rev. B* **2010**, *81*, 195442
[17] *Handbook of Nanoscience, Engineering and Technology*; Goddard, W., Brenner, D., Lishevski. S., Iafrate. G.J., Eds.; Taylor and Francis-CRC Press: Boca Raton, FL, 2007
[18] Persson, B.N.J.; Ueba, H. *Europhys. Lett.* **2010**, 91, 56001.
[19] Chen, D.Z.A.; Haman, R.; Soljacic, M.; Joannopoulos, J.D.; Chen, G. *Appl. Phys. Lett.* **2007**, 90, 181921
[20] Shen, S.; Narayanaswamy, A.; Chen, G. *Nano Lett.* **2009**, 9, 1883-1888.
[21] Rousseau, E.; Siria, A.; Jourdan, G.; Volz, S.; Comin, F.; Chevrier, J.; Greffet, J.J. *Nature Photon.* **2009**, *90*, 181921
[22] Joulain, K.;Mulet, J.P.;Marquier, F.;Carminati, R.; Greffet, J.J. *Surf. Sci. Rep.* **2005** , *57*,59-112
[23] Wunscvh, B.; Stauber, T.; Sols, F.; Guinea, F.; *New J.Phys.* **2006**, *8*, 318.
[24] Hwang, E.H.; Sarma, S.D.; *Phys. Rev. B* **2007**, *75*, 205418
[25] Chen, D.Z.A.; Hamam, R,; Soliači c, M.;Joannopoulos, J.D. *Appl. Phys. Lett.* **2007**, *90*, 181921
[26] Chen, J.H.; Jang, C.; Xiao, S.; Ishigami, M.; Fuhrer, M.S. *Nat. Nanotechnol.* **2008**, *3*, 206
[27] Freitag, M.; Steiner, M.; Martin, Y.; Perebeinos, V.; Chen, Z.; Tsang, J.C.; Avouris, P. *Nano lett.*, **2009**, *9*, 1883-1888
[28] Pendry, J.B. *J. Phys.C*, **1997**, *9*, 10301-10320
[29] Philbin, T.G.; Leonhardt, U. *New J. Phys.* **2009**, *11*, 033035
[30] Pendry, J.B. *New J. Phys.* **2010**, *12*, 033028
[31] Volokitin, A.I.; Persson, B.N.J. to be published
[32] Pogrebinskii, M.B. *Fiz.Tekh.Poluprov.* **1977**,*11*, 637-641 [*Sov.Phys. Semicond.* **1977**, *11*, 372-376].
[33] Price, P. J. *Physica B+C*, **1983**, *117*, 750-752
[34] Gramila, T.J.; Eisenstein, J.P.; MacDonald, A.H.; Pfeiffer, L.N.; West, K. W. *Phys. Rev. Lett.* **1991**, *66*, 1216-1219
[35] Sivan, U.; Solomon, P.M.; Shtrikman, H. *Phys. Rev. Lett.* **1992**, *68*, 1196-1199
[36] Bønsager, M.C.; Flensberg, K.; Hu, B. Y.K.; Macdonald, A.H. *Phys.Rev. B*, **1998**, *57*, 7085-7102
[37] Meyer, J.C.; Geim, A.K.; Katsnelson, M.I.; Novoselov, K.S.; Booth, T.J.; Roth, S. *Nature*, **2007**, *446*, 60-63

Journal of Physics: Conference Series **291** (2011) 012019 doi:10.1088/1742-6596/291/1/012019

Phenomenological model of an abnormal thermal field electron emission from the 2D nanoheterostructured surfaces

V E Ptitsin

Institute for Analytical Instrumentation of the Russian Academy of Sciences, Russia,

e-mail: vptitsin@yandex.ru

Abstract Phenomenological model of an abnormal thermal field electron emission from the 2D nanoheterostructures of a kind: a conductor – a thin dielectric film, is offered and developed.

1. Introduction

As is known [1], a phenomenon of an excitation of a self sustaining electron emission from the thin oxide dielectric films was revealed and systematically investigated by Malter [1]. "*Malter emission*" is the electron emission from the thin high-resistivity materials films with the high-secondary-electron coefficients, induced by a primary electron beam directed at the film front surface [1, 2].

Now it is established that the basic features of the *Malter emission* phenomenon takes place for a wide range of the 2D nanoheterostructures (NHS) of a kind: a conductor- a thin dielectric films [1-5].

The basic features and laws of the *Malter emission* phenomenon[1] or, other words, ATFE phenomenon consist in the following.

1. ATFE occurs for various NHS, formed by the thin ($1 \div 10^4$ nm) dielectric films (in particularly, Al_2O_3 + Cs_2O, MgO, B_2O_3, KCl, SiO_2, CdS, ZrO_2, diamond, diamond like, carbon and polymeric films) on the conducting substrates surfaces. ATFE process properties depend from the film formation technological conditions and its thickness, physical-chemical properties of the NHS substance, temperature and also of the film doping degree by the impurities.
2. For ATFE excitation it is necessary to "activate" emission process by an external physical impact at NHS substance (by electron or ion bombardment, NHS heating).
3. Emission process functional dependence I_e (F_e), where I_e is an emission current, F_e is the normal component extracting field strength, does not submit to field emission (FE) and thermal field emission (TFE) theories [2, 6].

[1] *Malter emission* phenomenon we will name here as an *abnormal thermal field emission* (ATFE). In our opinion, the offered abbreviation more adequately and correctly reflects the basic features and anomalies of the low-macroscopic field (LMF) electron emission processes [2] activated impact on the NHS substance of various external physical factors.

Journal of Physics: Conference Series **291** (2011) 012019 doi:10.1088/1742-6596/291/1/012019

4. ATFE phenomenon takes place for the NHS with various geometry and superficial morphology. In particular, ATFE it is observed as for NHS samples on a basis of tip-shaped wide band gap semiconductors and for 2D NHS on the basis of diamond like films, carbon films and also for carbon nanotubes [5].
5. F_e numerical values necessary for activation and maintenance ATFE process approximately at 2-3 order is less than field strength necessary for field emitters on the homogeneous substances basis.

Listed ATFE properties and, in particularly, high NHS emission ability at rather low values F_e, represent considerable practical interest for realization of the some modern projects for vacuum microelectronics and also for high current electronics [2, 5].

However, despite rather long history of researches ATFE phenomenon has no adequate quantitative interpretation till now. The purpose of the present work consists in designing semi quantitative phenomenological model for ATFE phenomenon.

2. Phenomenological model of the ATFE phenomenon

The model of the ATFE phenomenon was constructed on the basis of generalization of the data obtained in the previous researches of thermal field emission properties of the low-dimensional NHS. [1-5]. To create ATFE model a real vacuum diode was modeled by a 2D NHS placed into vacuum in a homogeneous electrostatic field F_e (figure 1).

The model was based on following assumptions.

1. Electron injection from metal to dielectric through the metal-dielectric barrier, can involve both thermal field emission and thermionic emission (TE) mechanisms; the metal-dielectric junction is locking for metal electrons;
2. Electric field on the metal-dielectric interface is a superposition: of an external field *(F_e)*, field of the positive space charge (SC) formed as a result of ionization of the localized states in the dielectric film band gap and also SC field of electrons injected into the dielectric conduction band;
3. Ionization of the localized states in the dielectric film band gap may occur both by thermal field mechanism and, generally speaking, under action of the electromagnetic radiation absorbed by the film, including thermal radiation of NHS substance;
4. At small thicknesses of the dielectric film (about the free path length of low-energy electrons in relation to the process of electron-phonon interaction) electron transport in the film is occurs mainly by the quasi-ballistic mechanism;
5. The electron flux emitted into vacuum was accepted to be equal to the product of probability of electron tunneling through the potential barrier on the dielectric–vacuum interface (in WKB approximation [6]) by the electron flux injected into dielectric.

The space charge density distribution (ρ) formed in a result of ionization of the localized states in the band gap of the dielectric film, was approximated by the function

$$\rho = \rho_0 \exp\left[-x^2/(\lambda^2 \tau^2)\right], \qquad (1)$$

where x ($x \leq \tau$) is the coordinate counted from the metal – dielectric interface in a direction to the dielectric-vacuum interface; τ is the dielectric film thickness; λ is a dimensionless parameter; $\rho_0 = qN_t/\varepsilon\varepsilon_0$, here q is an absolute value of the electron charge; N_t is the density of the ionized localized states; ε is the dielectric film permittivity, ε_0 is an electric constant.

The Poisson equation solution for the potential distribution φ(x) in the film with boundary conditions

$$\varphi\big|_{x=0} = 0; \quad \frac{d\varphi}{dx}\bigg|_{x=\tau-0} = \frac{F_e}{\varepsilon},$$

where F_e is the absolute value of the field strength on the dielectric-vacuum interface (at x = τ +0), looks like

$$\varphi(x) = \frac{F_e}{\varepsilon}x + \frac{\sqrt{\pi}}{2}\lambda\tau\rho_0 \cdot erf\left(\frac{1}{\lambda}\right)\cdot x + \lambda^2\tau^2\frac{\rho_0}{2}\left[1-\left(\frac{x\sqrt{\pi}}{\lambda\tau}erf\left(\frac{x}{\lambda\tau}\right)+\exp\left(-\frac{x^2}{\lambda^2\tau^2}\right)\right)\right] \quad (2)$$

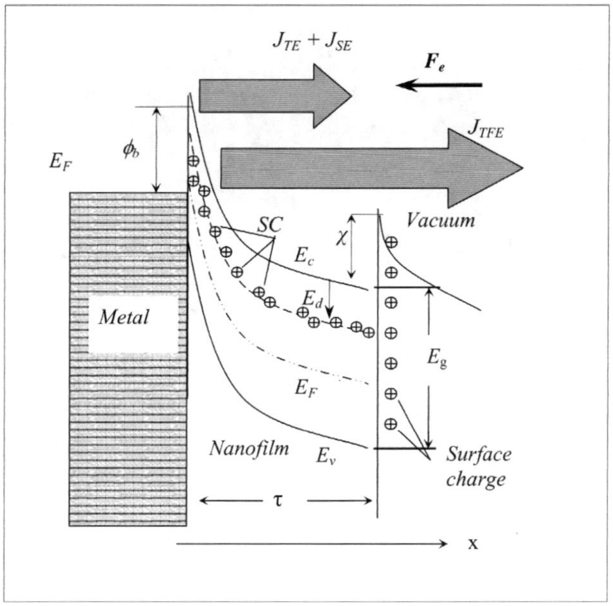

Figure 1. Two barrier energy band diagram for ATFE phenomenon modeling

Note that equation (2) describes the potential distribution φ(x) in the so-called "zero current" approximation. This approximation is valid when the inequality $N_t >> n_e$ holds, here n_e is a concentration of the electrons injected into the dielectric film.

The current densities of electrons injected from the metal into the film by the thermal field emission *(J_{TFE})* mechanism of and by the thermionic emission mechanism *(J_{TE})* above the Schottky barrier can be described by the following expressions [7]

$$J_{TFE} = A^* T_s^2 \left(\frac{\pi E_{00}}{k^2 T_s^2}\right)^{1/2} \left[q\varphi_s + \frac{\phi_b}{ch^2\left(\frac{E_{00}}{kT_s}\right)}\right]^{1/2} \times \exp\left(-\frac{\phi_b}{E_0}\right)\cdot \exp\left(\frac{q\varphi_s}{E_1}\right), \quad (3)$$

where A^* is a constant which, generally speaking, depends on electron effective mass (m^*) and includes a quantum-mechanical factor of electron reflection from the interface, T_s is the NHS material temperature, φ_s is the surface potential of the film on the dielectric – vacuum interface,

$$\phi_b = \phi_m - \chi - \left[\frac{q^6 \cdot (\phi_m - \phi_d + q \cdot \varphi_s) \cdot N_t}{8\pi^2 \varepsilon^3 \varepsilon_0^3} \right]^{1/4},$$

here ϕ_b is the potential barrier height at the metal-vacuum interface, ϕ_m is the work function of an electron leaving the metal, χ is the electronic affinity of the dielectric, ϕ_d is the work function of an electron leaving the dielectric surface,

$$E_{00} \equiv \frac{qh}{4\pi} \left(\frac{N_t}{m^* \varepsilon \varepsilon_0} \right)^{1/2}, \; E_0 \equiv E_{00} \cdot cth\left(\frac{E_{00}}{kT_s} \right),$$

$$E_1 \equiv E_{00} \cdot \left[\frac{E_{00}}{kT_s} - th\left(\frac{E_{00}}{kT_s} \right) \right]^{-1},$$

where h is the Planck constant, k is the Boltzmann constant, m^* is the effective electron mass,

$$J_{TE} = A^* \cdot T_s^2 \cdot \exp\left(-\frac{\phi_b}{kT_s} \right). \qquad (4)$$

The values N_t (T_s, F_e) were defined from the known semiconductor physics relations between the concentration of the ionized localized states, substance temperature and electric field strength.
To perform numerical calculations with equations (3) and (4), the initial parameters describing physicochemical and structural properties of modeled NHS were taken from the data of [8 - 16]. For carrying out of numerical calculations and estimations the concrete NHS was used, namely, ZrO_2/W<100> needle-shaped microcrystal [17]. In view of these remarks, the calculations were carried out using the following set of parameters: $\phi_m \approx 4.0$eV (for W(111)); $\phi_m \approx 4.5$eV (for W(100)); $\phi_d \approx 4.0$eV (for ZrO_2); $m^* \sim 0.5 m_e$, $\chi \approx 3.0$eV (for ZrO_2); the thickness of a dielectric film was accepted equal to $\tau \approx 10$nm; the density of non-ionized localized states (N_{t0}) in the ZrO_2 film (from different sources [11 - 13]) can make up from $\approx 5.0\ 10^{24}$м$^{-3}$ to $\approx 5.0\ 10^{25}$м$^{-3}$; the maximum density of the localized states should be at a depth $E_d \approx 0.8$eV under a bottom of the ZrO_2 conductivity band [10]; the average free path length of electrons in relation to electron-phonon interaction in various thin oxide films is from ~ 3nm up to ~ 100 (and more) nm. The probability of tunneling through dielectric-vacuum potential barrier for partly thermalized electrons was accepted equal to the transparency of the potential barrier calculated in the WKB - approximation [6].

Results of modeling are given on figure 2.

From the curves resulted in figure 2, in particular, it is easy to see that emission ability of ZrO_2/W <100> NHS is significantly (~ 3-4 orders of magnitude) higher than for the Schottky - cathode (see figure 2; curves 1 and 2). In the area of rather low of external electric field intensities (F_e < 50V/μm) the emission flux from the surface of ZrO_2/W<100> NHS is formed mainly by electrons injected into ZrO_2 film as a result of thermionic (TE) process (and extended Schottky emission (SE) process) above the Schottky barrier from the surfaces of planes {100}, {110} and {111} of the W<100> microcrystal. At $F_e \geq 100$V/μm (when the transparency of the barrier becomes equal to ≈ 1 (both for "TE", and for "TFE" injected electrons)) the emission current rather weakly depends on F_e (see figure 2).

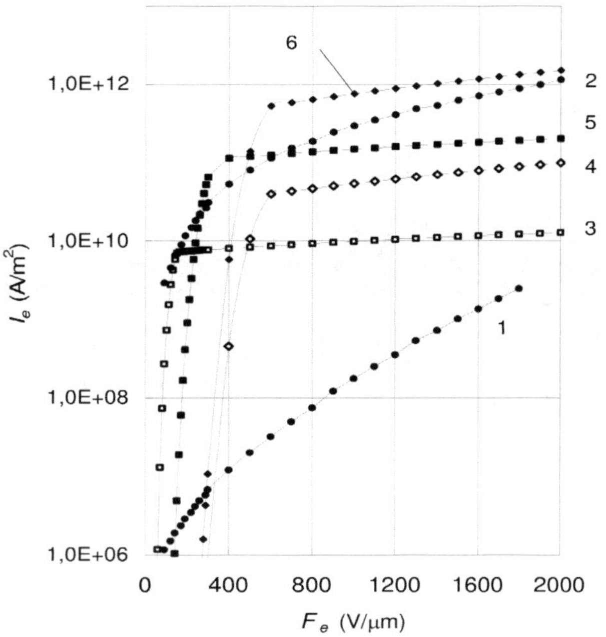

Figure 2. Emission current density versus field strength on the surface of ZrO_2/W <100> NHS (experimental (curves 1, 2) [17] and modeling (curves 3-6) data)
1. T_s = 2000K; the thickness of ZrO_2 layer is equal to one monolayer; 2. T_s = 2000K; the thickness of ZrO_2 layer is equal to ≈ 5nm; 3. For electrons injected into ZrO_2 film from planes W{100} by the TE mechanism; 4. For electrons injected into ZrO_2 film from planes W{100} by the TFE mechanism;
5. For electrons injected in ZrO_2 film from planes W{111} by the TE mechanism; 6. For electrons injected into ZrO_2 film from planes W{111} by the TFE mechanism.
(For curves 3 - 6: T_s = 2000K; $N_{t0} \approx 10^{25}$м$^{-3}$; the thickness of ZrO_2 layer is equal ≈ 5nm)

3. Conclusions

The above-stated results, as it is represented, allow making the following conclusions.

1. The main factor, which defines the unusual features (or anomalies) of the ATFE process from the surface of the low-dimensional NHS, is formation of the high density bounded space charge on the localized states in the band gap of the thin dielectric film.
2. In the ATFE conditions the emitted in vacuum electron flux is formed by electrons injected into a dielectric film by the mechanism of TE (above Schottky barrier) and by the mechanism of TFE. In other words, the physical mechanism of ATFE is complex and is not reduced only to one of the known classical TE and TFE phenomena.
3. ATFE properties: super high emission ability and stability of process, rather low level of F_e and also high density of an emission current (up to 10^8A/cm^2 and more) can be used in different areas of micro - and nanoelectronics, including high power electronics for produce high-intensity stationary electron sources [5].
4. ZrO_2/W and ZrO_2/Mo NHSs does not seem to be unique in its physicochemical and structural properties. To form new NHS - effective ATFE electron sources (based both on 1D or 2D structures

and ordered ensembles NHS) having *a priori* given emission properties (in particular, capable of ATFE of high current density at T = 300K) additional special researches using methods of modeling layer-by-layer epitaxial thin dielectric film growth processes and also methods of "zone engineering" are necessary [18, 19].

Acknowledgements
The work has been supported by the Russian Academy of Sciences (Saint-Petersburg Scientific Center Grant) and the Russian Foundation for Basic Researches.

References
[1] Malter L 1936 Thin film emission *Phys. Rev.* **50** 48-58
[2] Forbes R G 2001 *Solid-State Electronics* **45**, 779
[3] Baskin L M, Egorov N V, Ptitsin V E and Fursey G N 1979 *Pisma v ZhTF*, **5** 1345 (in Russian)
[4] Cirlin G E, Tonkikh A A, Ptitsin V E, Dubrovsky V G, Masalov S A, Evtikhiev V P, Denisov D V, Ustinov V M and Werner P 2005 *Physika Tverdogo Tela*, **47** 58
[5] Gulyaev Yu V 2003 *Vestnik RAN* **73**, No.5, 389 (in Russian)
[6] Modinos A 1990 Field-Thermo- and Secondary Electron Emission Spectroscopy, Moscow, 320
[7] Kao K, Hwang B 1984 *Electrical transport in solids,* Moscow, Mir, part 1, 350 (in Russian)
[8] Fomenko V S 1981 Emission Properties of Materials, A Handbook, Kiev, 340 (in Russian)
[9] Sayan S, Ngyen N V, Ehrstein J 2005, *Applied Physics Letters*, **86**, 152902
[10] Houssa M, Stesmans A and Heyns M M 2001 *Semicond. Sci. and Technol.* 427
[11] Sayan S, Bartynsky R A, Zhao X, Gusev E P 2004 *Phys. Stat. Sol.* **241** 2246
[12] Chang J P, Lin Y-S, Berger S, Kepten 2001, *J. Vac. Sci. Technol. B* **19** (6), Nov./Dec., 2137
[13] Chatterjee S, Nandi S K, Maikap S, Samanta S K and Maiti C K 2003 *Semicond. Sci. Technol.* **18** 92
[14] Goossens A, Vazquez M, Macdonald D 1996 *Electrochemica Acta* **35**
[15] Chakraborty S, Bera M K, Dalapati G K and Patramanik D 2006 *Semicond. Sci. Technol.* **21** 467
[16] Christenko V A, Meerson E E and Morokov Yu N 1997 *Phys. Rev.* B **57** R2081
[17] Ptitsin V E 2006 *Nauchnoie priborostroenie* ***16*** 3-10 (in Russian)
[18] Alferov Zh I 2002 *UFN*, **172** 1068 (in Russian)
[19] Velikhov E P 2003 *Vestnik RAN* **73** 395 (in Russian)

Journal of Physics: Conference Series **291** (2011) 012020 doi:10.1088/1742-6596/291/1/012020

Creation of Principally New Generation of Switching Technique Elements (Reed Switches) with Nanostructured Contact Surfaces

Karabanov S.M., Zeltser I.A., Maizels R.M., Moos E.N. Arushanov K.A.

Russia, Ryazan, 390027, Novaya Str., 51B,
Ryazan Metal Ceramics Instrumentation Plant JSC

E – mail: zeltseria@rmcip.ru

Abstract. The cycle of activities of the creation of principally new generation of reed switches with nanostructured contact surfaces was implemented. Experimental justification of the opportunity of reed switches creation with modified contact surface was given (instead of precious metals-based galvanic coating). Principally new technological process of modification of magnetically operated contacts contacting surfaces was developed, based on the usage of the ion-plasma methods of nanolayers and nanostructures forming having specified contact features.

1. Introduction

One of the basic manufacturing operations of the modern reed switches production is operation of the special (corrosion- and erosion-resistant) coatings deposition on working surfaces of permalloy contact springs of reed switches. Meanwhile, a galvanic technology is typically used for coating deposition, and precious and noble metals (gold, palladium, rhodium, ruthenium and so on) are used as materials. [1, 2].

The problems of existing electrodeposition methods of reed switch contact plating are a high precious metal consumption and stock loss thereof, long production time, difficulty and high cost of equipment, large expenditure of energy, complication in alloy deposition of the specified chemical and phase composition and given structure, complication in production of thin nonporous or thick films with low internal voltages and high adhesion to contact blade material [1, 2].

Non-usage of noble metals and methods of deposition thereof will allow producing goods with a competitive price-quality ratio.

The cycle of works [1 - 10] which common purpose is to create a principally new generation of switching technique elements (reed switches) with nanostructured contact surfaces, based on the usage of ion-plasma methods of nitriding and forming of nanolayers and nanostructures with the given contact features, was made.

Ion-plasma nitriding application for modification of surface properties and near-surface layers is known [11, 12]. Typically, such properties are hardness, durability, and corrosion stability. Different variations of the method of nitrided layers creation are known as well [11, 12]. However, in this case, nitriding was used for creation of the special-purpose surface structure having new special properties.

This paper is a logical follow-up of the cycle of works [1 - 10]. The purpose thereof is an experimental investigation of an opportunity of reed switches production with a modified permalloy contact surface (instead of galvanic plating made of precious metals).

The raised task solution is directly connected with study of the special new properties of surface structures being formed as the result of ion-plasma modification of reed switch contact surfaces.

2. Samples and Technique of Experiment

The samples made in a constructive way on the basis of MKA-14103 off-the-shelf devices [13] were MKA-14108 reed switches (Figure 1) [9]. The major distinctive feature of such reed switch design compared to off-the-shelf devices consisted in the absence of any special coatings on permalloy contacts. Contact springs were shaped from permalloy wire, degreased and fired in the hydrogen atmosphere. The wire was pulled out from permalloy (52% Ni, 48% Fe) of vacuum melting. During vacuum sealing a spectroscopically pure (99.999 %) nitrogen was used as a gas filling. Upon the vacuum sealing a gas pressure in reed switches was 250 – 290 Torr.

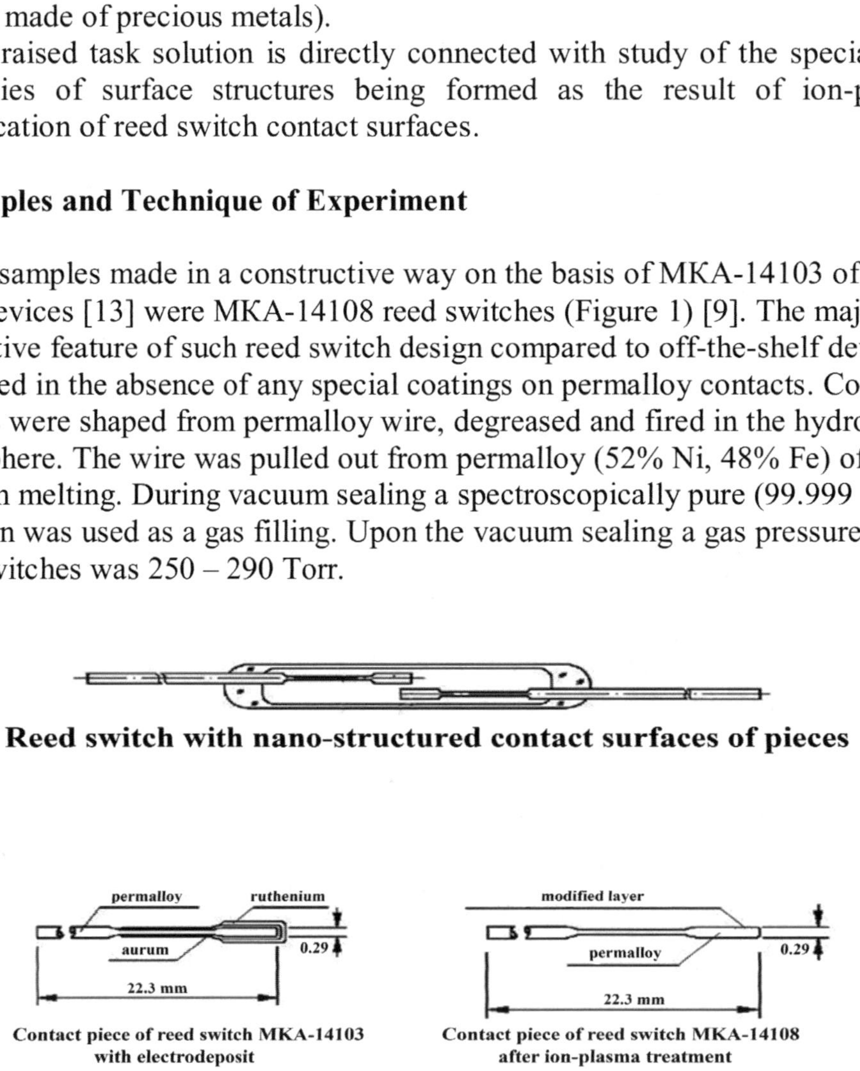

Figure 1. Reed switch design.

Ion-Plasma Treatment (IPT) of contact surfaces was conducted by high-voltage pulse discharges pursuant to the conditions specified in [10]. Discharges were initiated on reed switch open contacts (with a gap d = 20 – 30 mμ) with the help of ion-plasma treatment machine [4]. Under such conditions [10] the surface modification is implemented as a result of ion nitriding [1, 2, 6, 11, 12].

It is known [1, 2] that the major effect of corrosion and pollutions on electrical contacts consists in conductivity irregularity, so contact resistance to such processes can be characterized by a number of closings, whereby a conductivity changes tragically. Contact erosion resistance can be characterized by a number of closings, whereby conductivity irregularity or non-opening of contacts occurs.

Reed switch experimental samples after treatments by high-voltage discharges, and without such, were subject to the comparative switching tests.

It was tests for active load and in idle mode (under no-load).

The state of contact surface was judged by from diagnostics results by methods of AFM and optical microscopy, Auger-electron spectroscopy as well as integral measurement of electrical resistance R value of reed switch.

This allowed optimizing modes of contact spring surface modification of reed switch experimental samples, and studying physical processes occurring with all treatment stages and with reed switch operation.

3. Experimental Results and Discussion

3.1. SwitchingTests

Comparative switching tests of pilot MKA-14108 reed switches and MKA 14103 off-the-shelf devices showed that pilot MKA-14108 reed switch running time, pretreated by high-voltage pulse discharges, generally corresponds to requirements raised to durability of MKA 14103 off-the-shelf devices (Table 1).

Table 1. Results of Comparative Switching Tests

Mode of switching tests	Reed switch type	Without failing, %	Rmax before tests, Ohm	Rmax after tests, Ohm
50 mV - 5 μA, 100 Hz, 10^6 operations	MKA-14103	100	0.1	0.1
	MKA-14108, without treatment	21	0.25	100
	MKA14108, with treatment	100	0.07	0.08
5 V - 10 mA, 100 Hz, 10^6 operations	MKA-14103	100	0.17	0.11
	MKA14108, without treatment	27	0.35	22.7
	MKA14108, with treatment	100	0.08	0.08
20 V - 0.5 A, 50 Hz, $5\cdot10^6$ operations	MKA-14103	38	0.16	40
	MKA14108, without treatment	47	0.31	0.11
	MKA14108, with treatment	100	0.07	0.10
24 V - 400 mA, 50 Hz, $5\cdot10^5$ operations	MKA-14103	100	0.13	0.5
	MKA14108, without treatment	100	0.29	0.11
	MKA14108, with treatment	100	0.07	0.01
36 V - 15 mA, 50 Hz, $5\cdot10^6$ operations	MKA-14103	25	0.2	8.7
	MKA14108, without treatment	53	0.49	0.34
	MKA14108, with treatment	100	0.08	0.07
50 V - 50 mA, 50 Hz, $5\cdot10^5$ operations	MKA-14103	100	0.14	0.22
	MKA14108, without treatment	100	0.38	0.25
	MKA14108, with treatment	100	0.07	0.08
100 V -100 mA, 50 Hz, $5\cdot10^5$ operations	MKA-14103	2	0.17	0.19
	MKA14108, without treatment	100	0.27	0.23
	MKA14108, with treatment	100	0.07	0.07

For treatment mode optimization and identification MKA-14108 reed switch pilot samples tests before and after treatment under no-load (idle) were implemented. Switching quantity changed step-by-step from 0 to 10^7 collisions. At any stage of reed switch testing R value was measured (Figure 2).

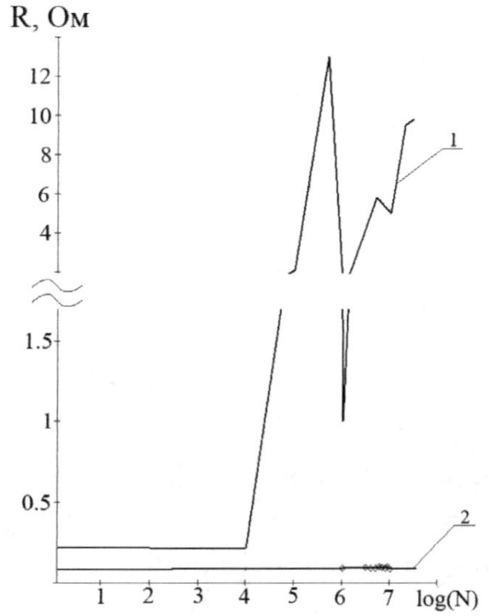

Figure 2. Relation between MKA14108 reed switch resistance (R) across median and operations number N in «dry» circuit (1 – untreated contact blades, 2 - contact blades after IPT). Reed switch quantity within a batch is 100 pcs.

Contact surface morphology changes in terms of switching number were studied under metallurgical and atomic-force microscopes (Figures 3 - 6).

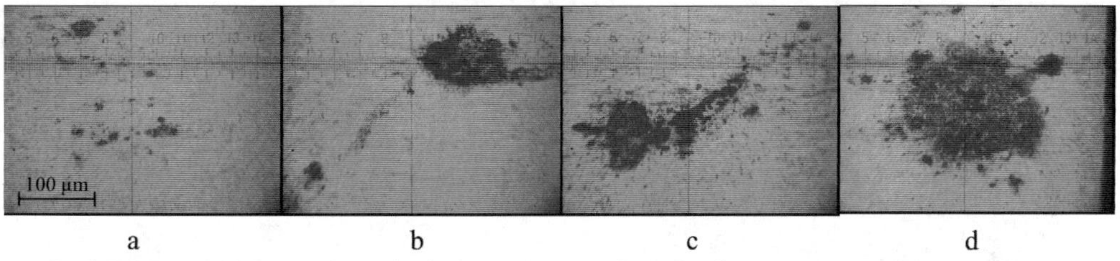

Figure 3. MKA-14108 reed switch contact surfaces after N operations under no-load, where N: a - 10^4; b - 10^5; c - 10^6; d - 10^7. Magnification is 300^X.

With operation number increasing a reed switch resistance pretreated by high-voltage pulse discharges remains stable and does not exceed 0.1 Ohm (Figure 2, Curve 2). On the contrary, with operation number increase an untreated reed switch resistance raises about 100 times, and with 10^7 operations it reaches 10 Ohm (Figure 2, Curve 1).

With an operation number of 10^4, independently in contacting area, some black-brown spots emerge on reed switch contact surface having no special coatings (Figure 3). It is thermal decomposition products of polymer films formed from adsorbed carbon, oxygen and hydrogen on a contact surface during switchings. With a switching number increase up to 10^5, size and degree of such spots blackening increase, and in step with this process R-value increases. Within range of 10^5 – 10^6 polymer films damage and R-value decrease are observed (Figures 2, 3). With a further increase of switching number (up to 10^7)

instead of films broken by collisions new polymer coatings start developing that is accompanied by R increase [1, 2, 6] (Figures 2, 3). Reasons for such island coatings origin were discussed in [1, 2, 6]. Here we will discuss an issue of contacts erosion.

From images of contact surfaces shown in Figures 4-6 it is evident that resistance of reed switches with untreated contact blades increases also due to surface erosion as a result of actual contact area decreasing. In Figure 4, one can clearly see areas of surface erosion of untreated contact blade. 2D, 3D- image of one of these areas (it is indicated by an arrow in Figures 4-6) is presented in Figures 5, 6, respectively.

Contact surface of nitrided contact blades, on the contrary, was found to be more resistant to erosion process due to higher hardness of nitrided layers, therefore resistance of reed switches while switching remains stably low (Figure 2, Curve 2). A similar stability is demonstrated by MKA-14103 reed switches having a gold-ruthenium coating while switching dry circuits [6].

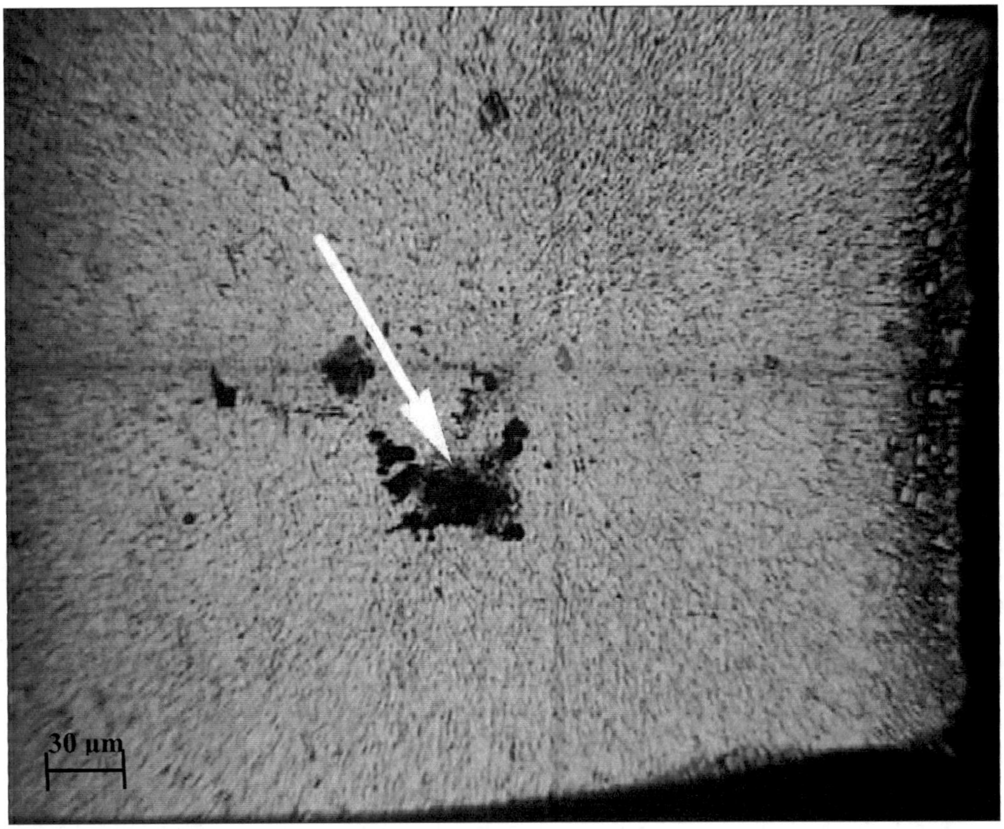

Figure 4. Image of untreated contact blade surface of MKA-14108 reed switch after $3{\cdot}10^7$ operations in a dry circuit. Magnification is 300^X.

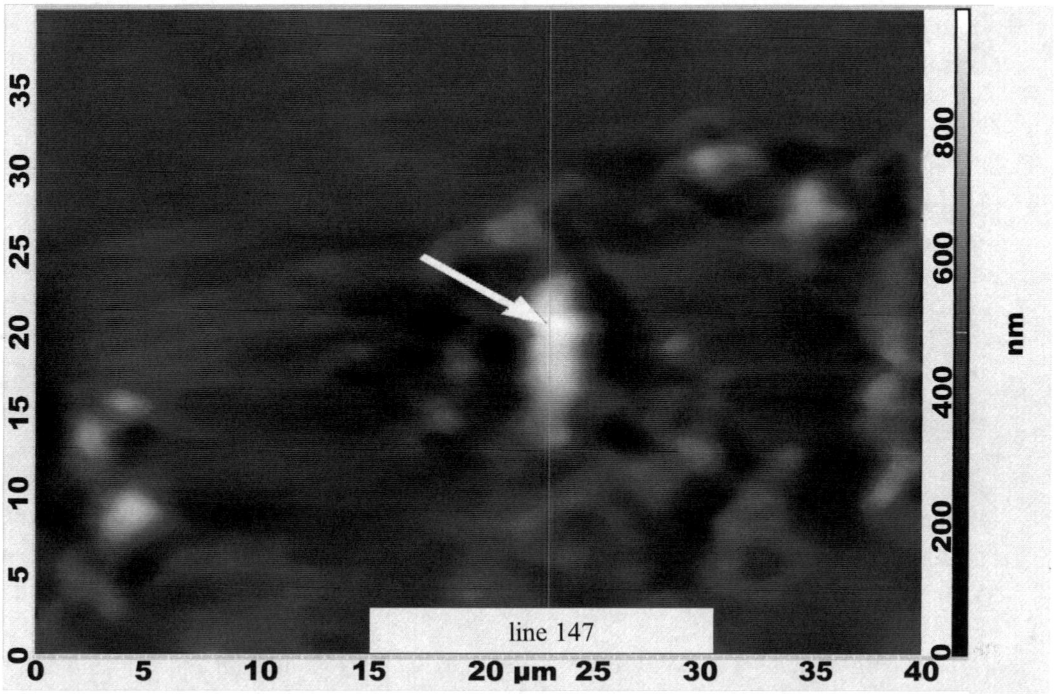

a b

Figure 5. 2-D-image of untreated contact blade surface of MKA-14108 reed switch after $3{\cdot}10^7$ operations in a dry circuit.

Figure 6. 3-D-image of untreated contact blade surface of MKA-14108 reed switch after $3{\cdot}10^7$ operations in a dry circuit.

Let's examine possible origins of erosion being observed.

Processes which take place in a gas filling, on surface and in volume of reed switch contacts, determine an evolution of contact blades surface of reed switches while switching, and in most cases, have a negative influence on functionability of devices.

A geometrical surface of contacts is such that an actual metal contact is realized only in separate points [13-15]. Even the smoothest metal surfaces have microasperities of 0.01-0.1 μm in height [16]. These asperities are located usually on some wavy surface, period of which in different conditions is 10^3 - 10^4 μm, and height of 1-10 μm [17]. Just because waviness and asperities are present, it results in that two surfaces come in contact only in separate points [17].

During switching, under the influence of pressure forces of contact surfaces, in places of asperities, metal will be deformed and they will be transformed into conductivity centres. The more pressure forces of contact surfaces occur the more asperities may be deformed: partly plastically and (or) partly elastically [14, 15].

If the contacts of reed switch are closed, which are being under some voltage, so with a critical approaching thereof, a vacuum breakdown occurs [17] which is similar to that discussed in works [18, 19]. After closing contacts, a force arises which may lead to the bounce thereof. This force is conditioned by resistance of contact springs [20] and by high pressure of vapour and metal plasma as a result of micro-explosions in places of contact areas [17].

While switching dry circuits (without load), contacts of reed switch are also under voltage of 10 - 30 mV. It has been found that the origins of this voltage (dynamic noise) are electro-induction processes in a ferromagnetic material during reconstruction of its domain structure, under the influence of mechanical stresses and magnetic field [21].

While current breaking, due to the surface irregularity, contacts diverge non-simultaneously. The whole circuital current will flow through separate spots. Therefore, the current constriction area will be melted, and a molten metal bridge is formed [17] which, being exploded, creates a specific surface relief of reed switch contacts, that has a negative impact on operability of device [21].

Thus, if the contact material is relatively soft and fusible (permalloy is exactly such a material compared to Ruthenium), as well as there are no corrosion- and erosion-resistant coatings, then while switching dry circuits (without load), it has to be expected that a bridge erosion will appear. This is observed in case of untreated contacts (Figures 5,6).

3.2. Auger Spectrum Investigations

Elemental composition of the contact surface (before and after treatment) was analyzed using an Auger electron spectrometer [22]. The device is equipped with a cylindrical mirror electron energy analyzer (resolution 0.25%) with an integrated electron gun having a beam current up to 1 μA at a beam diameter of 100 μm. To clean the surface of contact blades of reed switches and for layer-

Journal of Physics: Conference Series **291** (2011) 012020 doi:10.1088/1742-6596/291/1/012020

by-layer Auger analysis thereof, an ion gun, with a differential pumping of working gas (Ar) and ion current density up to 3,5 mA/cm^2, was used. All measurements were performed in vacuum of 2×10^{-7} Pa to eliminate effects of electron-stimulated adsorption of residual gas molecules on surfaces being researched.

Results of layer-by-layer Auger analysis of contact near-surface layers of reed switches after ion-plasma surface treatment thereof are shown in Figure 7. In the course of investigations, samples were subjected to sputtering by an ion probe with a simultaneous Auger analysis of primary element lines.

Etching was carried out using Ar^+ ions and 2.5 keV energy, in-beam current of 200 $\mu A/cm^2$. All that provides an ion-beam sputtering rate of 64.2 nm/min across permalloy.

Conclusions on the concentration and phase states as well as thickness of modified layer were drawn according to the conventional schedule: based on amplitude measurement of differentiated Auger lines of atoms, relation thereof, line profile stability in depth and shape of the lines which specify atomic Auger processes depending on atom coordination environment in various phase states.

Since a concentration stabilization of nitrogen atoms for depths from 14 up to 26 nm and from 58 up to 90 nm (Figure 7) is observed, when it is different in each case, and a shape of Auger line is typical for iron nitrides [21], it was assumed that all nitrogen atoms form a single-phase bonding only with iron atoms, but with various stoichiometry for every stabilization zone.

System solution: $C_{Fe} + C_N = 100$ [atom. %] and $C_{Fe} / C_N = K$ for two values obtained by calculation K = 2.7 and 4.07 (two stabilization zones) allows estimating a concentration of nitrogen atoms without regard to other present elements (Ni, C, and O), what is shown in Figure 8 in combination with nitrogen atomic concentration for stoichiometric compounds of Fe_3N and Fe_4N.

Due to the examination, a zonal, typical for intermetallic phases, distribution of atoms composing the sample, including doping nitrogen in depth, is revealed. This is the first transition-type zone about 14 nm in thickness (see fracture observed on atomic profiles) which is identified by a lateral nano-roughness of relief of surface atomic layers and by a crater effect known in probe ionic diagnostics (nonuniform ionic layer-by-layer etching across surface). All the above considered, the following zone specifies a stable stoichiometric relation of the main elemental components of heterogeneous sample, namely Fe and N, with a typical equidistant scaling. Regular breaks in concentration profile justify this assumption which comes into the open from transformations of shape of Auger lines that are typical for nitride phases, as shown below in Figure 7. In accordance to the data from scientific and technical literature, zone 2 is to be identified with nitride phase of Fe_3N type. It will be replaced (on the basis of typical breaks in profile of layer-by-layer ionic etching, line shape, and typical chemical displacements of Auger transitions) by Fe_4N phase which is regularly alloyed using interatomic additives of excess oxygen and carbon (Figures 7, 8).

Typical breaks and cophased nature of concentration decline on layer-by-layer atomic profiles (C, O) at a simultaneous relative stability of amplitudes of lines of Ni atoms allow ascertaining a heterogeneous lamination of phase state of examined samples which can be interpreted as nickel carbides and oxides. The findings conform to the working data [21].

Characteristics of Fe-Ni contact blades examined at atomic and molecular level, subjected to activation doping in a low-temperature plasma of nitrogen, specifically: transient resistance, potential surface relief (provided by attained values and lateral distribution of electron output work in a region of overlapping contact blades, and therefore, by a contact potential difference in conduction bands), effective mechanical and erosion wear-resistance, congruous stability of phase state of reed switch blades in bench and performance tests – all that is intentionally achieved due to reproduction processes of singular phase structure in range of nanoscaling.

Thereby, the results of integrated researches with corpuscular probing methods and received physical-chemical characteristics combined with reached results of comprehensive tests suggest on a purposeful reproduction of a new reed switch class unknown in world practice, commercial development thereof will permit to expand a market segment of the Russian science-intensive industrial products.

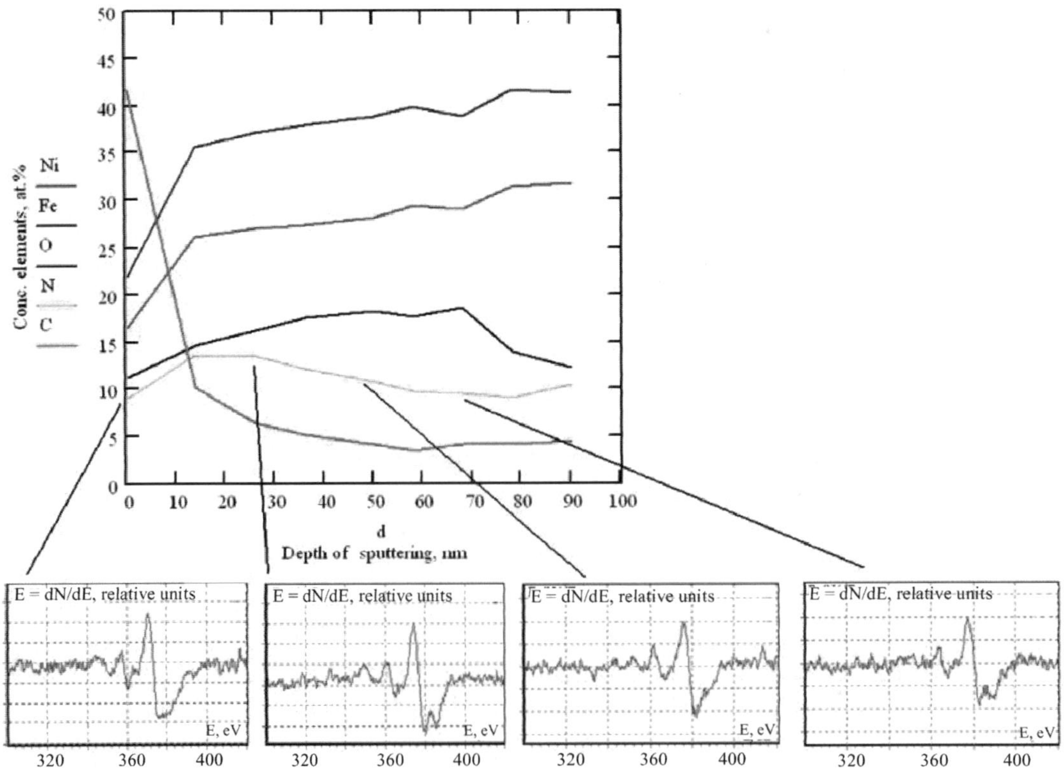

Figure 7. Elements concentration profile in surficial region of ion-plasma treated reed switch and shape of nitrogen Auger line, typical for nitride phase (E – kinetic energy of electrons; N – quantity of electrons with energy E).

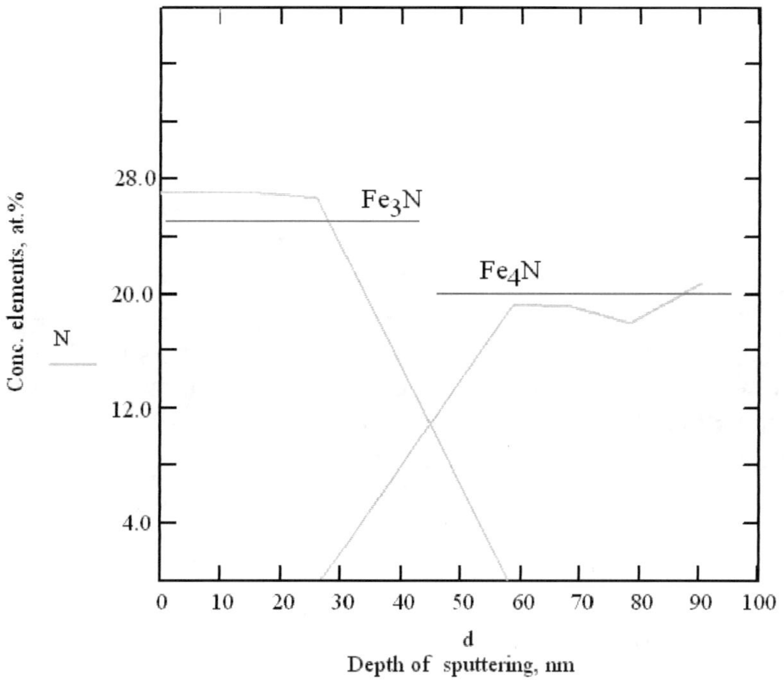

Figure 8. Regenerated phase distribution profile in depth specifying nitrogen concentration in stoichiometric compounds of Fe_3N and Fe_4N.

3.3. AFM- Research

In Figure 9 AFM-image of nitrided surface of reed switch contact spring is provided, that has grain boundaries and nanorelief thereon. One can see well interchange of dark and light round-shaped nanosized regions that points to the change of surface relief height. Profile of AFM-image and statistical analysis of height distribution, provided in Figures 10 b and 10 c , respectively, give additional and detailed information on the obtained relief.

Statistical analysis has revealed that an average peak height is 50 nm. Lateral dimensions of peaks at the bottom were 70-100 nm. On the average, width of peak at half height is 70 nm.

For the first time depictions of reed switch contact surface after ion-plasma treatment within the reflection of diffluence resistance were obtained (Figure 11), that is a positive evidence of the presence of special conductive properties of nanopeaks.

Before our researches there were no data about the formation of such nanostructures on a permalloy surface as a result of nitriding.

Contacting in reed switches always takes place in separate points on some areas because it is impossible to make and form a construction so, that contacting could be realized across the whole contact surface. In fact the number of contacting points is limited and their position changes during reed switch operation. With foreign dielectric particles are within contact region between two contacting planes, transient resistance increases because a gap is formed between contacting planes at once, and the number of contacting points may considerably decrease.

Our usage of special nitriding technology creates required properties of contacting surfaces of reed switch, improves stability its operation in different modes during long period of time, simplifies technology of contact creation, eliminates influence of foreign particles on contact properties and usage of comparatively "dirty", labour-intensive and complicated to use galvanic technologies of deposition of precious-metal-based coatings. Galvanic technology is one of the main sources of foreign particles in reed switch that leads to device failures in the process of operation. A new galvanic technology, that eliminates the usage of such main source of foreign particles, significantly improves reliability of reed switch operation. Level of failures (number of faulty reed switches under testing is 1 million pieces), with application of the new technology, decreases from 200 to 50 ppm and less (about 4 times).

Surface nanostructure (nanorelief with peaks of a higher conductivity) provides realization of reliability principle - multilevel echeloned protection of contact surface from corrosion, erosion and mechanical damages, increases amount of contacting points, improves noise immunity from foreign particles, and finally solves a problem of reed switches quality, substantially decreasing amount of service failures.

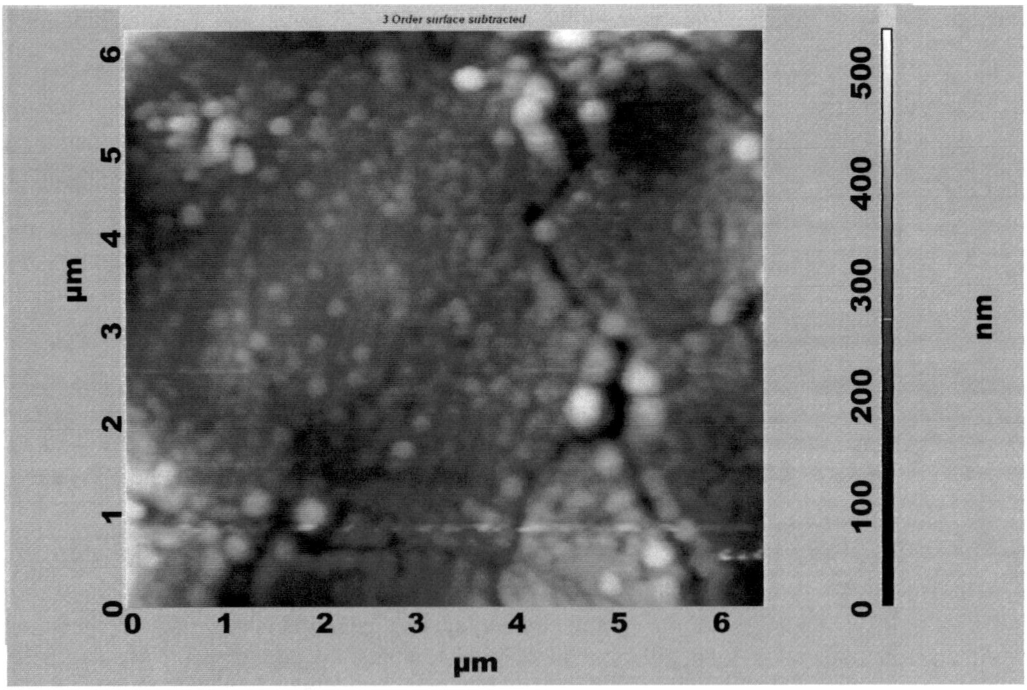

Figure 9. AFM – image of reed switch contact spring surface nitrided in discharge, field width 7 x 6 μm.

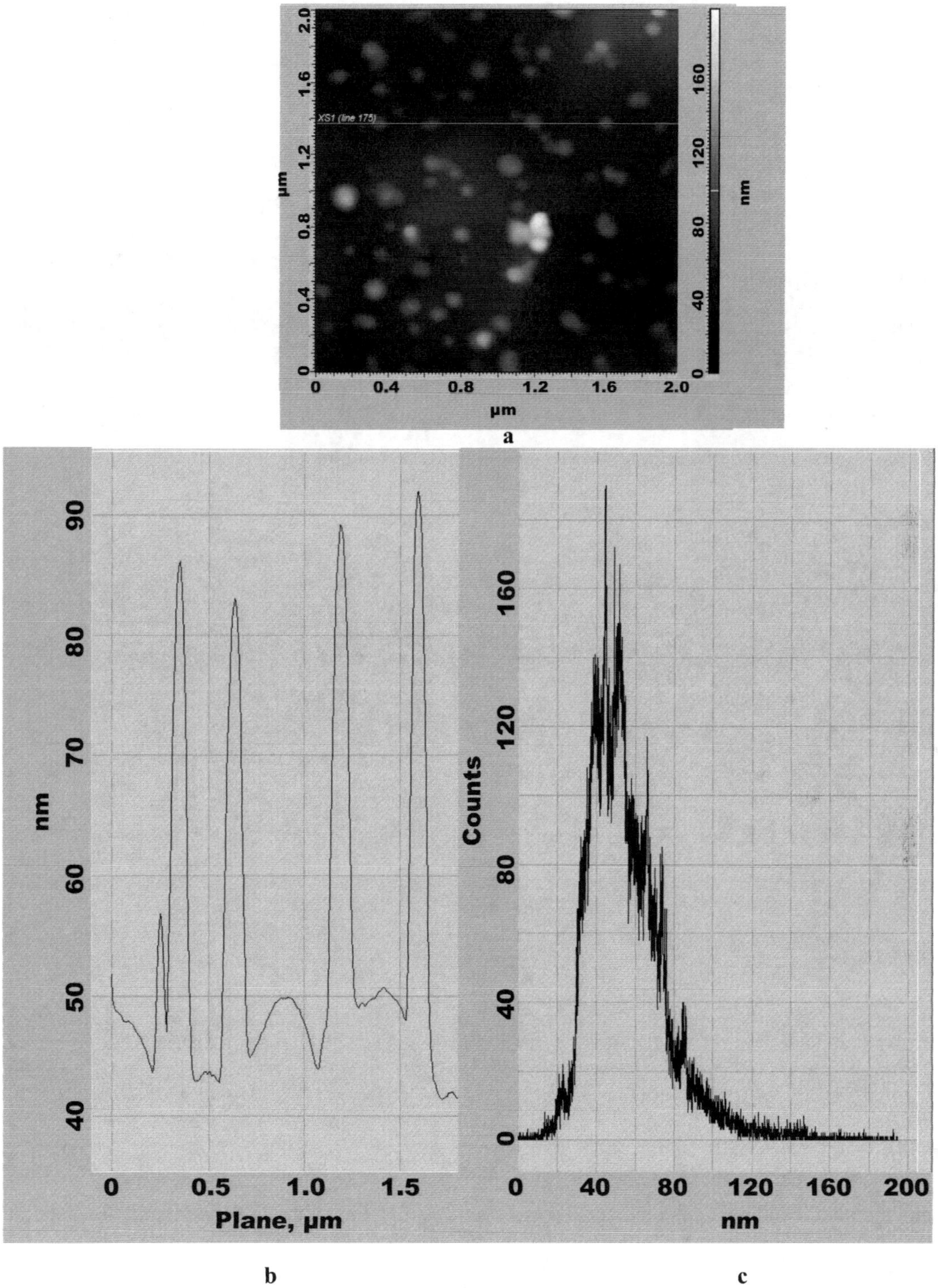

Figure 10. a) AFM – image of reed switch contact spring surface nitrided in discharge, field width 2x2 µm; b) profile of AFM - image in X-direction along a solid line 175 (see a); c) statistical distribution of relief height on surface area of reed switch contact spring (see a).

Journal of Physics: Conference Series **291** (2011) 012020 doi:10.1088/1742-6596/291/1/012020

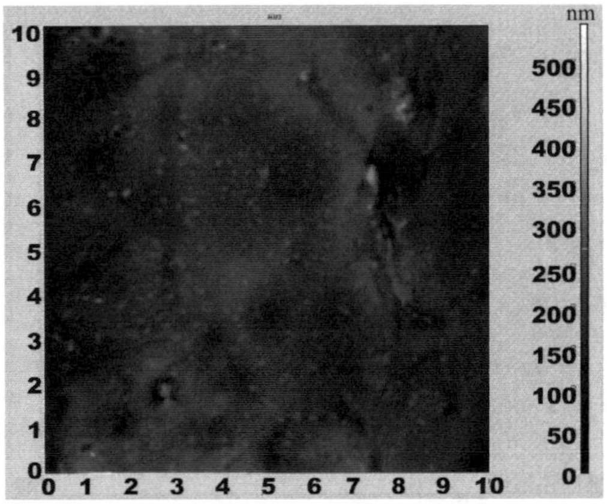

a) 10 x 10 μm

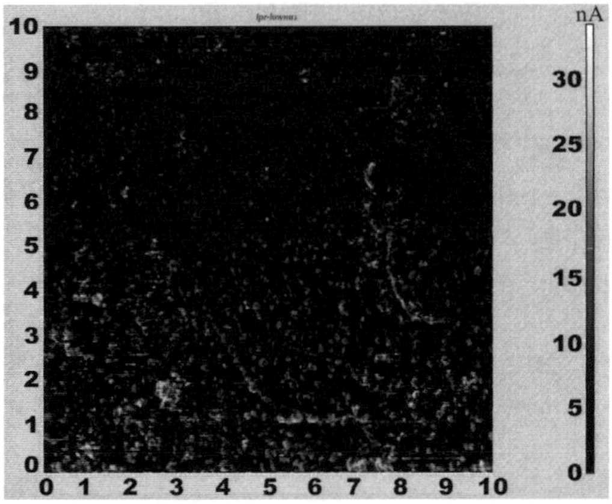

b) 10 x 10 μm

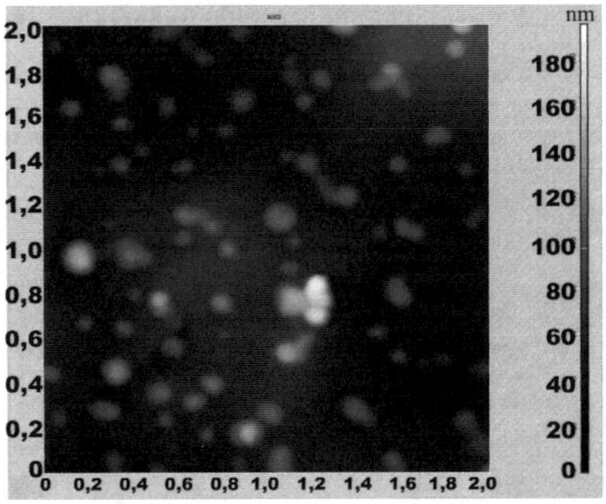

c) 2 x 2 μm

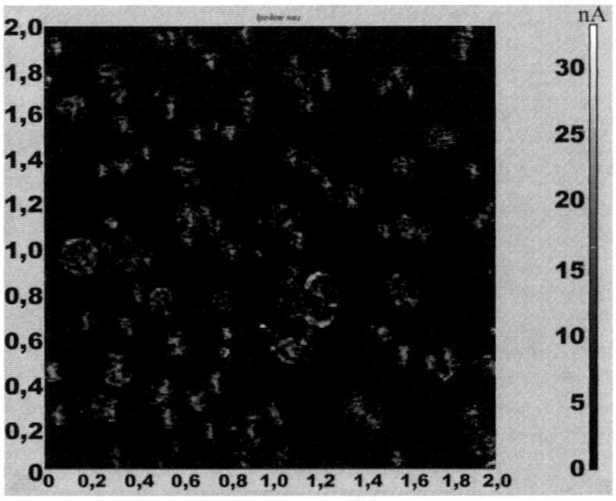

d) 2 x 2 μm

Figure 11. AFM-image of nitrided surface area of reed switch contact spring: a, c – method of constant effective force; b, d - method of reflection of diffluence resistance.

Origination of defective-deformative instability is shown in a number of researches devoted to investigation of results of energy deposition on surface of condensed matters, namely, laser irradiation, ion fluxes, and gradient fields [24, 25]. This stipulates realization of critical conditions for display of synergistic effect that leads to the development of surface structures of relief. Particularly, in our case, combination of temperature of permalloy surface, irradiation intensity and duration by nitrogen ions, forms a surface relief in the form of system of nanostructured peaks, as a result of atomic synergistic self-assembly.

4. Conclusion

Principally new technological process of modification of Reed Switch (RS) contacting surfaces was developed, that is after sealing of contact springs into a glass tube filled with nitrogen, across RS being opened, current impulses are passed which cause formation of nanosized layers with the specified contact features in a near-surficial region of RS.

As a result of these researches we have managed to create such a discharge electrophysical condition at which corrosion- and erosion-resistant nanosized high-conductivity layers are formed in the near-surficial region of permalloy contacts of reed switches. In our opinion, it enables not to use the galvanic technology of the special precious-metal based coatings.

Profile of element concentration distribution in near-surficial region and images of nanostructured contact peaks (50nm) (Figures 9-11) on new heterogeneous Fe_3N (30nm) and Fe_4N (40nm) base (Figures 7, 8), obtained by atomic-force microscope and electronic Auger spectrometer, indicate the formation of nanostructured surface contact layer. Moreover, cardinal change of contact layer characteristics also indicates the achievement of nanostructured states of near-surfacial atomic layers.

Surface nanostructure (nanorelief with conductive peaks) provides realization of high reliability principle - multilevel echeloned protection of contact surface from corrosion, erosion and mechanical damages, increases amount of contacting points, improves noise immunity from foreign particles, and, finally, solves a problem of reed switches quality, substantially decreasing amount of service failures.

So, one of the main problems of modern technology of reed switch production – relatively high level of service failures – has been overcome, that will not only improve reed switch reliability but also significantly enlarge the range of its application, competitive capacity and enable to expand the segment of world market of Russian science-intensive commercial nanoproducts.

The developed modification process was tested during the commercial manufacture of the principally new reed switch sample with nanostructured contact surfaces (MKA 14108 type) (see Figure 1).

References

1. Zeltser I.A, Karabanov S. M, Maizels R. M, Moos E. N, Sablin V.A. // Collected papers of the Second International Theoretical and Practical Conference «Reed switches and production on their basis», Ryazan: Poligraph, 2009. p. 174 (http:// www.rmcip.ru -in Russian).
2. Zeltser I.A, Karabanov S. M, Maizels R. M, Sablin V.A. // Collected papers of the Second International Theoretical and Practical Conference «Reed switches and production on their basis», Ryazan: Poligraph, 2009. p. 184 (http:// www.rmcip.ru -in Russian).
3. Zeltser I.A, Karabanov S. M, Kuznetsov A.A, Maizels R. M, Sablin V.A, Chernyak E.Y. // Collected papers of the Second International Theoretical and Practical Conference «Reed switches and production on their basis», Ryazan: Poligraph, 2009. p. 178 (http:// www.rmcip.ru -in Russian).
4. Karpov A.S., Maizels R. M., Shishkina L.V., Shkutenko L.N. // Collected papers of the Second International Theoretical and Practical Conference «Reed switches and production on their basis», Ryazan: Poligraph, 2009. p. 169 (http:// www.rmcip.ru -in Russian).
5. Kuznetsov A.A., Vasiliev E.V., Zeltser I.A, Chernyak E.Y. // Collected papers of the Second International Theoretical and Practical Conference «Reed switches and production on their basis», Ryazan: Poligraph, 2009. p. 84 (http:// www.rmcip.ru -in Russian).
6. Arushanov K.A., Zeltser I.A. // Herald of the Ryazan State Radio Engineering University. № 3 (issue 29). Ryazan, 2009. p. 93 (in Russian).
7. Zeltser I.A, Karabanov S. M, Maizels R. M, Moos E. N, Sablin V.A. // Theses of the reports of the X-th International workshop "Common frameworks of material modification by methods of alternative technologies", (MAT-X). Obninsk. 2009. p. 58 (in Russian).
8. Zeltser I.A, Karabanov S. M, Maizels R. M, Moos E. N, Sablin V.A. // In Abstract Book of International Conference "Information and Structure in Nanoworld" Saint-Petersburg, 1-3 July, 2009. P. 48.
9. Patent RF № 2391733. Reed switch. / Karabanov S. M, Zeltser I.A, Maizels R. M, Trunin E.B.
10. Patent RF № 2393570. Method of production of reed switches with nitrided contact blades. / Karabanov S.M., Maizels R.M., Arushanov K.A., Zeltser I.A., Provotorov V.S.
11. Lahtin Y.M. The theory and technology of nitriding.- M.: Metallurgy, 1991. - 319 p (in Russian).
12. Arzamasov B.N., Bratuchin A.G., Eliseev Y.S., Panayoty T.A. Ion beam surface machining of alloys in gas environment. - M: Publ. Bauman Moscow State Technical University 1999. – 400 p (in Russian).
13. Karabanov S.M., Maizels R.M., Shoffa V.N. Reed switches and production on their basis. - M.: Intellect, 2010. – 470 p (in Russian).
14. Merl B. Electric contact. - M.: SEI, 1962. – 490 p (in Russian).
15. Holm R. Electric contacts. -Stockholm, 1946.

Journal of Physics: Conference Series **291** (2011) 012020 doi:10.1088/1742-6596/291/1/012020

16. Rahovsky V. I. Basic physics of current switching in vacuum. - M.: Science, 1970. – 378 p (in Russian).
17. Mecyaz G.A. // Progress in Physical Science, 1995. V.165, №6. p 601.
18. Zeltser I.A., Kukushkin S.A, Sablin V.A., Moos E.N, "Common frameworks of material modification by methods of alternative technologies", (MAT-IX). Obninsk. June, 12-16, 2007. Theses of the reports p. 38 (in Russian).
19. Zeltser I.A., Moos E. N, Sablin V.A., Trunin E.B. // Collected papers of the Second International Theoretical and Practical Conference «Reed switches and production on their basis», October 1–3, 2008. / Under the editorship of Doctor Technics. Karabanov S.M. - Ryazan: Poligraph, 2009. p. 111 (http://www.rmcip.ru -in Russian).
20. Razumichin M.A. Erosion resistance of low-power contacts - M.-L.: Energy, 1964. – 80 p (in Russian).
21. Solomin Y.M. // Electronnaya Technica, series. 4, issue.7, 1975 (in Russian).
22. Kuznetsov A.A., Abramova S. Yu., Potapova T.E., Protopopov O.D. //Journ. of Electron Spectr. and Rel. Phenom., 1994, V. 68, pp. 407-412.
23. Vanden Berghe R.and Vlaeminck R.// Surf. Interface Anal. 10, 316 (1987).
24. Emelyanov V.I., Ruhlyada N.Y. // Vyusokiye Technologii . 2009. V. 10, № 6. p.3 (in Russian).
25. Zeltser I.A., Karabanov A. S., Moos E. N. // Izvestiya Academii Nauk, series phys., 2004. V.68, №3. p. 408 (in Russian).

Local structure of high-coercivity Fe-Ni-Al alloys

A P Menushenkov[1], V P Menushenkov[2], R V Chernikov[1,3], T A Sviridova[2], O V Grishina[1], V V Sidorov[1]
[1] National Research Nuclear University "MEPhI", 115409 Moscow, Russia
[2] National University of Science and Technology «MISIS», 119049 Moscow, Russia
[3] HASYLAB at DESY, D-22603 Hamburg, Germany

E-mail: menushen@htsc.mephi.ru

Abstract. Results of hard magnetic Fe-Ni-Al alloys after various thermal processing local structure researches by method of EXAFS-spectroscopy with use of synchrotron radiation at temperature 77 K are presented. It is established, that during cooling a firm solution with critical speed reorganization of a local environment of nickel relative to quickly tempered sample owing to stratification of a firm solution is observed. The subsequent aging at 780°C practically restores local structure, characteristic for quickly tempered sample, keeping thus rather high coercitive force.

1. Introduction

Hard magnetic Fe-Ni-Al alloys belong to a group of alloys with insolubility field at state diagram. On cooling below the solubility curve solid solution decomposes to a mixture of two isomorphic cubic phases $\beta \rightarrow \beta + \beta_2$ [1]. Maxima of coercitive force H_c achieved at intermediate decomposition stage is related with nonequilibrium microstructure which is formed at aging of homogeneous solid solution (heat treatment I), or at solid solution cooling with critical rate (heat treatment II). H_c value reached by heat treatment II is 1.5 times as large as H_c value reached by heat treatment I.

X-ray studies of Fe-Ni-Al solid solutions decomposition (see [2] and references therein) do not reveal any changes in shapes and positions of diffraction lines of spectra for samples prepared by quenching or cooling with critical rate. This is probably connected with the proximity of the lattice parameters of precipitating phases or strong diffusion of the β-phase diffraction lines. Available data on the structure of high-coercivity alloys are very contradictory. According to electron microscopic studies hetero-phase structure is formed by $\beta \rightarrow \beta + \beta_2$ decomposition, this structure consists of fine-grained (10-100 nm) β-phase precipitations enriched by iron and matrix β_2-phase enriched by Ni and Al. In fact no direct measurements of phase composition have been made. β-phase equated particles were observed in high-coercivity sample. On the contrary it was observed in [3] that microstructure consists of intermittent plates of β- and β_2 phases. Anisotropic shape of single-domain particles seems to be the main reason for high coercitive force in Fe-Ni-Al alloys. Improvement of magnetic insulation of β-particles increases H_c due to the process of second decomposition below 750°C related with Ni and Al precipitation from β-phase [2].

Thus data on structural transformations in Fe-Ni-Al alloys are contradictory and need to be specified. This work summarizes our studies of crystal structure and local environment of iron and

Published under licence by IOP Publishing Ltd

nickel in Fe-Ni-Al alloy after different heat treatment by means of X-ray absorption spectroscopy (EXAFS).

2. Experimental

The samples with nominal composition of $Fe_{46.4}Ni_{24.5}Al_{26.7}Si_{2.4}$ were prepared by induction melting in Ar atmosphere followed by casting in a copper mould. Three schemes of heat treatment were used: sample N1 was aged at 1200°C for 20 min and quenched in water ($H_c < 4$ Oe); sample N2 was aged at 1200°C for 20 min and cooled in air atmosphere with critical speed V = 2°/s ($H_c = 650$ Oe) and sample N3, heat treated in similar manner as N2, was aged at 780°C for 10 min and then quenched in water ($H_c = 320$ Oe).

Phase identification was carried out at powder samples by X-ray diffraction using Co-Kα radiation. EXAFS measurements above iron and nickel K-edge absorption were performed at E4 beamline of HASYLAB synchrotron centre (DESY, Hamburg, Germany) at temperatures of 77 K and 300 K. Spectra were measured for powder samples (d < 100 microns) in transmission mode. EXAFS spectra processing and modelling was performed in VIPER [4] software environment. The background in the experimental spectra was removed as described in [5], taking care to remove the low-frequency oscillations. The EXAFS functions $\chi(k)k^2$ obtained from the absorption spectra were Fourier transformed (FT) in the range of wave-number k from 2 to 12 Å^{-1}, using the Kaiser-Bessel windowing function. Fourier back-transformation (BFT) was carried out using a Hanning window over the range of 1.2 – 2.0 Å for K-Fe and 1.1 – 2.9 Å for K-Ni edges. BFT modelling was performed according to formula:

$$\chi(k) = -S_0^2 \sum_j \frac{|f_j(k,\pi)|}{kR_j^2} \exp(-2k^2\sigma_j^{\ 2}) \sin(2kR_j + \varphi_j(k,\pi)) \quad (1)$$

where R_j - mean radius of j-th sphere, σ_j^2- Debye-Waller factor (mean-square deviation of interatomic distance), S_0^2 - many-body reduction factor, $f_j(k,\pi)$ and $\varphi_j(k,\pi)$ - back-scattering amplitude and phase shift calculated by FEFF [6].

3. Results

Quenching from 1200°C in water (N1) fixes solid solution with low coercitive force ($H_c < 4$ Oe). X-ray structure analysis showed the superposition of two phases: β_2-phase which is partly ordered in CsCl-manner structure (B2) and β-phase with bcc-structure (A2), which can be partly ordered in Cu_2MnAl-like structure ($L2_1$). Both phases have practically equal lattice parameter a = 0.2877 nm. The cooling with critical rate (N2) transforms the alloy to the high-coercivity state ($H_c = 650$ Oe). The diffraction pattern shows peaks broadening (for example, integrated width of (220)-line changed from 2.83° to 3.48°) and increase of intensity of additional superstructural reflections belonging to $L2_1$ phase. The mean lattice parameter of alloy is a ≈ 0.2879 nm (or twice of it as regards $L2_1$ phase). Aging of sample N2 at 780°C for 10 min (N3) reduces coercitive force to $H_c = 320$ Oe and leads to additional broadening of peaks (integrated width of (220)-line increased from 3.48° to 4.28°) together with decrease of the intensity of superstructural reflections of $L2_1$ phase. The mean lattice parameter of alloy is a ≈ 0.2878 nm. It seems that aging leads to disordering of $L2_1$-phase and its transformation into A2 phase.

Transmission electron microscopy (TEM) image of the quenched sample N1 (fig. 1a) reveals two-phase nanostructure constituting the fine mixture of black and white precipitates of size 5÷10 nm, forming at the initial stage of solid solution decomposition. After cooling with critical rate (N2) the features of nanostructure becomes noticeably coarser (fig. 2a). The sample contains lamellar precipitates with an average size of 80÷150 nm and pronounced preferred orientation inside a single grain. The diffraction pattern (fig. 2b,c) reveals the superposition of reflections of two phases: β-phase with bcc-structure (A2) and β_2-phase with CsCl structure (B2). Figure 1b shows TEM image of the sample aged at 780°C for 10 min (N3), which indicate a visible increase of precipitates size up to 200÷250 nm.

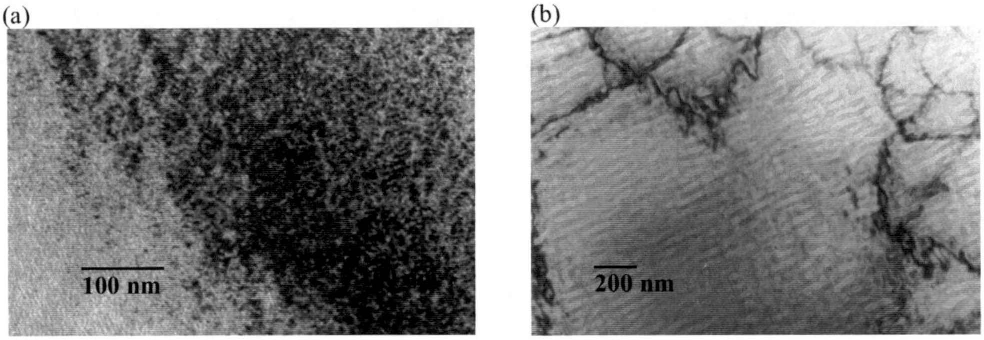

Figure 1. TEM images of quenched sample N1 (a) and sample N3 aged at 780°C for 10 min (b).

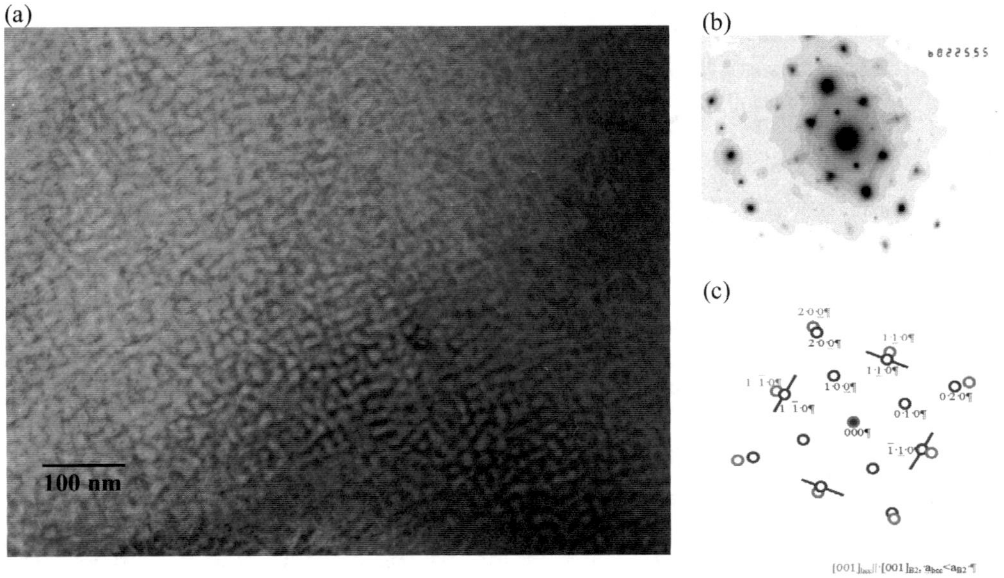

Figure 2. (a) TEM image of sample N2. (b,c) Diffraction pattern for the region depicted in (a).

According to EXAFS spectra analysis, iron local environment is the same for both $L2_1$ and B2 types of ordering, so parameters of coordination shells around iron atoms are practically identical in all samples N1-N3. However analysis of the EXAFS-function measured above K-Ni edge showed the strong difference of the nickel local environment in sample N2 from the environment in sample N1 and N2. One can observe similar picture for BFT of EXAFS functions measured above K-Ni absorption edge for N1 and N3 samples (fig. 3a,b). EXAFS spectra modelling showed that local environment corresponds to B2 type of ordering in both N1 and N3 samples.

At the same time modelling of high-coercivity sample N2 EXAFS functions showed that it is impossible to fit experimental spectrum with single phase (B2 or $L2_1$) model due to samples heterogeneity (fig. 3c). Good results were obtained by modelling the spectrum assuming the simultaneous existence of two ordering types. Thus the local structure analysis confirms appearance of $L2_1$ phase in addition to B2 phase in the structure of sample N2 after cooling with critical rate. Apparently this transformation is the main reason of the high-coercivity state in Fe-Ni-Al alloys.

According to X-ray diffraction data aging of sample N2 at 780°C (N3) returns the alloy to A2+B2 phase state. EXAFS data analysis shows that parameters of coordination shells around Fe and Ni as well as BFT shapes are restored to values corresponding to sample N1 (fig. 3b). It seems that aging at 780°C leads to disordering of $L2_1$-phase and its transformation into A2 phase [2]. As a result the coersivity decreases.

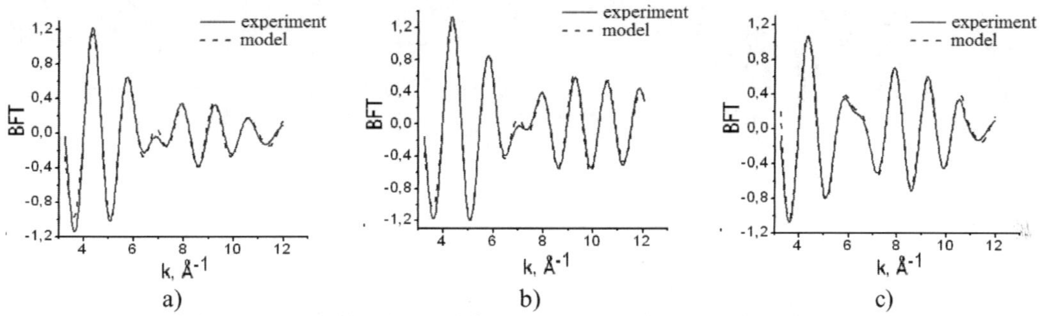

Figure 3. BFT of the nearest shells, obtained from treatment of EXAFS function measured above K-Ni absorption edge for samples N1 (a), N3 (b) and N2 (c).

4. Conclusions

Quenching from 1200°C (N1) fixes initial stage of solid solution decomposition into two phases: β_2-phase which is partly ordered in CsCl-manner structure (B2) and β-phase with bcc-structure (A2). The cooling with critical rate (N2) transforms the alloy to high-coercivity state with β-phase partly ordered in Cu_2MnAl-like structure ($L2_1$). Annealing at 780°C (N3) transfers the alloy to initial phase state: parameters of coordination spheres around Fe and Ni are restored to values corresponding to sample N1 (A2+B2 structures) due to disordering of $L2_1$-phase and its transformation into A2 phase.

Acknowledgments

This work was supported by Federal Program"Scientific and scientific-pedagogical stuff of innovative Russia" and Russian Foundation for Basic Researches (grant 11-02-01174-a). The authors thank the HASYLAB Program Committee for providing beamtime.

References

[1] Bradley A J, Taylor F 1938 *Proc. Roy. Soc.* (A) **166** 353
[2] Menushenkov V P, Sviridova T A, Shelekhov EV 2010 *Proceedings of the 19-th Int. Metallurgical and Materials Conference "Metal-2010" (Roznov pod Radhostem, Czech Republic, 18-20 May 2010)* pp 807-811
[3] Cahn J W 1962 *Acta Met.* **10** 179
[4] Klementev K V 2001 *J Phys D: Appl. Phys* **34** 209
[5] Newville M, Līviņš P, Yacoby Y, Rehr J J and Stern E A 1993 *Phys. Rev.* B **47** 14 126
[6] Rehr J J, Mustre de Leon J, Zabinsky S I et al. 1991 *J. Am. Chem. Soc.* **113** 5135

Journal of Physics: Conference Series **291** (2011) 012022 doi:10.1088/1742-6596/291/1/012022

The structure peculiarity of perfluorocarbon nanodispersions

I N Kuznetsova

Russian Research Institute for Haematology and Transfusiology (Rus Rsch Inst Hem/Trans), 2-d Sovietskaya Street, 16, 191024 Saint Petersburg, Russia

e-mail: kuz-ir@ inbox.ru

Abstract. Nano-dispersions of perfluorocarbon (PFC) or PFC emulsions are substitutes of erythrocytes. They perform the function of the gas transport on condition when the particle structure will be preserved. The invariable values of the size and surface properties of the spherical PFC particles characterize the stability of their structure. The stable nano-dispersion on base PFC and phospholipids (PL) as the basis of the oxygen carrier preparation have been prepared. This PFC/PL nano-dispersion conformed to the essential requirements: the heat sterilization of the final product and the structure preservation of particles: size (~100 nm) and surface properties, during one year of storage in the unfrozen state. The obtained PFC/PL nano-dispersion conformed to medicine-biology employment.

1. Introduction

Blood transfusion may cause risks of emerge of infection diseases. The creation of artificial blood substitutes with effective gas-transport function is actual task of modern medicine.

Perfluorocarbons (PFCs) contain large amounts of various gases including oxygen and carbon dioxide. There are many publications [1-4]about chemical and physico-chemical properties of PFCs that are used in biology and medicine for elaboration of different emulsions. The authors pay particular attention to the peculiarities of molecular structures of these compounds. That molecular structure brings about increased physical solubility of gases in PFC without the formation of intermolecular chemical bonds PFC - gas.

Nano-dispersions of perfluorocarbon (PFC) or PFC emulsions are basis of oxygen carrier blood substitutes. There are several elaborated PFC preparations, in which nature phospholipids as emulsifier have been used: – Oxygent, Oxyfluor, Oxycyte. Differ adverse reactions were observed during preclinical trails of the preparations according to GMP demands [5]. The reasons of these side effects are not fully understood. There is question about improving the quality of the final product – the PFC emulsion preparations [2]. The assertion have been stated that progress in studying of the quality and stability of PFC emulsions was related to the increase of notions about their structure [6].

It is more correctly to represent PFC nano-dispersions as red cell substitutes because they do not replace all components and functions of blood (coagulation factors and white cells) [5]. PFC particles are a model of erythrocyte indeed because they have the structure of two-layer sphere [7]. In the sphere centre is PFC – the nucleus of particle, where dissolves blood gases. On the surface of nucleus there is the thin layer of surfactant, the shell. Surface properties and size of two-layer sphere particles characterize stability of their structure.

PFC emulsions (nano-dispersions) are the basis of pharmacology preparations. The stability of PFC emulsions and their behavior in a vascular bed depend on a surface layer firmness of surfactant around the particles. The study of solely a particle size of these dispersing preparations is not enough to judge of their stability and quality. It is essential to have simultaneously information about the particle size and structure changes, taking place in these media.

2. Materials and methods

The nano-dispersions were prepared by high pressure method. Fluorocarbon phase was perfluorodecalin (PFD) with additions of other known PFCs (perfluorotripropylamine and perfluoroctylbromide) in an optimal ratio. The PFC mixture was dispersed at a high pressure Gaulin-tipe homogenizer. Soybean (or egg yolk) phospholipids in the form of some complex aqueous dispersion was used as surfactant.

The particle size was found by turbidity spectrum. Surface properties of particles were characterized using indirect physico-chemical parameter [8] – the coefficient of PFC nanodispersion interaction with blood serum: $K_\tau = \tau_1/\tau_2$, where τ_1 – turbidity of the mixture of blood serum / nano-dispersion (experience), τ_2 – turbidity of a mixture of blood serum / saline (control). Ratio of components in both mixtures "experience and control" was the same.

3. Results and discussion

PFC molecules do not form complexes with blood gases. Therefore, the preservation of the corpuscular nature of PFC particles (the structure) under the circulation in the bloodstream is the main condition to fulfill gas transport function. The decisive role to ensure the PFC emulsion stability *in vitro* and *in vivo* belongs to the surface surfactant layer around the particles. This layer became loosened as a result of molecular diffusion (Ostwald ripening or isothermal distillation), which is the main mechanism of destruction of PFC emulsions during storage [9]. Molecule PFC passage through the disperse medium from small particles to larger ones. At the same time the particle size can vary a little, but the surface properties change significantly. Into the bloodstream, the particles with the loosened surfactant shell will be more intensive to interact with the plasma macromolecules as compared to the original emulsion (Figure 1).

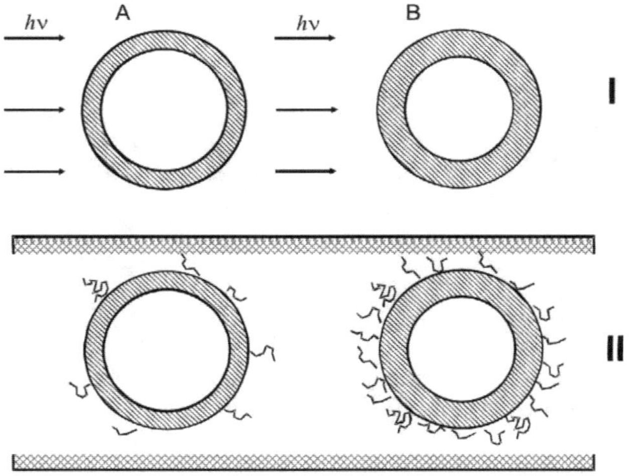

Figure 1. Schematic representation of behavior of PFC particles with strong (A) and loosen (B) surfactant layer when they are infused into the bloodstream.
I – state *in vitro*; II – status *in vivo* (bloodstream). A – adsorption of plasma macromolecules is negligible. B – absorption of macromolecules is significant.

It was shown, that the turbidity of blood serum /PFC emulsion mixtures is not additive in contrast to the turbidity of blood serum/saline and saline/ PFC emulsions mixtures. These differences indicate the interactions between two systems – blood serum and PFC emulsion. We proposed to evaluate the change of surface properties of PFC nano-particles using indirect physico-chemical parameter K_τ – the coefficient of PFC nano-dispersion interaction with blood serum as a model system [8, 10]. This parameter reflects the response reaction extent of nano-particles with plasma macromolecules.

Values K_τ for the emulsions of different compositions was different and the relationship between the amount of bound protein on the particle surface and K_τ parameters was established (Figure 2) [10].

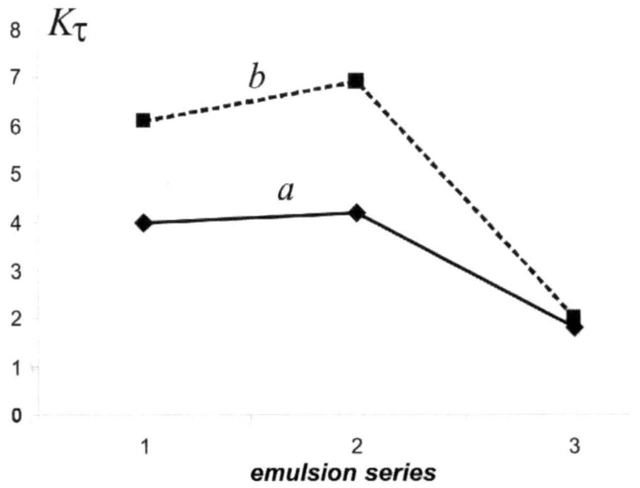

Figure 2. Values of coefficients of interaction of PFC emulsions of different composition with blood serum.
a – serum / emulsion ratio = 1 / 0.05, *b* – serum / emulsion ratio = 1/ 0.10

The stable nano-dispersions on base of PFC and phospholipids (PL) have been prepared. These PFC/PL nano-dispersions conformed to the essential requirements for model of pharmacological preparation: the heat sterilization of the final product and the structure preservation of particles: size (~100 nm) and surface properties (K_τ) during one year of storage in the unfrozen state. The relative viscosity (relative to water) prepared nano-dispersions did not exceed the critical values of plasma viscosity (1.5 – 1.75) [11].

On a large experimental material is shown that the K_τ values for different series nano-dispersions PFC / PL of the same composition fluctuate within a narrow range and did not change after one year storage in the unfrozen state – unpublished results [12] (Figure 3).

The obtained PFC/PL nano-dispersions conformed to medicine-biology employment. The constancy of surface properties and particle size of PFC nano-dispersions *in vitro* reduces significantly the probability of manifestation of adverse reactions *in vivo*.

We did not encounter such a detailed study of the physico-chemical properties of different PFC emulsion preparations on the stage of pre-clinical study. Data on the study of PFC particle surface properties for the known blood substitutes – Oxygent, Oxyfluor, Oxycyte are not described in the literature.

Obviously, it is necessary before the clinical study of such complex PFC emulsion preparations have advanced information about their physical-chemical parameters characterizing their quality: the average diameter of particles and particle size distribution, integrity of PFC particle structure, viscosity, stability and others. The date about the structure integrity of particles and the use of indirect physico-chemical parameter Kτ can help to solve technological problems of manufacture of new effective PFC emulsions.

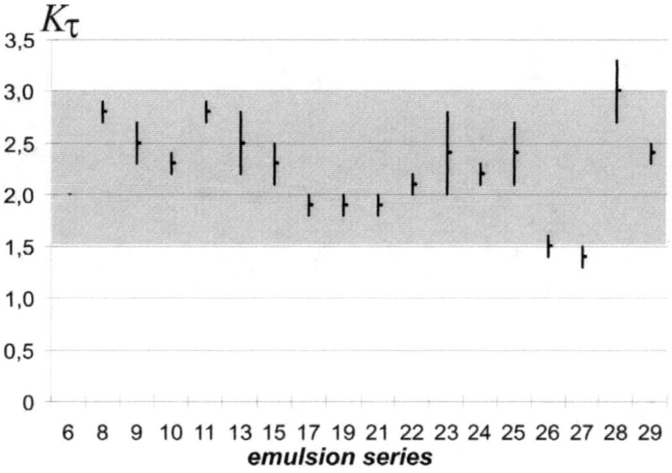

A

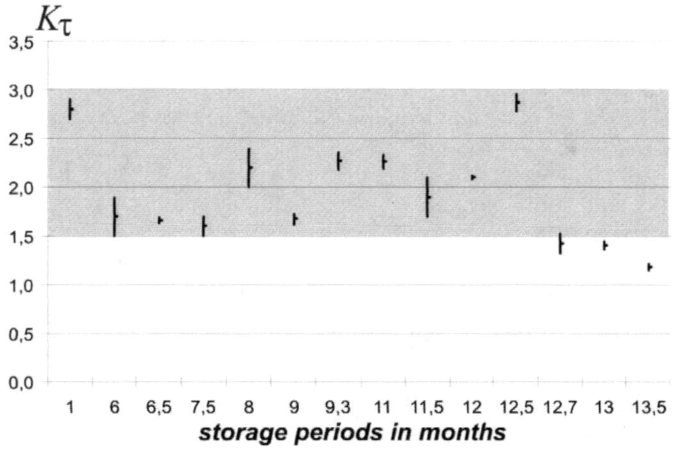

B

Figure 3. Values of coefficient of interaction with blood serum (K_τ) for PFC/PL nano-dispersions.
A – K_τ values for different series of nano-dispersions of the same composition. B – K_τ values for the same series of nano-dispersions during different periods of storage in unfrozen form.

4. Summary

Evaluation of the integrity of the structure of pharmaceutical preparations based on PFC nano-dispersions (PFC emulsions) is a prerequisite for the conclusion about their stability. The structure is characterized by size and surface properties of particles.

The indirect physico-chemical parameter K_τ – coefficient of PFC emulsions interactions with blood serum have been proposed. K_τ reflects the dynamic of change or constancy of the surface properties of PFC emulsified particles.

The nano-dispersions on base PFC and phospholipids have been prepared. These nano-dispersions conformed to the necessary requirements for a model of pharmacological preparations: heat sterilization of the final product and preservation of the particle structure integrity (size and surface properties) during to one year of storage in the unfrozen state.

Acknowledgements
I wish to express my appreciation to all my co-authors.
This work was supported in part by NPO "BioMed-Nd" Samara.

References

[1] Riess J G 2001 Oxygen carriers ("blood substitutes") - Raison d'Etre, chemistry, and some physiology *Chemical Reviews* **101** 2797-2919

[2] Riess J G 2005 Understanding the fundamentals of perfluorocarbons and perfluorocarbon emulsions relevant to in vivo oxygent delivery *Art. Cells, Blood Subs., and Immob. Biotech.* **33** 47-63

[3] Riess J G 2006 Perfluorocarbon-based oxygen delivery *Art. Cells, Blood Subs., and Immob. Biotech.* **34** (6). P. 567-80

[4] Krafft M P, Riess J G 2009 Chemistry, Physical Chemistry, and Uses of Molecular Fluorocarbon-Hydrocarbon Diblocks, Triblocks, and Related Compounds-Unique "Apblar" Components for Self-Assembled Colloid and Interface Engineering *Chemical Reviews* **109** 1714-92

[5] Cohn C S and Cushing M M 2009 Oxygen therapevtic: perfluorocarbon and blood substitute safety *Crit. Care Clin.* **25** 399–414

[6] Kuznetsova I N 2003 Perfluorocarbon emulsions. Stability in vitro and in vivo.(View) *Russian Chemico-Pharmaceutical J.* **37** 20-5

[7] Kuznetsova I N, Bezrukova A G, Lopatin V N and Parshin A V 1988 On determination of refractive index and shell thickness of the perfluorocarbon particles of dispersed blood substitutes *Russian Biophysic* **33** 126–9

[8] Kuznetsova I N, Gokhman N Sh and Lavrova Y V 1993 Interaction of perfluorocarbon emulsion particles with blood serum *Russian J. Physical Chem* **67** 1884–8

[9] Trevino L, Sole-Violant L, Daumur P, Davallet B, Postel M and Riess J G 1993 Molecular diffusion in concentrated fluorocarbon emulsion and its effect on emulsion stability *New J. Chem.* **17** 275–8.

[10] Kuznetsova I N and Rybakova L P 2007 Blood serum protein sorption on the surface of perfluorocarbon emulsion particles *Russian Chemico-Pharmaceutical J.* **41** 40–3

[11] Kuznetsova I N, Yurchenko V S and Kochetkova G A 2009 Physico-chemical parameters of perfluorocarbon emulsions of different osmolarity *Russian Chemico-Pharmaceutical J.* **43** 34-40

[12] Kuznetsova I N, Yurchenko V S and Kochetkova G A (in print) The stability of nano-dispersions on base of perfluorocarbon and phospholipids *Russian Chemico-Pharmacetical J.*

Journal of Physics: Conference Series **291** (2011) 012023 doi:10.1088/1742-6596/291/1/012023

Fabrication of porous nanostructured lead chalcogenide semiconductors for modern thermoelectric and optoelectronic applications

S P Zimin[1,4], E S Gorlachev[1,2], V V Naumov[2], E Yu Buchin[2] and H Zogg[3]

[1] Microelectronics Department, Yaroslavl State University, Sovetskaya Street 14, Yaroslavl, 150000, Russia
[2] Yaroslavl Branch of the Institute of Physics and Technology, Russian Academy of Sciences, Universitetskaya Street 21, Yaroslavl, 150007, Russia
[3] Thin Film Physics Group, Laboratory for Solid State Physics, Swiss Federal Institute of Technology, ETH Technopark, CH-8005, Zurich, Switzerland

E-mail: zimin@uniyar.ac.ru

Abstract. Porous structures of lead chalcogenides with porosities of 20-68% and pore dimensions as small as 7-20 nm were fabricated using an anodic electrochemical etching technique applied to epitaxial PbTe and PbSe films on Si. Anodized lead selenide demonstrated two basic types of porous morphology: quasiporous noncontinuous layers and hierarchical porous layers. Lead telluride had a more typical mesoporous morphology, with pores propagating at an angle of 35° to surface, which corresponds to the <100> directions in the epitaxial films and is promising for the fabrication of photonic crystals with Yablonovite-like structure. The sizes of PbTe nanocrystallites in porous layers with high porosity were calculated to be 26 nm, which indicates that quantum confinement conditions are realized in this material. Such low-dimensional morphology of porous lead telluride is prospective for the fabrication of microscale thermoelectric devices with high *ZT*.

1. Intoduction

The processes of pore formation in solids present a unique and diverse approach to the fabrication of novel nanomaterials with improved properties. As of late, the top-down anodic electrochemical etching process resurged as an extremely promising technique for porous modification on nanoscale. Porous materials with high porosity and nanometer-sized pores are known to exhibit unique physical properties due to specific and rather complex nanostructured morphologies, which result in an extremely large surface to volume ratio and the quantum confinement effects. The possibility to employ new properties of high-porosity materials, which take place due to the quantum-size effects, is widely studied for porous silicon and an array of other semiconductor materials. However, one glaring omission is the study of the binary compounds of lead chalcogenides PbX (X = Te, Se, S). The main attractive property of PbX materials is the large value of exciton Bohr radius (20-50 nm), which gives a unique opportunity for the fabrication of the nanostructured materials with outstanding properties [1]. Currently there are two major fields of the practical application of lead chalcogenides: infrared

[4] To whom any correspondence should be addressed.

optoelectronic devices (quantum dot lasers, detectors, superlattice-based Bragg mirrors) [2], and thermoelectric devices (microcoolers, microthermogenerators) [3]. PbTe is one of the most studied and widely used intermediate temperature (450-800 K) thermoelectric materials. It is well known, that the modern development of thermoelectric devices strongly requires an increase in the value of the thermoelectric figure of merit *ZT* so that it exceeds 1 at 300 K. A scientific approach to solving this problem exists and it is associated with the creation of low-dimensional systems on the basis of existing thermoelectric materials. From a physical point of view this is to be expected from the reducing of thermal conductivity by suppressing the phonon thermal conductivity due to the scattering of phonons at the nanoscale boundaries and inhomogeneities. PbTe is currently receiving a significant amount of attention due to the demonstration of $ZT > 1$ by employing nanogranular powder materials or creating nanosized inclusions in bulk crystalline material [3-6]. For example, for thin film systems with PbTe quantum dots in a PbSe matrix $ZT = 2.5$ at 550 K was reported [5]. A completely novel approach for thermoelectric devices fabrication is to employ porous materials with a nanometer-scaled morphology. Application of nanostructured porous lead chalcogenide materials can lead to a remarkable improvements and new devices, such as monolithically integrated cooling systems for IR-detectors and microprocessors, microrefrigerators, microthermogenerators, which seem to become more and more in demand considering the new alternative energy sources. Therefore, it is extremely important to fabricate, investigate and apply new nanostructured porous PbX materials. In our work we present our latest results in porous lead chalcogenide layers study based on the recently developed method of PbX anodic etching in a Norr solution [7].

2. Experimental details

In this work we have applied a traditional anodic electrochemical etching widely used for porous material fabrication for lead chalcogenide film structures. Initial samples were PbTe and PbSe epitaxial films with 2-3.5 μm thickness grown with molecular beam epitaxy on Si(111) wafer substrates with 2 nm thin CaF_2 buffer layers.

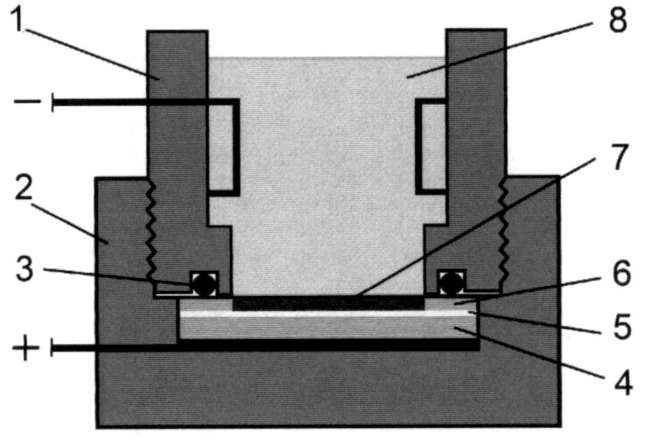

Figure 1. Schematic image of electrochemical cell. 1 - upper cell unit with a circular Pt cathode, 2 – lower cell unit with flat anode, 3 – elastic O-ring, 4 – sample's silicon substrate, 5 – sample's buffer layer, 6 – sample's epitaxial PbX film, 7 - modified area of the PbX film, 8 – electrolyte.

A vertical type teflon electrochemical cell with a circular platinum cathode and a flat metallic anode was used to perform anodic electrochemical etching of $PbX/CaF_2/Si(111)$ structures. A schematic of the cell is given in figure 1. The prefabricated teflon base of the cell consisted of two main parts: upper unit with a circular platinum cathode and lower unit with flat metallic anode. Experimental samples were placed directly on the anode and were pressed to it by elastic O-ring to ensure no leaks. Since lead chalcogenides have a significantly lower resistivity (~0.1 Ohm × cm) than Si substrate (1000 Ohm × cm), there was no necessity to create a highly doped layer at the bottom of the Si substrate in order to successfully carry out anodic electrochemical etching of the PbX layers. Anodic etching of lead chalcogenides was carried out in a potassium hydroxide based Norr electrolyte

[7, 8]. Anodizing current density j_{an} was in the range of 2-8 mA/cm^2. The etching duration time t_{an} was 10-20 min. Morphological properties of the PbX samples were studied by high-resolution scanning electron microscopy (SEM) using Zeiss Supra-40, the structural parameters of the porous samples were investigated with the specifically developed x-ray reflectometry and triple crystal diffractometry methods [9].

3. Results and discussions

3.1. *Anodic electrochemical etching of lead selenide films*

Lead selenide films were divided into two groups depending on their initial surface morphology. While some films had a flat surface with 2-3 nm high nanoterraces typical for PbX epitaxial films on Si(111) [10], another group had a developed faceted surface morphology. During anodic electrochemical etching a strong surface modification took place for PbSe films in both cases. For the PbSe/CaF_2/Si(111) structures with an initial flat nanoterraced PbSe surface, anodic etching in a Norr electrolyte resulted in a non-continuous surficial film morphology formation, when interconnected maze-like pattern consisted of structures with 80-400 nm lateral sizes and 90-100 nm thickness. The formation of such porous network structures took place only in the near-surface region of the film and did not depend on the film's entire thickness and anodizing conditions. This allowed us to fabricate isolated PbSe nanoislands directly on the Si substrate for the initial epitaxial films with lower thickness using higher current densities [11]. On the other hand, for the PbSe films with a developed faceted initial surface a developed hierarchical surficial porous network layer formation was observed after anodic etching (figure 2). Characteristics of the porous nanostructured morphology were as following: large open macropores (100-250 nm), smaller secondary macropores (50-75 nm), and mesopores (15-20 nm).

Figure 2. SEM image of the surface of the anodized porous PbSe epitaxial layer (j_{an} = 8 mA/cm^2 and t_{an} = 10 min) with hierarchical porous surficial layer.

The most possible reason for the observed complex pore formation in this case is the non-uniformity of the electrical field for the anodized films with prominent initial faceted surface relief. At applied low current densities the surface is preferably etched at defects and surface irregularities, and facet boundaries serve as nucleation sites for the pores of the nanonetwork. Hence, the mechanism of the modified layer formation, which results in the variety of the obtained PbSe porous structures, is purely electrochemical. Therefore, during subsequent research it is necessary to optimize conditions of the anodic etching process (firstly, by varying the electrolyte solution) for the fabrication of a defined pore morphology in PbSe samples.

3.2. *Anodic electrochemical etching of lead telluride films*

The processes of the modification of PbTe layers during anodic etching were significantly different from previously described PbSe layers. The first unique feature of the anodized $PbSe/CaF_2/Si(111)$ structures was the presence of a domain-like surface morphology with inclined exits of the pores clearly visible on the surface (the initial surface was flat with nanoterraces [10]). The lateral dimensions of the domains were 40-300 nm with an average value of 130 nm and their density was ~1 $\times 10^{10}$ cm^{-2}. Any striking dependence for their geometrical parameters versus the anodizing conditions for the series of PbTe samples was not observed. On the surface of each large domain there were exits of the pores seemingly propagating at a certain angle to the surface due to the (111) orientation of the lead telluride epitaxial films [12]. Cross-sections of the anodized PbTe films obtained with a cleavage of substrate along natural direction [110] displayed uniform arrays of columnar mesopores inclined to the surface at an angle of 35° with a pore diameter of 7-26 nm (figure 3). Remarkably, the porous layer occupies only the top region (150-180 nm) for all the anodized PbTe films. Just like for PbSe, an important feature of the anodic etching of PbTe films was the surficial material removal. Hence, we were able to develop a technique to fabricate completely mesoporous PbTe layers with 150-300 nm thickness on Si [12].

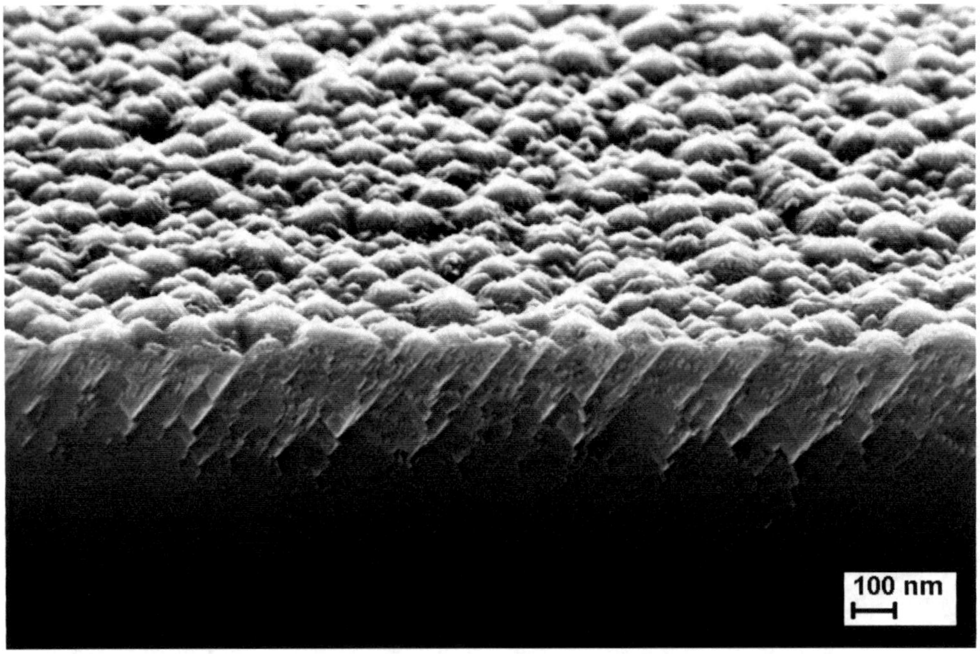

Figure 3. SEM image of the cross-section of the anodized porous PbTe epitaxial layer (j_{an} = 4 mA/cm^2 and t_{an} = 10 min) with 35° inclined mesopores (SEM tilt view at the angle of 70°).

In order to analyze the physics of the specific porous morphology formation in epitaxial (111)-oriented PbTe films we have to contemplate that the characteristic property of the epitaxial lead chalcogenide films on Si(111) is the presence of the threading dislocations and of a large amount of point defects. The process of the fragmentation of the epitaxial films into domains during anodic etching is therefore influenced by these structural inhomogeneities. It is well known [13] that the threading dislocations in lead chalcogenide (111)-oriented films have about the same propagation angle of 36° (angle between [111] and [101] directions). Evaluations showed that the density of the pores is on several orders of magnitude higher than the density of dislocations in the lead telluride films and therefore the observed picture can not be identified with the threading dislocations in the film. It is widely known that in (111)-oriented semiconductors preferential pore propagation corresponds to two main directions: <100> or <113> [14]. In the former case the inclination angle from the surface normal of the pore array is 35°, while in the latter case – 29°. Considering the close obtained value of the pore growth vector angle for the studied PbTe anodized films, we can assume that the crystallographic <100> directions play a determining role in the pore propagation in PbTe films on Si(111) substrates.

Average pore sizes in a lateral and vertical directions calculated from a triple crystal x-ray diffractometry measurements [9] were in a perfect correspondence with the SEM data. The porosity values (P) calculated using high-resolution x-ray reflectometry in the angle interval of total external reflection for anodized PbTe films were P = 41-68% with the highest porosity values corresponding to the longer anodizing durations [9, 12]. It is very important that using our etching technique the high porosity values can be relatively easily achieved.

In order to investigate the modification of the optical properties of PbTe films during pore formation we have measured the transmission spectra of PbTe/CaF_2/Si(111) structures before and after electrochemical treatment using Perkin Elmer 16PC Fourier transform spectrometer in the wavelength range of 2.2-16.7 micron. Before and after anodic etching the spectra had the classic form of interference associated with the phenomenon of interference in the film system. As a result of the anodic electrochemical etching of the PbTe films with high porosity P = 65% an increase of energy gap value from 0.31 to 0.34 eV took place. This result indicates the formation of low-dimensional structures in the sample. To estimate the size of the nanoparticles the hyperbolic model [15] was used, which has recently been widely applied for the PbTe nanoparticles of spherical shape. The exact form of nanoparticles in our case is unknown, but we would assume that PbTe nanocrystallites have a spherical shape, as is generally the case for other mesoporous semiconductors. For the anodized PbTe material with the nanoparticles with a spherical shape the energy gap E_g^{an} can be calculated according to the following equation:

$$E_g^{an} = \sqrt{E_g^2 + E_g \frac{\pi h}{m^* R^2}}, \tag{1}$$

where E_g is the band gap of bulk monocrystalline PbTe, h – the Planck constant, R – average radius of a nanoparticle, m^* – effective exciton mass. According to the isotropic model [16]

$$\left.\begin{aligned} \frac{1}{m_e^*} &= \frac{1}{3}\left(\frac{2}{m_{et}} + \frac{1}{m_{el}}\right) \\ \frac{1}{m_h^*} &= \frac{1}{3}\left(\frac{2}{m_{ht}} + \frac{1}{m_{hl}}\right) \end{aligned}\right\} \tag{2}$$

where m_{et} and m_{ht} are transverse and m_{el} and m_{hl} are longitudinal effective masses of electrons and holes in PbTe. Using data from [16], at 300 K m_0/m_{et} = 11.6, m_0/m_{ht} = 10, m_0/m_{el} = 1.2, and m_0/m_{hl} = 0.7, where m_0 is electron rest mass. Using equation (1), we have determined that for E_g^{an} = 0.34 eV, the radius of nanoparticles is 13 nm. Therefore, in nanostructured porous PbTe films with high porosity, the conditions of strong quantum confinement effects take place, when the particle size becomes smaller than the exciton Bohr radius (50 nm for PbTe). These results indicate that it is extremely promising to apply the method of nanostructured porous lead telluride layers fabrication for the practical aims. The nanoscale morphology is known to result in a strong phonon scattering and consecutive *ZT* gain, which is why PbTe porous layers with high porosity can be used for thermoelectric devices. An important positive attribute of our results is the capability of PbX modified layers formation directly on silicon substrates, providing an effective way for the fabrication of Si chip integrated devices. The mesopore propagation angle of 35°, on the other hand, is promising for the fabrication of photonic crystals with Yablonovite-like structure.

4. Conclusions

In summary, porous nanostructured lead telluride and lead selenide layers on silicon substrates were fabricated. SEM investigations showed the formation of material specific porous morphology. For PbSe layers two types of modified layers were distinguished, one of them being a complex hierarchical porous layer. For PbTe layers inclined mesopore ensemble formation was observed. Energy gap value increase for lead telluride with high porosity, determined from transmission spectra, allowed us to calculate the nanocrystallite radius confirming the presence of quantum size effects in nanostructured samples.

Acknowledgments

Authors are thankful to Vasin V M (Yaroslavl State University) and Petrakov A P (Syktyvkar State University) for their valuable contributions. This work was performed with the financial support of the "Scientific potential development of higher school" Program (projects 2.1.1/466 and 2.1.1/13083).

References

[1] Zimin S P, Gorlachev E S, Amirov I I and Zogg H 2009 *J. Phys. D: Appl. Phys.* **42** 165205
[2] Rahim M, Khiar A, Felder F, Fill M and Zogg H 2009 *Appl. Phys. Lett.* **94** 201112
[3] Heremans J P, Jovovic V, Toberer E S, Saramat A, Kurosaki K, Charoenphakdee A, Yamanaka S and Snyder G J 2008 *Science* **321** 554
[4] Cao Y Q, Zhu T J and Zhao X B 2009 *J. Phys. D: Appl. Phys.* **42** 015406
[5] Harman T C, Taylor P J, Walsh M P and LaForge B E 2002 *Science* **297** 2229
[6] Dmitriev A V and Zvyagin I P 2010 *Phys.-Usp.* **53** 789
[7] Zimin S P, Bogoyavlenskaya E A, Buchin E Yu, Petrakov A P, Zogg H and Zimin D 2009 *Semicond. Sci. Technol.* **24** 105008
[8] Norr M K 1962 *J. Electrochem. Soc.* **109** 433
[9] Zimin S P, Vasin V M, Gorlachev E S, Petrakov A P and Shilov S V 2010 *Proc. SPIE* **7521** 752114
[10] Zimin S P, Gorlachev E S, Amirov I I, Gerke M N, Zogg H and Zimin D 2007 *Semicond. Sci. Technol.* **22** 929
[11] Zimin S P, Vasin V M, Gorlachev E S, Buchin E Yu and Naumov V V 2011 *Phys. Stat. Sol. (c)* (in print)
[12] Zimin S P, Vasin V M, Gorlachev E S, Naumov V V, Petrakov A P and Shilov S V 2011 *Phys. Stat. Sol. (c)* (in print)
[13] Zogg H, Maissen C, Blunier S, Teodoropol S, Overney R M, Richmond T and Tomm J W 1993 *Semicond. Sci. Technol.* **8** S337
[14] Christophersen M, Carstensen J, Rönnebeck S, Jäger C, Jäger W and Föll H 2001 *J. Electrochem. Soc.* **148** E267
[15] Wang Y, Suna A, Mahler W and Kasovski R 1987 *J. Chem. Phys.* **84** 7315
[16] Tudury G E, Marquezini M V, Ferreira L G, Barbosa L C and Cesar C L 2000 *Phys. Rev. B* **62** 7357

Journal of Physics: Conference Series **291** (2011) 012024 doi:10.1088/1742-6596/291/1/012024

Pulse-reverse treatment of carbon nanomaterials for the electrochemical storage of energy

V V Zhylinski[1], V B Drozdovich[1], A V Krauklis[2], S A Zhdanok[2] and I M Zharski[1]

[1]Belarusian State Technological University, Minsk, Sverdlov Str., 13a, 220006, Belarus
[2]A V Lyekov Heat and Mass Transfer Institute of NAS Belarus, Minsk, P. Brovki Str., 15, 220072, Belarus

E-mail: zhilinski@yandex.ru, elchemdvb@mail.ru

Abstract. The paper dwells upon the investigation of carbon nanomaterials synthesized by an electric arc-method from methane-air mixture under atmospheric pressure in presence of Ni catalyst. These materials may be used for the electrochemical storage of energy. According to the data of transmission electron microscopy the carbon nanomaterials are generally represented by carbon nanotubes with the diameter of 20-80 nm. The pulse-reverse treatment of carbon nanomaterials in concentrated sulphuric electrolytes with addition of fluoride and lithium ions is carried out to increase their discharge capacity in 1 M H_2SO_4 and 1 M KOH. The properties of carbon nanomaterials are studied using the Brunauer-Emmett-Teller method, chemical energy-dispersive X-ray analysis, potentiostatic charge method of electrochemical saturation of hydrogen and potentiodynamic discharge method. The maximum discharge capacity of 940 C g^{-1} in 1 M H_2SO_4 is revealed for CNMs subjected to pulse-reverse treatment in concentrated sulphuric electrolyte containing fluoride and lithium ions.

1. Introduction

Carbon nanomaterials (CNM), their production, purification and certification are intensively investigated [1–6] to be used for electrochemical storage of energy. Their unique electrical, thermal, and chemical properties of these materials are due to the carbon nanotubes that usually constitute big part of CNMs. Individual carbon tubes have properties of both microparticle and single nanosize molecule and belong to the class of quasi-one-dimensional nanoscale objects [1]. Especially, the carbon nanotubes have a large specific surface, high electrical conductivity and relative chemical inertness. These materials are potential additives to the electrodes of lithium-ionic and lithium-polymeric accumulators and as electrode materials of electrochemical supercapacitors and hydrogen electric accumulators. However they may be used only as nanostructured bulk CNMs in which carbon nanotubes can potentially outperform other forms of carbon (amorphous carbon, parts of graphite etc.) [2-6].

The traditional synthesis of CNMs such as electric arc discharge, laser ablation, thermal and plasma-enhanced chemical deposition generally give carbon powders with significant amount of amorphous carbon and metal-catalyst impurity [2–4]. The carbon nanotubes can be purified by high-temperature annealing, plasma treatment and chemical methods. In most cases purification of carbon nanotubes lead to an uncontrolled destruction of carbon nanotubes and assisted oxidation. Moreover, the oxidation is accompanied by the formation of a large number of carboxylic, hydroxyl, ketone groups [1–7] that can substantially change the change chemical properties of CNMs.

An alternative method of purification is electrochemical treatment of CNMs which makes it possible to control over oxidation-reduction processes of carbon nanotubes [2-5, 7]. Moreover, many investigations are aimed at improving of these treatments by injecting intercalation ions (Li^+, Na^+ etc.) and activators of electrode oxidation processes [3–7].

One of these methods is the potentiostatic treatment of CNMs under potential exceeding thermodynamic potential of oxygen evolution in the concentrated sulphuric acid electrolytes [7]. This treatment of CNMs can significantly increase the index of electrochemical hydrogen sorption in 1 M H_2SO_4 and 1 M KOH. This effect is attained by the oxidation of layered structures of CNMs under the potentiostatic polarization in concentrated sulphuric-acid electrolyte. It has been established [7] that the electrochemical hydrogen sorption in CNMs increases due to pulse-reverse treatment of CNMs using potentiostatic pulse-reverse modes with a long anodic pulse and shot cathodic pulse at potentials which exceed thermodynamic potential of oxygen and hydrogen evolution respectively. This article focuses on investigation of pulse-reverse CNMs treatment in concentrated sulphuric-acid electrolyte containing fluoride and lithium ions.

2. Materials and Methods

We investigate the CNMs synthesized from methane-air mixture by the electric-arc method at atmospheric pressure in the presence of the Ni catalyst. The aqueous suspension (0.1 wt.%) of synthesized carbon masses undergoes ultrasonic treatment to separate them [8]. The fraction of CNMs is observed on the surface of the suspension after ultrasonic treatment at specific power of 7 kJ ml^{-1} and frequency of 20 kilohertz. It can be seen (figure 1) that this fraction of CNMs consists of carbon nanotubes 20-60 nm in diameter and small fibrils whereas the initial carbon material mainly contains large particles of amorphous carbon and graphite (figure 1).

According to the data of the TEM investigation, the CNMs obtained by the ultrasonic treatment are black powder with bulk density of 0.45 g cm^{-3} consisting of amorphous carbon (less than 30 %), carbon fibrils 100-1000 nm in diameter and 1-15 μm in length and carbon nanotubes 20-80 nm in diameter and 1-10 μm in length [8].

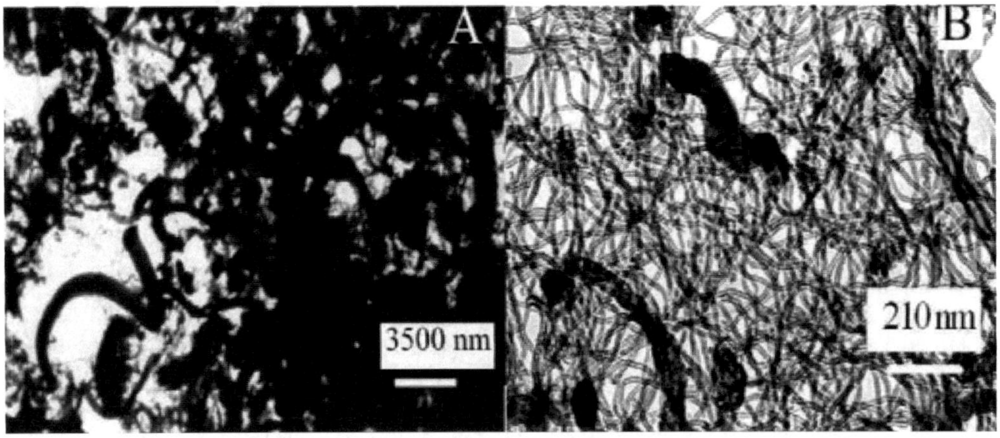

Figure 1. TEM images of the initial carbon material (a) synthesized from methane-air mixture by electric-arc method and the CNMs (b) obtained by ultrasonic treatment.

The electrochemical treatment of CNMs in the concentrated sulphuric acid (96 wt.%) containing lithium and fluorine ions is carried out at the temperature of 298 K. The concentration of Li^+ and F^- ions is varied from 2 to 20 g l^{-1}. The electrochemical treatment is carried out in potentiostatic and pulse-reverse modes [8]. Table 1 displays other parameters of electrochemical treatment of CNMs. After electrochemical treatment the CNMs are washed by distilled water.

Table 1. Parameters of electrochemical treatment of CNMs.

Current mode	Anode potential (V)	Anode pause (s)	Cathode potential (V)	Anode pause (s)	Time (h)
Potentiostatic	+3.5[a]	–	–	–	1
Pulse-reverse	+3.5[a]	20	–1.3[a]	10	1

[a] Potentials are referenced by the normal hydrogen electrode at temperature of 298 K.

Charge-discharge characteristics of initial and electrochemically treated CNMs were investigated in 1 M H_2SO_4 and 1 M KOH. The charge was carried out in potentiostatic mode under –0.2 V and –1.0 V in 1 M H_2SO_4 and 1 M KOH, respectively. The discharge was carried out in potentiodynamic mode, the potential rate being 0.01 V s^{-1} using the voltammetric system IPC Pro. The absorbance infrared spectra of solid CNMs and KBr mixture are detected by NexusTM ESP (Thermo Nicolet). The mass ratio of CNMs and KBr is 1:500. The specific surface of CNMs was determined by the nitrogen BET method using NOVA 2200. The chemical energy-dispersion X-ray analysis of CNMs was carried out by the system EDX JED-2201 JEOL.

3. Results and Discussion

The electrochemical properties and discharge capacity of CNMs were estimated by discharge characteristics (figure 2). It was established that the discharge currents of the initial CNMs reach 1.2 A g^{-1} and 1.0 A g^{-1} in 1 M H_2SO_4 and 1 M KOH, respectively. The discharge capacity was automatically calculated by the voltammetric system IPC Pro is 150 C g^{-1}. This capacity of CNMs can be accumulated by the formation of double electric layer and sorption of hydrogen on a large specific surface which was 140 m^2 g^{-1} according to the nitrogen BET method.

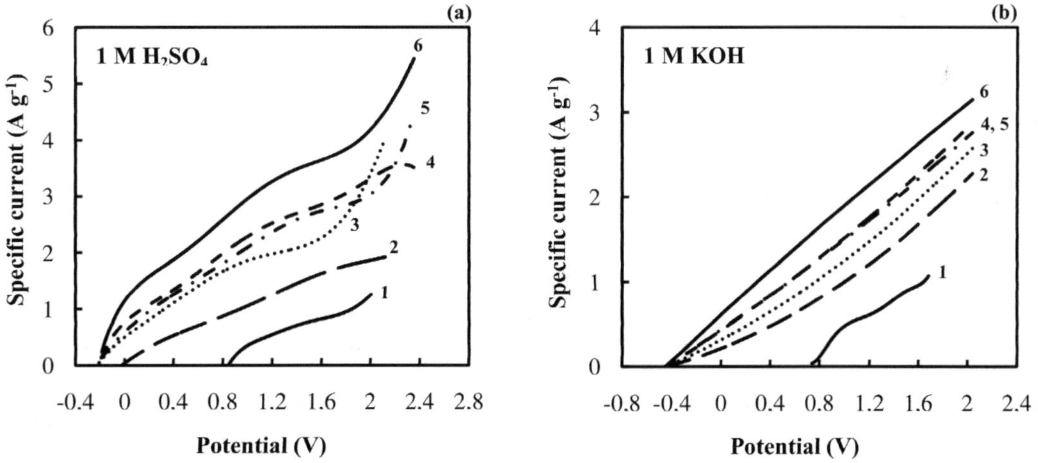

Figure 2. (a, b) Discharge potentiodynamic curves of (1) initial CNMs, (2) CNMs after potentiostatic treatment in concentrated H_2SO_4, (3–6) CNMs after pulse-reverse treatment (3) in concentrated H_2SO_4, containing (4) fluoride ions, (5) lithium ions, and containing (6) lithium and fluoride ions.

The pulse-reverse treatment of CNMs in concentrated sulphuric acid resulted in some oxidation of carbon materials [7]. According to the energy-dispersion X-ray analysis (table 2), the mass fraction of carbon decreased from 87 to 78.5 wt.% while the fraction of oxygen increased from 6 to 20.4 wt.% after pulse-reverse treatment in concentrated sulphuric acid. In addition, the presence of the sulfur fraction indicated active intercalation of CNMs in the solution of H_2SO_4. This effect and increasing oxygen fraction may lead to increase in electrochemical activity of treated CNMs in 1 M H_2SO_4 and 1 M KOH (figure 2).

Table 2. Results of chemical energy-dispersion X-ray analysis of the CNMs after pulse-reverse treatment in concentrated sulphuric acid.

Element	Energy (keV)	W (wt.%)
C	0.294	78.5±0.3
O	0.582	20.4±4.0
S	2.307	0.6±0.3
Other elements (K, Na)	–	0.3

Types of oxygen-containing groups can be qualitatively analyzed using absorbance infrared spectra of CNMs. Marked peaks are observed at 2923 cm^{-1} and 2850 cm^{-1} that corresponds to the fluctuations of C-C and C-OH bonds (figure 3).

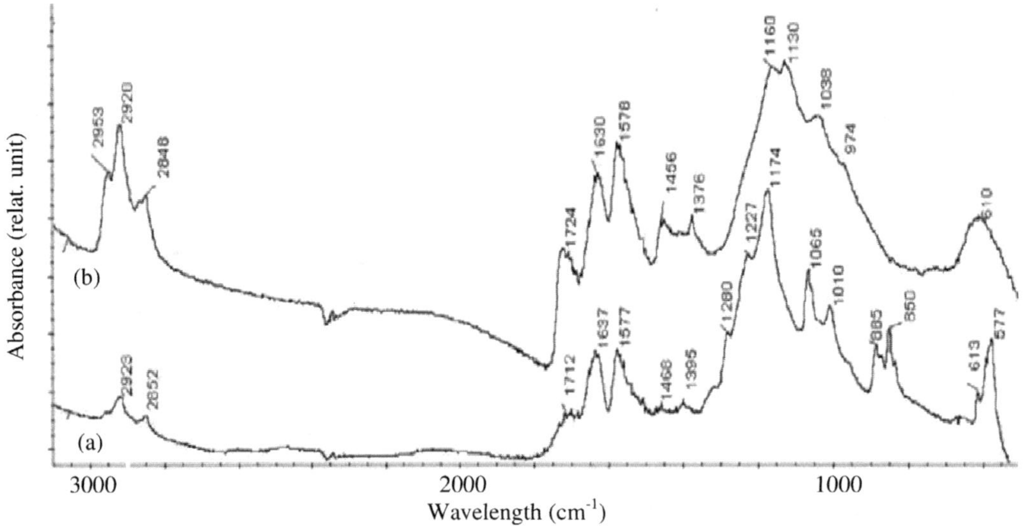

Figure 3. Infrared absorbance spectra of initial CNMs (a) and CNMs (b) after pulse-reverse treatment in concentrated sulphuric acid.

The peak at 1716 cm^{-1} corresponds to C=O bond which is characteristic of ketonic, carboxylic and lacton groups. They can catalyze processes of electrochemical sorption and reduction of hydrogen [9]. Marked peaks at 1578 cm^{-1} and 890 cm^{-1} are caused by the deformations of aromatic structure. The maximal peak at 1200 cm^{-1} corresponds to oxygen of phenol type. The initial CNMs in contrast to the treated ones are characterized by the peak at 1630 cm^{-1} due to the presence of significant amount of C=C bond. Thus treated CNMs demonstrate more clearly marked peaks of C=O and C-OH bonds what indicates an increased number of oxygen-containing groups after pulse-reverse treatment in concentrated sulphuric acid.

The additive of 8 g l^{-1} Li^+ in concentrated sulphuric electrolyte of electrochemical treatment corresponds to higher discharge capacity of treated CNMs which achieved 640 and 530 C g^{-1} in 1 M H_2SO_4 and 1 M KOH, respectively (table 3).

Table 3. Influence of ion additions in concentrated sulphuric acid on specific capacity of treated CNMs.

		Specific capacity of CNMs (C g^{-1})	
Type of addition	Concentration of ion (g l^{-1})	in 1 M H_2SO_4	in 1 M KOH
Li^+	2.0	510	385
	4.0	580	415
	6.0	590	460
	8.0	640	530
	10.0	580	480
F^-	4.5	530	280
	9.0	580	290
	13.5	600	305
	18.0	590	300
	22.5	585	300

In this case the discharge currents reached 6.5 A g^{-1} and 3.0 A g^{-1} in 1 M H_2SO_4 and 1 M KOH respectively (figure 2). This influence of the Li^+ additive was obviously caused by active diffusion and electromigration of Li^+ in CNMs structures during the cathode polarization. The lithium ions are known [10] to promote intercalation of interlaminar spaces of CNMs and increase of amount of CNMs defects. This order-disorder transformation can lead to the formation of turbo-stratified carbon nets of AB or ABC types [10].

The additive of 13.5 g l^{-1} F^- in concentrated sulphuric acid resulted in better efficiency of electrochemical treatment that corresponded to a higher value of discharge capacity of CNMs activated by this method in comparison with other values of discharge capacity of CNMs activated at other concentrations of F^- (table 3). This influence of the fluoride anions could be caused by their specific adsorption on the anode. They partially block the oxygen evolution that can cause some processes such as ozone evolution, synthesis of peroxysulphate acids and peroxide carbon groups [11].

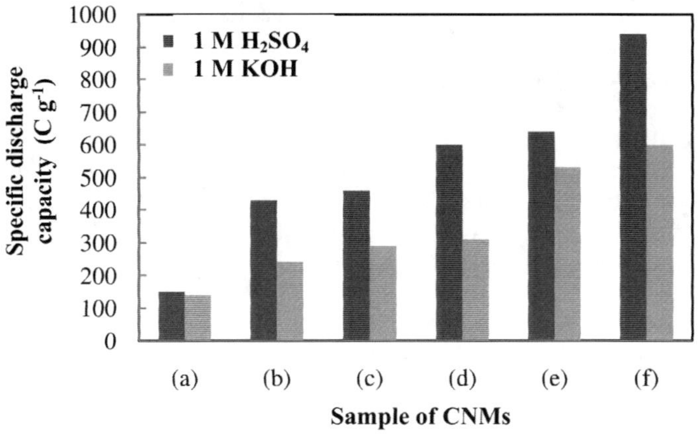

Figure 4. Specific discharge capacity of (a) initial CNMs, (b) CNMs after potentiostatic treatment in concentrated H_2SO_4, (c-f) CNMs after pulse-reverse treatment (c) in concentrated H_2SO_4, containing (d) fluoride ions, (e) lithium ions, and containing (f) lithium and fluoride ions.

The maximum discharge currents in 1 M H_2SO_4 and 1 M KOH are reached for the CNMs after pulse-reverse treatment in concentrated sulphuric electrolyte wich contains 8 g l^{-1} Li^+ and 13.5 g l^{-1}

fluoride ions (figure 2). The discharge current of these CNMs under the potential of oxygen electrode is more than 3 times higher than that of the initial CNMs in 1 M H_2SO_4 and 1 M KOH. The maximum capacity of activated CNMs reaches 940 C g^{-1} in 1 M H_2SO_4 and 600 C g^{-1} in 1 M KOH (figure 4). This effect may be caused by the complex influence of fluoride and lithium ions on the efficiency of pulse-reverse treatment of CNMs in concentrated sulphuric acid. The fluoride ions activate oxidation and dispersion of CNMs under anode polarization. The lithium ions may lead to increasing number of CNMs defects by the processes of intercalation and de-intercalation of interlaminar spaces. As a result we can observe increased discharge capacity of CNMs subjected to pulse-reverse treatment in concentrated sulphuric electrolyte containing fluoride and lithium ions.

4. Conclusions

The electrochemical treatment of CNMs in the pulse-reverse mode results in intercalation of carbon materials by H_2SO_4 and oxidation that is indicated by increase of oxygen and sulfur contents in the treated CNMs. This oxidation is accompanied with the formation of phenol, ketonic, carboxylic and lacton groups. Moreover, the discharge capacity of the treated CNMs increases in 4 times as higher as discharge capacity of initial CNMs. The introduction of lithium or fluoride ions in concentrated sulphuric acid increases the discharge capacity of the treated CNMs. However, the maximum capacity of 940 C g^{-1} in 1 M H_2SO_4 is achieved in the CNMs subjected to pulse-reverse treatment when concentrated sulphuric electrolyte contains 8 g l^{-1} Li^+ and 13.5 g l^{-1} F^- ions. This effect is caused by the complex influence of fluoride and lithium ions. It may lead to morphologic changes of CNMs and increase of defects. These changes are made by Li^+ intercalation of layered structures of CNMs and formation of oxygen-containing groups in the presence of F^- ions.

Thus the CNMs treated by this method are potential materials for the electrochemical storage of energy. They may be used for electrochemical supercapacitor, hydrogen electric accumulator.

References

[1] Yun Y-H, Dong Zh, Shanov V, Heineman W R, Brian Halsall H, Bhattacharya A, Conforti L, Narayan R K, Ball W S and Schulz M J 2007 *Nanotoday* **2** 30.
[2] Dai K, Zhang D-S, Yu B-K, Fang J-H and Shi L-Y 2006 *J. Shanghai Univ.* **2** 557.
[3] Liua Ch G, Fangb H T, Lia F, Liua M and Chenga H M 2006 *J. Power Sources* **160** 758.
[4] Lin Ch-Ch and Hsien-Chieh H 2009 *J. Power Sources* **188** 332.
[5] Wang Q, Li Ch, Yuan G and Gu Ch-Zh 2009 *Chinese Phys. Lett.* **26** 86.
[6] Kalb M, Kavan L and Dunsch L 2009 *J. Am. Chem. Soc.* **131** 4529.
[7] Zhylinski V V, Drozdovich V B, Zhdanok S A and Krauklis A V 2010 Possibility of using activated CNM in electrochemical energy accumulation devices *Electrochemical Technologies and Materials for XXI Century: Abstracts of 9th Int. Frumkin Symp. (Moscow, Russia, 25–29 October 2010) (Institut of Phisical Chemistry and Electrochemistry)* ed A Yu Tsivadze (Moscow: IPCE) p 145.
[8] Buyakov I F, Volzhankin V M, Krauklis A V, Samtsov P P and Solntsev A P 2005 Formation of Carbon Nanomaterials Under the Conditions Formed by Atmospheric Pressure High-Voltage Plasma *Int. Colloquium on Physics of Shock Waves, Combustion, Detonation and Non-Equilibrium Processes (Minsk, Belarus, 14–19 November 2005) (A V Lykov Heat and Mass Transfer Institute)* ed S A Zhdanok (Minsk: MIC) pp 54-56.
[9] Bandosza T J and Aniaa C O 2006 *Interface Science and Technology* **7** 159.
[10] Yang Zh-H and Wu H-Q 2001 *Solid State Ionics* **143** 173.
[11] Stein A, Wang Zh and Fierke M A 2009 *Adv. Mater.* **21** 265.

Peculiarities of TiNi-based shape memory alloys local crystalline structure

A A Yaroslavtsev[1], A P Menushenkov[1], R V Chernikov[1], O V Grishina[1], Y V Zubavichus[2], A A Veligzhanin[2], A V Shelyakov[1] and N N Sitnikov[1]

[1] National Research Nuclear University "Moscow Engineering Physics Institute", Kashirskoe sh., 31, Moscow 115409, Russia

[2] Russian Research Centre "Kurchatov Institute", Academic Kurchatov sq., 1, Moscow 123182, Russia

E-mail: yalex03@gmail.com

Abstract. The extended X-ray absorption fine structure spectroscopy (EXAFS) was applied to investigate the local structure peculiarities of TiNi-based shape memory alloys $Ti_{50}Ni_{25}Cu_{25}$ and $Ti_{39.2}Ni_{24.8}Cu_{25}Hf_{11}$. The phase composition of ternary alloy was examined with the additional X-ray diffraction. Our experimental results demonstrate that the most significant changes in local crystalline structure under martensitic transformation arise in the Ni-Cu sublattice. The static disordering of the Ti coordination shell is more significant for Ni than for Cu environment. Such a conclusion is consistent with the refined values of atomic ν-shifts from the planar center of symmetry in (010) layers which are greater for Ni atoms than for Cu.

1. Introduction

TiNi-based alloys are some of the most widespread shape memory alloys (SMA). While being heated they are able to fully restore their pre-deformation shape as a result of reverse martensitic transformation (MT). Under such a behavior these alloys demonstrate considerable mechanical force. Due to their unique properties SMAs are of great interest to find the unusual ways in development of various important technical applications, such as electronics, power and machine engineering, instrument-making, space industry, medicine etc. [1, 2, 3, 4]. However, the demand of micro- and nanodevices construction drives production of new types of SMAs as well as their further exploration on different scales.

Direct and reverse MTs occur in their own temperature ranges, therefore the material demonstrates hysteretic properties under phase transformation. Characteristic temperatures of MT depend on both the alloy grade and its chemical composition. The slight composition changes (intentional or as the result of flaw) lead to the shift of these parameters. The transformation properties undoubtedly have the connection with the fine effects taking place in the local crystalline structure upon cooling and heating the alloys. Therefore, SMA are also important for the understanding of chemical bonding and phase transitions in the intermetallics. In present work we made an attempt to investigate the peculiarities of Ni and Cu local environment in ternary $Ti_{50}Ni_{25}Cu_{25}$ (MT temp. region 43 – 65°C) and quaternary $Ti_{39.2}Ni_{24.8}Cu_{25}Hf_{11}$ (MT temp. region 168 – 200K) SMAs by means of extended X-ray absorption fine structure spectroscopy (EXAFS). Both alloys crystallize in cubic B2 structure

Journal of Physics: Conference Series **291** (2011) 012025 doi:10.1088/1742-6596/291/1/012025

in austenitic phase and orthorhombic B19 in martensitic phase [3, 5]. Such properties as the interatomic distances, local disordering Debye-Waller factors, atomic shifts to define the character of distortion in unit cell etc. were examined.

2. Experimental

In initial state the samples were thin bands obtained by components melt extrusion from the quartz crucible through the thin nozzle on the surface of rotating copper disk, where the sample's cooling took place with the rate of 10^6K/sec. The thickness of the initial amorphous bands was $\sim 40 - 45\mu$m. Then samples were crystallized by izotropic annealing in the air at $\sim 500°$C for 4 min. Finally the electron microscope TESLA BS-540 was used to estimate the size of crystalline grains which turned out to be $300 \div 500$nm [6].

The EXAFS spectra were collected above K-Ni (8333eV) and K-Cu (8979eV) absorption edges. Series of measures were carried out upon consistent sample heating and cooling through the temperature regions of martensitic and austenitic phase transformations. Spectra of quaternary $Ti_{39.2}Ni_{24.8}Cu_{25}Hf_{11}$ alloy were collected at E4 beamline of HASYLAB (DESY, Hamburg, Germany) in transmission mode. Energy resolution of the double-crystal Si (111) monochromator (detuned to reject 50% of the incident signal in order to minimize harmonic contamination) with a 0.3mm slit at 7keV was about 1.2eV. Low-temperature measurements were carried out using a liquid helium continuous flow cryostat with a temperature control of $\pm$1K at 300K and 0.1K at 5K. Spectra of ternary alloy $Ti_{50}Ni_{25}Cu_{25}$ were collected at station "Structural Materials Science" of KCSR and NT (Russia, Moscow) in transmission mode. The sample was fixed in the *in situ* chamber and did not move during all the time of heating and cooling. The processing of spectra was performed using VIPER [7] software with standard procedure of EXAFS-oscillations function $\chi(k)$ deconvolution and Fourier-analysis. Magnitudes and phases of back scattering were computed using the FEFF-8.20 code [8] within a model of martensitic phase crystalline structure, $Pmmb$ symmetry group.

Additional X-ray diffraction (XRD) was carried out at STM station at KCSR and NT to examine the phase composition of ternary alloy $Ti_{50}Ni_{25}Cu_{25}$. In order to eliminate the resonance effects the wavelength of 1.54Å (Cu-$K\alpha$) was used, which is far from the K-Ni and K-Cu absorption edges. XRD patterns were measured with the linear position-sensitive detector, situated at the exit window of the *in situ* chamber. The study was carried out in transmission geometry using Debye-Scherrer method in order to examine the bulk structure of alloy.

3. Results and discussion

Fig. 1 represents two characteristic experimental XRD patterns at room temperature and 80°C as well as theoretically calculated for the B2, B19 and B19' type structures. The experimental intensities are nomalized at the integrated K-Ti fluorescence measured with the detector set in the *in situ* chamber. There is a pronounced difference between the present experimental XRD patterns. The one at room temperature demonstrates the series of B19 structure Bragg reflections, while the other at 80°C coincides with the pattern calculated for the B2 structure. Bragg reflections of B19' structure aren't visible in any of two experimental patterns. Thus, XRD results demonstrate that $Ti_{50}Ni_{25}Cu_{25}$ alloy really has *B2 type structure in austenitic phase* and *B19 in martensitic phase*. No direct signs of frequently noted monoclinic B19' structure [9, 10] appear in this alloy. Up to $\sim$ 80% of material undergo the phase transformation which occur in rather narrow temperature range $50 - 55°$C. By analyzing the width of the peaks at the XRD patterns the size of crystallites was estimated: $\sim$ 32nm in the austenitic phase and $\sim$ 20nm in martensitic phase. Besides, the position of large peak at the 80°C XRD pattern $2\theta \sim 41.7°$ allowed us to refine the unit cell parameter of the cubic B2 phase $a \sim 3.060$Å. This value was useful in the subsequent EXAFS study.

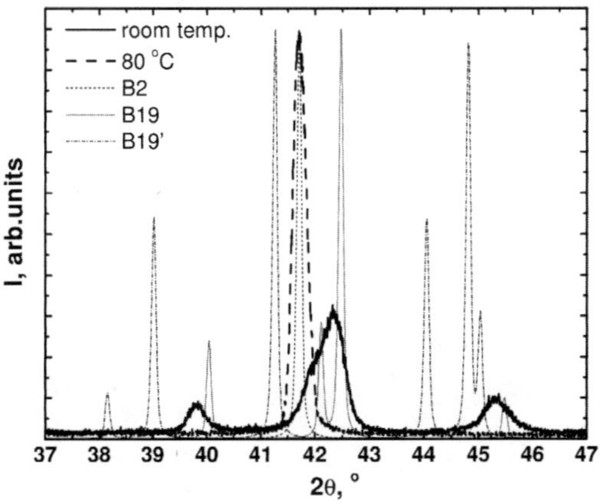

Figure 1. Comparision of the experimental XRD patterns of $Ti_{50}Ni_{25}Cu_{25}$ alloy and theoretically calculated for the crystalline phases possible in this system.

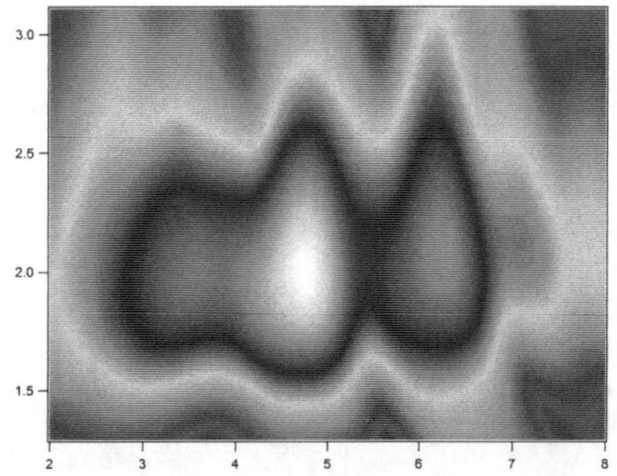

Figure 2. Wavelet transform of K-Ni-EXAFS-functions for the $Ti_{50}Ni_{25}Cu_{25}$ alloy at 30°C. Wave vector k, Å^{-1} is set along the horizontal axis; distance R, Å is set along the vertical axis.

Qualitative analysis of spectra was performed using wavelet transform of EXAFS oscillation functions $\chi(k)$ with IGOR PRO (WaveMetrics, Lake Oswego, OR, USA). Results of wavelet transform for the $Ti_{50}Ni_{25}Cu_{25}$ alloy K-Ni spectrum are shown in Fig. 2. They demonstrate three clear resolved maxima, i.e. three scattering acts are registered on different elements situated in approximately same distances from the absorbing atom. Thus, the closest local environment of Ni and Cu atoms consists of 3 atom types.

Ni and Cu atoms occupy equivalent positions in lattice. In ternary alloy $Ti_{50}Ni_{25}Cu_{25}$ the closest coordination shell contain 8 Ti atoms and 6 atoms which can be Ni or Cu (in our study we implied there were 3 Ni and 3 Cu atoms in these shells). Parameters of Ni and Cu atoms local environment in ternary alloy were investigated in temperature range 30 – 70°C. EXAFS-functions of spectra above K-Ni edge at the temperature of martensitic phase 30°C, austenitic phase 70°C, and in the region of phase transition at 58°C are shown in Fig. 3. There is a notable difference in the region 4.2 – 4.8Å^{-1}, where two oscillations maxima are situated. Amplitudes ratio of these maxima changes with temperature, and besides, their shape at 58°C is not the same under heating and cooling. Modules of Fourier transforms (FT) of EXAFS-functions above K-Ni absorption edge at various temperatures are shown in Fig. 4a. The first large peak corresponds to the first Ti shell and doesn't change much with temperature. Whereas several smaller peaks in the region 3.5 – 5Å beyond the first one change significantly and their shape depends on circumstances the spectra was measured under. To support current observation the back Fourier transform (BFT) of FT region is also resented in Fig. 4b. It is clearly seen that in case of sample heating at 58°C the oscillation shape is close to one at 30°C. On the contrary, under cooling it is much similar with the shape at 70°C, hence we can consider the change of EXAFS line shape hysteretic. Discussed FT region beyond the first Ti shell peak corresponds to the variety of shells containing Ni and Cu atoms, few shells of Ti atoms in the neighbouring cells and the multiscattering path Ni-Ti-Ti. However, magnitude of latter is rather low and the Ni-Ti bonds beyond one unit cell are more relevant to long-range order that obviously changes with the symmetry upon phase transformation, than to the local structure. Therefore, we suppose that the general reversible changes namely in the local crystalline structure upon MT occur in the sublattice of Ni and Cu atoms.

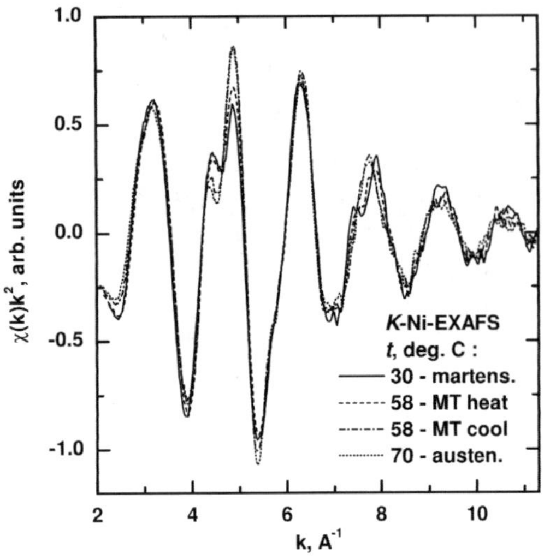

Figure 3. K-Ni-EXAFS-functions for the $Ti_{50}Ni_{25}Cu_{25}$ alloy at various temperatures.

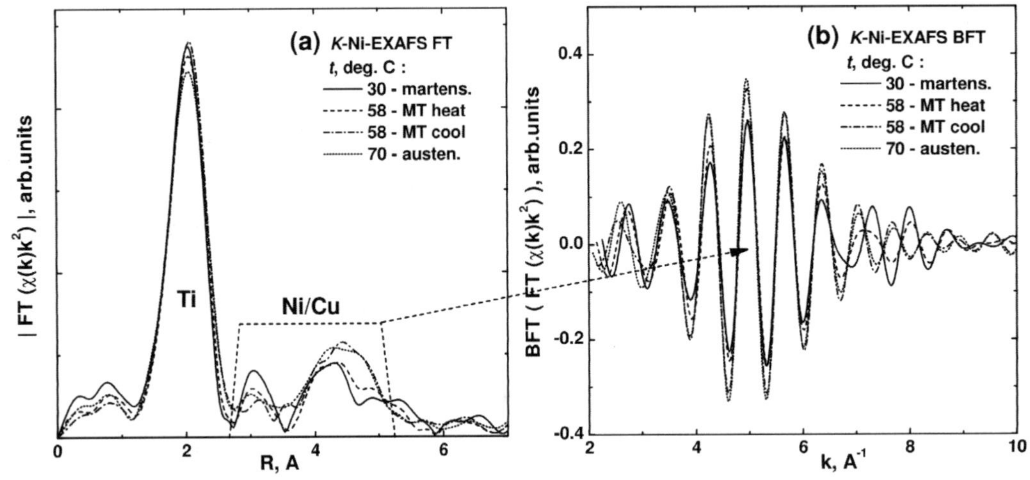

Figure 4. (a) Fourier transforms of K-Ni-EXAFS-functions for the $Ti_{50}Ni_{25}Cu_{25}$ alloy at various temperatures; (b) Back Fourier transforms of K-Ni-EXAFS-functions FT region beyond Ti-shell peak.

EXAFS-functions fitting reveal that the interatomic distances Ni-Ti and Cu-Ti don't change much with temperature, but the average degree of their local disordering expressed by the Debye-Waller factor σ^2 increases. Besides, Ti atoms in the Ni local environment are characterized by considerably higher static disordering than in the Cu environment. Situation with the Ni-Ni, Ni-Cu and Cu-Cu distances is more complicated. Their average values at room temperature are $R_{Ni-Ni} \sim 2.80$Å, $R_{Ni-Cu} \sim 3.00$Å, $R_{Cu-Cu} \sim 3.05$Å and reversibly increase by $0.04 - 0.05$Å upon heating and cooling. The character of this change is approximately the same if we consider Ni or Cu local environment. Disordering of these distances also increases with temperature and within the error of technique employed might be considered the similar for heating and cooling. It is notable, that changes upon reversible MT in short-range order of our SMA crystalline structure occur gradually in compare with drastic changes in long-range

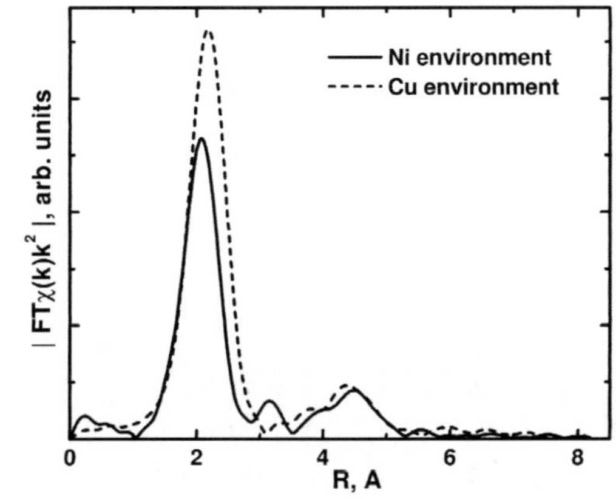

Figure 5. Modules of K-Ni and K-Cu EXAFS-functions Fourier transforms for the $Ti_{39.2}Ni_{24.8}Cu_{25}Hf_{11}$ alloy at temperature 5K.

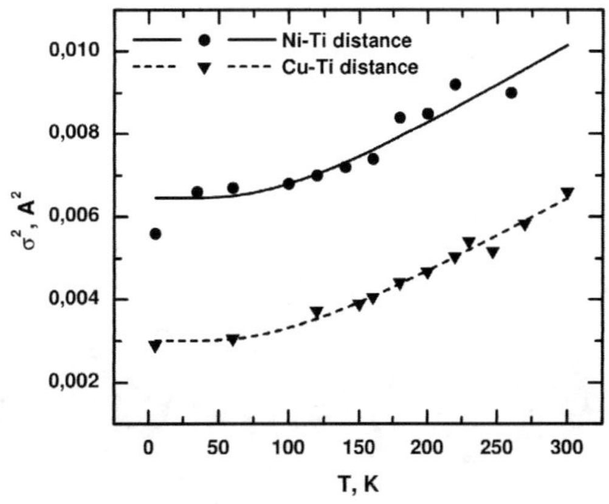

Figure 6. Temperature dependences of Ni-Ti and Cu-Ti interatomic distances local disordering Debye-Waller factors in the $Ti_{39.2}Ni_{24.8}Cu_{25}Hf_{11}$ alloy.

order examined by the X-ray diffraction.

Parameters of Ni and Cu atoms local environment in quaternary alloy $Ti_{39.2}Ni_{24.8}Cu_{25}Hf_{11}$ were also investigated in a broad temperature range $5 - 300$K, covering the direct and reverse MTs occuring at $168-200$K. However, unlike the case of the ternary alloy, analysis of heating and cooling spectra didn't yield any signs of hysteretic effects near the transformation temperature. Only monotonous increase of EXAFS-oscillations magnitude with temperature is observed. The spectral compositions of EXAFS-oscillations are notably different around Ni and Cu atoms (Fig. 5) but do not change considerably with temperature. The lack of visible hysteretic effects in EXAFS-spectra can be caused either by small amount of material undergoing MT or by insufficiently precise temeperature following over MT region during the experiment.

The temperature dependences of local disordering Debye-Waller factors σ^2 for the first titanium coordination shells around Ni and Cu atoms (Fig. 6) were treated within the harmonic Einstein model, the characteristic Einstein temperatures Θ_E were found for the average distances Ni-Ti ($\Theta_E \sim 303$K) and Cu-Ti ($\Theta_E \sim 300$K). The Θ_E values as well as line shapes in Fig. 6 are almost similar, hence Ni-Ti and Cu-Ti interatomic bonds are approximately of the same rigidity. Whereas, the curve for the Ni-Ti bond lies above the one for the Cu-Ti bond, and therefore the static disordering of the first titanium shell is more significant for Ni than for Cu environment. This effect is the same as in the ternary alloy.

The values of Cu-Ti and Ni-Ti distances allowed us to refine the positions of Ni and Cu atoms in unit cell of $Ti_{50}Ni_{25}Cu_{25}$ alloy. The first Ti shell around Ni or Cu atoms in martensitic phase, space group $Pmmb$, is divided into 3 "subshells" containing 2, 4 and 2 Ti atoms respectively. The first 2 Ti and the third 2 Ti atoms lie in (010) layer while the second 4 Ti atoms lie in orthogonal (001) layer. In paper [5] authors reveal that the structure differs from the standard B19 type and demonstrates a ν-shift of Ni and Cu atoms from the planar centro-symmetric position along c-axis inside the (010) layers similar to that observed in the B19' monoclinic martensite structure. Our examinations also demonstrate this shift, moreover, for Ni and Cu atoms it is not similar. The distances between absorbing atom and Ti atoms in these 3 subshells obtained at room temperature turned out to be $R_1\ _{Ni-2Ti} \sim 2.51$Å, $R_2\ _{Ni-4Ti} \sim 2.63$Å, $R_3\ _{Ni-2Ti} \sim 2.86$Å around Ni atom and $R_1\ _{Cu-2Ti} \sim 2.61$Å, $R_2\ _{Cu-4Ti} \sim 2.61$Å, $R_3\ _{Cu-2Ti} \sim 2.75$Å around Cu atom (the error of technique is ~ 0.02Å). ν-shift determined from these values is -1.9% for Cu atom and

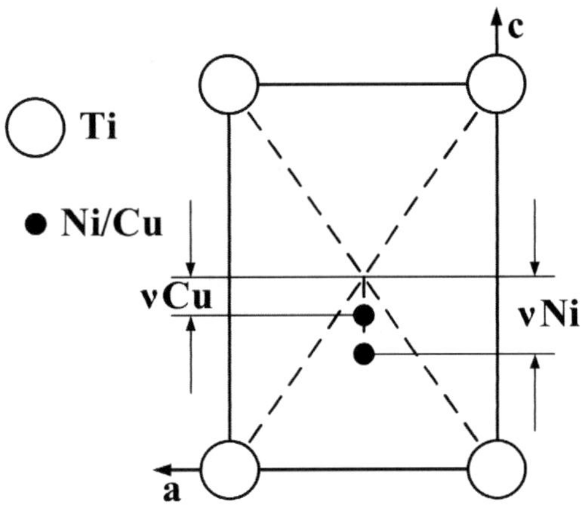

Figure 7. Schematic of the atomic shifts in (010) layers in *Pmmb* space group for the $Ti_{50}Ni_{25}Cu_{25}$ alloy.

−4.6% for Ni atom in coordinates of conventional unit cell (Fig. 7), while in [5] the "universal" shift for both types of atoms is −2.7%. The difference between values is obtained due to the EXAFS's selectivity to the type of chemical element in contrast to XRD employed in [5]. Such an inhomogeneity of Ni and Cu atomic shifts in (010) layers is consistent with our conclusion that Ti atoms are more disordered in Ni local environment.

4. Conclusion

Results of EXAFS spectroscopy study demonstrate that the changes in short-range order of investigated TiNi-based SMAs crystalline structure, unlike the long-range order, occur rather monotonously and can be modeled using B19 structure or *Pmmb* symmetry group both in austenitic and martensitic phases. XRD results demonstrate that $Ti_{50}Ni_{25}Cu_{25}$ alloy has B2 structure in austenitic phase and B19 in martensitic phase without any signs of monoclinic B19' structure. Hysteretic discrepancies in X-ray absorption spectra of ternary alloy $Ti_{50}Ni_{25}Cu_{25}$ generally originate from the changes in coordination shells of Ni and Cu atoms around the absorbing atom. Apparently, the most significant changes in local crystalline structure under MT occur in the Ni-Cu sublattice. No hysteretic effects in the region of phase transformation were found in quaternary alloy $Ti_{39.2}Ni_{24.8}Cu_{25}Hf_{11}$. However, in both alloys average static disordering of Ti atoms in coordination shell around Cu atom is less than around Ni. This assertion is consistent with the refined values of the Ni and Cu atomic ν-shifts from the planar centro-symmetric position in the (010) layer. The value of ν-shift, which in itself is not typical for general B19 orthorhombic structure, for Ni atoms turned out to be greater than for Cu atoms. This distortion may drive the further local deformation and affect the parameters of the phase transition under certain conditions. Such an impact might become the subject of further experimental and theoretical investigations aiming to explain the processes in shape memory alloys concerning their local crystalline structure.

Acknowledgments

This work was partially supported by RFBR (grant No. 11-02-01174-a) and Federal target program "Scientific and scientific-pedagogical stuff of innovative Russia" (State Contracts No. P39 and 16.740.11.0139)

References

[1] Yongqing F, Hejun D and Weimin H 2004 *Sensors and Actuators* A **112** 395–408

[2] Van Humbeeck J 1999 *Mat. Sc. & Eng.* A **273-275** 134–48
[3] Otsuka K and Ren X 1999 *Intermetallics* **7(5)** 511–28
[4] Shelyakov A V, Antonov V A and Bykovsky Yu A 1995 *Proc. SMST-94 (Pacific Grove, CA (USA))* (Monterey: MIAS) pp 335–40
[5] Potapov P L, Kulkova S E, Shelyakov A V, Okutsu K, Miyazaki S and Schryvers D 2003 *Phys. IV France* **112** 727–30
[6] Sitnikov N N, Shelyakov A V 2009 *Proc. Scientific Session of MEPHI - 2009 (26-30 January 2009, Moscow)* vol 4 (Moscow: MEPhI) pp 107–110
[7] Klementev K V 2001 K. V. Klementiev, VIPER for Windows, freeware *J. Phys. D: Appl. Phys.* **34** 209–17
[8] Ankudinov A L, Bouldin C, Rehr J J, Sims J and Hung H 2002 *Phys. Rev.* B **65** 104107-1-11
[9] Kudoh Y, Tokonami M, Miyazaki S and Otsuka K 1985 *Acta Metall. Mater.* **33**(11) 2049–56
[10] Sanati M, Albers R C and Pinski F J 1998 *Phys. Rev.* B **58**(20) 13590–3

Synthesis of precious and non-ferrous metals in hydrothermal medium

E V Fesik[1], V V Grebnev, V I Zarazhevsky, T N Lykova, V S Paveljev, G D Malchikov

Department of Nanoengineering, Samara State Aerospace University, Moscow highway 34, Samara, 443086, Russia

[1]E-mail: 1707-fesik@mail.ru

Abstract. This paper contains an overview of our results on platinum group metals disperse particles synthesis in water solutions in hydrothermal conditions. Metal phases (metal sols, powders and coatings) are formed in halogen and halogenammin complexes solutions in autoclave conditions at 110-230°C. The received materials are promising for the practical applications (catalysts, sensor controls, fuel elements, photonics and etc.).

1. Intoduction

Nanoparticles are the basic material used in creating classical materials: powders, sols, films, coatings, composites, and other with unexpected properties [1]. To solve practical tasks it is necessary to ensure continuity of the entire sequence of synthesis' stages of nanomaterial, namely:

- nanoparticles production,
- dimensional stabilization,
- insertion to the target system with non-fading and reproducible properties,

where relaxation processes of nanoparticles growth would have been kept idle.

According to academician M.V. Alfimov [2], it is the production of nanopowders that is one of the most rapidly growing areas of nanotechnology in the world.

From this point of view among all possible methods for obtaining nanoparticles, in our opinion, chemical methods have greater advantage over the physical ones. They essentially allow the influence of external environments to control the phase transition of condensation in the system and simultaneously influence the rate of small particles coagulation or coalescence. In chemical methods it is possible to deal with metastable states stabilization, for example, by adding SAS, cladding, microencapsulation by viscous dispersed medium, porous materials and other structurally-constrained medium.

The present report reviews our results on platinum group metals dispersed particles synthesis in aqueous solutions in autoclave conditions.

Journal of Physics: Conference Series **291** (2011) 012026 doi:10.1088/1742-6596/291/1/012026

2. Experimental details

Experiments were carried out in fluoroplastic digester according to the methods previously described in papers [3-5].

3. Results and discussions

Reaction (1) of a reversible scrambling according to degrees of platinum oxidation occurs in acidic medium [3]:

$$2\,[PtHal_4]^{2-} \rightarrow Pt^0 + [PtHal_6]^{2-} + 2Hal^- \qquad (1)$$

Reaction reversibility is the key characteristic of its use for platinum recrystallization in aqueous solution, for example, to change platinum particles dispersion. When various metals are inserted into the reaction system [3, 7] we can observe a reversible process of dissolution - crystallization, cementation (solid-phase exchange) and other (2, 3):

$$\begin{aligned} Pd^0 + Pt^{2+} &\rightarrow Pt^{4+} + Pd^{2+} + Pt^0 \\ Pd^0 + Ir^{4+} &\rightarrow Ir^{3+} + Pd^{2+} + Ir^0 \\ Rh^0 + Pd^{2+} &\rightarrow Pd^0 + Rh^{3+} \end{aligned} \qquad (2)$$

$$M^0_{disp.}\ M^{n+} \rightarrow M^0_{comp.}\ (sol, powder) \qquad (3)$$

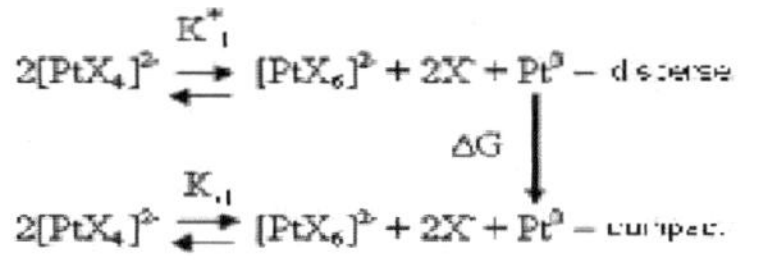

(4)

The reaction products are binary metal deposits Pd-Pt, Pd-Ir, Rh-Pd in the form of solid solutions or mixtures. Dispersed deposits can spontaneously coarsen by equation (4) according to the small particles thermodynamics. This happens quite quickly at elevated temperatures in contact with the liquid phase.

For reaction (1) equilibrium with ultra compact and phases of platinum can be offered:

Figure 1(A) presents a graph of the function $Q = \frac{C^2_{Pt^{2+}}}{C_{Pt^{4+}}}$ characterizing pseudoequilibrium (metastable) state of the system, depending on time. The graph resembles the classical dependence (Fig. 1(B)) of free energy change during the new phase particles (nuclei) formation.

For spherical particle formation with radius R the total change in free energy ΔG is presented like [1]:

$$\Delta G = 4/3\pi \Delta g R^3 + 4\pi\sigma R^2,$$

where Δg - the difference between the specific free energies of two phases, σ - surface tension coefficient.

The maximum value of ΔG (in our case, 7.5 kJ/mol) is the energy of critical nucleus formation. In the subcritical particle size nucleus formation has ΔG> 0, and according to classical thermodynamics, it can not be spontaneous, new-phase particles dissolve. In the region where the particle sizes are more than Rcr., spontaneous heterogeneous process of particle growth proceed up to the balanced compact state. Such arguments are only of qualitative sense, since the macroscopic approaches of classical thermodynamics are not used here.

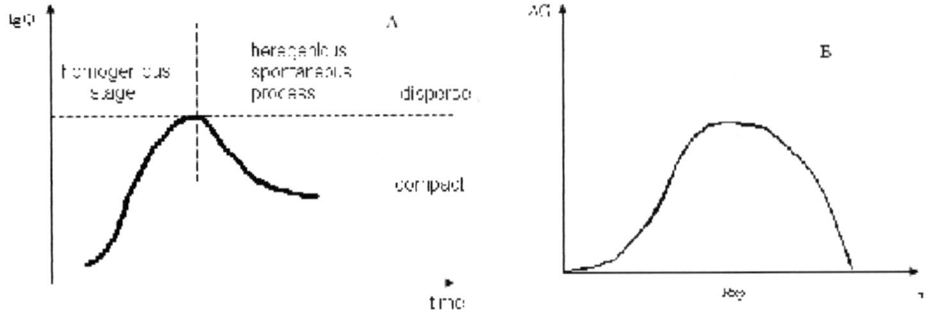

Figure 1. Variation of Q and ΔG over dispersed system evolution coordinate in the parameters: time (A) or particle size (B)

New phase particles nucleation in the subcritical (in size) conditionally homogeneous region has random nature stages; this must be shown in results non-reproducibility according to the allocated solid phase amount. Herein it is homogeneous nucleation stage, obviously, that is limiting in speed. Such facts were observed, for example, in paper [3]: only 1-2 mg of mohr platinum were extracted in 6 hours, and in the next 2.5 hours - 71.6 mg.

For the scrambling reaction we observed wave changes in time of platinum concentrations in various oxidation states. Figure 2 shows the experimental results for reaction (1) in the change of Pt (II) and Pt (IV) concentrations in process at 190° C and 210° C. This confirms the random nature of solid phase nucleation, which was discussed above.

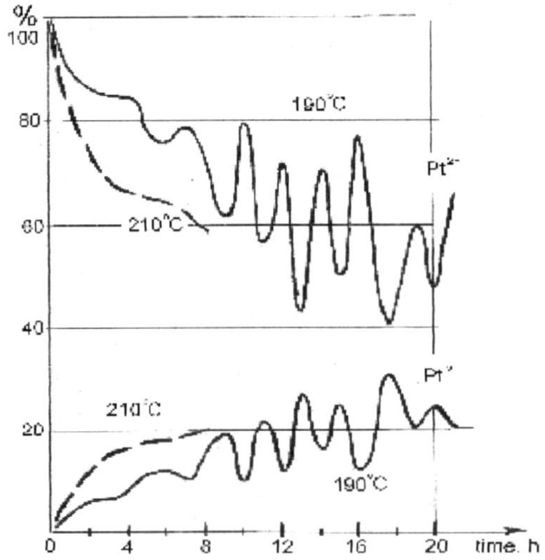

Figure 2. Wave nature of Pt (II) and Pt (IV) concentrations changes.

For the following reactions:

$$2\,[PtCl_4]^{2-} \rightleftarrows Pt^0 + [PtCl_6]^{2-} + 2Cl^- \qquad \text{atm. } CO_2,\ pH = 7$$

$$Pt^0_{UPD} \xrightarrow{190^0 C} Pt^0 \text{ compact}$$

$$Pt^0 \xrightarrow{210^0 C} Pt^0 \text{ compact}$$

wave behavior allows to hypothesize about scrambling reaction mechanism. Figure 2 illustrates the mechanism of the heterogeneous process of scrambling in time at 190 ° C, which has a wave character. At 210 ° C fluctuations are smoothed out. This means that platinum quickly becomes compact.
Wave process explains the given hypothesis.

$$\begin{cases} Pt^{2+} \rightleftarrows Pt^{4+} + Pt^0_{UDP} \\ Pt^{4+} + Pt^0_{UDP} \rightleftarrows Pt^{2+} \\ Pt^0_{UDP} \rightarrow \underline{Pt^0_{comp.}} \downarrow \end{cases} \qquad (5)$$

The first two reactions are linked: the basic component of the first is the output of the second. Basic components of the second reaction are the outputs of the first. The third stage is the irreversible consumption of one component. This combination of stages fully complies with the conditions of oscillatory processes time, such as the famous Belousov-Zhabotinsky's reaction. We consider important to point out a curious fact. At Chernyaevsk Chemistry, Analysis and Technology of platinum metals Conference, (2006), the report made by Distler V.V. [8] on ore genesis of grains of platinoids and gold ingrained in copper-nickel ores massive pieces in direct experiments the course of platinum and gold scrambling in the volcanic magma is proved. Currently, the study of scrambling autoclaving according to oxidation and crystallization degrees in the bimetallic system is being successfully developed in the ICCT SB RAS under the guidance of O.V. Belousov [9-11].
Autoclaving acquires a different character in alkali solution of halogen-ammonia complexes. The main reaction (6) - irreversible metal reduction with inner-ammonia:

$$[M(NH_3)_{4-i}Cl_i]^{2-i} + 2OH^- \rightarrow M^0 + 1/3N_2 + + (10/3\text{-}i)\ NH_3 + iCl^- + 2H_2O, \qquad (6),$$

where M - metal: Pt, Pd, Rh, Ir, Ru, Au, Ag, Cu, Ni, Co, etc.

State of solid metal phase (powder, mirror, sponge, weighed in solution in small portion, loose or dense coating on the substrate) depends on the composition and structure of the basic complexes, the composition and pH solution, temperature, time, agitation, etc. The dominating routes, which lead to the metallic phase, appear in the system depending on the conditions. Ammonia complexes irreversible restoration is realised for platinum, palladium, rhodium, ruthenium, cobalt, nickel, etc. [12-14]. The same processes (6) of autoclave thermolysis proceed in amin complex of platinum, where Amin is ethylenediamine, pyridine, glycine and other [15].
Carrying out the processes in a dispersion medium for metallic phase separation in structure-constrained conditions, which can control particle size, is of basic interest. Methods of sol-gel technology are widely used for the synthesis of hydroxides and oxides of amphoteric metals. A combination of sol-gel technology methods, autoclave methods producing metal dispersions in gelified medium of hydroxides allows to widen preparation possibilities for new catalyst systems, combining all the consistent technological operations. For example, boehmite crystalline modification of aluminum hydroxide is formed in autoclave conditions. Under the same conditions platinammine are restored to a metallic state. Combining these two processes, one can produce a suspension or coating

on a thick substrate containing boehmite and platinum group metals. Drying and calcination will lead to the desired combination of metal on γ-Al_2O_3. With such a synthesis method even distribution of metal in hydroxide and then in aluminum oxide should be expected. Figure 3 shows micro photos of surface images of samples in the characteristic emission of ruthenium with its different content in the samples, which show that ruthenium is evenly distributed on the surface. In addition, metal reduction and crystallization of the metallic phase proceed in a "microreactor" of gel, which contributes to the high dispersion of platinum group metals particles.

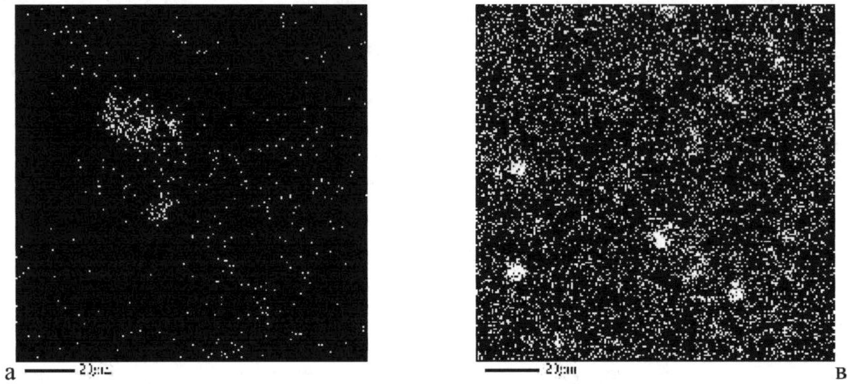

Figure 3. The image of the sample surface in a characteristic radiation:

a) Ruthenium 5% Ru /AlO (OH); b) ruthenium, 10% Ru /AlO (OH)

We studied the reaction of the ions interaction $[Pt(NH_3)_4]^{2+}$, $[Pd(NH_3)_4]^{2+}$ with NH_4ReO_4 in alkaline solutions under hydrothermal conditions.

In paper [16] the stoichiometry of tetraamminplatinic (II) and ammonium perrhenate interaction is studied and proved:

$$3[Pt(NH_3)_4]Cl_2 + 6NH_4ReO_4 + 6KOH \rightarrow 3Pt^0 + + 6Re^0 + 8\ N_2\uparrow + 2NH_3\uparrow + 6KCl + 30H_2O \quad (7)$$

Table 1 shows the experimental and theoretical data on XPA of solid product reaction product (7), which is a solid solution or a mixture of metallic phases of platinum and rhenium.

Table 1. X-ray characteristics of the solid product of an autoclaved thermolysis mixture of $[Pt(NH_3)_4]Cl_2$ and NH_4ReO_4.

Product of autoclave thermolysis - Pt phase				Product of autoclave thermolysis – Re phase			
2θ, deg	I, %	d/n, Å (exp.)	d/n, Å (ASTM)	2θ, deg	I, %	d/n, Å (exp.)	d/n, Å (ASTM)
50.7	100	2.261	2.27	47.4	30	2.408	2.388
–	–	–	–	51.5	30	2.228	2.226
59.3	85	1.957	1.956	54.7	100	2.107	2.105
88.8	84	1.383	1.385	56.5	11	1.254	1.626
110.2	100	1.180	1.179	67.8	22	-	1.380
117.8	60	1.130	1.130	82.1	20	-	1.173

Experiments have been carried out to prove the stoichiometry of this reaction and to quantify the free ammonia by the method described in the experimental part of this paper.

Data from table 2 show good agreement between the experimental and theoretical values of free ammonia. Thus, sals interaction in autoclave conditions proceeds according to equation (7).

Table 2. The results of the NH_3 content determination in solutions after autoclave experiments completion (150 min, 190° C).

Weighed portion, g		m NH_3, (theor.), g	m NH_3, (exp.), g	% of theoretical value
$[Pt(NH_3)_4]Cl_2{\bullet}H_2O$	NH_4ReO_4			
1.036 (0.55 Pt)	0.789 (0.55 Re)	0.101	0.0969	95.9
1.036 (0.55 Pt)	0.789 (0.55 Re)	0.101	0.0942	93.3
0.776 (0.43 Pt)	1.181 (0.82 Re)	0.025	0.0241	96.0

Similarly, the interaction of $[Pd(NH_3)_4]Cl_2$ and NH_4ReO_4 mixture in platinum-rhenium system has been studied. The phase composition of the solid reaction product was determined by XFA. The results given in Table 3 show that the solid product of autoclave thermolysis of this mixture contains three phases - the metal palladium, the metal rhenium and rhenium dioxide.

Table 3. X-ray characteristics of the solid product of an autoclaved thermolysis mixture of [Pd $(NH_3)_4$] Cl_2 and NH_4ReO_4.

Product of autoclave thermolysis - Pd phase				Product of autoclave thermolysis - phase ReO_2				Product of autoclave thermolysis - phase Re			
2θ, degr	I, %	d/n, Å (exp.)	d/n, Å (ASTM)	2θ, degr	I, %	d/n, Å (exp.)	d/n, Å (ASTM)	2θ, degr	I, %	d/n, Å (exp.)	d/n, Å (ASTM)
40.25	100	2.238	2.23	26.3	40	3.386	3.401	37.1	30	2.421	2.388
46.7	50	1.943	1.94	31.4	80	2.846	2.865	40.25	30	2.238	2.226
68.4	30	1.37	1.371	46.7	50	1.943	1.946	42.7	100	2.116	2.105
82.3	30	1.17	1.17	53.0	30	1.726	1.701	68.4	20	1.37	1.38
86.8	10	1.121	1.12	76.4	20	1.245	1.262	76.4	20	1.245	1.262
				81.4	30	1.181	1.203	82.3	20	1.17	1.173

Table 4 summarizes the experimental results of free ammonia determination, which show that its quantity in the system corresponds to 24-30% of the total.

Table 4. The experimental values of the basic amount of reactants (for metal and ammonia) and of free ammonia in the reaction products.

Pd, g	Re, g	Pd:Re (atomical)	$\sum NH_3$, (basic quantity), g	Quantity of free NH_3	
				g	%
0.057	0.1	1:1	0.0459	0.0140	30.5
0.057	0.1	1:1	0.0459	0.0113	24.6
0.057	0.1	1:1	0.0459	0.0121	26.4
0.057	0.1	1:1	0.0459	0.0121	26.4
0.057	0.066	1.5:1 (3:2)	0.0425	0.0090	21.2
0.1	0.1	1.75:1	0.0728	0.019	26.1
0.1	0.1	1.75:1	0.0728	0.0224	30.8
0.1	0.1	1.75:1	0.0728	0.020	27.5

Specified as the reaction products, metallic states of palladium, rhenium and rhenium dioxide allow to assume a significant variety of equations for the recovery processes of mixture $[Pd(NH_3)_4]Cl_2$ - NH_4ReO_4 in an alkaline environment in the autoclave conditions. Here are the atomic ratios of metals, the total number of ammonia molecules before the reaction, the number of molecules of free ammonia and its percentage of the total number of molecules for the corresponding reactions:

Table 5. Presumptive variants of recovery processes of mixture $[Pd(NH_3)_4]Cl_2$ - NH_4ReO_4.

Possible reactions	Pd:Re	$\sum NH_3$	Quantity of free NH_3		
$2NH_4ReO_4+2[Pd(NH_3)_4]Cl_2+4KOH \rightarrow 2Pd^0+2Re^0+3N_2+4NH_3+4KCl+12H_2O$	2:2 (1:1)	14	4	28.6	(8)
$12NH_4ReO_4+6[Pd(NH_3)_4]Cl_2+12KOH \rightarrow 6Pd^0+10Re^0+13N_2+8NH_3+12KCl+2ReO_2+54H_2O+2NO$	6:12 (1:2)	36	8	22.2	(9)
$7NH_4ReO_4+4[Pd(NH_3)_4]Cl_2+8KOH \rightarrow 4Pd^0+6Re^0+8N_2+6NH_3+8KCl+ReO_2+33H_2O+NO$	4:7	23	6	26.1	(10)

Data given in table 5 shows that the most close to the description of the experimental results are hypothesis, described by equations (8), (9) and (10) as the quantity of free ammonia (%) agrees well with experimental data obtained after titration of free ammonia. Based on the above experimental data, in our opinion, we should give preference to the reaction equation (10).

4. Conclusion

On an example of binary metal systems results of researches of the processes proceeding in hydrothermal conditions which products are disperse phases are resulted;
The observable reactions aren't typical for processes of coordination chemistry under normal conditions it is noticed.
For the first time it is experimentally proved that under hydrothermal conditions redox interaction processes of tetraamines platinum (II) and palladium (II) with perrenatum of ammonia proceed;
For the first time results of research of autoclave thermolysis processes ruthenium ammines chlorides in the depressive environment –of hydroxide aluminum suspensions are resulted.
In the products of these reactions metallic phases of platinum, palladium and rhenium were discovered;
Studied processes in the platinammine (II) and palladium (II) systems with perrenatum of ammonia liberalize the autoclave thermolysis method and allow to deliberately perform the dispersed bimetallic materials synthesis in solutions and the dispersion medium.

References

[1] Kukushkin S A, Slezov V V Dispersion systems on solid surfaces (evolutionary approach): thin films formation mechanisms 1996 St-Pb. Science 304
[2] Alfimov M V 2008 *Russian nanotechnologies* **3** № 01-02 http://nanorf.ru/educ.aspx
[3] Kovalenko N L 1982 Scrambling and platinum halogen complexes aquation in aqueous solutions at elevated temperatures: synopsis of a PhD thesis 02.00.01. Inorganic chemistry Krasnoyarsk 20Kovalenko N L 1982 Scrambling and halogen complexes of platinic aquation in aqueous solutions at elevated temperatures: *synopsis of a PhD thesis 02.00.01. Inorganic chemistry* Krasnoyarsk 20
[4] Kovalenko N L, Malchikov G D, Vershkov A V 1987 *Coord. chem..* **13** 4 554-557
[5] Kovalenko N L, Rogin N J, Malchikov G D 1985 *Coord. Chem.* **11** 9 1276-1280
[6] Krasheninnikov S A, Kuznetsova A G, Sultanov V P, Surkov E L, Yakhontova E L 1968 Technical analysis and control in inorganic compounds production M "Higher School" 208 - 210
[7] Belousov O V 2000 Influence of highly dispersed state of platinum metals on the reaction of scrambling and cementation processes *synopsis of a PhD thesis* 02.00.01. Inorganic chemistry Krasnoyarsk 17
[8] Distler V 2006 Nanoblock structure of natural mineral phases of platinum-group metals: genetic and technological effects *XVIII Chugaevsk meeting on the chemistry of complex compounds* Abstracts Moscow 13
[9] Belousov O V, Kovalenko N L, Dorokhov L I, Chumakov V G 2001 *J. Inorg. Chem.* **4** 684-688
[10] Belousov O V, Saltykov Yu V, Dorokhov L I, Solovyov L A, Zharkov S M 2008 *J. Phys. Chem.* **82 4** 749-753

[11] Belousov O V, Dorokhov L I, Solovjev L A, Zharkov S M, 2007 *J. Phys. Chem.* **81 8** 479-1482

[12] Malchikov G D, etc 1997 Aqueous solutions of precious metals halogen-amine complexes behavior in autoclave conditions [Text] Malchikov G D [And others] // Production and use of precious metals products Sat. articles Ekaterinburg URO RAN 125-133

[13] Kovalenko N L, Malchikov G D, Vershkov A V 1987 *Coord chem.* **13 4** 554-557

[14] Kovalenko N L , Vershkov A V, Malchikov G D 1986 *Coord. Chem.* **11** 1546-1549

[15] Fomenko L V 1985 Decomposition of glycine complexes of platinum (II) in aqueous solutions [Text] Fomenko L V [and others] XIV Chugaevsk meeting on the chemistry of complex compounds Abstracts Kiev 542

[16] Malchikov G D, Zarazhevsky V I, Grebnev V V, Fesik E V 2008 *J. Gen. Chem..* **78** 1761-1763

Journal of Physics: Conference Series **291** (2011) 012027 | doi:10.1088/1742-6596/291/1/012027

The effect of particle size on the toxic action of silver nanoparticles

L S Sosenkova, E M Egorova

Laboratory of nanopathology,
Institute of General Pathology and Patophysiology of RAMS,
Baltijskaya st., 8, 125315 Moscow, Russia
Science-technology Company "Nanomet", Moscow, Russia

E-mail: lubasya@inbox.ru

Abstract. Silver nanoparticles in AOT reverse micelles were obtained by means of the biochemical synthesis. Synthesis of nanoparticles was carried out with variation of the three parameters of reverse-micellar systems: concentration of silver ions, concentration of the stabilizer (AOT) and hydration extent w = $[H_2O]/[AOT]$. The combinations of varied parameters have been found, allowing to prepare micellar solutions of spherical silver nanoparticles with average sizes 4.6 and 9.5 nm and narrow size distribution. From micellar solution the nanoparticles were transferred into the water phase; water solutions of the nanoparticles were used for testing their biological activity. Our assay is based on negative chemotaxis, a motile reaction of cells to an unfavorable chemical environment. Plasmodium of the slime mold *Physarum polycephalum* used as an object is a multinuclear amoeboid cell with unlimited growth and the auto-oscillatory mode of locomotion. In researches of chemotaxis on plasmodium it is learned that silver nanoparticles of smaller size exhibit a higher biological activity (behave as stronger repellent) and this correlates with the literary data obtained in studies of silver nanoparticles interaction with other biological objects.

1. Introduction

It is known that in the course of interaction with living organisms, metal nanoparticles can both render curative effects and cause various pathologies. Silver nanoparticles exhibit the widest spectrum of applications, the fact that stimulated intensive studies of their biological effects. The main goal of such researches is accumulation of the data on the toxicity of these nanoparticles for living organisms, that allows to define conditions of their safe usage. Therefore, the necessity arose to determine, first, the range of nanoparticle concentrations where their toxic action is dangerous for life and health and, second, to obtain as much information as possible about the effect of various nanoparticle parameters (size, form, structure, surface charge) on their biological activity.

In studies of the biological effects of silver nanoparticles experimental evidence was gained on their toxicity towards bacteria and viruses [1-7], algae [8], animal and human cultured cells [9-11], fish embryos [12], animal organisms [13, 14]. Apart from the determination of lethal concentrations

Journal of Physics: Conference Series **291** (2011) 012027 doi:10.1088/1742-6596/291/1/012027

and threshold concentrations, which mark the appearance of the toxic effects, it was found also that the extent of toxicity depends of the particle size [3, 7] and shape [4].

For the studies of biological effects of silver nanoparticles it is necessary that they were stable on air in solution, and homogeneous in sizes. The method of biochemical synthesis in reverse micelles [15, 16] allows to prepare silver nanoparticles, stable on air in solution for a long time. In the present work the biochemical synthesis has been used for preparation of silver nanoparticles of different average diameter with narrow size distribution. This allows to study the dependence of their biological effects on the particle size. It seemed important to investigate the area of small sizes (less than 20 nm), where the so-called "size effects" are most clearly expressed [e.g.17]. To obtain micellar solutions achieve our goal we varied three parameters of the reverse-micellar system: the concentration of stabilizer (AOT), that of silver ions and the hydration extent w = $[H_2O]/[AOT]$. As reported earlier in a series of works devoted to the studies of metal nanoparticles formation in reverse micelles (e.g. [18-20]), these parameters affect the sizes of water micelles and (or) nanoparticles. As shown earlier [21], we managed to find the combinations of parameters providing the formation of at least two populations of spherical silver nanoparticles with rather narrow distribution and large enough difference in average sizes, the maximum average size not exceeding 10.5 nm.

From their micellar solutions, water dispersions of silver nanoparticles were prepared by the specially developed procedure [22]. These water dispersions may be used, in particular, for the studies of the nanoparticles' interaction with biological objects. In this work, two water dispersions obtained from the nearly monodispersed micellar solutions of silver nanoparticles were tested on plasmodium of the acellular slime mold Physarum polycephalum.

Physarum is a classical object for studying cell motility, and characteristic features of its chemotactic behavior are well established [24–26]. In particular, substances causing negative taxis (repellents) were shown to increase the period of contractile oscillations and to decrease the area of spreading when applied at spatially uniform concentrations [25, 26]. Due to this characteristic features, our experiments with *Physarum* allowed to register not only the lethal events, as with the majority of other objects, but also the changes in its behavior taking place in the mild conditions, at sublethal nanoparticle concentrations. Hence, this chemotaxis – based assay may be regarded as a highly sensitive tool for the studies of the biological action of silver nanoparticles.

The experiments performed with pairs of *Physarum* preparations on agar plates [23] showed that the chemotactic tests were sensitive enough for recognizing the difference in their efficiency, the fact that allowed to use such tests for the estimation of the particle size effect in the repellent action of silver nanoparticles.

2. Experimental

Silver Nanoparticles (SNP) were obtained by the biochemical synthesis.

Silver nitrate (analytical grade) and deionized water were used to prepare the $AgNO_3$ aqueous solution. The $[Ag(NH_3)_2]NO_3$ solution was obtained by adding 27 % aqueous ammonium hydroxide until silver hydroxide sediment was fully dissolved with creation of $[Ag(NH_3)_2]^+$-ions. To obtain reverse micelles, AOT (sodium bis(2-dioctyl)sulphosuccinate) was used as surfactant, and isooctane (analytical grade), as solvent. The natural flavonoid quercetin (Qr or 3,5,7,3',4'-pentahydroxyflavone) was applied as reducing agent.

The standard procedure for preparation of Qr micellar solution with the known concentration of the flavonoid is described in detail elsewhere [16, 22, 27].

For the synthesis of silver nanoparticles, $[Ag(NH_3)_2]NO_3$ water solution was added to the quercetin micellar solution to the given silver salt concentration (C_{Ag}) and hydration extent. The AOT concentration (C_{AOT}) in micellar solution did not exceed 0.135 M. The concentration of silver nanoparticles (C_{SNP}) in micellar solution was determined from the measured optical densities in absorption band maximum and the extinction coefficient ($\varepsilon = 1.03 * 10^4$ l/mol *sm) found by us as described in [27]. The C_{SNP} in solution is estimated as equivalent concentration either of silver salt (usually in millimols/liter) or as weight concentration of metallic silver (most often in µg/ml).

The absorption spectra of micellar or water solutions were recorded on spectrophotometer Helios-α (Thermo Electronics, GB) in 1 mm quartz cell at room temperature. Either isooctane or distilled water was used as reference solution. Particle sizes in micellar solutions were determined by transmission electron microscopy on LEO912 AB OMEGA microscope (production of Carl Zeiss, Germany) at 120 kV accelerating voltage. From electron micrographs particle size distributions were found for no less than 350 particles. Average sizes and standard deviations were determined using the Gauss approximation.

Plasmodium ***Physarum polycehallum*** is capable of unlimited growth as a single multinuclear cell. The migrating plasmodium looks like a fan-shaped protoplasmic film, which is smooth and continuous at the front and transforms into a tree-shaped network of individual protoplasmic strands in more caudal regions. Migration of plasmodium occurs due to the more intensive or more prolonged protoplasm streaming toward the leading edge. Any fragment of plasmodium restores the integrity of the plasma membrane [28] and resumes the contractile and motile activities. By this property, strands excised from a plasmodium can be used for force measurements, and film fragments of standard size and shape can be used in chemotactic assays based on the sensitivity of plasmodium growth to its physiological state and external influence.

For the studies of size effects the agar strips have been used, containing the nanoparticles added as water solution. The plasmodium film fragment was placed on the boundary between the two agar strips with the SNP of different size introduced to the equal total silver concentration. The strips were incubated in standard conditions and the difference in the direction of plasmodium growth was recorded, showing the preference of Physarum to the nanoparticles of one of the two sizes studied.

3. Results and discussion

3.1. Synthesis of Silver nanoparticles with different average size

As shown earlier [21], we managed to find the conditions which allow to receive solutions of silver nanoparticles with the average size in the range from 4.6 to 9.3 nm and with small degree of polydispersion. The nanoparticles are approximately spherical, stable in the solution for a long time (for several years).

From these micellar solutions water dispersions were obtained for the studies of biological effects. It is known that, during their transfer to the water phase, the sizes of nanoparticles do not change, and their concentration appears to be equal or close to that in micellar solution [22, 29]. Typical change of the optical spectra at the transfer of silver nanoparticles from micellar solution to water phase in the process of preparation of AgNPw water solution is shown in Figure 1.

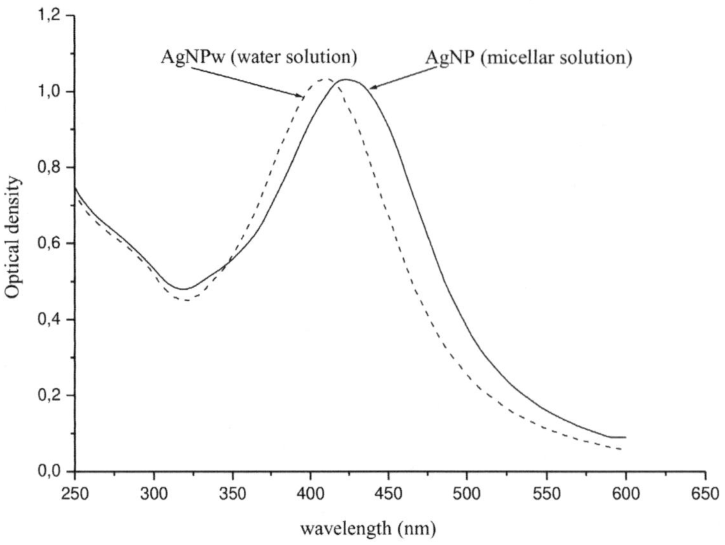

Figure 1. Typical change of the optical spectra at the transfer of silver nanoparticles from organic (micellar solution) to water phase in the process of preparation of SNP water solution.

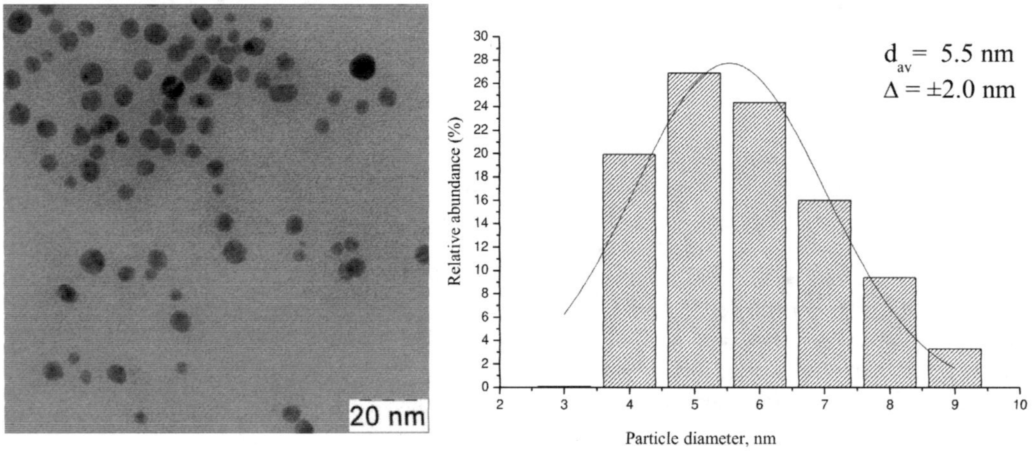

Figure 2. Electron micrograph (a) and histograms of the sizes (b) of silver nanoparticles in water solution, average diameter 5.5 nm.

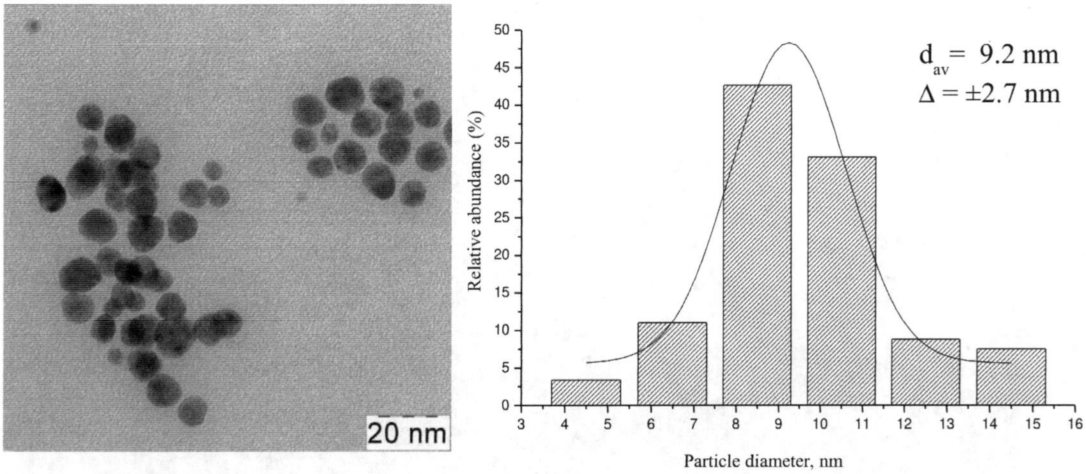

Figure 3. Electron micrograph (a) and histograms of the sizes (b) of silver nanoparticles in water solution, average diameter 9.2 nm.

Comparison of the TEM micrographs and hystograms for water solution (Figures 2, 3) and the corresponding micellar solution [21] shows that there is almost no change both of average size and size distribution.

3.2. Toxic Effects of Silver nanoparticles

Water dispersions presented in Figures 2, 3 have been used were tested on the Physarum plasmodium. Testing was made on the agar strips containing 3 μg-ion/l of SNP1 (d_{av}=5.5±2.0 nm) and SNP2 (d_{av}=9.2±2.7 nm). The result of testing is shown in Figure 4. It is seen that almost all plasmodia are oriented towards SNP2, and consequently, the SNP1 behave as more strong repellent. This correlates with the data reported in literature on the more strong toxicity of smaller nanoparticles [3, 7]. It seems probable that the effect of particle sizes is connected mainly with the difference in their number per unit volume: for a given overall concentration of silver, the number of smaller nanoparticles will exceed that of the nanoparticles of larger size. This shows probably to the significance of the numerical concentration of nanoparticles per unit volume of the medium: at the equal silver concentration, numerical concentration of the smaller particles will be higher, thus providing the possibility of the more intensive adsorption onto the cell surface.

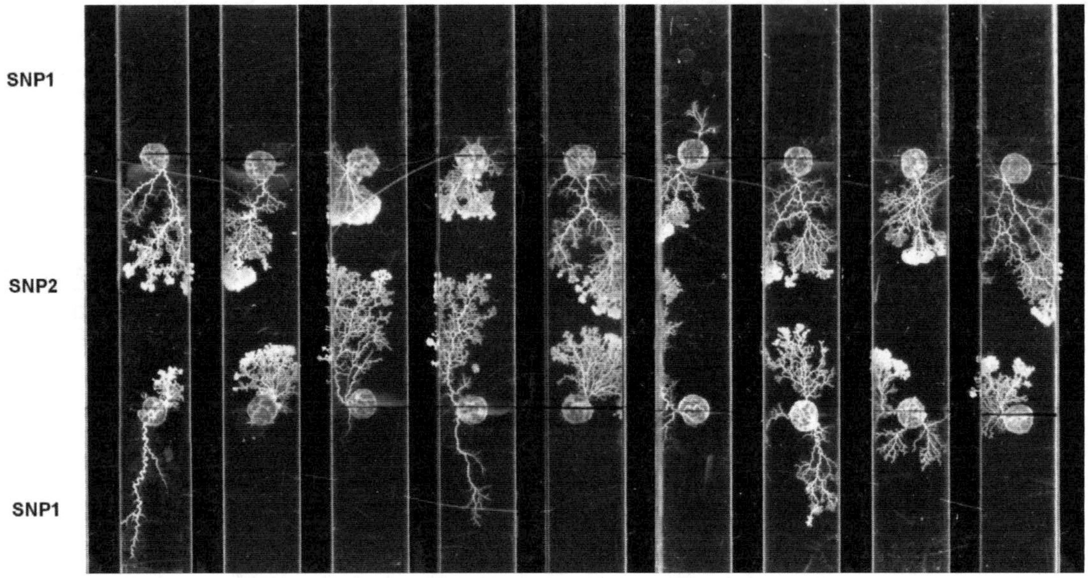

Figure 4. Comparison of the repellent action of SNP1 (d_{av}=5.5±2 nm) and SNP2 (d_{av}=9.2±2.7 nm) at equal concentration of 3 μg-ion/l. Directional movement of all plasmodia toward SNP2 containing strips (in the center of chamber) proves the higher repellent efficiency of SNP1. Scale bar, 1 cm.

4. Conclusions

The conditions are found allowing to obtain stable silver nanoparticles of average size in the range 4.6 - 9.3 nm with narrow size distribution. The nanoparticles are approximately spherical, stable in solution for a long time (for several years). For the studies of the nanoparticles interaction with living organisms, water dispersions of silver nanoparticles different in size were obtained from their micellar solutions. Investigation of the effect of water dispersions on chemotactic behavior of plasmodium of the slime mold *Physarum polycephalum* showed that nanoparticles of smaller size demonstrate a higher biological activity (are stronger as repellent). This observation agrees with the literary data obtained in studies of the interaction of SNP with some other biological objects.

References

[1] Karamushka V I, Gadd D M, Grusina T G, Ulberg Z R 1998 *Kolloidn. Zh.* **60** 775(Colloid. J. 1998 **60** 775)
[2] Sondi I, Salopek-Sondi B 2004 *J. Colloid Interface Sci.* **275** 177
[3] Lok C N, Ho C M, Chen R, et al. 2007 *J. Biol. Inorg. Chem.* **12** 527
[4] Pal S, Tak Y K, Song J M 2007 *Applied and Environmental Microbiology* **73** (6) 1712
[5] Kim J S, Kuk E, Yu K N, et al. 2007*Nanomedecine: nanotechnology, biology and medicine* **2**, 95
[6] Elechiguerra J L, Burt J L, Morones J R, et al. 2005 *Journal of nanobiotechnology* 3:6 doi:10.1186/1477-3155-3-6.

[7] Neal A L 2008 Ecotoxicology. **17** 362
[8] Ulberg Z R, Marochko L G, Savkin A G, Pertsov N V 1998 *Kolloidn. Zh.* **60** 836 (Colloid. J. 1998 **60** 775)
[9] Braydich-Stolle L, Hussain S, Schlager J J, Hofmann M C 2005 *Toxicological sciences* **88** 412
[10] Hussain S, Hess K, Gearhart J, et al. 2006 *Toxicological sciences* **92** 456
[11] Sheikpranbabu S, Kalishwaralal K, Venkataraman D, et al. 2009 *Journal of nanobiotechnology* 7:8 doi:10.1186/1477-3155-7-8.
[12] Asharani P V, Wu Y L, Gong Z, Valivaveettil S 2008 *Nanotechnology* **19** 255102
[13] Ji J H, Jung J H, Kim S S, et al. 2008 *Inhalation toxicology* **19** 857
[14] Kim Y S, Kim J S, Cho H S, et al. 2008 *Inhalation toxicology* **20**, 575
[15] Egorova E M, Revina A A, Kondratieva V S Patent RF N 2147487 priority from 01.07.1999.
[16] Egorova E M, Revina A A 2000 *Colloids Surf. A* **168** 87
[17] Sergeev G B 2003 *Nanochemistry* (Moscow: Publishing House of Moscow State University)
[18] Maitra A 1984 *J. Phys. Chem.* **88** 5122
[19] Robinson B H, Khan-Lodhi A N, Towey T, *Structure and Reactivity in reverse micelles* ed Pileni M P 1989 (Amsterdam: Elsevier,Oxford, New York) p 198
[20] Zhang W, Qiao X, Chen J, Wang H 2006 *J. Coll. Interf. Sci.* **302** 370
[21] Sosenkova L S, Egorova E M 2011 *Zn. Fiz. Khim.* **85** 317 (*Russ. J. Phys. Chem.* **85** 264)
[22] Egorova E M, Revina A A, Rumyantsev B V, et al. 2002 *Zn. Prikl. Khim.* **75** 1620 (*Russ. J. Appl. Chem.* **75** 1585)
[23] Egorova E M, Beylina S I, Matveeva N B, Sosenkova L S 2010 *In: Chemotaxis Types, Clinical Significance, and Mathematical Models* Nova Science Publishers (in press)
[24] Kincaid R L, Mansour T E 1978 *Exptl. Cell Res.* **116** 365
[25] Ueda T, Kobatake Y *Chemotaxis in Plasmodia of Physarum polycephalum* ed Aldrich H C, Daniel J W 1982 *Cell Biology of Physarum and Didymium: Organisms, Nucleus, and Cell Cycle* vol.1 (New York: Academic Press) p111
[26] Beylina S I, Matveeva N B, Teplov V A 1996 *Biophysics* **41** 137
[27] Egorova E M, Revina A A 2003 *Zh. Fiz. Khim.* **77** 1683 (*Russ. J. Phys. Chem.* A **77** 1513)
[28] Wohlfarth-Botterman K E, Stockem W 1970 *Entwicklungsmech. Org.* **164** 321
[29] Egorova E M 2010 *Biological effects of silver nanoparticles, in: Silver nanoparticles: properties, characterization and application* ed. Welles A E (New York: Nova Science Publishers) p 221

Cavitations synthesis of carbon nanostructures

S Voropaev
Vernadsky Institute of Geochemistry and Analytical Chemistry (GEOKHI RAS), Moscow, Kosygina 19, 119991 Russia

E-mail: svoropaev@rambler.ru

Abstract. Originally an idea of diamonds production by hydrodynamical cavitation was presented by academician E M Galimov. He supposed the possibility of nature diamonds formation at fast magma flowing in kimberlitic pipes during bubbles collapse. This hypothesis assumes a number of processes, which were not under consideration until now. It concerns cavitation under high pressure, growth and stability of the gas- and vapors bubbles, their evolution, and corresponding physical- and chemical processes inside. Experimental setup to reproduce the high pressure and temperature reaction centers by means of the cavitation following the above idea was created. A few crystalline nanocarbon forms were successfully recovered after treatment of benzene (C_6H_6).

1. Introduction

Mainly, nanodiamonds (ND) and other carbon nanostructures are produced by means of detonation of explosive containing carbon [1], [2] or by CVD [3], PVD [4] methods and laser ablation [5] now. Hydrodynamical cavitation at carbon contacting liquids [6], [7] provide alternative way, where a number of advantages are evident (control of impurities, defects structure, etc.). Additional interest to the cavitations concerns manipulation with the carbon nanotubes. Shock waves and jets, which are results of the bubbles collapsing, can deform effectively nanostructures providing possibility to investigate theirs physical- and mechanical properties [8]. So, hydrodynamical cavitation is unique process which strongly differs from acoustical cavitation (produced by sound wave with small amplitude in the liquid) concerning carbon nanostructures creation and control of their size and properties.

From the general point of view, the physics by cavitational bubbles collapsing and by shock-compression is very similar. But, the environment conditions, especially theirs that appropriate for the rapid cooling, are important as well. For example, detonation nanodiamonds production crucially depends on the synthesis condition and especially on the heat capacity of the cooling medium in the detonation chamber (water, air, CO_2, etc.). The higher the cooling capacity, the larger the detonation nanodiamonds yield, which can reach 90%. Various measurements, including X-ray diffraction and high-resolution TEM [9] revealed that the size of the diamond grains in the soot is distributed around 5 nm. The purpose of this study is to investigate a new rapid cooling technique under shock compression during hydrodynamical cavitation and to clarify carbon behavior under special range of P-T conditions. A few crystalline nanocarbon forms were found in the samples recovered from the benzene after treatment at the proper experimental setup. The results of characterization for the produced nanocarbon forms by means of the transmission electron microscope (TEM) and Raman spectroscopy are reported.

2. Experimental setup

At GEOKHI RAS experimental setup was constructed in collaboration with Moscow Bauman Technical University, group of Prof. V N Skorobogatskiy, to argue the idea of academician E.M. Galimov. Schema of the experimental setup is shown in the figure 1.

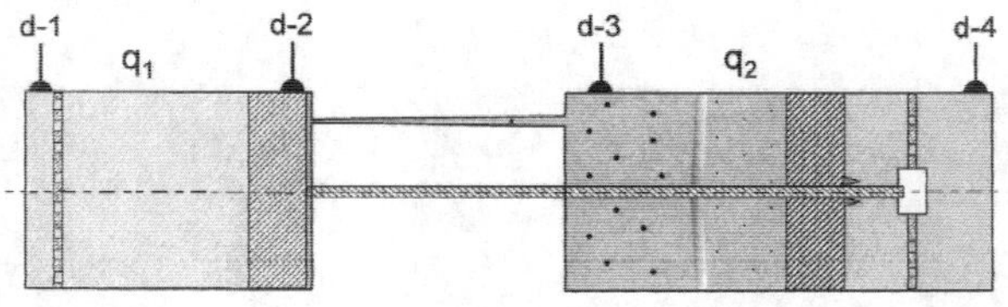

Figure 1. Experimental setup: two work cavities 2 and 3 are joined by Venturi tube; q1 and q2 are moving pistons; d-1, d-2, d-3 and d-4 are pressure sensors.

It includes the cavities (1) and (4), in which the ignition of the powder charges occurs and pistons q1 and q2 were moved. The cavities (2) and (3) and jet block accomplished in the form of the Venturi tube are filled with the working liquid (benzene). The profile of the Venturi tube in its critical section provides the decrease of pressure of flowing liquid lower than the pressure of its saturated vapors and its continuous flow in the diffuser part. The experimental installation operates in the following way: under action of pressure of the powder gases formed in the cavity 1 the piston q1 begins to move and drives the liquid from the cavity 2 into the cavity 3 through the Venturi tube. At that time the vapor bubbles are formed in the minimum section of the Venturi tube (the first stage of cavitations). These bubbles increase in their size and get into the cavity 3. The piston q3 moves in its extreme right position and pin out the cartridge primer, which initiate the powder cartridge in the cavity 4. Under pressure of the powder gases in the cavity 4 the piston q2 gets the impulse in the reverse direction that generate the shock wave in the cavity 3, which initiate the condition for the adiabatic collapse of the vapor bubbles and diamond producing.

Pressure measurements were made during experiments by means of number of pressure sensors d-1, d-2, d-3, d-4 at the cavities and Venturi tube. Pressure levels at benzene were at the interval from 80 MPa to 140 MPa, pressure drop at the cavity 3 (until initiation of the powder cartridge in the cavity 4) was from 0.002 MPa to 0.1 MPa, pistons velocities were about 2 m/sec. In some experiments, temperature sensor was installed instead pressure sensor at the cavity 3. Its value demonstrate that the initial temperature of benzene was not remarkable changed during the flowing, so the process was considered as isothermal in modeling.

After tuning of the setup experimental parameters, the second phase of the bubbles collapse was produced as planned. Namely, spherical-symmetric adiabatic bubbles collapse in the cavity 3 under proper external pressure jump.

3. Analysis of results

There are reliably registered sparkling grains of the nanodiamonds from 5 to 20 nanometers in size and aggregates of different forms of diamonds and graphite till 1-5 micrometers in size, see figure 3. Investigation of grains observed under transmission electron microscope was carried out by JEOL JEM100-C11 by Prof. A S Aronin (ISSP RAS, Chernogolovka). Renishaw InVia Reflex Spectrometer system for Raman spectral analysis using visible excitation at 785 nm was prepared by Dr. A Eliseev, Chemistry Department, MSU, see figure 4.

Figure 2. Industrial shock-compressed graphite with ND inside, size 1-10 μm.

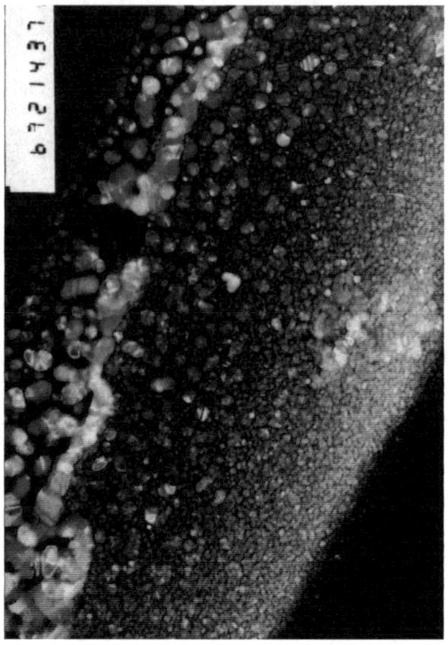

Figure 3. Collection of caviation's ND grains, size 5-20 nm, x100000.

Raman spectroscopy is particularly well suited to molecular morphology characterization of carbon materials. Every band in the Raman spectrum corresponds directly to a specific vibrational frequency of a bond within the molecule. The vibrational frequency and hence the position of the Raman band is very sensitive to the orientation of the bands and weight of the atoms at either end of the bond. So, when we compare the Raman spectra of two carbon allotropes – diamond and graphite – we can easily distinguish the two materials by their Raman spectrum even though both are composed entirely of C-C bonds. The graphite spectrum has several bands in the spectrum and the main band has shifted from 1332 cm^{-1} in cubic diamond to 1582 cm^{-1} in graphite [11]. The reason for this is that graphite is composed of sp2 bonded carbon in planar sheets in which the bond energy of the sp2 bonds is higher than the sp3 bonds of diamond. The higher energy of the sp2 bonds in graphite pushes the vibrational frequency of the bonds and hence the frequency of the band in the Raman spectrum to higher frequency. The presence of additional bands in the graphite spectrum indicate that there are some carbon bonds with different bond energies in the graphite sample and this is in fact the case, as graphite is not quite as uniform in structure as diamond.

Another good example to show the remarkable sensitivity of Raman to molecular morphology is to compare the Raman spectra of produced nanocrystalline diamonds with that of industrial nanodiamond grains (see figure 2 and figure 4) recovered from the shock-compressed graphite (SCG) at 60-70 GPa, which present intermediate and metastable phase on a way to cubic diamond. The small crystal size of nanocrystalline diamond results in a finite-size effect in which the lattice is somewhat distorted. This is manifested in the Raman spectrum by a slightly downshifted tetrahedral sp3 band. The additional band at 1340 cm^{-1} and the shoulders on the 1350 cm^{-1} and tetrahedral sp3 band are also indicative of sp2 bonded carbon that represents surface defect modes, which would be insignificant in larger diamond crystals.

Finally, Raman can reveal a lot of information about more complex carbon structures. So, fullerenes are essentially hollow carbon shells of various sizes. The most well known of these is a 60-carbon unit called Buckminster fullerene or C60. There are many other fullerenes, from a few to many hundreds of carbon atoms. The main feature in the C60 spectrum is a relatively sharp line at around 1462 cm^{-1}, known as the pentagonal pinch mode. This tells us several things. Firstly, it tells us that

C60 is composed of sp2 bonded carbon. The sharpness of the band also tells us that the bonds are for the most part very uniform in nature. In fact, the carbon atoms in C60 are equivalent and indistinguishable. In contrast, the spectrum of the produced nanocarbon forms (see in figure 4) is littered with numerous bands. This is due to a reduction in molecular symmetry which results in more Raman bands being active. So, we can suppose complex fullerenes forms or nanosized crystalline graphite as resulting nanocarbon forms after cavitational bubbles collapsing.

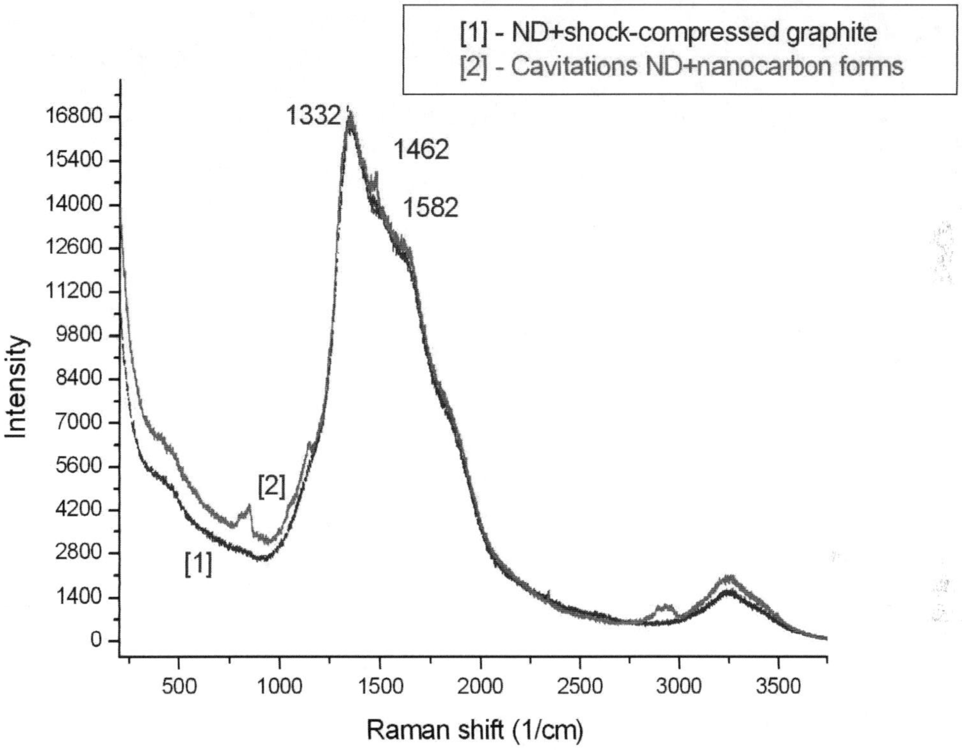

Figure 4. Raman spectrum of different samples: [1] – ND and shock-compressed graphite mixture; [2] – nanocarbon forms after hydrodynamical cavitation at benzene.

4. Conclusions

The experimental setup description and the results of the experiments with the hydrodynamical cavitation at benzene are presented. Grains of diamonds in size of 5-20 nm and aggregates of different carbon nanostructures are obtained. The transmission electron microscopy and Raman spectroscopy methods were used to compare produced nanocarbon forms with industrial nanodiamond grains from the shock-compressed graphite. These results support the idea academician E M Galimov that the thermodynamic conditions inside of cavitational bubbles during collapse are appropriate for the diamond phase formation and provide new alternative method of carbon nanostructures synthesis.

Acknowledgments

The authors would like to thank Prof. E Goodilin for his valuable help with Raman spectral analysis and Dr. V Shkinev for his help with chemical treatment of benzene.

References

[1] Lyamkin A, Petrov E, Ershov A, Sakovich G, Staver A and Titov V 1988 *Dok.Akad. Nauk SSSR* **302** 611
[2] Aleksenskii A, Baidakova M, Vul' A Ya and Siklitski V 1999 *Phys.Solid.State* **41** 668
[3] Whittaker A 1978 *Science* **200** 763
[4] Vora H and Moravec T 1981 *J. Appl. Phys.* **52** 6151
[5] Guo T, Nikolaev P, Thess A, Colbert D and Smalley R 1995 *Chem. Phys. Let.* **243** 49
[6] Galimov E M 1973*Nature* **243** 389
[7] Galimov E, Kudin A, Skorobogatskiy V and Aronin A 2004 *Dok. Acad. of Sciences* **395** 187
[8] Quinto-Su P 2010 *Physical Review Letters* **104** 14501
[9] Dnestrovskiy A, Voropaev S and Ponomareva E 2011 *Doklady Physics* **56** 78
[10] Iakoubovskii K, Mitsuishi K and Furuya K 2008 *Nanotechnology* **19** 155705
[11] Zerda T, Xu W, Zerda A, Zhao Y and Von Dreele R 2000 *Carbon* **38** 355

Journal of Physics: Conference Series **291** (2011) 012029 doi:10.1088/1742-6596/291/1/012029

Superstrength of nanostructured alloys produced by SPD processing

R Z Valiev*[,1], N A Enikeev[1], X Sauvage[2]
[1]Institute of Physics of Advanced Materials, Ufa State Aviation Technical University, 12 K. Marx. Str., Ufa 450000, RUSSIA

[2]University of Rouen, Groupe de Physique des Matériaux, CNRS (UMR 6634), Avenue de l'Université – BP 12, 76801 Saint-Etienne du Rouvray, France

E-mail: rzvaliev@mail.rb.ru

Abstract. The metals and alloys subjected to severe plastic deformation (SPD) can possess not only ultrafine-grained structure but also specific nanostructural features, such as non-equilibrium grain boundaries, nanotwins, grain boundary segregations and nano particles. The present work considers the role some of these features in exhibition of high strength of nanostructured metals and alloys. In particular, it is demonstrated that the presence of grain boundary segregations and non-equilibrium boundaries can result in yield stress values that considerably exceed those predicted from the Hall-Petch relation for the given materials.

1. Introduction

Nanostructuring is the new and promising way to enhance the properties of metals and alloys for advanced structural and functional application [1]. To date, it is well established that bulk nanostructured materials (BNM) can be produced successfully via grain refinement using severe plastic deformation (SPD), i.e. heavy straining under high imposed pressure [2]. SPD processing is an attractive procedure for many advanced applications, as it allows enhancing significantly properties of commonly used metals and alloys. Since the pioneering work on producing UFG materials by SPD processing [3], two SPD techniques have attracted close attention and have lately experienced further development [2]. These techniques are high-pressure torsion (HPT) and equal-channel angular pressing (ECAP). For the last 10-15 years there appeared a wide diversity of new SPD techniques: for example, accumulative roll bonding (ARB), multi-axial forging, twist extrusion and others (see for example [2] for a more comprehensive review). Nevertheless, processing by HPT and ECAP has remained the most popular approach and recently this has acquired a new impulse for development through the modification of conventional die-sets and demonstrations that new opportunities are now available for involving these procedures in industrial processing, e.g. using ECAP with parallel channels and continuous ECAP-Conform [4].

Grain refinement is well known to result in strength enhancement of metals and alloys, with the experimental relation between yield strength σ_y and a mean grain size d described by the Hall–Petch relationship: $\sigma_y = \sigma_{y0} + k_y\, d^{-1/2}$, where σ_{y0} and k_y are material's constants [5,6]. However, recently the

[1] Corresponding author

Journal of Physics: Conference Series **291** (2011) 012029 doi:10.1088/1742-6596/291/1/012029

results were reported that ultrafine-grained (UFG) alloys produced by SPD processing can exhibit a considerably higher strength than the Hall–Petch relationship predicts for the range of ultrafine grains [7-9]. The nature of such superstrength may be associated with another nanostructural features (dislocation substructures, nanophase particles and segregations, nanotwins et al) which could be observed in the SPD-processed metals and alloys [7,9].

In recent years in our laboratory in close collaboration with colleagues and partners there was performed a number of investigations of unusual mechanical performance of SPD-processed Al and Ti alloys as well as in several steels [7-11]. The aim of present study is to consider and investigate the manifestation and nature of high strength observed in these nanostructured materials.

As it is known for SPD processing by ECAP and HPT it is typical to form ultrafine-grained structures with mean grain sizes within the submicrometer range so that, typically, the grain sizes are ~100-500 nm [4]. However, in the process of SPD the formation of other nanostructural elements takes place as well like dislocation substructures, twins, grain boundary segregations and precipitations. They may also significantly influence the materials' properties after processing [9, 12-15].

An example of remarkable contribution of nanostructured elements formed in the UFG materials processed by SPD to the change of yield stress and flow stress was recently studied in detail for the case of UFG Al alloys [7,8,13]. Namely, 1570 alloy (Al, 5.7 Mg, 0.32 Sc, 0.4 Mn, wt.%) and 6061 alloy of Al-Mg-Si system were nanostructured by HPT-processing, no subsequent heat-treatment was applied.

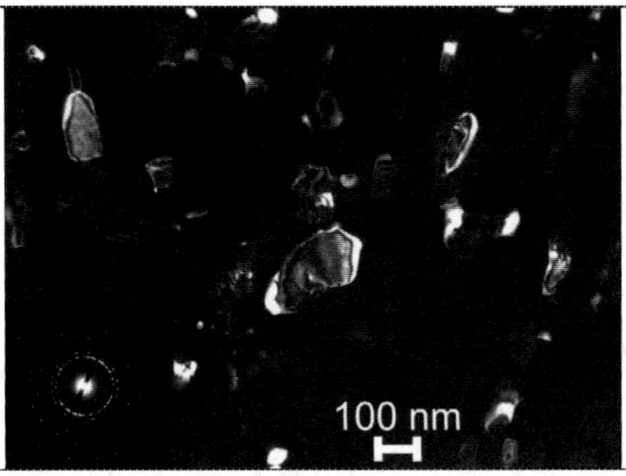

Figure 1. A dark-field TEM image of the UFG 1570 alloy with a corresponding SAED.

2. Results and Discussion

TEM observations showed that the HPT processing of the alloys leads to complete refinement of their initial coarse-grained structure. The dark field image presented in Fig. 1 illustrates the homogeneous UFG structure formed in the HPT 1570 alloy with the mean grain size of 97±2.0 nm. The SAED pattern (Fig.1 inset) testifies that the ultrafine structures are characterized predominantly by high-angle grain boundaries. The 6061 alloy demonstrated similar microstructure features [13].

Table 1 displays the mechanical tests data for the 1570 and 6061 alloys in coarse-grained (CG) and HPT-processed states. These data demonstrate that UFG 6061 and 1570 alloys exhibit significantly increased strength as compared to CG states.

One can analyze the obtained data thanks to a Hall–Petch plot where the yield stress ($\sigma_{0.2}$) is plotted against the inverse square root of the grain size ($d^{-1/2}$) (Fig. 2). For comparison, data reported by Tsuji for a UFG 1100 Al alloy obtained by ARB [16] and data reported by Furukawa et al for the UFG Al-3%Mg alloy obtained by ECAP [17] were also added on the plot. The data for 1100 Al alloy highlight the Hall-Petch slope resulting from the SPD-induced refinement while the Al–3%Mg alloy exhibits the additional contribution of solid solution hardening.

Table 1. Mechanical properties of the 1570 and 6061 alloys in UFG and CG states

Alloy	CG 1570	UFG 1570	CG 6061	UFG 6061
	(solid soluted)		(solid soluted+quenched)	
σ_{02} (MPa)	190	905	150	660
UTS (MPa)	376	950	275	690

From the presented data it may be seen that the $\sigma_{0.2}$ values for 6061 and, especially, 1570 alloy are notably higher than expected from an interpolation of the AA1100 or the Al–3%Mg Hall-Petch slopes.

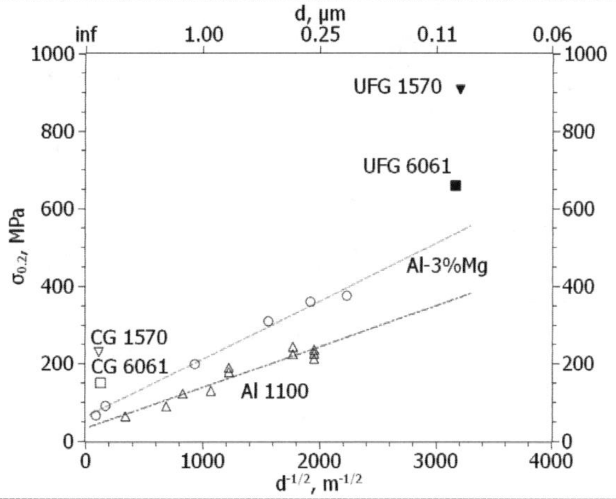

Figure 2. The Hall–Petch relation for the Al alloys: 1100 [17], Al–3%Mg [18] and the data on the yield stresses of CG and UFG Al alloys 1570 [7] and 6061 [13].

Since the solid solution strengthening component is not a function of the grain size, one can suppose that this manifestation of superior strength exceeding Hall-Petch predictions may result from specific microstructure features of the HPT-processed alloys. The atom probe tomography technique was applied to investigate the distribution of solute elements in the alloys processed by HPT.

Fig. 3 shows some data collected on the UFG 1570 alloy [7]. The 3D reconstructed volume (Fig. 3a) clearly exhibits a planar segregation of Mg. Since Al atomic planes are resolved on the right part while they disappear on the left, it is thought that there is a significant disorientation between these two regions and that Mg atoms have segregated along a grain boundary. Such segregations give rise to a significant decrease of Mg in solid solution (from 7.0 ± 0.2 down to 6.4 ± 0.2 at.%), also confirmed by X-ray diffraction data [10]. The composition profile computed across the boundary (Fig. 3b) reveals that the local concentration is up to 30 at.% Mg within a layer of about 6 nm width. Recently, we observed a very similar behavior of segregations at grain boundaries also in a number of Ti alloys as well as low-carbon steel subjected to SPD and these data correlated with displaying by these materials of very high strength [9, 11].

It is natural to suppose that grain boundary segregations may influence the mechanical behavior of the UFG alloys. It is well established that deformation of UFG materials (with grains larger than 30–50 nm) is mainly associated with intragranular movement of dislocations [12]. Dislocations are generated at grain boundaries and move through a grain to be captured by an opposite boundary. In this case the rate-controlling mechanism is "dislocation-grain boundary" interaction. Elevated concentration of solutes in grain boundaries can suppress emission of dislocations due to solute drag.

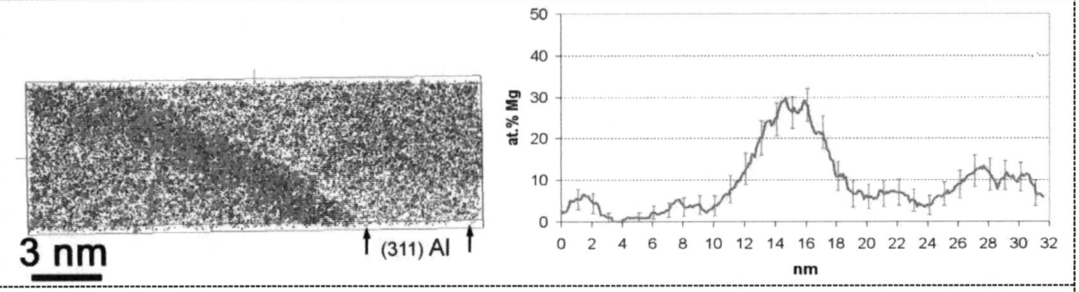

(a) (b)

Figure 3. 3D image of an analyzed volume in the UFG 1570 alloy [7]: (a) a planar segregation of Mg (Al atoms are displayed as dots and Mg atoms as bubbles). (311)Al atomic planes on the right of the planar segregation are displayed; (b) concentration profile computed across the segregation.

3. Conclusion

The conducted investigations testify to the fact that yield stress of UFG metals and alloys produced by SPD may be considerably higher than it is predicted by Hall-Petch relation for their grain size range. The observed superstrength of these materials is attributed to nanostructural features of their grain boundaries – their non-equilibrium state and segregations. Thus, the problem of detailed investigations of grain boundaries in UFG materials produced by SPD is topical for the achievement of high strength in nanomaterials. Besides, it is important to study the role of intragranular nanostructural elements – nanotwins, nanoparticles, which can lead to additional strengthening of the materials.

References

[1] Gleiter H 2000 *Acta Mater.* **48** 1
[2] Valiev R Z, Estrin Y, Horita Z, Langdon T G, Zehetbauer M J, Zhu Y T 2006 *JOM* **58** (4) 33
[3] Valiev R Z, Krasilnikov N A, Tsenev N K 1991 *Mater. Sci. Eng.* A **137** 35
[4] Valiev R Z, Langdon T G 2006 *Prog. Mater. Sci* .**51** 881
[5] Hall E O 1951 *Proc. Phys. Soc. Lond.* **64b** 747
[6] Petch N J 1953 *J. Iron Steel Inst.* **174** 25
[7] Valiev R Z, Enikeev N A., Murashkin M Yu, Kazykhanov V U, Sauvage X 2010 *Scripta Mater.* **63** 949
[8] Valiev R Z, Enikeev N A, Murashkin M Yu, Aleksandrov S E, Goldshtein R V 2010 *Doklady Phys.* **55**(6) 267
[9] Valiev R Z, Langdon T G 2010 *Adv. Eng. Mater.* **12** 677
[10] Murashkin M Yu, Kilmametov A R, Valiev R Z 2008 *Phys. Met. Metallgr.*, **106**(1) 93
[11] Semenova I, Salimgareeva G, Da Costa G, Lefebvre W, Valiev R Z 2010 *Adv. Eng. Mater.* **12** 803
[12] Valiev R Z, Islamgaliev R K, Alexandrov I V 2000 *Prog. Mater. Sci.* **45**103
[13] Nurislamova G, Sauvage X, Murashkin M, Islamgaliev R, Valiev R Z 2008 *Phil. Mag. Lett.* **88** 459
[14] Sha G, Wang Y B, Liao X Z, Duan Z C, Ringer S P, Langdon T G 2009 *Acta Mater.* **57** 3123

[15] Horita Z, Smith D J, Furukawa M, Nemoto M, Valiev R Z, Langdon T G 1996 *J. Mater. Res.* **11** 1880
[16] Tsuji N 2006 In Zhu Y T and Varyukhin V (Eds.) *Nanostructured materials by high-pressure severe plastic deformation* (Springer Netherlands, 2006) p 227
[17] Furukawa M, Horita Z, Nemoto M, Valiev R Z, Langdon T G 1998 *Phil. Mag. A* **78** 203

Antituberculous effect of silver nanoparticles

G N Kreytsberg[1], I E Gracheva[1], B S Kibrik[2], I V Golikov[3]

[1]Limited liability Company "Scientific and Production Association (NPO)"Likom", Russia, 150049, Yaroslavl, Magistralnaya str., 32.

[2] The Yaroslavl State Medical Academy
Russia, 150000, Yaroslavl, Revolutsionnaya str., 5

[3] The Yaroslavl State Technical University
Russia, 150023, Yaroslavl, Moskovskiy avenue, 88

e-mail: likomm@yaroslavl.ru

The in vitro experiment, involving 1164 strains of the tuberculosis mycobacteria, exhibited a potentiating effect of silver nanoparticles on known antituberculous preparations in respect of overcoming drug-resistance of the causative agent. The in vitro experiment, based on the model of resistant tuberculosis, was performed on 65 white mice. An evident antituberculous effect of the nanocomposite on the basis of silver nanoparticles and isoniazid was proved. Toxicological assessment of the of nanopreparations was carried out. The performed research scientifically establishes efficacy and safety of the nanocomposite application in combination therapy of patients suffering from drug-resistant tuberculosis.

Tuberculosis problem

The current trait of the problem is expressed in extremely low efficacy in treatment of patients, particularly, concerning those who are diagnosed on a result of their visit to a doctor. In many cases the efficacy falls below 30%, which proves low efficiency of treatment rate of the modern strategy in tuberculosis treatment [5, 11]. The main reason lies in drug resistance of the tuberculosis mycobacterium to the existing numerous drugs, which to a large degree have lost their potency [7]. In Russia milliards of roubles are spent on free treatment of tuberculosis every year. Due to drug resistance considerable part of the sponsored funds, before the treatment starts, can be regarded and planned as useless and futile expenses. Ineffective treatment promotes formation and occurrence of "new" tuberculosis, called chemo-resistant. The research objective is to evaluate efficacy, safety and tolerance of the silver nanoparticles in isolated form as well as in joint application with the antituberculous preparation isoniazid in vitro and in vivo research during treatments of experimental drug-resistance cattle tuberculosis

Journal of Physics: Conference Series **291** (2011) 012030 doi:10.1088/1742-6596/291/1/012030

Treatment and nanoparticles of silver

One of the ways out of the existing situation might be the search for the means to reverse the known antituberculosis preparations directed at overcoming the tuberculosis mycobacteria resistance during their interaction with the chemi-pharmacological agents. The nanoparticles of silver are discussed to be such an agent . The history of the scientific research of silver proves its pronounced antiseptic properties. The history of silver research studies established that the daily diet of a man should contain on average 80 micrograms of silver, since the latter is a microelement essential for maintaining normal metabolism as well as regular functioning of his immune system. The same research proved that the doses of silver in the volume of 50 – 250 mg per liter are physiological and during prolonged consumption have no negative effect on the human organism. The "Sanitary Regulation and Standards 2.1.4.1074-01", existing in the Russian Federation, state that the Maximum Concentration Limit (PDK) of silver cations in drinking water should be within 0,05 mg per litre. According to a number of researches the doses of silver in volumes of 50 – 200 mg per litre eliminate possibility of argyria. The effect of silver exceeds similar concentrations of chlorine, chlorinated lime and sodium hypochloride as well a number of other potent oxidizers. The research established bacteriostatic and bactericidal action of ions of silver on drug-resistant strains of staphylococcus, Proteus vulgaris, blue pus bacillus and colon bacillus. It was established that ions of silver are able to inactivate vaccine of variolovaccine, strains of the A-1 and B flue, of entero and adenoviruses. The dose-dependent action of ion solutions of silver towards a number of causative agents was proved. For instance, complete elimination of colon bacillus bacteriophage N 163, the Coxsackie virus of the serotype A-5, A-7, A-14, requires a higher concentration of silver (500 – 5000 mg per litre) than elimination of Escherichia, Salmonella, Shigella and causative agents of other enteric infections (100 – 200 mcg). [2, 8, 10].

Mechanism of bactericidal action of nanosilver

The main function of the silver nanoparticles is a continuous accumulation of silver ion s in the vicinity of the affected area. This is to protect the silver ions from the interaction with the anions of salts in the blood, and thereby increase the therapeutic effect of application composition .

The mechanism of antimicrobic action of silver has not been studied exhaustively. Adsorption theory focuses on derangement of electrostatic interactions, which occurs between the cells of bacteria, charged negatively, and positively charged ions of silver. Some researches explain the bactericidal action of silver by its catalytic properties, which trigger "the oxygen cascade" accompanied by destruction of bacteria protoplasm. Certain data is available, indicating that silver ions interact with the exterior peptidoglycanes of cell membrane, which promotes interruption of oxygen delivery to bacterial cell. A number of researchers suggest that the effect of the antimicrobic action of silver might be explained by the fact that it inhibits the transmembrane electrolyte transport. Other data draw attention to formation of complexes of nucleic acids with silver ions, which disturbs stability of the DNA and bacteria viability. Research of interaction of ions of silver with the cells of pro- and eukaryotic microorganisms indicated that the biocidic effect of silver substrates is determined by its attachment to the membrane-associated protein and lipid stroma of membranes, which leads to changes of the transmembrane potential and necrocytosis. [2, 9].

Technologies of nanosilver preparation have raised topicality of its application to a new level. [6, 12, 13, 14, 15, 16, 17]. Utilization of silver in its nanodimensional range makes it possible to reduce its concentration hundredfold and at the same time multiplying its biocidic properties. Study of bactericidal properties of silver nanoparticles indicated that they possess a more pronounced biocidic effect in comparison with ionic silver [1, 3, 4, 9,]. Comparative study of antimicrobial properties in ionic and cluster form proved that application of silver nanoparticles on suspension of yeast cells results in destruction of the membrane surfaces, but utilization of ionic silver is not accompanied by destruction of the cells. [4].

Materials and methods

For the first time Russian researchers demonstrated the antituberculous effect of silver nanoparticles with isoniazid. [18].

The water solution of particles of the silver, transferred to tests, is received by a method in detail described in the patent application [19], are characterized by the spectroscopy and electron microscopy method. Results are presented in the *table* and the *figures 1*, (conducted by the TsKP DMIS (Common User Facilities "Diagnostics of micro- and nanostructures"of Yaroslavl).

Table 1. Key characteristics of silver nanoparticles.

Determined characteristics	Test results
Silver nanoparticles size, nm	From 5 to 50
Number of silver nanoparticles in 1 mcm²	120-270
Size of the stabilizer-formed membrane	From 2 to 5

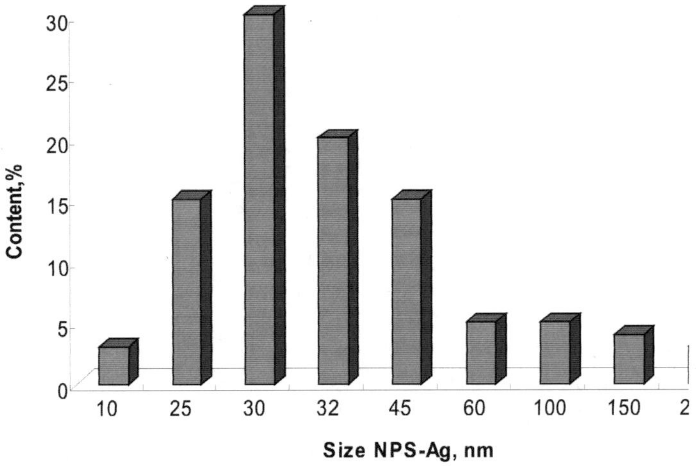

Figures 1. Distribution of the nanoparticles according to sizes

Water solutions of silver nanoparticles were analysed in concentration 2,5 mg/l, 5 mg/l, 25 mg/l, 50 mg/l and as antitubercular activity of a composition of silver nanoparticles in concentration 5 mg /l, 25 mg /l, 50 mg /l in a combination to 1 mg of an isoniazid is studied are taken [20].

Research has been performed according to the recommendations stated in «the Management on experimental to studying of new forms of pharmacological substances» [21].

Statistical data processing has been performed using the standard methods of the statistical analysis at medical and biological researches. Reliability level was estimated as sufficient at $p>0,05$.

Results and their discussion

We performed bacteriological research on 1164 clinical isolates of Koch's bacillus (MTB), bacteriological research, on 65 white mice, and evaluation of general toxicity on 83 non-linear white mice as well as on 146 white rats. We carried out comparative evaluation of antituberculous activity of the isolated nanoparticles in concentrations of 5, 25 and 50 mg per l and the nanocomposite, which contained the specified concentration of nanosilver in combination with antituberculous preparations that inhibited growth of the MTB strains, featuring initial resistance to the used preparations [20].

The efficacy study of various silver nanoparticles concentration in isolated mode of application (*figures 2*) indicated that at concentration of 5 mg/l complete growth inhibition of the MTB occurred 3,5 times more often, than at concentration of 25 mg/l and 7 times more often than at concentration of 50 mg/l.

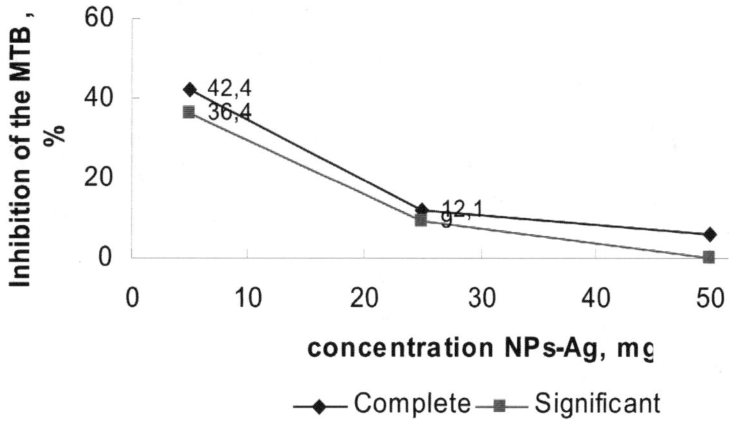

Figures 2. Inhibition of the MTB growth by the nanoparticles of silver solutions of various concentrations in isolated mode of application.

Results of the research, performed to establish efficacy of the silver nanoparticles in concentration of 2.5, 5, 25 и 50 mg/l in combination with individual antituberculous preparations, is represented in the chart of the figures 3.

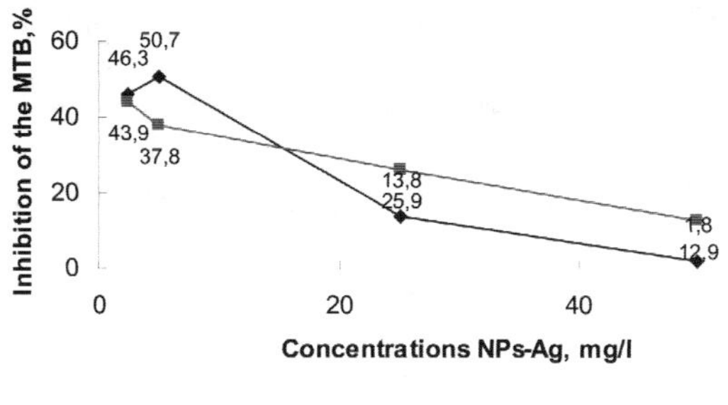

Figures 3. Inhibition of the MTB growth by the nanocomposites with various concentrations of the silver nanoparticles.

Efficacy of combined application of the nanoparticles with chemopreparations significantly exceeded isolated mode of the nanoparticles application. It is worth noting that maximal inhibiting efficacy was exhibited at concentration within the range of 2.5 – 5.0 mg/l.

The research data as obtained in vitro indicate the possibility of potentiating effect of silver nanoparticles on the total spectrum of known antituberculous chemopreparations in respect of overcoming drug-resistance of the Koch's bacilli. The highest bacterial growth-inhibitory activity was displayed by the silver nanoparticles with isoniazid 95,4%, with rifampicin – 93,3%, ethionamide – 100%, levofloxacin - 100% and ofloxacin - 100%. The lowest efficacy was observed (61,2%) in case the silver nanoparticles were used in combination with kanamycin. *(figures 4).*

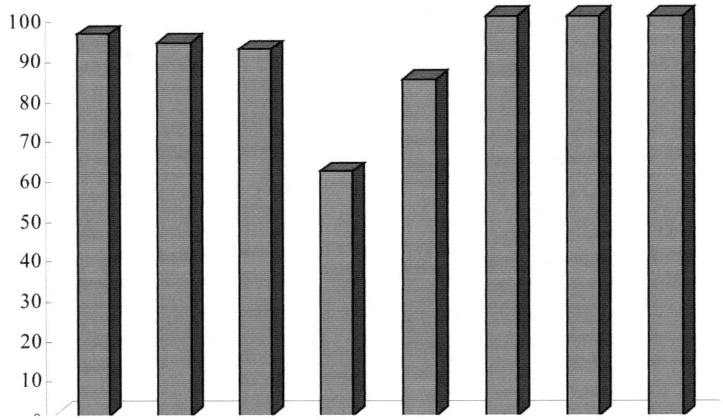

Figures 4. Bacterial growth-inhibitory activity of nanocomposites on the basis of silver nanoparticles and chemopreparations (H - isoniazid, R-rifampicin, S – streptomycin, K – kanamycin, E - ethambutol, Ea - ethionamide, Lft – levofloxacin, Ofl – ofloxacin, Cs - cycloserine.)

The results of the bacteriological research in vitro significantly indicate existence of potentiating bacteriostatic and bactericidal effect of silver nanoparticles towards the drug-resistant MBT strains that reaches 100 % in case of reserve preparations (drugs).

Research of the chemotherapeutic activity was performed in the in vivo experiment on 65 white mice. The laboratory animals were infected with a two-week virulent culture of the clinical M. tuberculosis strain, isolated from a tuberculous patient, resistant to traditional antituberculous drugs. The infective dose of the white mice amounted to $5x10^6$ CFU (colony-forming units), which was contained in 0,5 ml of the obtained suspension. The experimental animals were infected via the intravenous route, using the aforementioned dose in the area of retroorbital sinus.

When survival rate of the animals was estimated, it was revealed that at the moment of death of all 100% of the mice from the control group that didn't receive the treatment, the survival rate of the animals that were treated with nanoparticles in combination with isoniazid, amounted to 95%. Survival rate of the animals that received isolated nanoparticles of silver amounted to 35% *(figures 5).*

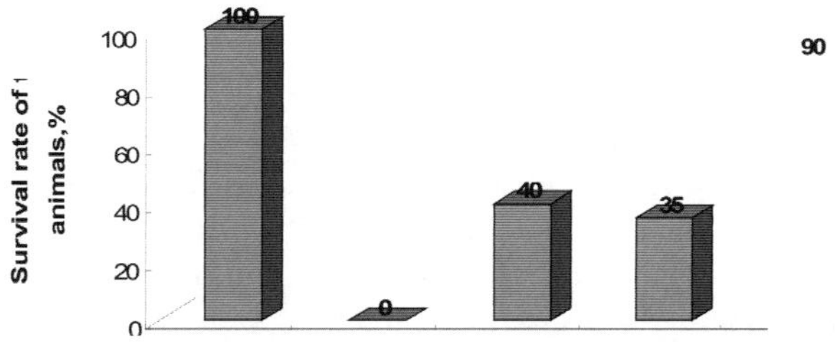

Figures 5. Survival rate of the animals by the 45th day of treatment (1– healthy, 2 – without treatment, 3 – isoniazid, 4 – nanoparticles, 5 – phthizarg)

Conclusions

The data, obtained in the course of the in vivo experiment made it possible to conclude with statistical validity that the antituberculous activity of the silver nanoparticles in isolated variant as well as used jointly with isoniazid possess a dose-dependent character. The maximal antituberculous activity on the model of the experimental drug-resistant tuberculosis was observed when the silver nanoparticles where used in the dose of 25 mg/kg and isoniazid in the dose of 50mg/kg.

The results of the therapeutic action of the silver nanoparticles on the models of generated tuberculosis of the white mice in terms of clinical doses of the preparation for a man, obtained in the course of the preclinical research, allow establishing a minimal therapeutic dose that equals 2 mg/kg.

The results of the performed research revealed a significant potentiating effect of silver nanoparticles on known antituberculous preparations (isoniazid, rifampicin, streptomycin, kanamycin, ethambutol, ethionamide, levofloxacin, ofloxacin, cycloserine) with possibility to overcome the drug resistance of the pathogenic agent.

List of bibliography

[1] Baranova E, Revina A, Voyno L and Gorbatiuk V 2003 *Comparison of effect of silver ions and silver nanoparticles on the yeast cells and colon bacillus (E Cili)* Moscow: Materials of the 1st Russian Scientific Methodological Workshop 53-60

[2] Blagitko E, Burmistrov V, Kolesnikov A, Mikhailov Y and Rodionov P 2004 *Silver in Medicine* Novosibirsk: Nauka-Center 254

[3] Gmoshinski I, Khotimchenko S and Vernikov V 2009 *Nanoparticles of Silver in nature, industry, packing materials, intended for food products: characteristics of possible risks* Issues of Nutrition **6** 13-20

[4] Yegorova E, Revina A, Rostovschikova T and Kisiliova O 2001 *Bactericidal and catalytic properties of stable metallic nanoparticles in inverted micelle* Moscow University Newsletter series 2: chemistry **42** 332 – 338

[5] Kibrik B and Chelnokova O 2005 *Acute progressive destructive forms of pulmonary tuberculosis* Yaroslavl 191

[6] Kuzmin S, Privalov L, Katsnelson B, Nikolaeva E, Degtiariova T, Minigalieve I, Valamina I, Kireeva E, Sutunkova M, Shur V, Yereminko O, Khodos M, Glazirina Y, Kozitsina A, Malakhova N and Shishkin E 2010 *Experimental data to evaluation of pulmonary toxicity and resoptive toxicity of magnetite particles (Fe304) of nano- and micrometer range* Toxicological bulletin **2** 17 – 24

[7] Mishin V 2009 *Pulmonary tuberculosis with drug resistant of the causative agent* Moscow: GEOTAR-Media 208

[8] Odegova G, Burmistrova V and Rodionov P 2004 *Study of silver state in silver-containing antibacterial preparations argovit and argogel* Novosibirsk: Collection of scientific papers, connected with research and practice conference "New chemical systems and processes in medicine" 58 – 63

[9] Parkhomenko N, Yurkova I and Riabushko V 2008 *Antibacterial and antifungal effect of water-soluble silver-based nano-bio formulations and sea biopolymers* Works of Taurida National University named after V.I. Vernadskiy. Biology and Chemistry Series **21** 106 – 112

[10] Blagitko E 2004 *Applications of silver preparations in medicine* Novosibirsk: Collection of scientific papers, connected with research and practice conference "New chemical systems and processes in medicine" 115

[11] Shilova M and Khruliova E 2005 *Efficacy of tuberculosis patients treatment at the present stage* Problems of tuberculosis and lung disease **3** 3-11

[12] Samsonova M 2008 *Nanomedicine: modern approach to diagnostics and treatments of diseases, problems of safety* Pulmonology **5** 5 – 13

[13] Shvetsh V, Gelperina C, Maksimenko O, Vanchugova L, Bondarenko G, Shipulo E and Shandruk G 2010 *Influence of toxicological parameters on physicochemical properties of poli-lactide nanoparticles, containing rifampicin* Yaroslavl 42 – 47

[14] Sahoo S, Parveen S and Panda J 2007 *The present and future of nanotechnology in human health care* Nanomedicine: Nanotechnology/ Biology and Medicine **3** 20 -31

[15] Alt V, Bechert Th, Steinrücke P, Wagener M, Seidel P, Dingeldein E, Domann E and Schnettler R 2004 *An in vitro assessment of the antibacterial properties and cytotoxicity of nanoparticulate silver bone cement* Moscow: Biomaterials **25** 4383-4391

[16] *Safety evaluation of nanomaterials* 2008 Moscow: Federal Service on Surveillance for Consumer rights protection and human well-being, approved on 12 oct 2008 **280**

[17] Shahverdy A, Fakhimi A and Minaian S 2007 *Synthesis and effect of silver nanoparcles on the antibacterial activity of different antibiotics against Staphylococcus and Escherichia coli* Nanomedicine:Nanotechnology/ Biology and Medicine **3** 168–171

[18] Russian Agency for Patents and Trademarks Registration № 2008142461 of 28 oct 2008, decision of grant of 22 apr 2010, № 151410

[19] Russian Agency for Patents and Trademarks patent № 2390344 for the invention *Ways of generation of silver nanoparticles in aqueous medium*

[20] Kibrik B, Zaharov A and Kreytsberg G 2010 *Nanocomposite "isoniasidi+Ag" in the treatment of tuberculosis in vitro, in vivo* Yaroslavl: "YarMediaGrup" 97 – 195

[21] Habriev R 2005 *Management on experimental to studying of new pharmacological substances* Moscow

Multiscale modelling of damage and fracture processes in nanomaterials

V Lepov[1]
V.P. Larionov's Institute of Physical-Technical Problems of the North SB RAS
1, Oktaybrskaya str., Yakutsk, 677980, Russia

E-mail: lepov@iptpn.ysn.ru

Abstract. The common multiscale evolution approach to structural modelling of the damage accumulation and fracture processes of submicrocrystalline materials is considered. The method of scanning probe microscopy has been applied to describe the damage evolution on nanoscale. The small-sized plane deformation specimens have been loaded in the tensile device integrated in the scanning probe microscope. The multifractal analysis has been used for the quantitative characterization of *in situ* deformation of the surface by means of scaling properties. To reveal the mechanism of low-temperature fracture the atomic-force microscopy has been applied to characterize the surface morphology of the stretching deformations of samples fractured at different temperatures.

1. Introduction

The multiscale modelling approach has been actively developed last years [1-6]. First of all, there are many conferences in material and computer sciences taking place where this problem has been discussed [1,3,7]. Some studies have very specific nature like a multiscale modelling of a quantum dot in a semiconductor solid containing a free surface [4], and here the lattice-statics and continuum Green's functions integrated with classical molecular dynamics are used. The same approach has been used to show significant differences in the deformation behaviour of nanocrystalline nickel with low and high angle grain boundaries [5]. Such works analyse the dislocation activity mainly and are restricted by the method on nanoscale only. Some possibility of multiscale approach was shown as realization of structural and statistical approach [6,7]. In general case the multscale evolution modelling approach for the damage nucleation and crack propagation in structural submicrocrystalline materials could be presented (see Figure 1). Here the electronic microfractography data means the analysis of micrographs taken by scanning probe or electron microscope, the gage theory means the calibration theory of defects in solids where the Lagrangian of quantum electrodynamics in form $L=\frac{i}{2}(\bar{\psi}\gamma^{\mu}\partial_{\mu}\psi-\partial_{\mu}\bar{\psi}\gamma^{\mu}\psi)-m\bar{\psi}\psi$ has been used [8]. The procedure of global symmetry localization of this Lagrangian has been used on place to obtain the partial calibration theory [9].

But one of the key feature of the multiscale modelling approach is the obtaining of data of damage nucleation on nanoscale during the deformation in situ. So the device and method of scanning probe microscopy for the in situ deformation of plane specimens has been used. The further quantitative estimation of the scaling features of surface images has been carried out by the multifractal analysis.

---

[1] To whom any correspondence should be addressed.

There is only a small number of similar studies known [10], and none of them is applied for the structural alloys and steels.

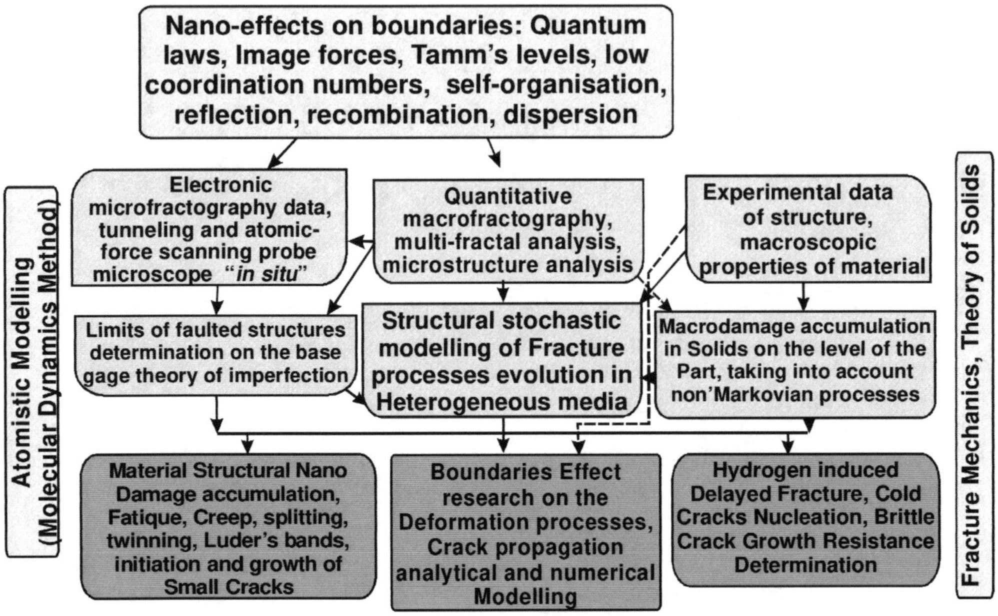

Figure 1: Concept scheme of Multilevel Modelling of Fracture

Evolution approach means taking into account the history of the each element on the certain level of structure. The ultimate state of the current level controls the transfer on the next level of structure. The main difficulty is the determination of the ultimate state because it is guided by the number of parameters. But in most cases it is possible to restrict the state to only one control order factor. Here a defect modelling by the calibration theory, not the atomistic simulation, is preferable. The main restriction of the molecular dynamics method is the small size of simulation volume, a necessity to simplify the quantum dependences for electron configuration and difficulty in connection to upper scale level [1-3]. The defect modelling on different structural levels allows simulating behaviour of material under the wide range of external influences without abovementioned faults. On the concept scheme the main points and methods used have been shown.

2. Materials and methods

As it has been mentioned beforehand it is the necessary to obtain the data of damage nucleation on nanoscale during the deformation in situ. So the device of scanning tunnelling and atomic-force microscopy for the in situ deformation of plane specimens has been developed and used. The first specially designed device has been produced by "Proton-MIET" company (Zelenograd, Moscow Distr.) as a combination of a scanning tunnelling microscope with a strain gauged loading setup. Recent experiments were performed using the second device containing the loading setup and the removable scanning platform with atomic-force probe suitable for structure experiments in different conditions (see Figure 2). Both microscopes have scanning platform (SP-STM and SP-AFM) construction. However the massive antivibration table and removable scanning platform in SP-AFM construction provide higher resolution and images of high quality are to be obtained much easy.

For the small-size specimens the ultrafine-grained steel produced by the method of intensive plastic deformation (equal-channel angular pressing) has been used [11].

In addition to revealing the mechanism of metal fracture under low temperatures, the fracture surfaces of construction elements and sample models are being studied. Optical and scanning tunnelling microscopy of fractured surfaces of samples under different external environments and after different types of influences on the metal has been realized, and a complex of metallographic and microstructure studies has been carried out also. So the small-size specimens were saturated at high temperature and high pressure by hydrogen. It is necessary to define the microdefects size distribution function, its shapes and average concentration depending upon the static and corrosion influence, for the obtaining the probabilistic damage estimations.

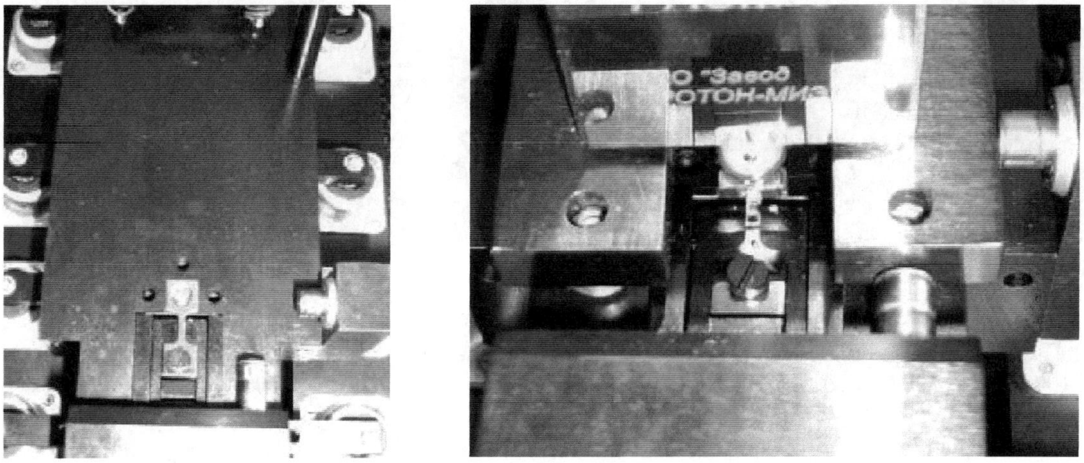

a b

Figure 2. View of (a) loading device and fixed plane tensile specimen on antivibration table and (b) SP-AFM probe placed on the table with specimen.

Journal of Physics: Conference Series **291** (2011) 012031 doi:10.1088/1742-6596/291/1/012031

3. Results and Discussion

Using the method of scanning tunnelling microscopy the fracture of low-alloys steel samples was explored. The samples were saturated preliminarily by the hydrogen at the 400˚C temperature and 10 MPa pressure conditions and tested for the delayed fracture sensibility. As the size of investigation area is 2,6×2,6 µm, the height of microrelief is achieved 0,2 µkm. The typical sizes of the secondary microcracks – depth is 0,2 and width is 0,3 µm. The indication that the cracks are propagating by the mechanism of nucleation and coalescence of such microdefects as micropores and microcracks is obtained here. It is agree with the known visualisation model of fracture [12].

The study of ultimate states of material is directly connected to the lifetime numerical estimation of constructions manufactured from widely used and new materials. The last investigations in this directions are linked to the study of influence of grain boundaries and interfaces, particularly, impurity segregations on grain boundaries, on the deformation and fracture of polycrystalline and nanostructural materials. By means of scanning tunnelling microscopy the "in situ" investigation of the evolution of surface damage during the deformation of small-sized specimens of the experimental cold-resistant steel was conducted. The fitted 4,5 kN loading device and the electronic tunnelling probe with 20×20×2 µm scanning field has been used. The results of scanning of the grain boundary during the deformation (the deformation curve is indicated on figure 3) have been shown on figure 4.

It is seen that before the extensive macroscopic plastic deformation the surface relief change and the microcracks is appeared on the 3D-images and profiles. The damaging has been numerically estimated by the methods of multifractal analyses of the scanned images [7].

The calculation of parameters of the multifractal spectrum has been performed for the equal areas (100 × 100 pixels or 580 × 580 nm) in grain boundary zone for the (a) initial state, and for the steps on deformation curve - (b) 1500, and (c) 1650 N (see fig.4). According to the initial surface image and multifractal parameters obtained the initial roughness is high and during the deformation is growing (see the table 1). The changing of fractal dimensions – the Hausdorff D_0, the correlative D_1 and the informational D_2 – is same for the deformation surfaces.

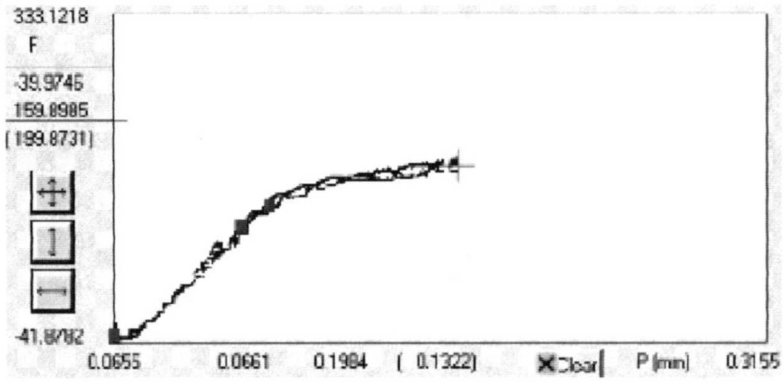

Figure 3. View of in situ loading control curve in electron probe scope.

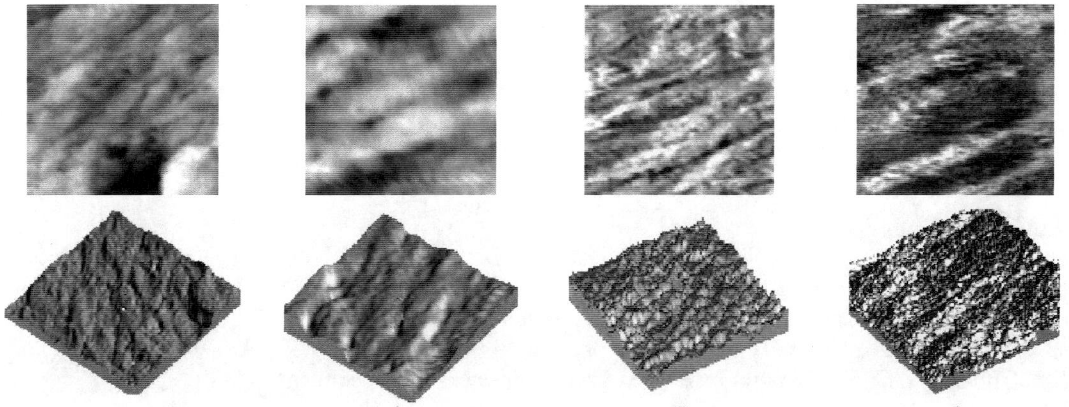

a b C d

Figure 4. The 2D and 3D-images of the probed surface at grain boundary during the deformation in situ, (a) in initial state, and at the steps on deformation curve - (b) 1350, (c) 1500, and (d) 1650 N.

Table 1. The multifractal spectre parameters of deformation surfaces in situ.

Parameter	On initial surface	On loaded at 1500 N	On loaded on 1650 N
D0	2,389	2,455	2,376
D1	2,277	2,297	2,258
D2	2,206	2,218	2,179
K	1,505	2,518	1,799
Δ	0,325	0,462	0,433
f(40)	0,528	0,107	-0,292

Concerning the structure of the investigated surfaces, it has been the hidden periodicity measure $K=D(-40) - D(40)$ that estimate the regularity of surfaces, and measure of order $\Delta=D(1) - D(40)$. The higher values of thus measures mean the more periodic components the surface contains. The last parameter of the multifractal spectrum reveals the porosity of surface, and during the deformation it is monotonic decreasing.

The atomic-force microscopy of the morphology of surface stretching deformations of samples fractured at different temperatures has been carried out also. Microstructure peculiarities of damage accumulation and the character of nanostructured steel fractures at low temperatures were revealed. The microstructural aspects of deformation and fracture of nanostructured steel at low temperature (–60°C) are considered. By the methods of optical and atomic-force microscopy the deformation surfaces of the fractured tension probes was studied. The fracture character changes (from a brittle cleavage to a ductile shear) at low temperatures, it has been shown to be associated, on the one hand, with a fuzzing of grain boundaries during angular pressing with solid structural phases and, on the other hand, by the replacement of the prevalent plastic deformation carriers [13].

4. Summary

According to the common multiscale evolution approach to structural modelling of the damage accumulation and fracture processes of submicrocrystalline materials the method of scanning probe microscopy has been applied to subscribe the damage evolution on nanoscale. The small-sized plane specimens have been loaded in the tensile device integrated in scanning probe microscope. For the quantitative characterization of the damage of in situ deformation surface the multifractal analysis has been used. To reveal the mechanism of low-temperature fracture the atomic-force microscopy of stretching deformation surfaces has been applied. The fracture character changes have been shown to be associated with a fuzzing of grain boundaries and by the replacement of prevalent plastic deformation carriers.

Acknowledgements
The support of Russian Foundation for Basic Research (Grant 09-01-98511) and Presidium of Russian Academy of science (Project 27.40) is gratefully acknowledged. Author thanks Loginov B.A., Head of Laboratory of atomic modification and semiconductor surface analysis of Moscow Institute of Electronics for the co-operation and help in experimental work, and Ivanov A.M., senior researcher of IPTPN SB RAS to help in preparation of the ultrafine-grained steel specimens.

References

[1] Sih G C, editor 2007 *Multiscaling in Molecular and Continuum Mechanics: Interaction of Time and Size from Macro to Nano* (London: Springer)
[2] Loehnert S, Belytschko T 2007 *Int. J. Num.. Meth. Eng* **71** 1466
[3] Meguid S A, Wernik J Multiscale modelling of the nonlinear behaviour of nano-reinforced composite interfaces 2009 *Proceedings of the International Conference IRF'*2009 ed J F Silva Gomes and Shaker A. Meguid pp 5-6
[4] Tewary V K, Read D T 2005 *Nanomechanics of Materials and Structures* **342** 89
[5] Vaithyanathan V, Wolverton C, Chen L Q 2002 *Phys.Rev.Lett.* **88** 125503
[6] Lepov V, Ivanova A, Achikasova V, Lepova K 2007 *Key Engineering Materials* **345-346** 809
[7] Lepov V V, Ivanova A A, Achikasova V S, Lepova K Ya 2008 Structural-statistical Aspects of the Damage Accumulation and Fracture Modelling *ICF Interquadrennial Conference. Fracture Mechanics in Design of Fracture Resistant Materials and Structures* ed R V Goldstein
[8] Abers E.S., Lee B.W. 1973 *Phys. Reports (Section C of Phys. Lett.)* **9** 1
[9] Yang C N, Mills R L 1954 *Phys.Rev.* **96** 191.
[10] Bagrov D, Yaminskiy I 2008 *Nanoindustry* **5** 32 (in Russian)
[11] Ivanov A M, Lukin E S, Petrova N D, Vashenko S S 2010 *Reviews on advanced materials science* **25** 203
[12] Broberg, K B 1997 *Computational Mechanics* **19** 447
[13] Lepov V V, Ivanov A M, Loginov B A, Bespalov V A, Achikasova V S, Zakirov R R, Loginov V B 2008 *Nanotechnologies in Russia* **3** 734

Journal of Physics: Conference Series **291** (2011) 012032 doi:10.1088/1742-6596/291/1/012032

Study of structure and vibrational spectra of silicon and titanium dioxide nanoclusters for technology of composite materials

V A Astapenko and L P Sukhanov[1]

Moscow Institute of Physics and Technology, 9 Institutsky Side-Street, Dolgoprudny of Moscow Region 141700, Russia

E-mail: sukhanov1955@mail.ru

Abstract. The formation of structure and IR spectra of the $(SiO_2)_n$ (n = 1-4) and $(TiO_2)_n$ (n = 1-3) clusters with the increasing of their size n are simulated using the B3LYP/6-31G* method of density functional theory. It is shown that the structural fragments of the low-lying isomers of even small $(SiO_2)_n$ and $(TiO_2)_n$ clusters are exhibited in IR spectra of the SiO_2 and TiO_2 films. The IR absorption bands calculated by us for structures of the $(SiO_2)_n$ and $(TiO_2)_n$ clusters are in good agreement with the bands observed in IR spectra of the SiO_2 and TiO_2 films.

1. Introduction

Nanoclusters of silicon and titanium oxides are building blocks in preparation technology of a number of nanomaterials. The interest to silicon oxide clusters has increased highly after synthesis of silicon nanowires [1,2], in which the nucleation and growth of nanowires have been achieved by either laser ablation or thermal evaporation of Si powder mixed with SiO_2. A high yield of nanowires has been obtained when the elemental composition of Si:O $\approx$ 1. The important role of oxide participation in the synthesis of nanosilicon has thus been demonstrated.

Over the past two decades, the essential efforts have been made to prepare, characterize, and employ nanocrystalline forms of TiO_2, as a result of which the films consisting of TiO_2 nanoparticles in the 1-100 nm size range are now known (see, for example, [3]). Compared with the bulk, the nanocrystalline forms of TiO_2 are characterized by a dramatically increased surface-to-bulk ratio, which makes them attractive for photocatalytic and photoelectrochemical applications (see, for example, reviews [4-6]).

Up to now the pure and composite SiO_2 - TiO_2 films were also prepared by the sol - gel technology, and their optical and structural properties were investigated by IR spectroscopy method [7], and in paper [8] by atomic force microscopy, ellipsometry, thermal gravimetric analysis, UV - visible spectroscopy, Fourier transform IR spectroscopy, x-ray photoelectron spectroscopy and Raman spectroscopy.

The database on IR absorption spectra of the SiO_2 and TiO_2 nanoclusters is necessary for the monitoring of the technological processes of production of nanostructured films of composite materials. In this paper on the basis of density functional theory [9] by the GAUSSIAN98 computer

[1] Author to whom any correspondence should be addressed.

code [10], the structure and vibrational spectra of the $(SiO_2)_n$ (n = 1-4) and $(TiO_2)_n$ (n = 1-3) clusters are calculated with the aim of the database formation. The Becke's hybrid three-parameter exchange functional [11] along with the Lee-Yang-Parr's correlation functional [12], i.e., B3LYP functional, was used. The Pople's 6-31G* basis set [10] was chosen as the optimal one in all our calculations. It consists of the contracted $[5s4p2d1f]_{Ti}$ + $[4s3p1d]_{Si}$ + $[3s2p1d]_O$ basis functions. The basis set 6-31G* is more complete and flexible than the LANL2DZ basis set used in paper [13] under study of electronic structure and stability of $(TiO_2)_n$ clusters since the external polarization functions of *f*- type at Ti atoms and *d*- type at O atoms are absent in the LANL2DZ basis set.

2. Results and discussions

The information on IR spectra of the $(SiO_2)_n$ clusters is practically absent in the literature. Only zero-point energies of the $(SiO_2)_n$ clusters with n = 1-5 are given in paper [14]. The paper [15] contains more information. It is shown in it that the molecular rings (see figure 1) have the characteristic IR bands distinguishing them from molecular chains. The calculated 12-membered chain harmonic frequencies have the following modes 866, 915, 838 cm^{-1}, and the 12-membered ring has modes 902, 886, 933 cm^{-1}. The corresponding experimental data 908, 888, 932 cm^{-1} obtained for the silica surfaces are in remarkable agreement with the above-indicated calculated values of IR frequencies for molecular rings. This surfacelike signature of rings contrasts in properties with chain structures, which also display modes from terminal oxygen atoms at frequency 1321 cm^{-1}.

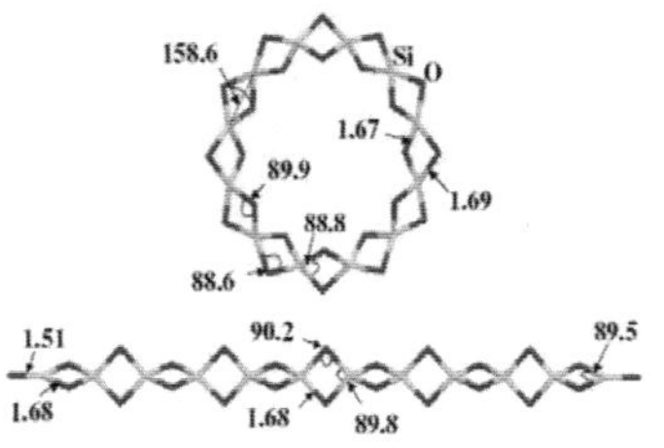

Figure 1. Structures of the $(SiO_2)_{12}$ molecular ring and chain. Bond lengths are in Å, angles in degrees.

Because of the information deficit on IR spectra for the $(SiO_2)_n$ clusters, there is a necessity to carry out our studies of the formation of their IR spectra absorption with the increasing of the size n of clusters. One of the examples of the structures and IR spectra, calculated by us for the $(SiO_2)_n$ clusters, is shown in figure 2.

The analysis of IR spectra [16] in the 750-1650 cm^{-1} range for the SiO_2 films (see figure 3) has revealed a number of absorption bands. The absorption band at ~ 800 cm^{-1} agrees with the band at 799 cm^{-1} observed and assigned in [7] to transverse optical component of the Si - O - Si mode of in-plane symmetric stretching vibrations. Our calculations of IR spectra of the $(SiO_2)_n$ clusters by B3LYP/6-31G* method also have shown that the sufficiently strong band at 765-778 cm^{-1} is characteristics for in-plane vibrations of the SiO_2Si cycle. In our calculations, the out-of-plane vibrations of the SiO_2Si rhombus fragment at 915 - 923 cm^{-1} correspond to the absorption band between 850 and 950 cm^{-1} and agree with the mode 915 cm^{-1} from theoretical paper [15]. According to our calculations, the six-atomic ring structure is distinctly displayed at 1036 - 1044 cm^{-1} that correlates with the band near 1050 cm^{-1} (figure 3) and also with the band at 1076 cm^{-1} detected in paper [7] for transverse optical component of Si - O - Si asymmetric stretching vibrations. In the same region at frequency of 1048 cm^{-1}, the fundamental band due to Si - O stretching vibrations of silica network is identified in Fourier transform IR spectra [8].

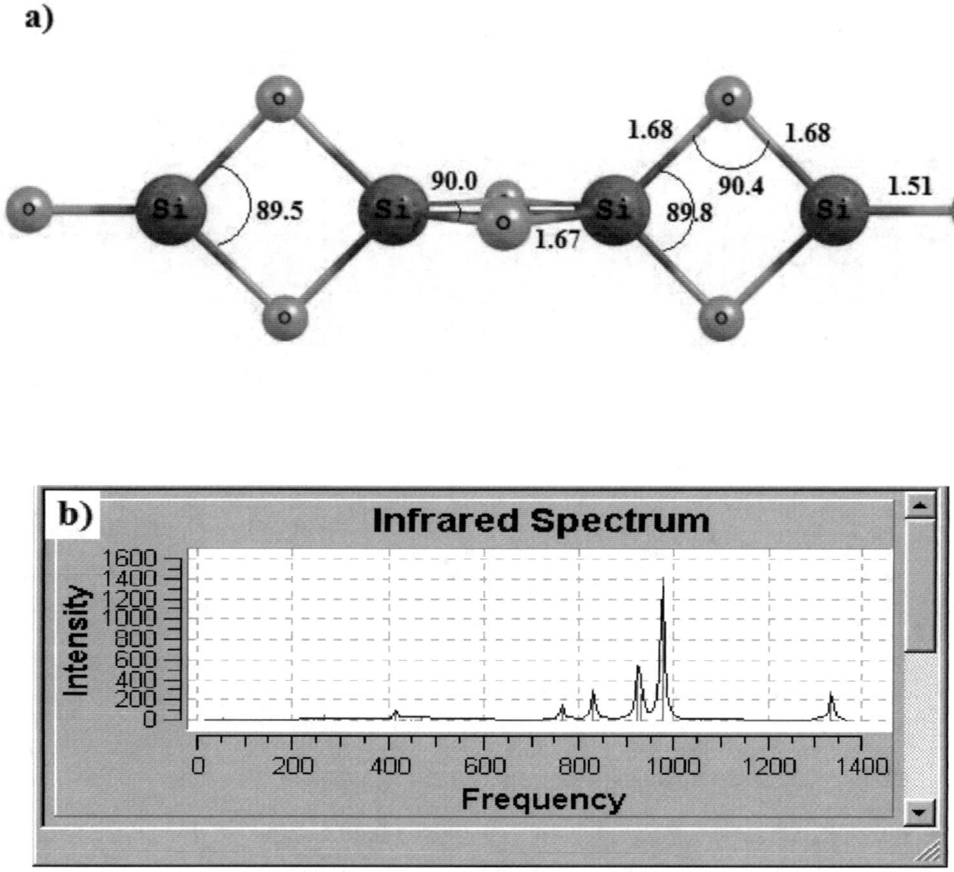

Figure 2. Optimized by B3LYP/6-31G* minimum energy chain structure of the $(SiO_2)_4$ cluster (D_{2h} symmetry) **(a)** and its IR spectrum **(b)**. Bond lengths are given in Å, angles are in degrees **(a)**. The frequencies are given in cm^{-1} and the intensities are in km/mole **(b)**.

In IR spectra of figure 3 between 1150 and 1250 cm^{-1}, it is presented the shoulder that is assigned in [7,8] to longitudinal optical component of the Si - O - Si asymmetric stretching vibrations at ~ 1200 cm^{-1}. In our cluster calculations, such mode is not found, but there is the very strong band at 1321-1334 cm^{-1} caused by stretching vibrations of terminal Si - O bonds (see figure 2), and revealed besides in paper [15] under calculations of 12-membered chain structure of the $(SiO_2)_n$ clusters. However, the corresponding absorption band is absent in IR spectra of the SiO_2 films that means a small concentration of terminal Si - O bonds in the film samples. The absorption band between 1550 and 1650 cm^{-1} of figure 3 is attributed to bending vibrations of H_2O molecule adsorbed on the surface of SiO_2 films at 1632 cm^{-1} [7] and 1627 cm^{-1} [8]. In IR spectra of films [16] there is a wide absorption band between 3000 and 3700 cm^{-1} that is assigned (at 3390 cm^{-1} [7] and 3400 cm^{-1} [8]) to fundamental stretching vibrations ν(OH) of hydroxyl groups as products of water hydrolysis on the SiO_2 surface.

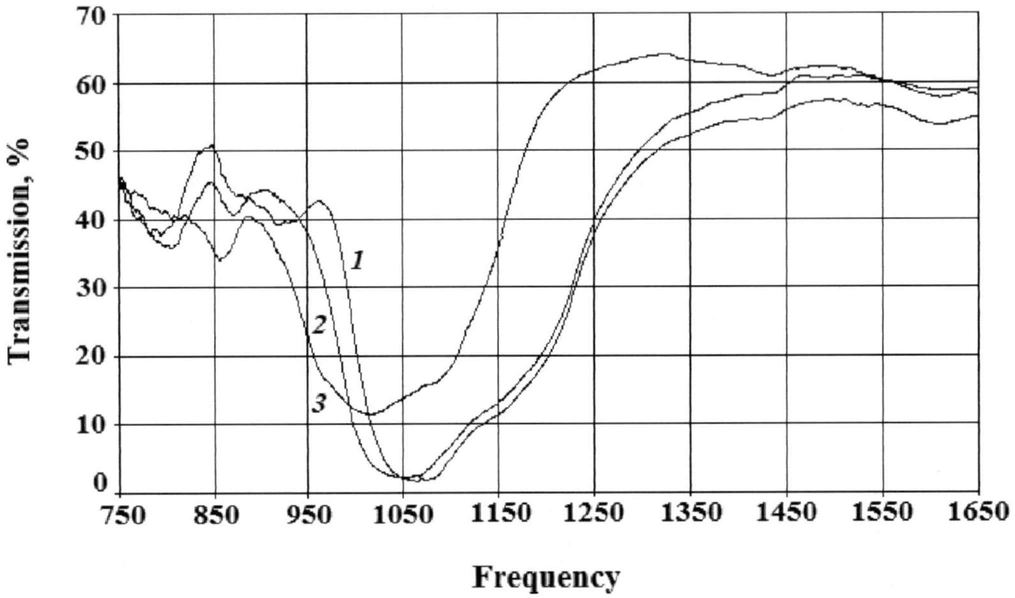

Figure 3. IR spectra in the 750-1650 cm^{-1} range for the SiO_2 films [16] at different conditions **1 – 3** of their preparation (initial sample **1** is prepared by gun, **2** is the sample **1** after exposure to water vapour, initial sample **3** is prepared by magnetron reactive sputtering).

If to compare our calculated IR spectra for $(TiO_2)_n$ (n = 2, 3) with IR spectra [7] measured for the TiO_2 films, the following features are become clear. The IR absorption bands calculated by us for compact bulk structures, C_{3v} for $(TiO_2)_2$ at 275 cm^{-1} and C_s for $(TiO_2)_3$ at 286, 298 cm^{-1}, are consistent with the band at 260 cm^{-1} observed in [7] for TiO_2 films. In ours theoretical IR spectra, the bands at 430 cm^{-1}, 421 cm^{-1} correlate with the band at 435 cm^{-1} [7], and the bands at 892 cm^{-1}, 854 and 902 cm^{-1} are in accordance with the band at 839 cm^{-1} [7]. Whereas the predicted in theoretical paper [13] characteristic bands in the 530-760 cm^{-1} range were not observed in IR absorption spectra [7] of the TiO_2 films, that puts under doubt the validity to use the B3LYP/LANL2DZ method under study of the $(TiO_2)_n$ clusters (n = 1-9) [13] and (n = 10-16) [17] because of the LANL2DZ basis set limitations (see above). It is worth to note that the strong IR absorption bands near 1100 cm^{-1} due to the terminal Ti-O bonds are not found in IR spectra of TiO_2 films that means a small concentration of the appropriate structural fragments in samples of films.

3. Conclusions

The B3LYP/6-31G* method of density functional theory may be recommended for study of the size and elemental composition nanoclusters influence on the IR absorption spectra of nanostructured films of composite materials. Using this method, we have established for the first time the tendencies in IR spectra formation of the $(SiO_2)_n$ clusters with the increasing of their size *n*. The IR absorption bands calculated by us for structures of the $(SiO_2)_n$ and $(TiO_2)_n$ clusters are in good agreement with the bands observed in IR spectra of the SiO_2 and TiO_2 films. Therefore, the obtained in this work database on IR absorption spectra of the SiO_2 and TiO_2 nanoclusters may be applied to control the technological processes of fabricating nanostructured films of the ceramic and composite materials.

Acknowledgments
Work was supported by Russian ministry of education and science within framework of program "Development of scientific potential of high school" (Grant No. 2.1.1/2637).

References

[1] Zhang Y F, Tang Y H, Wang N, Yu D P, Lee C S, Bello I and Lee S T 1998 *Appl. Phys. Lett.* **72** 1835
[2] Wang N, Zhang Y F, Tang Y H, Lee C S and Lee S T 1998 *Appl. Phys. Lett.* **73** 3902
[3] Lundquist M J, Nilsing M, Persson P and Lunnel S 2006 *Int. J. Quant. Chem.* **106** 3214
[4] Argios A G and Pichat P 2005 *J. Appl. Electrochem.* **35** 655
[5] Hoffmann M R, Martin S T, Choi W and Bahnemann D W 1995 *Chem. Rev.* **95** 69
[6] Chen X and Mao S S 2007 *Chem. Rev.* **107** 2891
[7] Portal S and Almeida R M 2004 *Phys. Stat. Sol.*(a) **201** 2941
[8] Que W, Sun Z, Lam Y L, Chan Y C and Kam C H 2001 *J. Phys. D: Appl. Phys.* **34** 471
[9] Kohn W and Sham L J 1965 *Phys. Rev.* **140** A1133
[10] Frisch M J, Trucks G W, Schlegel H B, Scuseria G E, Robb M A, Cheeseman J R, Zakrzewski V G, Montgomery J A Jr, Stratmann R E, Burant J C *et al.* 1998 *Gaussian 98. Revision A.3* (Pittsburgh, PA: Gaussian, Inc.)
[11] Becke A D 1988 *Phys. Rev.* A **38** 3098; 1993 *J. Chem. Phys.* **98** 5648
[12] Lee C, Yang W and Parr R G 1988 *Phys. Rev.* B **37** 785
[13] Qu Z -W and Kroes G -J 2006 *J. Phys. Chem.* B **110** 8998
[14] Lu W C, Wang C Z, Nguen V, Schmidt M W, Gordon M S and Ho K.M. 2003 *J. Phys. Chem.* A **107** 6936
[15] Bromley S T, Zwijnenburg M A and Maschmeyer Th 2003 *Phys. Rev. Lett.* **90** 035502-1
[16] Astapenko V A, Sukhanov L P and Dorofeeva E V 2008 *Summary Report on Scientific Research Work on Theme No. 205/07 (The Fourth Stage)* (Moscow: Moscow Institute of Physics and Technology) [in Russian]
[17] Qu Z -W and Kroes G -J 2007 *J. Phys. Chem.* C **111** 16808

A study of structure formation in Ti-Nb-Zr shape memory alloys for medical application

M R Filonov [1], V Brailovski [2], S D Prokoshkin [1], Y S Zhukova [1,*] and S M Dubinsky [1,2]

[1] National University of Science and Technology "MISIS", 4 Leninsky prospect, Moscow, Russia 119049

[2] Ecole de Technologie Superieure, 1100 Rue Notre-Dame Ouest Montréal, QC H3C 1K3 Monreal, Canada

zhukova@misis.ru

Abstract. A study of structure formation in Ti-Nb-Zr shape memory alloys was carried out by means of transmission electron microscopy. After 450 °C annealing of cold-worked specimens a very high dislocation density (10^{11} cm^{-2}) remained; after 600 °C annealing a mixed structure consisting of submicron sized grains and subgrains formed. Therefore, a nanosubgrained structure is expected to form after 500 – 550 °C annealing.

1. Introduction

Biocompatible shape memory alloys (SMA) are in the number of the most advanced metallic implant materials [1]. Titanium based and titanium intermetallic based alloys, especially Ti-Ni, are well studied. However, on account of nickel carcinogenicity, development of nickel-free alloys, e.g. solid solution Ti-Nb-(Ta, Zr) based alloys, is in a progress [2 – 4].

As the SMA functional properties are defined by the alloy structure, studies of the structure formation rules under the thermomechanical treatment comprising cold rolling + post-deformation annealing is of interest. The work was aimed at a search for the annealing temperature range to obtain nanosubgrained or nanocrystalline structures which provide the best combination of SMA functional characteristics, in particular, superelasticity [5 – 8].

2. Experimental

The study of structure formation in Ti-20.8Nb-5.5Zr (at. %) alloy was carried out by transmission electron microscopy after cold rolling and subsequent annealing under the following conditions: (1) 85 % reduction (true strain e = 2); (2) 85 % reduction + 450 °C, 1 hr annealing; (3) 85 % reduction + 600 °C, 1 hr annealing.

For structure investigation JEM 200CX transmission electron microscope was used with magnification x 15000 and x 50000. Electron diffraction patterns were obtained from selected areas with 0.5 μm in diameter. The studied samples were thin foils cut from thermomechanically treated samples and subsequently thinned by electropolishing.

* To whom any correspondence should be addressed.

3. Results and discussion

In as-deformed state after deformation e = 2, a non-homogeneous deformed structure of β-phase is observed (figure 1). The experiments show that β-phase exhibits a very high dislocation density (not less than 10^{11} cm^{-2}). The α- or α''-phase plates are also present, their width is about 100 nm. Dark field images show that very fine (below 10 nm) round shaped ω-phase particles are distributed through the whole volume. No features of amorphous or nanocrystalline structure are detected.

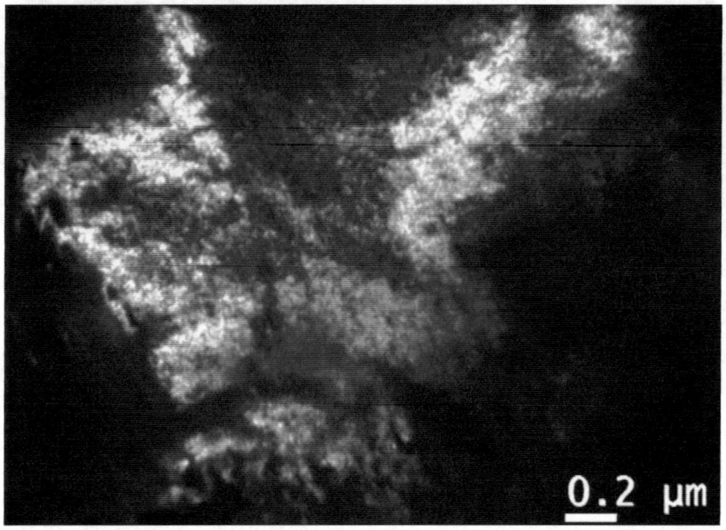

Figure 1. Dark-field image under diffraction conditions corresponding to deformed β-phase; e=2.

After 450 °C annealing, the alloy structure does not change significantly, but becomes more uniform. Mainly the regions with deformed β-phase and the regions with alternate α- and β-phase plates are observed; in all cases dislocation density remained very high. There are fine ω-phase particles inside β-plates. There are also very fine (less than 20 nm) lamellar crystals inside α-plates which morphology is similar to α''-phase crystals (figure 2). Obviously, annealing temperature of 450 °C corresponds to recovery stage, a polygonization in dislocation substructure does not yet develop.

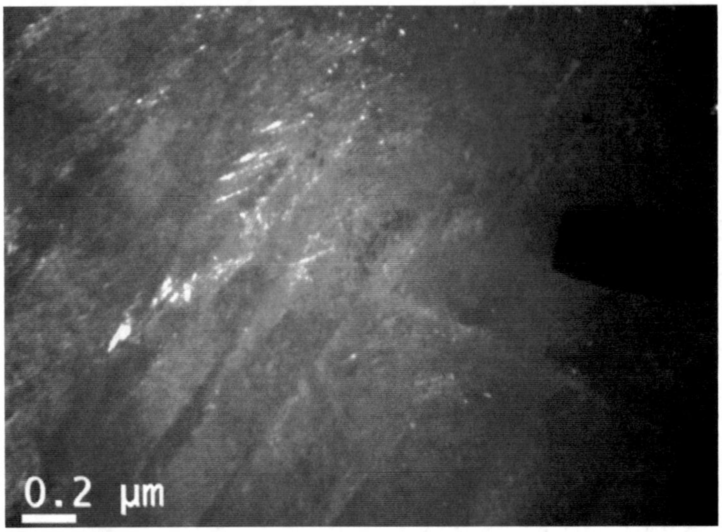

Figure 2. Dark-field image of plate-like α''-phase crystals after post-deformation annealing at 450 °C.

After annealing at 600 °C, an ultrafine-grained structure is observed (figure 3). Dark-field images obtained under diffraction conditions corresponding to α- and β-phase indicate that some grains belong to α-phase, while the greater fraction of grains, including the largest ones, belong to β-phase. According to these electron diffraction patterns, the visible structure elements (size 0.1 – 0.5 μm) are separated by high-angle and low-angle boundaries (see azimuthal misorientations between reflexes in figure 4). Obviously, annealing at 600 °C corresponds to transition to recrystallization. At this stage grain structure and polygonized substructure with submicro-sized structure elements (subgrains and grains) coexist.

Figure 3. Bright-field image of structure after 600 °C annealing.

Figure 4. "Ring" diffraction pattern corresponding to fine-grained structure after 600 °C annealing.

The results of the TEM studies indicate that to obtain nano-sized (less than 100 nm) structure elements, the post-deformation annealing should be carried out at 500 – 550 °C; the obtained structure elements will rather be subgrains (surrounded by low-angle boundaries) than grains (surrounded by high-angle boundaries).

A preliminary study of electrochemical behavior of Ti-Nb-Ta alloy which is analogous with Ti-Nb-Zr was carried out. It was shown that the alloy exhibits inclination for self-passivation in artificial saliva solution, with an electrode potential more positive than those for pure titanium that is known to be corrosion resistant material. Passivation rate of Ti-Nb-Ta is relatively low compared to those for titanium; it may indicate lower concentration of defects in passive film on Ti-Nb-Ta that may influence the film's resistance in the biofluid positively.

It is well known that electrochemical behavior of an alloy depends on its structure and, consequently, on thermomechanical treatment conditions. It was clearly shown for Ti-Ni [9] and Ti-6Al-4V [10] alloys. Therefore among the directions of the further biomedical Ti-Nb-Zr alloys research there is study of structure, including nanostructure, influence on the alloys' electrochemical behavior, as well as optimization of their treatment conditions concerning combination of mechanical, functional and electrochemical properties.

4. Conclusions

A study of structure formation in biomedical Ti-Nb-Zr shape memory alloy (SMA) was carried out by means of transmission electron microscopy after thermomechanical treatment under different conditions.

- Cold deformation of Ti-Nb-Zr SMA with true strain e = 2 forms dislocation substructure with very high dislocation density (not less than 10^{11} cm^{-2}) in β and α"-phases. High dislocation density remains after annealing at 450 °C, 1 hr, that denotes the preservation of significant strain hardening.
- After annealing at 600 °C, 1 hr in the case of initial strain e = 2 a mixed β-phase structure consisting of submicron sized (0.1 – 0.5 μm) grains and subgrains is observed. A nanosubgrained structure is expected to form after during post-deformation annealing at 500 – 550 °C.

Acknowledgements

The present work was carried out with the support of Ministry for Science and Education of Russian Federation (ADTP "Development of scientific potential of the higher school (2009-2010)", project 2.1.2.7129, FTP "Scientific and scientific-educational personnel of innovative Russia", projects 16.740.11.0014 and 14.740.11.0465), Ministry of Economic Development of Quebec, Natural Science and Engineering Research Council of Canada, and the Programme of Creation and Development of the National University of Science and Technology "MISIS".

References

[1] Gunter V E, Itin V I and Monasevich L A 1992 *Shape memory effects and their application in medicine* (Novosibirsk: Nauka) p 741 (in Russian)
[2] Miyazaki S, Kim H Y and Hosoda H 2006 *Mater. Sci. Eng.* A **438–440** 18–24
[3] Niinomi M, Hanawa T, Narushima T 2005 *JOM* **57** 4 18–24
[4] Brailovski V, Prokoshkin S, Gauthier, M, Inaekyan K, Dubinskiy S, Petrzhik M and Filonov M 2010 *Mater. Sci. Eng.* C, (in press)
[5] Prokoshkin S D, Brailovski V, Khmelevskaya I Yu, Dobatkin S V, Inaekyan K E, Turilina V. Yu, Demers V and Tat'yanin E V 2005 *Met. Sci. Heat Treat.* **47**, 5-6, 182-7
[6] Brailovski V, Prokoshkin S D, Khmelevskaya I Yu, Inaekyan K E and Demers V 2006 *Mater. Trans. JIM* **47** 795-804
[7] Prokoshkin S D, Brailovski V, Inaekyan K E, Inaekyan K E, Demers V, Dobatkin S V and Tatyanin E V 2008 *Mater. Sci. Eng. A* **481-482**, 114-8
[8] Brailovski V, Prokoshkin S, Inaekyan K and Demers V 2010 *J. Alloys Comp.*, (in press)
[9] Vandenkerckhove R, Chandrasekaran M, Vermaut P, Portier R and Delaey L 2004 *Mater. Sci. Eng.* A **378** 532–6
[10] Vasilescu E, Drob P, Raducanu D, Cinca I, Mareci D, Calderon Moreno J M, Popa M, Vasilescu C and Mirza Rosca J C 2009 *Corr. Sci.* **51** 2885–96

Journal of Physics: Conference Series **291** (2011) 012034 doi:10.1088/1742-6596/291/1/012034

Self-oscillations of the atomic structure of $Nd_2Fe_{14}B$ nanoclusters

S K Godovikov[1], S M Nikitin[1] and E A Nikitina[2]

[1] Skobeltsyn Institute of Nuclear Physics, Moscow State University. Leninskie Gory, Moscow 119992, Russia

[2]Institute of Applied Mechanics Russian Academy of Sciences, Leninsky prospect 32A, Moscow 119991, Russia

E-mail: godov@srd.sinp.msu.ru

Abstract. The phenomenon of self-oscillations of the magnetic material atomic structure, occurring at intervals of about 30 days having duration of about 150 days, was observed at the first time. The oscillations are initiated by single field pulse with duration of 1 μs. The essence of the fluctuations was found due to the redistribution of the iron atoms over the six crystallographic non-equivalent positions. To assess the role of vacancies and boron atoms in the oscillations the vibration spectrum of substance has been calculated.

1. Introduction

The collective synergetic behavior of atoms and molecules in condensed media was discovered long ago and continues to be studied today. The place and role of Mössbauer spectroscopy in this area are still not sharply defined. The unique possibilities of the nuclear gamma-resonance phenomenon are either little-known or virtually unknown. This work is devoted to considering some aspects of the above problem.

The self-organization and self-oscillation concerns a very vast class of phenomena of quite different nature. The phenomenon of self-oscillations in liquids was observed in a reaction with periodic (over several minutes) color changes [1]. Oxidation took place in a malonic acid solution with the presence of cerium sulphate and potassium bromate. The color change (from red to blue) was due to fluctuating changes in the cerium ion valence. The reaction was also accompanied by the appearance of spatial ordering in the system (concentric waves on the liquid's surface). The first researchers of this reaction were amazed the synchronism of molecular behavior in macroscopic volumes of the matter.

Another typical example is so-called Benar cells in a layer of oil heated on a planar surface [2]. Hexagonal cells of $\sim 10^{-1}$ cm shape the liquid's surface into a stable spatial structure. The liquid flows in neighboring cells are directed oppositely one another.

We should note that periodic reactions in chemistry are observed quite often, but such processes rarely occur in a solid body. This is apparently connected with the sharp difference between atom mobilities in liquids and solid bodies. Diffusion in crystals and amorphous bodies is a slow process. Weeks, months, and even years are sometimes required to notice any periodicity. The question therefore arises: which method from the arsenal of experimental physics is better suited to this problem? It is evident that nuclear γ-resonance is an extraordinarily appropriate and promising method in this case.

2. Experimental

An object for the synergetic phenomenon study was carefully chosen. According to the previous research experience [3] the following requirements were taken into account.

1. It had to be a metallic one with a large number of atoms per unit cell and with multiple pathways for the atoms migration in the crystal.
2. The crystal should possess both stoichiometric vacancies, which ensure high migration speed, and light atoms, which can be the initiators of self-organization process.
3. The initial magnetic structure of the crystal had to be rather simple, allowing properly assess the form of its deformation in fluctuation process.
4. It had to be energy intensive structure, ensuring the creation of a long living non-equilibrium state due to external influence (electric or magnetic). Energy intensity is associated with high values of magnetic and magnetoelastic energy.

$Nd_2Fe_{14}B$ crystal meets all of these requirements. Its crystalline structure is shown in figure 1.

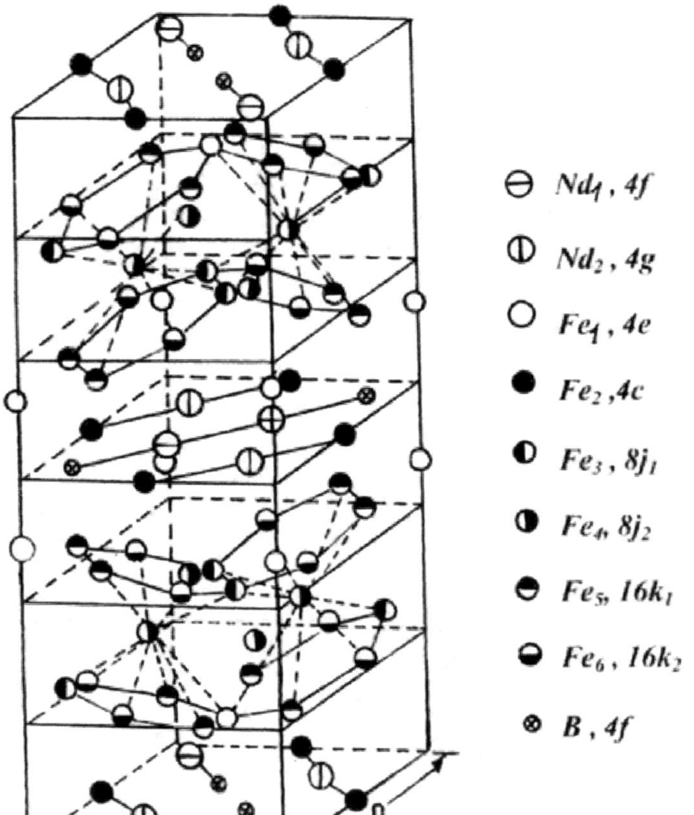

Figure 1. Crystalline structure of $Nd_2Fe_{14}B$ compound.

This is a tetragonal structure (α = 8.8 Å and c = 12.19 Å), including 68 atoms per unit cell [4]. There are six crystallographically nonequivalent Fe atom positions: $16k_1$, $16k_2$, $8j_1$, $8j_2$ and $4e$ and $4c$; two nonequivalent Nd atom positions; and one position of B atoms. Its magnetic structure at T = 293K is quite simple: it is a collinear ferromagnetic. $Nd_2Fe_{14}B$ compound is used in the fabrication of permanent magnets with record high magnetic energy of ~35 MGs Oe. Its magnetic anisotropy is ~10^7 erg·cm^{-3}, and its magnetostriction constant is ~10^{-3}. Our sample, prepared by the agglomeration method, contained a certain number of stoichiometric vacancies (~ 2 vacancies per one unit cell).

The sample was activated using the method traditionally applied in earlier experiments: an electric pulse, $dE/dt \approx 10^8$ kV·cm^{-1}·s^{-1}, emitted when the plates of the capacitor containing the sample were closed. A ^{57}Co source in Pd was used. A selection of different models was employed in spectrum processing: a model of a continuous distribution of hyperfine fields $p(H)$; calculations of the density function of probability distribution $\rho(V)$ for resonance line centers of gravity according to a scale of rates; and a model of discrete distribution by four, five, and six sextets. A four-sextet model with variable parameters of hyperfine field H_{hf}, isomeric shift δ, quadrupole splitting Q, line width Γ, and relative weight A of a partial sextet turned out to be the best suited to our purpose, and the most reliable one. Typical processing parameters are shown in the Table 1. Mössbauer spectra were analyzed by "Univem" program.

Table 1. Processing parameters of the $Nd_2Fe_{14}B$ Mössbauer spectrum.

Sextet	H_{hf}, kOe	A, %	δ, mm/s	Q, mm/s	Γ, mm/s
1	333 – H_1	10	0.45	-0.16	0.5
2	290 – H_2	30	-0.33	-0.02	0.5
3	285 – H_3	40	-0.16	0.41	0.5
4	250 – H_4	20	-0.17	0.08	0.5

On the basis of the known data [5], it was established that field H_1 corresponded to the position j_2; H_2 corresponded to the position k_2; and H_4 corresponded to the position c. Field H_3 corresponded to the sum of positions $(k_1 + j_1 + e)$, which differ only slightly in their parameters of hyperfine interactions. In our study the behaviour of parameters A_2 and A_3 (i.e., the behaviour of the occupancies of positions k_2 and $(k_1 + j_1 + e)$) was the most interesting. Average values of parameters A_2 and A_3 are about 30% and 40%, respectively. About 10% and 20% corresponds to positions j_2 and positions c. These values are calculated from the spectra in accordance with partial weights of Zeeman's sextets with different H_{hf} values.

3. Results and discussion

The behaviour of an activated sample exhibits the emergence of a clear time order. It was found that strong fluctuation processes are observed in parameters H_2, H_3, δ_2, δ_3, A_2, and A_3 (i.e., in the second and third sextets). Figure 2 shows the time dependence of parameters A_2 and A_3. Explicit fluctuations with a period of ~30 days and a lifetime of ~150 days are observed. Amplitude $\Delta = A_3 - A_2$ is very large: –5% to +20%. The places where the curves converge and intersect may be labeled inversion points, since here $A_2 \geq A_3$. The lower curve in figure 2 is an internal reference point, as it shows the stability in the occupancies A_1 of the j_2 positions. Fluctuations in parameters $\Delta H = H_2 - H_3$ and $\Delta\delta = \delta_2 - \delta_3$ were also observed. These fluctuations were quite regular as well, but had a period of 8–12 days and a concurrent antiphase (i.e., a reduction in ΔH corresponds to an increase in $\Delta\delta$). Let us now consider the microscopic mechanism of the observed phenomenon.

For this purpose we calculated the $Nd_2Fe_{14}B$ infrared spectrum. The X-ray data [4] was used for the initial spatial structure of $Nd_2Fe_{14}B$. The structure was optimized by molecular mechanics force field MM2 [6]. The optimization was fulfilled with well-known Polak-Rebiery approach with RMS gradient value limit 0.001 kcal/unit. The optimized structure was used for the calculation of the vibration spectrum represented in figure 3a. The calculation of the vibration spectrum and the normal coordinate analysis were performed within approximation of the valence force field. The basis of this approach is the fact that Fourier transform of velocity autocorrelation function defines the density of vibration states. In the proposed work an analytical expression of the autocorrelation function and its Fourier transform have been found in approach of the cyclic boundary conditions. The curve of the Fourier transform includes a number of sharp peaks, whose positions define the frequency of the normal vibration modes and their areas correspond to the masses that are vibrating with these modes.

Then, one of $8j_2$ iron atoms was removed from the original unit cell and the same sequence of calculations was repeated. The resulting spectrum is shown in figure 3b.

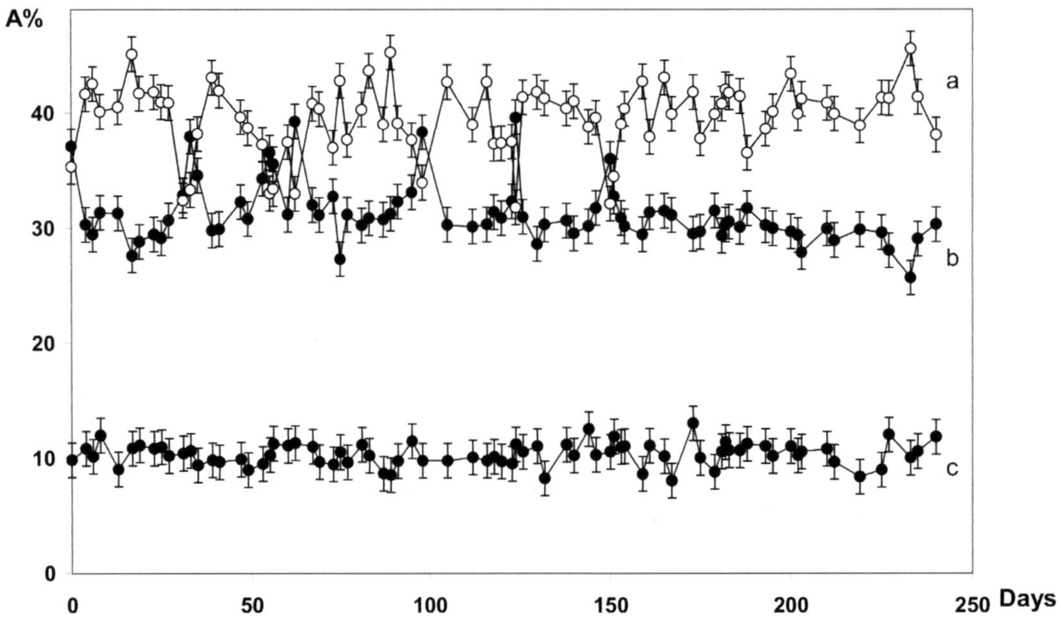

Figure 2. The time dependence of A parameters of $Nd_2Fe_{14}B$ compound perturbed by an electrical impact: (a) A_3, (b) A_2, (c) A_1.

Figure 3. Calculated IR spectra of (a) $Nd_2Fe_{14}B$ and (b) $Nd_2Fe_{13.75}B$ compounds.

The spectra have significant difference, both in position and quantity of the bands. The lowest frequencies of the vibration spectra correspond to boron atoms and to the ensemble of boron and the surrounding iron atoms.

We can conclude that:

1. The presence of stoichiometric vacancies has the influence on the vibration bands. In the low frequency part of spectrum Fig.3b bands of vibration combines in regions of the discontinuous spectrum.

2. It means that the energy difference between atoms in different positions is not great. For example, k_2 easily transferred to ($k_1 + j_1 + e$). It is sufficiently to slightly change the distance between atoms and bond angles, which is provided by electron impact.

3 Moving the light atoms of B between vacancies distributes wave of transformations $k_2 \rightarrow (k_1 + j_1 + e)$ on the whole sample. This is the reason of the macroscopic time needed for the process.

Synchronic movement of atoms between vacant positions in the whole volume of the sample occurs, by all means, for the following reasons.

1. One start time for the whole sample - the time of electric shock.

2. Mutual influence of cells on each other by electrical and magnetic exchange interaction.

The duration of the process of moving atoms (it takes several days) is obviously due to diffusive nature of the process. The energy barrier for this type movement is at about 1 eV.

Filling of vacant positions j_2 within 15 days causes a change in the whole crystal structure of the substance, which becomes high energetic, and than slowly relaxes to its original state (figure 2). However, the magnetoelastic interaction energy of the sample, which was increased due to the electric shock, is not being spent completely. This causes a second round of oscillation caused by internal rather than external reasons. After the 5th round the energy is completely spent and the oscillation process breaks down.

4. Conclusions

Self-oscillations of the local atomic structure of the $Nd_2Fe_{14}B$ crystal were observed by Mössbauer spectroscopy. Calculations by molecular mechanics method suggest the mechanism of the phenomenon, based on filling the vacant position by moving of boron atoms.

References

[1] Belousov B P 1959 *Collection of Papers on x-ray medicine* Moscow: Medgiz
[2] Prigozhin I S 1985 *From Existing to Emergence* Moscow: Nauka
[3] Godovikov S K and Lagutina E S 2009 *Phys. Metals Metallogr.* **108** 67
[4] Herbst J F, Croat J J and Pinkerton F E 1984 *Phys. Rev.* **29** 4176
[5] Niarchos D 1989 *Hyper. Interact.* **49** 367
[6] Allinger N L 1977 *J. Am. Chem. Soc.* **99** 8127

Journal of Physics: Conference Series **291** (2011) 012035 doi:10.1088/1742-6596/291/1/012035

Electro Mechanical Scanning Probe Lithography of Carbon Nanostructures

V D Frolov[1], V I Konov[1], S M Pimenov[1], S V Kosheleva[2]

[1] A.M. Prokhorov General Physics Institute of RAS, Vavilov str. 38, 119991 Moscow, Russia
[2] Moscow State Institute of Radiotechnics, Electronics and Automation, Vernadsky pr-t 78, 119454 Moscow, Russia

E-mail: frolov@ran.gpi.ru

Abstract. Carbon nano-processing as one of the most promising methods for creation of nanostructures and various types of electronic and optical systems is analyzed. A new method of electro-induced scanning probe lithography (SPL) based on the formation of ordered carbon nanostructures by combining electrical and mechanical actions on the original material in a scanning probe microscope (SPM) is proposed and tested.

1. Introduction

In recent years, many materials have been subjects for electric field-induced modification in a scanning probe microscope (SPM) (see for example the review [1]). The nanoprotrusion formation was demonstrated in experiments with silicon [2-5] and some metals [6-9] that was considered as a result of their local oxidation at positive voltages, i.e., anodization of the sample surface under the probe actions in the presence of adsorbed water vapors.

Scanning probe lithography (SPL) is highly effective and essentially the only method for precision processing of carbon compounds in the nanometer scale. However, in the case of carbon the surface modification shows more complex character compare to inorganic materials. For instance, either protrusions or cavities were resulted from the SPL of amorphous carbon films [10-13], mostly depending on the applied voltage and the environmental conditions (air or vacuum). For highly oriented pyrolytic graphite (HOPG), the only type of modification (nanocavity) was observed in similar nanoprocessing regimes. In the case of hydrogenated diamond surfaces it was found that the SPL induced protrusions as well as cavities, the formation (and registration) of which depends not only on the bias voltage but also on the mode (contact or non-contact) of registration in SPM [15-18].

Experimental evidences were done about local oxidation of carbon films whilst prevailing the SPM-actions threshold that lead to the removal of the uppermost surface layers (i.e. the appearance of nanocavities). In order to explain the formation of nanoprotrusions on diamond-like carbon (DLC) films, it was proposed that the carbon phase transitions occur in carbon, being the result of local heating of the material [11,13,18]. Another explanation related to the properties of hydrogenated diamond surface was suggested in works [15,16], where nanoprotrusions were described as artifacts of

Published under licence by IOP Publishing Ltd

SPM mapping, namely, the congestions of water adsorbate on the oxidized area which are manifested in the form of appropriate modifications of the surface topography.

Thus, a variety of reactions of carbon materials under the electric field actions is of great interest and importance. At the same time, such a variety causes considerable difficulties in the search of regimes and conditions, providing stable and reproducible modifications of carbon material.

In this paper, a new method of electro-mechanical SPL of carbon materials is described. Experimentally, the SPL-based processing is explored for nanostructuring of thin graphite-like films.

2. Experimental

The origin samples were prepared by annealing the 10 nm thick diamond-like carbon films on Si substrates at 620° C under vacuum conditions. According to X-ray photoelectron spectroscopy, under these conditions there is a complete graphitization of the films with the formation of graphite nanoclusters, which is manifested in the considerable narrowing of the C 1s line to values of 1.0 - 1.1 eV [17]. Experiments were carried out in SPM Solver P47 (NT-MDT). To implement the mechanical effects (SPM-nanoengraving), the SPM contact mode was used. As instruments of nanoengraving, silicon cantilevers (force constant is ~ 10 N / m, tip radius is of 5-10 nm) were used. The reason is that such cantilevers have sufficient mechanical strength and toughness (~ 2.4 GN/m^2) for the deformation of graphite layers (~ 0.4 GN/m^2). The scan speed was 100 nm/sec. The designed pressure of the probe on the surface was 10 μN. In order to realize electric field actions on the film surface, the voltage of positive polarity was additionally applied between the sample and the grounded probe.

3. Results and discussion

In Fig. 1, the original surface topology is presented. The right of Fig. 1 shows statistical data on the surface relief. As follows from the relief map, the film surface has a granular structure with a characteristic lateral size of grains ~50 nm. The surface roughness is of 0.15 nm, the relative spread of the film thickness is ~20%, thus indicating the films completely cover the substrate.

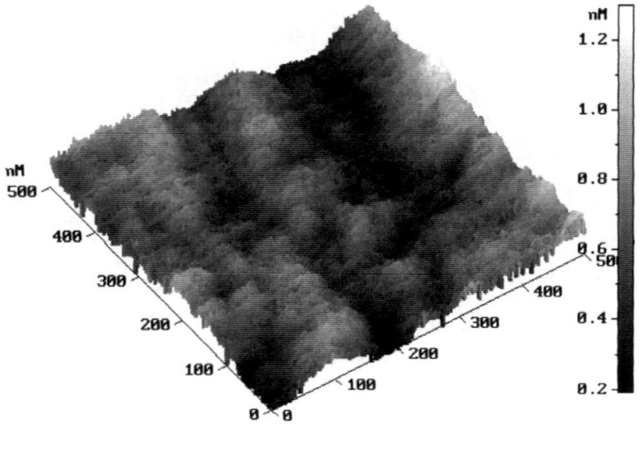

Figure 1. The original surface topology.

Fig. 2 illustrates the surface topography after 3 cycles of the mechanical actions of the probe (the cycle duration was – 5 s).

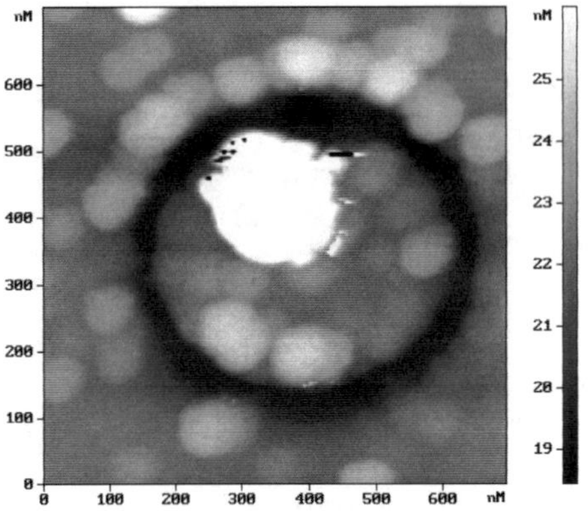

Figure 2. Topography of the sample surface after the mechanical treatment along the circle.

As can be seen from Fig. 2, scanning the probe leads to the formation of cavities in the film surface. (the cutting speed is of 0.2 nm/s), Additionally, the ledges the film surface appears that can be obviously related to ejections of particles of matter from the area of the cut.

The surface topography after the combined electro-mechanical treatment is shown in Fig. 3. It is established that the cutting speed increases to ~0.5 nm/s, leading to the cut about 5 nm in depth. We propose that this process develops due to the transformation of graphite particles to the soot [18]. (The chemical content of soot in this study was not the subject of investigation, but we plan to continue to further explore this aspect).

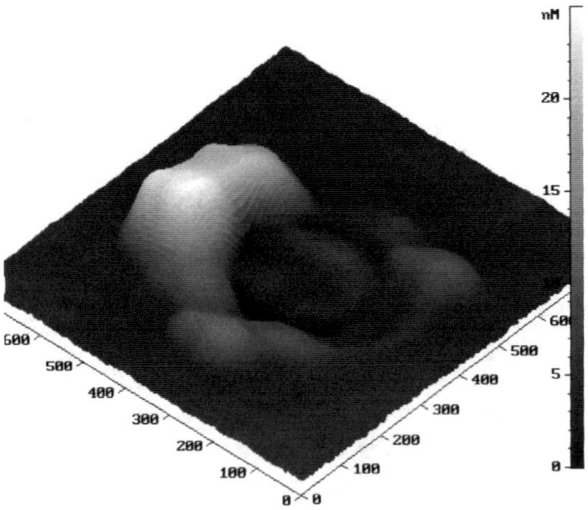

Figure 3. Topography of the sample surface after the electro-mechanical treatment..

It is well known that the soot is rather soft material hence can be effectively removed by scratching. To confirm this circumstance, the treated film was scanned in a contact mode regime. The result is presented in Fig. 4. It is found that the nanoparticles are effectively removed that is clearly seen from comparison of Fig. 3 with Fig. 4.

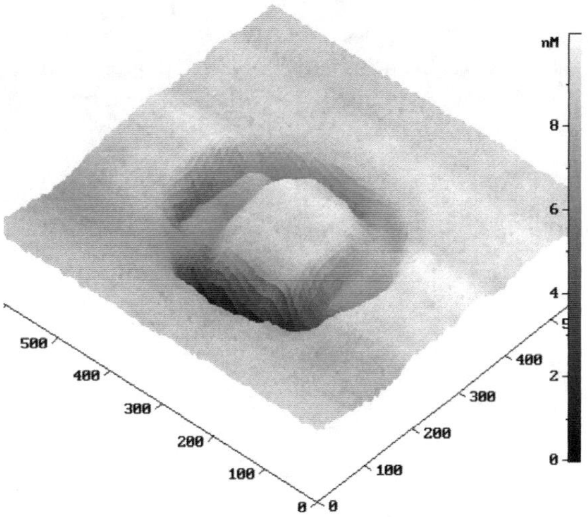

Figure 4. Topography of the sample surface after the electro-mechanical treatment..

It allowed us to form nanoobject arrays with the density of ~10^{12} elements/cm^2.

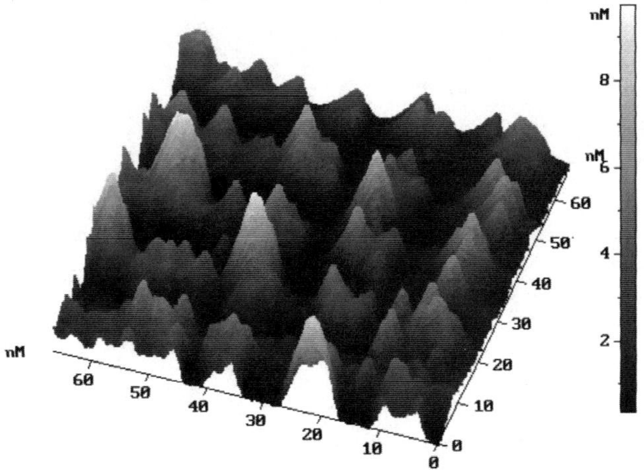

Figure 5. The nano-object array with the density of ~10^{12} elements/cm^2, formed on the graphite film surface.

It is distinctive that the proposed method can be used for the formation of nanostructures in different carbon films, including mechanically strong and chemically resistant materials such as DLC films.

4. Conclusion

The experience on electro-induced SPL of carbon nanostructures enables us to conclude that this process can be effectively realized in air. At once this process allows us to solve the problem on removing the remnants of reaction which during the only mechanical action become apparent as particles of matter removed from the cut, and during the additional electric actions are transformed to the soot. Owing to the small tip radius, it provides the formation of narrow grooves.

Acknowledgement

This work was supported by RFBR under grant No 09-02-01029.

References

[1] Saavedra H M, Mullen T J, Zhang P, Dewey D C, Claridge S A and Weiss P S 2010 Rep. Prog. Phys. **73** 036501 (40pp)

[2] Dagata J A, Schneir J, Harary H H, Evans C J, Postek M T, Bennett J 1990 *Appl. Phys. Lett.* **56** 2001

[3] Day H C, Allee D R 1993 *Appl. Phys. Lett.* **62** 2691

[4] Snow E S, Campbell P M 1994 *Appl. Phys. Lett.* **64** 1932

[5] Avouris P, Hertel T, Martel R 1997 *Appl. Phys. Lett.* **71** 285

[6] Sugimura H, Uchida T, Kitamura N, Masuhara H 1993 *Appl. Phys. Lett.* **63** 1288

[7] Gwo S, Yeh C-L, Chen P-F, Chou Y-C, Chen T T, Chao T-S, Hu S-F, Huang T-Y 1999 *Appl. Phys. Lett.* **74** 1090

[8] Snow E S, Park B, Campbell P M 1996 *Appl. Phys. Lett.* **69** 269

[9] Wang D, Tsau L, Wang K L, Chow P 1995 *Appl. Phys. Lett.* **67** 1295

[10] Muhl T, Bruckl H, Weise G, Reiss G 1997 *J. Appl. Phys.* **82** 5255

[11] Mercer T W, DiNardo N J, Rothman J B, Siegal M P, Friedmann T A, Martinez-Miranda L J 1998 *Appl. Phys. Lett.* **72** 2244

[12] Eagle S C, Fedder G K 1999 *Appl. Phys. Lett.* **74** 3902

[13] Myhra S 2005 *Appl. Phys., A* **80** 1097-1104

[14] Albrecht T R, Dovek M M, Kirk M D, Lang C A, Quate C F, Smith D P E 1989 *Appl. Phys. Lett.* **55** 1727

[15] Tachiki M, Fukuda T, Sugata K, Seo H, Umezawa H, Kawarada H 2000 *Jpn. J. Appl. Phys.* **39** 4631

[16] Rezek B, Sauerer C, Garrido J A, Nebel C E, Stutzmann M, Snidero E, Bergonzo P 2003 *Appl. Phys. Lett.* **82** 33360

[17] Frolov V D, Pimenov S M, Konov V I, Lubnin E N 2008 *Nanotechnologies in Russia* **3 Nos. 11–12** 688

[18] Frolov V D, Konov V I, Pimenov S M, Zavedeev E V 2004 *Diam.Relat. Mater.* **13** 2160

Carbonate substituted hydroxyapatite (CHA) powder consolidated at 450°C

Ya Yu Filippov[1], E S Klimashina[1], A B Ankudinov[2], V I Putlayev[1]

[1] Department of Materials Science, Lomonosov Moscow State University, Leninskie Gory, Moscow, Russia

[2] Baikov Institute of Metallurgy and Materials Science, Leninskii Prospect 49, Moscow, Russia

filippovya@gmail.com, alenakovaleva@gmail.com, a-58@bk.ru, valery.putlayev@gmail.com

Abstract. CHA powder was consolidated without noticeable changing of its carbonate content using low-melting Na-Ca phosphate glass with the composition referred to 54% $Na_4CaP_6O_{18}$ and 46% $Na_2CaP_2O_7$ as a binder. A temperature window suitable for obtaining the composite was assessed as 450 - 475°C. The composites obtained at 450°C and 400 MPa for 1 hr demonstrate compressive strength up to 25 MPa.

1. Introduction

Hydroxyapatite $Ca_{10}(PO_4)_6(OH)_2$ (HA) is widely used as clinically approved material for bone tissue replacement since its chemical composition is quite close to inorganic part of a human bone. HA implants are biocompatible, although the rate of their remodeling in vivo is too low to induce a massive formation of a new bone, and, thus, the HA implant serves as a substituent of the lost part of a bone. However, prominent trend in orthopedic and maxillofacial surgery is related to a regenerative approach which implies rather fast resorption of the implant material and simultaneous filling of respective resorption lacunas with newly formed bone. Pure HA has a limited potential in the frame of regenerative approach due to its low bioresorption rate [1]. Materials based on carbonate-substituted hydroxyapatite $Ca_{10-x}Na_x(PO_4)_{6-x}(CO_3)_x(OH)_2$ (CHA) are more prospective since their chemical composition approximates that one of inorganic part of a bone tissue more accurately [2], and its rate of bioresorbtion is high enough [3] due to structural distortions occurring under carbonate anion incorporation into the crystal lattice.

Choosing CHA as a material for bone implants immediately requires special technique for its consolidation into bulk bodies. Conventional method of solid state sintering cannot be applied for the case of CHA since it starts to decompose at temperatures higher than 700°C into pure (stoichiometric) HA and CaO with CO_2 release [4,5]. This problem can be overcome by selection of an appropriate binder with low melting point. Thereby, the process of consolidation can be described in terms of liquid-state sintering running essentially in liquid-state environment. Naturally, the binder has to be biocompatible, i.e. has to contain calcium, phosphorus and/or other biogenic elements. The next requirement is connected with binder melting point that should be reasonably low, less than the decomposition point of CHA, but high enough to provide the melt with low viscosity, otherwise molten binder cannot fill pores and capillary channels between CHA particles in a green body. From

Journal of Physics: Conference Series **291** (2011) 012036 doi:10.1088/1742-6596/291/1/012036

these viewpoints, quasi-binary section $Na_4CaP_6O_{18}$ - $Na_2CaP_2O_7$ of CaO – Na_2O - P_2O_5 ternary system looks very attractive as one can find there eutectics (54% $Na_4CaP_6O_{18}$ and 46% $Na_2CaP_2O_7$) melted at 473°C [9]. Pyro- and polyphosphate species composing the eutectic facilitate supercooling of the melt and, thus, enable to quench the melt into glass which can be soften at temperature even lower than the melting point.

In the present paper, we report on the possibility to consolidate CHA powder without noticeable changing of its carbonate content using low-melting Na-Ca phosphate glass with the above composition as a binder.

2. Experimental

2.1. Preparation of CHA powder

CHA powder was prepared by precipitation from aqueous as described in [6-8] according to reaction (1). Chemically graded $(NH_4)_2HPO_4$, $Ca(NO_3)_2{\cdot}4H_2O$, $NaHCO_3$ and 30 wt. % ammonia solution were used as raw materials.

$$8.5Ca(NO_3)_2 + 4.5(NH_4)_2HPO_4 + 1.5NaHCO_3 + 8NH_4OH \rightarrow Ca_{8.5}Na_{1.5}(PO_4)_{4.5}(CO_3)_{1.5}(OH)_2 + 17NH_4NO_3 + 6H_2O, \quad (1)$$

Stoichiometric amount of $NaHCO_3$ was added to 0.3 M $(NH_4)_2HPO_4$ solution, and pH was adjusted to 11 ± 1 by adding ammonia. 0.5 M $Ca(NO_3)_2$ was added to the above solution drop wisely under vigorous stirring and temperature T=40°C. The CHA-precipitate was maturated for 24 hours in mother liquor followed by filtration at Buchner funnel. This precipitate was thoroughly washed with distilled water (more than 1 litter) to remove by-products, and, then, was dried at T=60°C for 24 hours. Thermal treatment at 260°C for 1 hour was applied for complete elimination of NH_4NO_3 by-product.

2.2. Fabrication of low-melting phosphate glass

Phosphate glass composed of 54% $Na_4CaP_6O_{18}$ and 46% $Na_2CaP_2O_7$ was obtained from $CaHPO_4$ (obtained by precipitation from $Ca(NO_3)_2{\cdot}4H_2O$ and $(NH_4)_2HPO_4$ solutions), NaH_2PO_4, $Na_2HPO_4{*}12H_2O$, $Ca(H_2PO_4)_2$ precursors according to reactions (2-7) followed by mixing $Na_4CaP_6O_{18}$ and 46% $Na_2CaP_2O_7$, subsequent melting at 950°C for 6 hours and quenching (by pouring the melt onto the cold steel plate).

$$2CaHPO_4 \rightarrow Ca_2P_2O_7 + H_2O\uparrow \quad (2)$$
$$NaH_2PO_4 \rightarrow NaPO_3 + H_2O\uparrow \quad (3)$$
$$2Na_2HPO_4 \rightarrow Na_4P_2O_7 + H_2O\uparrow \quad (4)$$
$$Ca(H_2PO_4)_2 \rightarrow Ca(PO_3)_2 + 2H_2O\uparrow \quad (5)$$
$$4NaPO_3 + Ca(PO_3)_2 \rightarrow Na_4CaP_6O_{18} \quad (6)$$
$$Na_4P_2O_7 + Ca_2P_2O_7 \rightarrow 2Na_2CaP_2O_7 \quad (7)$$

2.3. Fabrication of CHA based composite materials

The CHA powders were mixed with 40 wt.% of the phosphate glass in a ball mill for 30 min under acetone (powder/balls ratio was set to 1:4). The resulted powder was dried at 100 °C for 24 hours. Thus obtained mixture was annealed at 450, 500, 600 and 700°C (t=24 h) in order to choose consolidation temperature (heating rate was about 5°C/min, after annealing the samples were quenched to room temperature in air). Two different ways were used to consolidate the composites – "warm" pressing at T=250°C and P=450 MPa for 6 hours (Fig. 1a) and hot pressing at T=450°C and P=200 MPa for 3 hours (Fig. 1b).

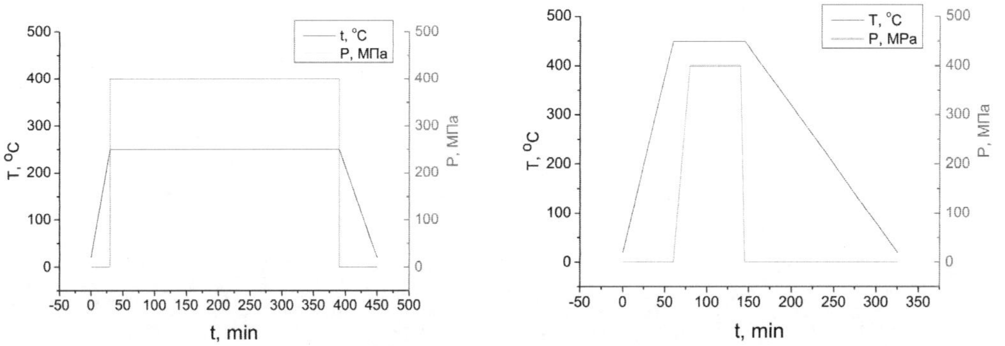

Fig. 1. Pressure and temperature profiles for (a) "warm" and (b) hot pressing.

3. Sample characterization

All samples were studied by X-ray diffraction analysis in the interval of angles 2θ=20–70° (Cu Ka radiation, Rigaku D/MAX 2500 with rotating anode, Japan). The micromorphology of the samples was examined by scanning electron microscopy (LEO SUPRA 50VP, Carl Zeiss, Germany, 5 kV). The carbonate content was analyzed via microvolumetric determination of CO_2 released from the samples under their dissolution in nitric acid. The overall content of CO_2 and H_2O was examined by TG/DTA (Diamond Pyris, Perkin Elmer, USA, quasi-isothermal heating to maintain decomposition rate at 10 μg/min; STA 409 PC, Netzsch, Germany, with IR-analysis of evolved gases). Distribution of particle sizes in powders was quantified by dynamic light scattering method (Analysette-22, Fritsch, Germany) in a window of 0.2-100 μm with preliminary ultrasonication for 20 s.

Compressive strength of the cylindrical samples was determined in the test of their axial loading (crosshead rate 1 mm/min) using electromechanical universal machine Instron 5982 (Instron, UK, ultimate load up to 100 kN). Height/diameter ratio of the cylindrical samples was adjusted to 2:3; both bases of the cylinders were made parallel by polishing.

4. Results and discussion

Details of variation of CHA properties with carbonate content can be found elsewhere [4]. Briefly, an increase of carbonate content gives rise to a particle size decrease as well as to a microstrain increase (and, therefore, solubility enhancement). Coming from this point, CHA samples with maximal carbonate content are prospective in sense of fabrication of bioresorbable implants.

The sequence of transformations occurring in the phosphate glass binder under its heating can be deduced from DTA-curve (Fig. 2). It is apparent that transition in glass connected with its devitrifying and softening takes place at about T ~ 450°C as it is evidenced by step-wise behavior of the DTA base line. Exothermal effect at T ~ 550°C has to be attributed to crystallization of the glass, and prominent endothermic effect at T ~ 750°C corresponds to global melting of the sample. It is of note that the last temperature points a maximum of the rate of melt formation, rather than the malt starting which seems to occur earlier, just after crystallization of the glass proceeds. Nevertheless, a contradiction on melting point of this composition between literature data and our experiment still exists. Feasible explanation of this fact can be made assuming that in our glassy samples melting is not an equilibrium process apart from [9], where attaining the equilibrium is a key factor to construct a phase diagram. However, it is quite clear that the temperature of composite processing should lie in the interval 450-700°C.

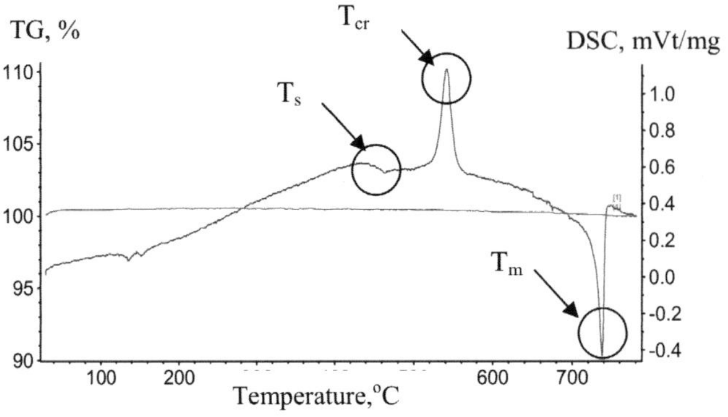

Fig. 2. DTA-curve for the phosphate glass under study.

The mixture CHA/glass (60% mass. CHA and 40% mass glass) was sintered at different temperatures, namely 450, 500, 600 and 700°C and according to XRD, chemical interaction of CHA with the glass starts at temperature T= 475±25°C (Fig. 3). This phenomenon narrows the interval suitable for the composite processing to 450-475°C.

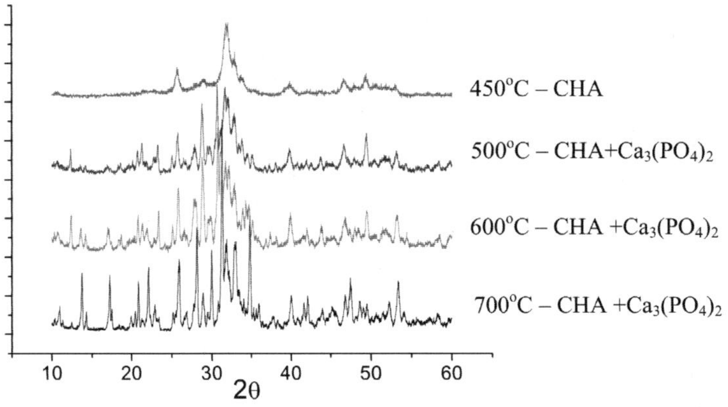

Fig. 3. XRD spectra of the CHAP/glass mixture annealed at 450, 500, 600 and 700°C for 24 hours.

Interestingly, XRD peaks of β-TCP ($Ca_3(PO_4)_2$) phase have considerable shift with respect to correspondent ICDD-data. Such an observation implies the fact of sodium incorporation into TCP lattice. Since β-TCP phase appears already at about 500°C, right in between devitrifying and crystallization, this temperature can be regarded as a starting one for CHA reaction with glass environment with the accuracy of a half of a temperature step in our annealing experiments (i.e. 25°C).

Our attempts to consolidate the mixture CHA/glass via conventional sintering at 450-500°C give no yield: thus "sintered" samples demonstrate rather weak strength (less than 5 MPa). It looks exactable since the main atomic mechanism of sintering glassy matrix is viscous or creep flow, the rate of which depends of matrix viscosity being the function of temperature. The driving force for sintering is either capillary or external pressure. Therefore, affective densification of the composite of CHA in a viscous thermoplastic matrix assumes elevated temperatures and pressures.

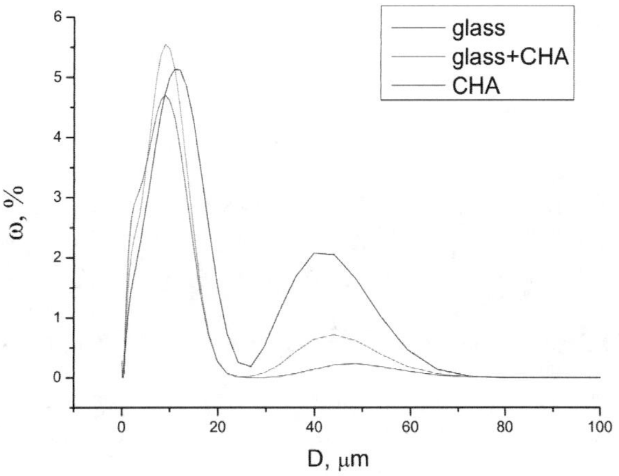

Fig. 4. Particle size distribution in CHA, phosphate glass and CHA/glass mixture powders after milling.

Another factor affecting consolidation is uniform distribution of CHA and glassy phases across the bulk of the samples. Apparently, that homogeneity of such distribution depends on proximity of particle sizes for both phases. Laser diffraction analysis showed that in the case of phosphate glass bimodal distribution with the modes of 11 μm corresponded to the average of primary particle aggregates, and of 44 μm - attributed by us to secondary aggregates. The cases of CHA and CHA/glass mixture look quite similar to the described above (Fig. 4). Although, the size of CHA crystals are less than 100 nm and, therefore, a large (10 μm in magnitude) particle of the glass is surrounded by a shell with comparable thickness composed of 100 nm particles of CHA.

According to SEM intrusion of CHA particles composing the shell into the glass particle takes place under "warm" pressing (Fig. 5). A viscosity of the glass is likely to be rather high under these conditions, and the glassy matrix cannot fill all pores and channels for the time less than 6 hrs. Extending the time of such procedure is hardly attractable from practical viewpoint.

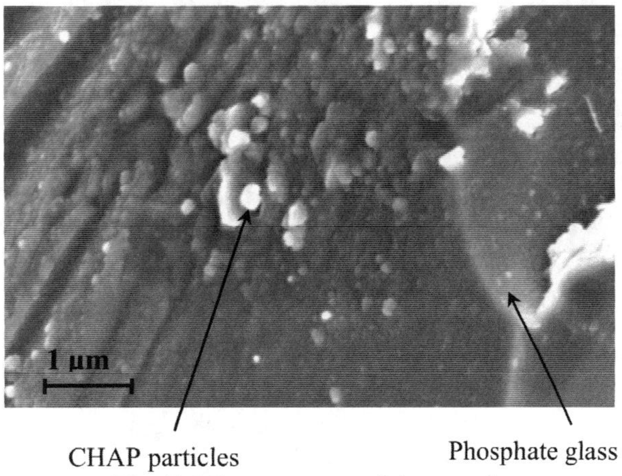

Fig. 5. Microstructure of the sample obtained by "warm" pressing.

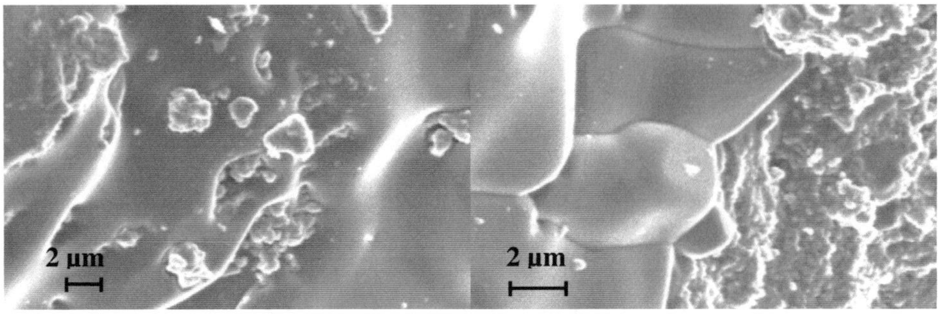

Fig. 6. Microstructures of the samples obtained by hot pressing.

Hot pressing at 450°C and pressure of 400 MPa leads to more effective creep of glass in comparison with the temperature of 250°C owing to considerable drop of glass viscosity behind the devitrifying region (Fig. 6). However, a decrease of viscosity is still not enough for effective densification in a time scale of 1 hour, since according to SEM data one can find some regions where CHA particles do not contact with the glass matrix. Perhaps, an increase of hot pressing time at temperature 450°C enables to obtain more homogeneous composite material with enhanced mechanical properties.

Thereby, formation of a composite material with phosphate glass as a matrix and CHA as a filler, combines two processes – intrusion of solid CHA particles into the glass matrix and a viscous flow of the glass into intergranular spaces by both capillary forces and external pressure. Since the creep rate of the glass is rather low, it needs definitive time for large glass particles to flow into the pores between CHA crystals (see Fig. 7.).

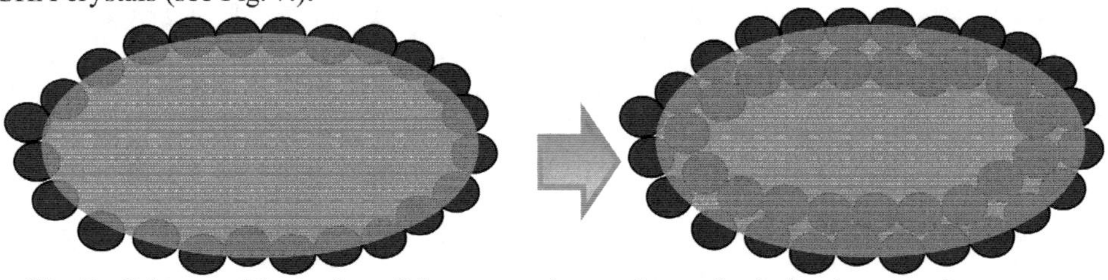

Fig. 7. Scheme of formation of the composites under study during hot pressing.

It is worth to note that according to chemical analysis both hot and "warm" PRESSED SAMPLES do not change significantly their carbonate content (Table 1); the content for both types of the samples as well as of starting CHA coincide within the accuracy of the analysis. This fact also supports that under chosen conditions neither interaction between CHA and glassy matrix occurs.

Table 1. Carbonate content in the CHA powder and the composites after high temperature pressing.

	CHA	CHA/glass (T=250°C)	CHAP/glass (T=450°C)
$\omega(CO_3^{2-})$, wt. %	7.5 ± 0.3	7.2 ± 0.5	6.9 ± 0.5

The samples fabricated by hot pressing technique were machined to shape necessary for correct mechanical testing (Fig. 8) (height/diameter ratio 3:2). It was found that the composites are easily machinable (via cutting and grinding) compared to HA ceramics obtained by solid-phase sintering due to moderate strength values.

Fig. 8. The composite fabricated by hot pressing machined for compressive strength test.

Evidently, the samples obtained by hot pressing at 450°C had better compressive strength values – 25 MPa compared to "warm"-pressed samples – 6 MPa. There is no doubt that the reason underlying such a behavior consists in significant drop in viscosity of the glass matrix at 450°C that leads to more effective intergranular space filling and better adhesion of filler particles to the matrix.

5. Conclusions

We have demonstrated that there is a possibility to consolidate CHA powder without noticeable changing of its carbonate content using low-melting Na-Ca phosphate glass with the composition referred to 54% $Na_4CaP_6O_{18}$ and 46% $Na_2CaP_2O_7$ as a binder. A temperature window suitable for obtaining the composite was assessed as 450 - 475°C since at 475±25°C the interaction between CHA and phosphate glass matrix starts. The main process contributing to overall densification of the samples is a viscous flow of the glass matrix under external pressure. The composites obtained at 450°C and 400 MPa for 1 hr demonstrate compressive strength up to 25 MPa.

This work was partially supported by RFBR grants № 09-03-01078-a, 10-03-00866-a, a call for grant 11-03-90463-Ukr_a. Funding by the state contract №-P403 from от 30.07.2009 in the frame of Federal Program "Scientific and Educational Staffs of innovative Russia, for the period 2009 -2013 years" is also acknowledged.

References

[1] LeGeros R Z 2002 *Clinical Orthopaedics and Related Research* **395** 81–98
[2] Kokubo T, Kim H M and Kawashita M 2003 *Biomaterials* **24** 2161–2175
[3] Torbenko N G and Karsavina V A 1977 *Nauka* 472
[4] Kovaleva E S, Shabanov M P, Putlayev V I, Filippov Ya Yu, Tretyakov Yu D and Ivanov V K 2008 *Mat.-wiss. u. Werkstofftech* **39** 1–8
[5] Barralet J, Best S and Bonfild W 1998 *Journal of Materials Science* **41** 79–86
[6] Gibson I R and Bonfield W 2002 *J. Biomed. Mater. Res.* **59** 697–708
[7] Nelson D G A and Featherstone J D B 1982 *Calcif. Tissue Int.* **34** 69–81
[8] Vignoles M, Bonel G and Young R A 1987 *Calcif. Tissue Int.* **40** 64–70
[9] Berak J and Znamierowska T 1972 *Rocz. Chem.* **46** 1697–1708

Syntesis of lanthanum zirconate hydrosols by the ion exchange method

E A Bovina[1], J V Tarasova, F Kh Chibirova

LLC "Superconductor", Varshavskoe shosse, 46, 115230, Moscow, Russia.

Email: bovina@rambler.ru

Abstract. Ion exchange of $LaCl_3$ and $ZrOCl_2$ aqueous solutions with anion-exchanger AV-17-8 was used to synthesize finely dispersed hydrosol of amorphous lanthanum zirconate $La_2Zr_2O_7$. Heat treatment of dried $La_2Zr_2O_7$ hydrosols at 700ºC and 1100ºC resulted in the formation of powders with fluorite and pyrochlore type structures, respectively. Epitaxial $La_2Zr_2O_7$ films were obtained on $SrTiO_3$ (001) single crystals. The substrate has an influence on the lanthanum zirconate crystal orientation, as well as strong inhibitory effect on sintering processes.

1. Introduction

Lanthanum zirconate $La_2Zr_2O_7$ has a cubic pyrochlore structure [1] and is highly thermostable. It is used in superconductors of the second generation as thin buffer layers since its lattice parameters match those of YBCO (0.5 %), and also because it provides a good barrier against oxygen diffusion [2,3].

Lanthanum zirconate (LZO) buffer layers are mainly prepared by the alkoxide sol-gel route [2,4], MOD [5], and CSD [3,6] methods from solutions of La and Zr organic compounds in organic solvents, for example, lanthanum isopropoxide and zirconium n-propoxide dissolved in 2-methoxyethanol [2,4] and lanthanum (III) and zirconium (IV) 2,4- pentadionate, that are transformed into respective propionate, when dissolved in the propionic acid [3,5]. Water-based solutions, containing metal acetates, acetic acid, triethanolamine, and ammonium hydroxide were proposed in order to avoid organic solvents [7].

We have developed the ion exchange method for synthesis of amorphous lanthanum zirconate hydrosol free from organic components. In this paper we report the synthesis route and properties of the product obtained.

2. Experimental details

The samples (La/Zr atom ratio=1/1) were synthesized by the ion exchange between aqueous solutions of lanthanum and zirconyl chlorides ($LaCl_3 \cdot 6H_2O$ and $ZrOCl_2 \cdot 8H_2O$ of the chemically pure grade) and anion-exchanger AV-17-8 in the OH^- form. Ion exchange was carried out by stirring aqueous solutions of lanthanum and zirconyl chlorides with anion exchanger AV-17-8 suspension with a mechanical mixer for a specified period of time at room temperature followed by separation of anion-exchanger by filtration through nylon-6 textile material. The anion exchanger amount was calculated from the given ratio of anion- exchanger to the sum of lanthanum and zirconyl chlorides (H) in terms of OH^- and Cl^- mg-equivalents. The sols were evaporated in a water bath, dried at 100°C in the oven and calcined in air for two hours at the given temperature. pH values of starting solutions and sols obtained were measured by a Sartorius Proffesional Meter PP-20 with the glass pH/ATC electrode. Lanthanum and zirconium concentrations were determined by titration with Trilon B [8]. Concentrations of chloride ions (C_{exp}) were measured nephelometrically on an ECOTEST 2020 instrument (λ=470nm), using silver nitrate $AgNO_3$. From these data, the degree of exchange was calculated: $\alpha=[1- (C_{exp}/C_{in})]$. The optical density of sols (D) was measured on an ECOTEST 2020 instrument (λ=595 nm; 1.0-cm cell). The size and morphology of sol particles were determined on a 200 kV JEM 2100-F/UHR (JEOL, Japan) transmission electron microscope (TEM). Dried sols (100-mg samples) were subjected to thermal analysis on a Q-1500 Paulik derivatograph at heating rate of 10°C/min. XRD patterns of dried and calcined samples were recorded on a Rigaku D/MAX 2500 diffractometer (CuK_α radiation, graphite monochromator). The specific surfaces of products were determined by measuring nitrogen desorption on an EMS-51 instrument. The size and form of the sample particles after heat treatment were estimated by a SEM Leo Supra 50 VP device.

3. Results and Discussion

Ion exchange between lanthanum and zirconyl chlorides solutions and anion-exchanger AV-17-8 gives clear colorless sols independent of the exchange conditions (concentrations of starting solutions in terms of $La_2Zr_2O_7$ were in the range of 0.00125-0.0050 mol/L; H=1.0-2.0 and time of stirring of 3 - 30 minutes). Concentrations of La and Zr in sols correspond to those of starting solutions. Hydroxyl groups of the anion-exchanger and chloride ions solutions of salts are exchanged rapidly, depending mainly on the value of the H parameter. At H=1.0, independent of the concentration of solutions, in 3 min, the degree of exchange reaches 0.85-0.87; when stirring time is 30 min the exchange degree increases up to 0.97-0.98. For H=1.5 and solution concentrations of 0.00125 and 0.0025 mol/L, after stirring for 3 min, the degree of exchange increases up to 0.95 and 0.97, respectively, and after 30 min stirring, it reaches to 0.99. For 0.00125 M sols at H=2.0 and 0.005 M sols at H=1.5, the degree of exchange reaches 0.98-0.99. pH values of starting lanthanum and zirconyl chlorides solutions are 2.0-2.5. pH values of the sols depend mainly on the H parameter. Independently of the solution concentration, at H=1.0 and 3 min stirring, pH values of sols are 8.1-8.5. At H=1.5 and 2, the further increasing of solution concentrations and stirring time increases pH of sol up to 10.3-10.9.
Optical density of sols (D) does not exceed 0.01. Sols have been analyzed by the transmission electronic microscopy (TEM). The typical TEM image is presented in Figure 1. It is obvious that the particles sizes are 2-5 nm and their form is close to spherical. The sols are stable for no less than two weeks.

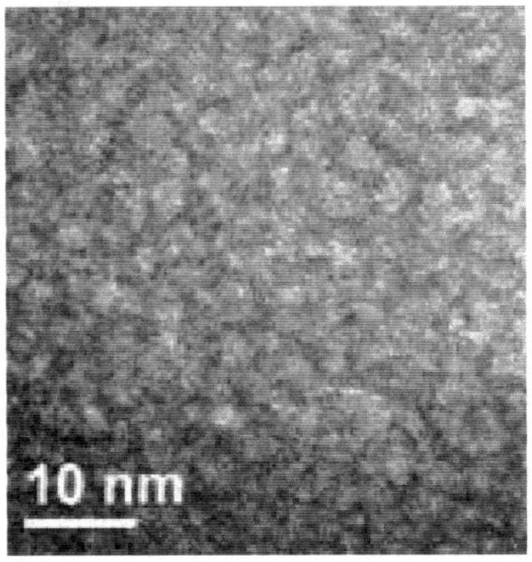

Figure 1. TEM image of a typical sol.

The dried sols are amorphous and have sufficiently developed specific surface of 150-161 m^2/g. The typical thermoanalytical curves for dried sols are presented in Figure 2. There is a blurred endotherm on the DTA curve near 130°C. Weight loss (26.6%) is also observed, that is, apparently, caused by removal of physically adsorbed water and, probably, hydroxyl groups. No other thermal and mass effects are seen in the curves.

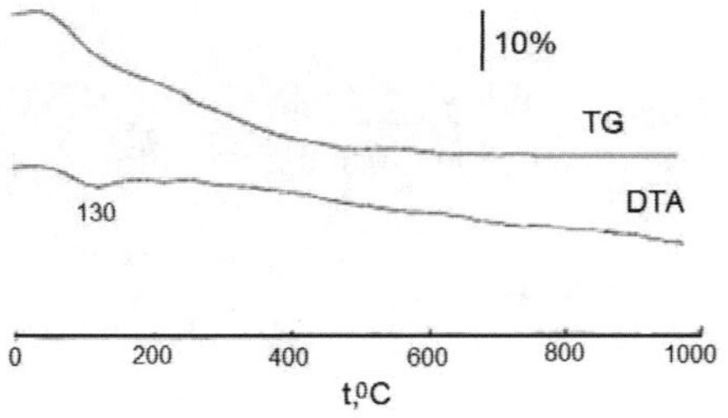

Figure 2. The typical thermoanalytical curves for dried sol.

XRD patterns of dried sols after heat treatment are shown in Figure 3. It is obvious that after the heat treatment at 700°C, reflections (220), (400), (440) and (622) characterizing the appearance of the fluorite crystal structure F-$La_2Zr_2O_7$ [9] are observed. Upon temperature increase up to 900°C, these reflections become narrower, and reflections (444) and (800) of lanthanum zirconate with the pyrochlore structure P-$La_2Zr_2O_7$ appear. In addition, very weak reflection (331) is observed, that characterizes lanthanum zirconate of the same structure. With further increase in temperature up to 1000°C, reflection (331) is increased and weak reflection (511) also related to P-$La_2Zr_2O_7$ is observed. At temperatures higher than 1100^{0}C, very weak reflection (111) also related to P-$La_2Zr_2O_7$ (JCPDS 73-444) appears. It is of note, that there are no reflections of lanthanum and zirconium oxides in the XRD, that proves the stoichiometric composition of lanthanum zirconate and the absence of admixtures in patterns.

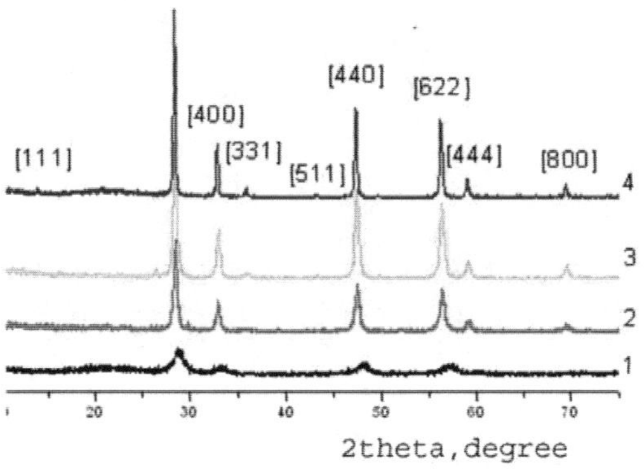

Figure 3. XRD pattern of lanthanum zirconate powder calcined at 700 (1), 900 (2), 1000 (3), and 1100°C (4).

In the SEM picture of LZO powder calcined at 900-1100^{0}C (Figure 4), transparent crystalline irregular plates of size up to several dozens of μm were observed.

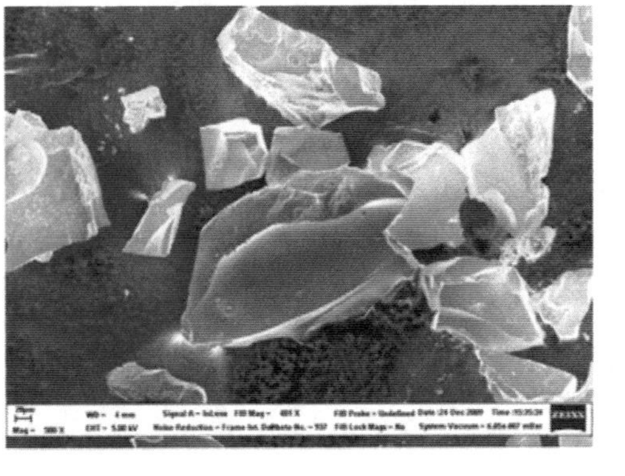

Figure 4. SEM picture of the $La_2Zr_2O_7$ powder calcined at 1000ºC (an increase of five hundred times).

Epitaxial LZO films on $SrTiO_3$ (001) single crystal were obtained at 1000ºC in air using the synthesized sols as precursor material. The $SrTiO_3$ single crystals (disks with diameter of 10 mm and thickness of 0.5 mm) was received from Moscow Power Engineering Institute. XRD pattern of a typical film is shown in Figure 5. The reflections (100), (200), (221) of the $SrTiO_3$ single crystal are observed. Besides, there are intensive reflection (400) at $2\theta = 33.25°$ and a less intensive reflection (800) at $2\theta = 69.70°$ of lanthanum zirconate. Their presence in the XRD pattern confirms epitaxial growth of the crystalline $La_2Zr_2O_7$ on the $SrTiO_3$ (001) single crystal.

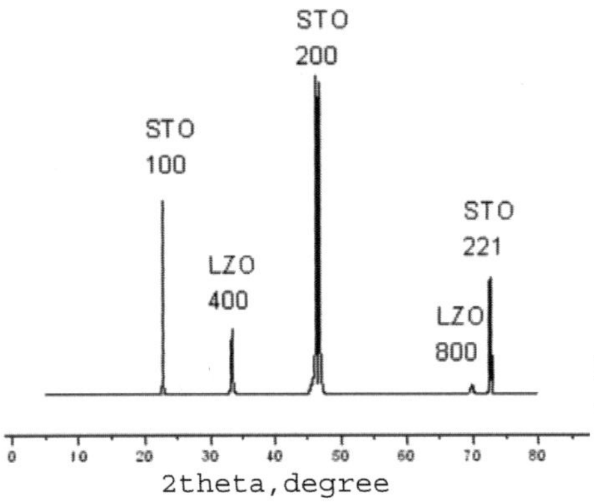

Figure 5. XRD pattern of the $La_2Zr_2O_7$ epitaxial film on $SrTiO_3$ (001) obtained at 1000ºC.

In the SEM-microphotograph (Figure 6), small grains of the $La_2Zr_2O_7$ film of size less than 100 nm are visible, that are much smaller than powder grains (Figure 4). The LZO film surface is in general homogeneous.

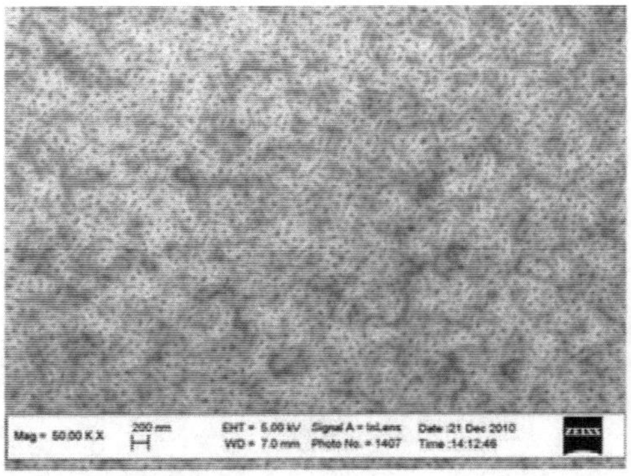

Figure 6. SEM picture of the epitaxial $La_2Zr_2O_7$ film on $SrTiO_3$ (001) single crystal obtained at 1000°C (an increase of one hundred thousand times).

Lanthanum zirconate sol formation may be represented as follows: OH^- groups appear in solutions as a result of rapid ion exchange of chloride ions of the dissolved lanthanum and zirconyl salts for hydroxide groups of the anion exchanger. Due to appearance of OH^- groups in the suspension, the acidic medium (pH= 2.0 -2.5) becomes alkaline (pH=8.0-10.9), and the processes analogous to those in coprecipitation, namely, hydrolysis of aquacations followed by polycondensation of forming aquahydroxy complexes occurs to produce nanoparticles of several nm in size. Thus, the sol is formed [10]. Taking into account the facts that according to the TEM picture (Figure 1), sol particles are homogeneous, the absence of any thermal effects except water loss in the DTA curve (Figure 2), and the formation of crystalline $La_2Zr_2O_7$ during the the heat treatment (Figure 3), it is reasonable to conclude that the dried sol is an amorphous lanthanum zirconate hydrate.

During the heat treatment of dried sols dehydration occurs, then, at 700°C, the amorphous compound begins to gradually crystallize to cubic lanthanum zirconate F-$La_2Zr_2O_7$ of the fluorite structure, consisting of LaO_8 and ZrO_8 cubes. However, because of the differences in charges and radii of La^{3+} (0.104 nm) and Zr^{4+} (0.082 nm) ions with increasing temperature, the disordered fluorite structure becomes unstable and begins to transform into the ordered pyrochlore structure that consists of the rigid frame of octahedrons ZrO_6 and a flexible sublattice of irregular cubes of LaO_8:

Amorphous compounds → F- $La_2Zr_2O_7$ → P- $La_2Zr_2O_7$

It should be noted that the ($La_2Zr_2O_7$) particle size in the epitaxial film is much less than that in the roughly dispersed massive sample, despite the fact that the epitaxial film and the massive sample are obtained at the same temperature (1000°C). This confirms strong effect of the textured substrate on process of formation of the $La_2Zr_2O_7$ epitaxial film.

The $La_2Zr_2O_7$ formation processes in the sol synthesis by the ion exchange method developed in the present work and those in the coprecipitation method [11, 12] are close. In both cases crystallization of lanthanum zirconate occurs during the heat treatment at temperatures above 600 °C (the structure of fluorite) and ≥1000°C (the structure of pyrochlore). In contrast to the coprecipitation method, the absence of the by-product (ammonium chloride) and lower solution concentrations used in this work prevent aggregation and coagulation of particles formed that allows to successfully synthesize sols by the ion exchange method.

4. Summary.

Nanosols of amorphous hydrated lanthanum zirconate were synthesized by the method of ion exchange of aqueous solutions of lanthanum and zirconyl chlorides with anion-exchanger AV-17-8. After heat treatment of dried sols, the crystal structure of $La_2Zr_2O_7$ obtained was determined. Calcination of $La_2Zr_2O_7$ sols at 700 and 1100°C on amorphous substrates resulted in the formation of films of fluorite and pyrochlore type structures, respectively. Upon sol deposition on a textured substrate at 1000ºC, epitaxial films were obtained.

References

[1] Deiseroth H-J, Müller-Buschbaum Hk 1970 *Zeitschrift für Anorganische und Allgemeine Chemie* **375** 152-156
[2] Chirayil TG, Paranthaman M, Beach DB, Goyal A, Williams RK, Cui X, Kroeger DM, Feenstra R, Verebelyi DT, Christen DK 2000 *Physica C* **336** 63-69
[3] Knoth K, Hühne R, Oswald S, Schultz L, Holzapfel B 2007 *Acta Materialia* **55** 517-529
[4] Parans Paranthaman M, Aytug T, Sathyamurthy S, Beach DB, Goyal A, Lee DF, Kang BW, Heatherly L, Specht ED, Leonard KJ, Christen DK, Kroeger DM 2002 *Physica C* **378-381** 1009-1012
[5] Carroff T, Morlens S, Abrutis A, Decroux M, Chaudouët P, Porcar L, Saltyte Z, Jiménez C, Odier P, Weiss F 2008 *Superconductor Science and Technology* **21** 075007-075015
[6] Zhu XB, Lei HC, Shi DQ, Zhang L, Wang L, Sun YP, Song WH, Yang ZR, Kim JH, Zeng R, Dou SX, Yang J, Gu HW 2007 *Physica C* **467** 73-79
[7] Cloet V, Feys J, Hühne R, Hoste S, Van Driessche I 2009 *Journal of Solid State Chemistry* **182** 37-42
[8] Kreshkov AP 1970 *The Base of the Analytical Chemistry* **2** Chemistry: Moscow 336 (in Russian)
[9] Kido H, Komarneni S, Roy R 1991 *Journal of the American Ceramic Society* **74** (2) 422-424
[10] Jolivet J-P 1988 *Metal Oxide Chemistry and Synthesis from Solution to Solid State* Chichester: Wiley J
[11] Nair J, Nair P, Doesburg EBM, Van Ommen JG, Ross JRH, Burggraaf AJ, Mizukami F 1998 *Journal of Materials Science* **33** 4517-4523
[12] Zwinkels MFM, Druesne S, Govind Menon P, Björnbom E, Järås SG 1998 *Industrial & Engineering Chemistry Research* **37** (2) 391-397

Journal of Physics: Conference Series **291** (2011) 012038 doi:10.1088/1742-6596/291/1/012038

Silica nanoparticles with a substrate switchable luminescence

O D Bochkova, A R Mustafina, S V Fedorenko and A I Konovalov

Institution of Russian Academy of Sciences A.E. Arbuzov Institute of Organic and Physical Chemistry of Kazan Scientific Center of RAS, Ak. Arbuzov Str., 8, 420088, Kazan, Russia

E-mail: o.d.bochkova@mail.ru

Abstract. Silica nanoparticles with visible (Tb and Ru doped), near IR (Yb doped) and dual visible-near IR luminescence (Ru-Yb doped) were obtained by reverse w/o microemulsion procedure. Plenty of luminescent complexes (from 4900 to 10000) encapsulated into each nanoparticle ensures the intensive luminescence of nanoparticles and their applicability as biomarkers. The silica surface decoration by definite anchor groups is the required step for the gaining to these nanoparticles marking and sensing functions. Thus covalent and non-covalent surface modification of these nanoparticles was developed to provide the binding with biotargets and sensing of anions. The dicationic surfactant coating of negatively charged Tb(III)-TCAS doped silica nanoparticles was chosen as the basis for the anion responsible system. The reversible insertion of the quenching anions (namely phenol red) into the surfactant based layer at the surface of luminescent nanoparticles switches off the Tb-centered luminescence. In turn the reversible reestablishment of the luminescence results from the competitive insertion of the non-quenching anions into the surfactant layer at the silica/water interface. The hydrophobic anions exemplified by dodecylsulfates versus hydrophilic ones (hydrophosphates) are preferable in the competition with phenol red anions.

1. Introduction

Smart colloidal systems, which are temperature, pH, or substrate responsive, gain much attention due to their possible application in bioanalysis, medicine and technology [1]. The appeal of silica nanoparticles results from the well-developed procedures of their synthesis with controlled size and uniformity, as well as nontoxicity [2]. The metal complex-doped nanoparticles have an advantage over a single complex, since the silica shell protects the metal complex from photodegradation, while the surface of the nanoparticle can be functionalized by various anchoring groups. Lanthanide complexes have gained a great deal of attention as dopants [3] owing to their unique spectroscopic characteristics, including long fluorescence lifetime, large Stokes shift, and sharp line-like emission bands [4]. These properties are particularly attractive because they enable temporal and spectral discrimination against background fluorescence often associated with commonly used fluorophores in chemical biology, leading to excellent detection sensitivity. The near-infrared (NIR)-emitting lanthanide complexes are of particular interest because biological tissues are transparent in this spectral range. Combination of luminescent and binding functions within nanoparticles is required for their application as biomarkers. The use of silica matrix, doped with luminophoric molecules and decorated by various anchor groups is a well known way of getting multifunctional nanoparticles for biolabelling [5]. The silica surface decoration is aimed to increase the colloidal stability of nanoparticles, as well as to develop their receptor properties towards biotargets. The decoration of the silica surface through the non-covalent interactions represents more convenient alternative to covalent anchoring due to the lack of multistep purification procedures [6]. For example the decoration of silica nanoparticles by adsorption of cationic surfactants enables to get non-aggregated positively charged silica nanoparticles [7] for a development of nanoparticles, exhibiting sensing function to anions. Anions play an important role in living organisms and even phospholipids membranes of cells are negatively charged.

The present report outlines the data concerning the development of highly luminescent and colloidal stable system with sensor function to anions on the basis of lanthanides complexes doped silica nanoparticles with the modified surface.

2. Materials and methods

2.1. Materials

Tetraethyl orthosilicate (TEOS) 98%, ammonium hydroxide (28-30%), *n*-heptanol 98%, cyclohexane 99%, tris(2,2'-bipyridyl)chlororuthenium(II) hexahydrate (99%) and ytterbium(III) nitrate hexahydrate (99.9%) from Acros; terbium(III) nitrate hexahydrate (99.9%) from Alfa Aesar, Triton X-100, cetyltrimethylammonium bromide (CTAB), Na_2HPO_4, sodium dodescylsulfate (SDS) and phenol red from Sigma-Aldrich were used as purchased without further purification. The synthesis of *p*-sulphonatothiacalix[4]arene tetrasodium salt (TCAS) was carried out according to the known procedure [8]. Dicationic surfactant gemini 16-6-16 (chart 1) was synthesized in analogy with the procedure [9].

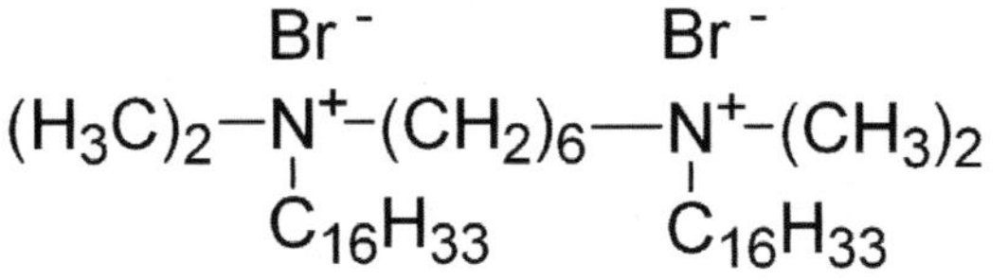

Chart 1. Gemini 16-6-16.

Luminescent Tb(III) and Yb(III) complexes with p-sulphonatothiacalix[4]arene and $[Ru(dipy)_3]Cl_2$ were used as dopants to silica nanoparticles. Synthesis of such silica nanoparticles has been performed according to reverse microemulsion procedure presented in the work [10]. Synthesis of silica coated Yb(III) and Yb(III)-Ru(II) nanoparticles at various Yb:Ru molar ratio has been performed with the addition to the reaction mixture of 7.8 mM of Yb(III)-TCAS and concentrations of $[Ru(dipy)_3]Cl_2$ varying from 3.9 mM to 15.6 mM.

Synthetic conditions of Yb, Ru and Yb-Ru doped silica nanoparticles.

SiO_2 Ru: Yb	$C_{Ru\ complexes}$, mM	$C_{Yb\ complexes}$, mM
0:1	0	7.8
1:0	7.8	0
2:1	15.6	7.8
1:1	7.8	7.8
0.9:1	7.02	7.8
0.5:1	3.9	7.8

Amino- and carboxy-modified silica nanoparticles were synthesized in analogy with the procedure [11].

2.2. Methods

The transmission electron microscopy (TEM) images were obtained using JEOL JEM 100 S microscope, Japan.

The dynamic light scattering (DLS) measurements were performed by means of the Malvern Mastersize 2000 particle analyzer. A He-Ne laser operating at 633 nm wavelength and emitting vertically polarized light was used as a light source. The measured autocorrelation functions were analyzed by Malvern DTS software and the second-order cumulant expansion methods. The effective hydrodynamic radius (RH) was calculated by the Einstein-Stokes relation from the first cumulant: D =kBT/6phRH, where D is the diffusion coefficient, kB is the Boltzmann constant, T is the absolute temperature, and h is the viscosity. The diffusion coefficient was measured at least three times for each sample. The average error in these experiments was approximately 4%. All samples were prepared from the bidistilled water with prior filtering through the PVDF membrane using the Syringe Filter (0.45 μm).

Zeta potential "Nano-ZS" (MALVERN) using laser Doppler velocimetry and phase analysis light scattering was used for zeta potential measurements. The average error in these experiments was approximately 10%.

The steady-state emission spectra were recorded on a spectrofluorometer FL3-221-NIR (Jobin Yvon) under 330 nm (slit 2), 320 nm (slit 14) and 450 nm (slit 2) excitation for recording Tb-centered, Yb-centered and and Ru-centered luminescence correspondingly.

All samples with nanoparticles were ultrasonicated within 30 min before measurements.

All measurements have been performed at least three times. The pH = 7.0, pH = 8.2 and pH=9.2 were adjusted by Tris (2.5 mM).

3. Results and discussion

3.1. Characterization of Tb(III)-TCAS and Yb(III)-TCAS doped silica nanoparticles

The nanoparticles were prepaired according to the known procedure by hydrolysis of TEOS with ammonium hydroxide in a W/O microemulsion [10], containing terbium(III) and ytterbium(III) complexes with *p*-sulphonatotiacalix[4]arene (Tb(III)-TCAS, Yb(III)-TCAS). The lanthanides complexes doped silica nanoparticles were characterized by transmission electron microscopy (TEM). As shown in figure 1 nanoparticles are spherical and uniform in size (40 ± 5 nm for Tb(III)-TCAS doped and 32±5 nm for Yb(III)-TCAS doped silica nanoparticles).

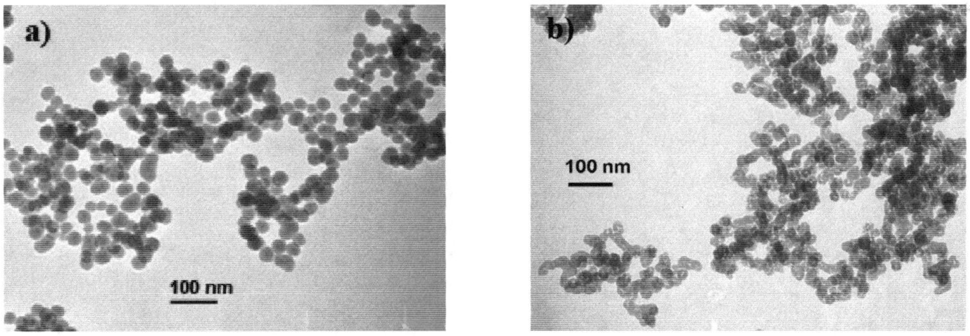

Figure 1. TEM images of Tb(III)-TCAS doped **(a)** and Yb(III)-TCAS doped **(b)** silica nanoparticles.

The aqueous dispersions of nanoparticles were analyzed by emission spectroscopy and dynamic light scattering (DLS) methods. These nanoparticles are negatively charged with ζ-value being about -33±5 mV in neutral aqueous solutions. DLS measurements performed for the aqueous dispersions of nanoparticles reveal the high monodispersity (polydispersity index is within 0.1-0.3) with the particle size being 223±4 nm for Tb(III)-TCAS doped and 262±4 nm for Yb(III)-TCAS doped silica nanoparticles, which stays unchanged within the concentration range from 0.0175 to 0.67 g L^{-1} (table 1). The gap between the sizes determined from the TEM (40±5 nm and 32±5 nm) and DLS data is rather high and can not be explained by the thickness of the hydration shell [7, 12]. Thus Ln(III)-TCAS doped nanoparticles are somewhat aggregated in aqueous dispersions.

Table 1. The averaged hydrodynamic diameters (d), ζ-potential values and polydispersity index (PDI) of Ln(III)-TCAS doped silica nanoparticles

Ln(III)-TCAS complex inside nanoparticle	C ($g\cdot L^{-1}$)	d (nm)	PDI	ζ (mV)
Tb(III)-TCAS	0.0175	237±2	0.190	-33
	0.028	223±4	0.243	-33
	0.07	232±2	0.186	-35
	0.67	300±5	0.115	-40
Yb(III)-TCAS	0.0175	234±7	0.265	-33
	0.028	262±4	0.287	-33
	0.07	273±9	0.315	-32
	0.67	268±2	0.292	-32

The aqueous dispersions of these nanoparticles exhibit very efficient Tb- (visible) and Yb-centered (near IR) luminescence, which confirms the formation of lanthanides doped silica nanoparticles (figure 2). The comparison of emission spectra of Tb(III)-TCAS and Yb(III)-TCAS complexes in aqueous solutions and within the silica core reveals that Ln(III)-TCAS in both above-mentioned cases show the same spectrum patterns [$^5D_4 \rightarrow {}^7F_6$(489nm), $^5D_4 \rightarrow {}^7F_5$(543nm), $^5D_4 \rightarrow {}^7F_4$(582nm), $^5D_4 \rightarrow {}^7F_3$(620nm) for Tb(III)-TCAS and $^2F_{5/2} \rightarrow {}^2F_{7/2}$ (976 nm) for Yb(III)-TCAS] without any emission band shift for Tb-doped and small one for Yb-doped nanoparticles (figure 2). The comparison of luminescence intensities of Ln(III)-TCAS doped silica nanoparticles with Ln(III)-TCAS complexes in aqueous solution reveals that each nanoparticle is doped with 4900 Tb(III)-TCAS complexes and 10000 Yb(III)-TCAS complexes, which ensure the intensive luminescence of nanoparticles and their applicability as biomarkers.

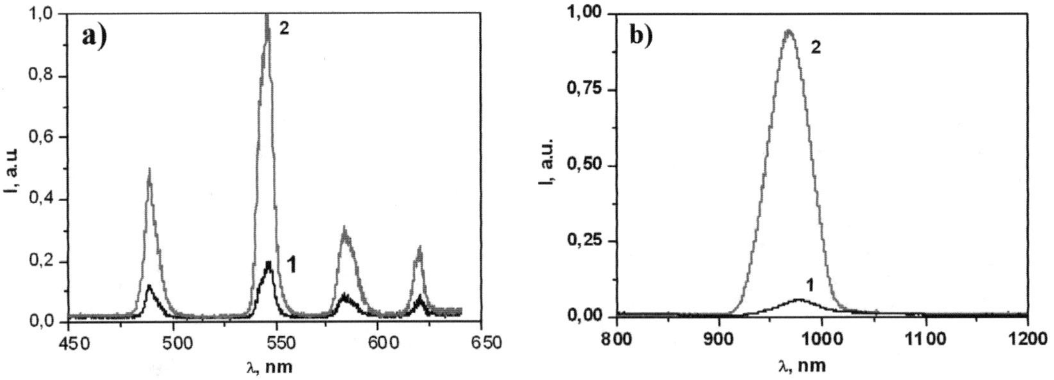

Figure 2. Emission spectra of **(a)** aqueous solutions of Tb(III)-TCAS **(1)** (C = 0.5 μM) and Tb(III)-TCAS doped silica nanoparticles **(2)** (C = 0.028 $g\cdot L^{-1}$) [10]; and **(b)** aqueous solution of Yb(III)-TCAS **(1)** (C = 0.01 mM) and Yb(III)-TCAS doped silica nanoparticles **(2)** (C = 0.28 $g\cdot L^{-1}$) at pH = 7.0.

It is also worth noting that 3 days of stirring of an aqueous dispersion of Tb(III)-TCAS doped nanoparticles (C = 0.028 $g\cdot L^{-1}$) provides no more than 10% decrease of Tb(III)-TCAS luminescence, while 2-fold decrease results from 7 days stirring of aqueous dispersions (figure 3). This decrease indicates the self-leakage of Tb(III)-TCAS complexes, which are rather unsteady in aqueous solutions due to photobleaching processes.

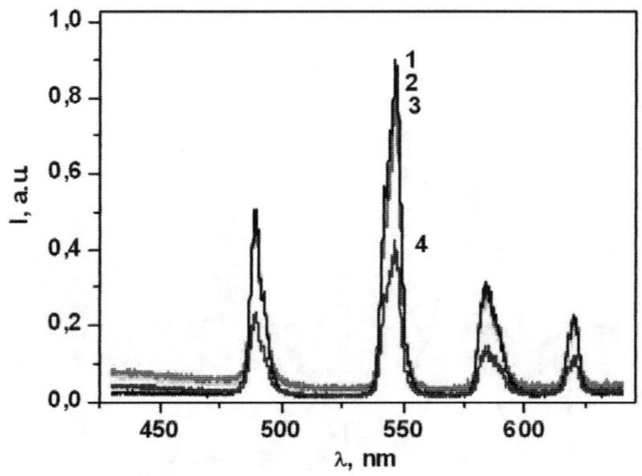

Figure 3. Emission spectra of Tb(III)-TCAS doped silica nanoparticles (C = 0.028 g·L^{-1}) in time: immediately after solution preparation **(1)**, after 1 hour of stirring **(2)**, after 3 days of stirring **(3)**, after 7 days of stirring **(4)** at pH = 7.0

3.2. Ru-Yb doped silica nanoparticles

The insertion of two luminophores into silica nanoparticles should provide more accurate recognition of the examined substrate through the ratiometric analysis. The variation of the ratio of luminescence intensities in the case of two luminophores can be measured more accurately than the change of the intensity of single luminophore system.

Silica nanoparticles exhibiting dual visible-NIR luminescence were obtained by the insertion of two type of complexes: tris(2,2'-bipyridyl)chlororuthenium(II) ([Ru(dipy)$_3$]Cl$_2$) and Yb(III)-TCAS into silica matrix using reverse W/O microemulsion procedure. The synthesized nanoparticles were doped with different ratio of Ru:Yb (2:1, 1:1, 0.9:1, 0.5:1, 0.3:1) and were analyzed by fluorimetry, electrophoresis, dynamic light scattering, microscopy and atomic emission spectroscopy methods [13].

The fluorometric measurements have been performed for single components [Ru(dipy)$_3$]Cl$_2$ and Yb(III)-TCAS, as well as their mixture at various molar ratio in aqueous solutions and in silica nanoparticles in order to compare the luminescent properties of luminophores in aqueous solutions with those within nanoparticles (figure 4). The obtained data indicate the quenching effect of [Ru(dipy)$_3$]$^{2+}$ on Yb(III)-centered luminescence of [Yb(III)-TCAS]$^{4-}$ in aqueous solutions (figure 4a) resulted from their ion pairing (chart 2).

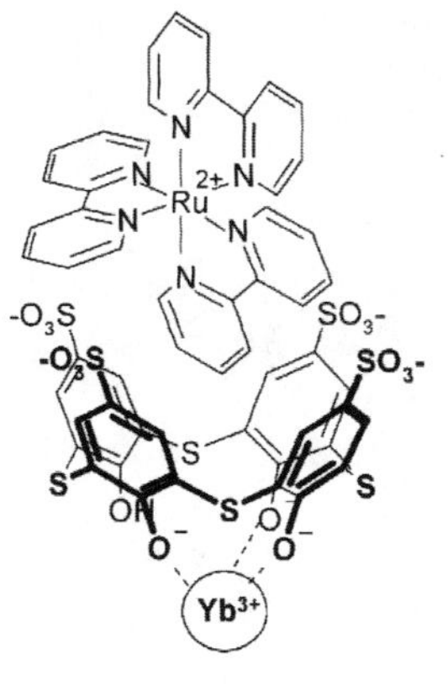

Chart 2. Schematic representation of the ion pair between [Ru(dipy)$_3$]$^{2+}$ and [Yb(III)-TCAS]$^{4-}$.

The analysis of Yb-centered emission intensity of nanoparticles indicates no correlation between the intensity of Yb-centered luminescence and Yb:Ru molar ratio (figure 4b). Moreover the deviation of the intensity of Yb-centered luminescence for nanoparticles with the mentioned above Yb:Ru molar ratio is not dramatic and does not exceed the experimentally observed deviation for Yb nanoparticles.

It should be concluded that Yb-centered emission in silica nanoparticles is not quenched by $[Ru(dipy)_3]Cl_2$ in these cases (figure 4b).

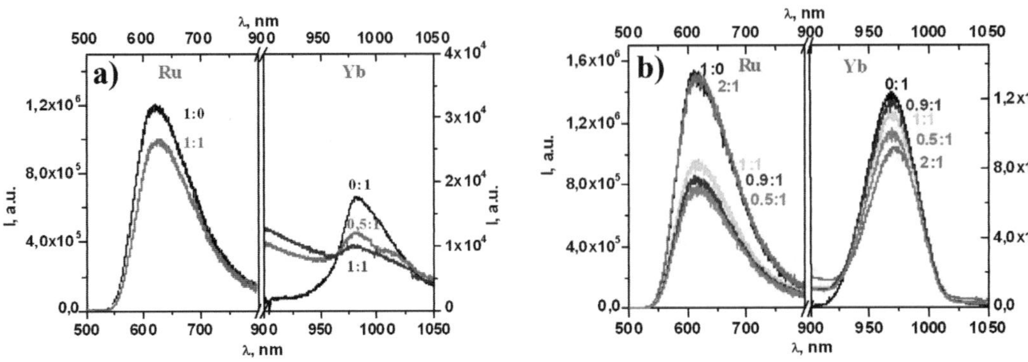

Figure 4. Emission spectra of aqueous solution of $[Ru(dipy)_3]Cl_2$ (C = 0.1 mM) and Yb(III)-TCAS **(a)** and Ru-Yb doped silica nanoparticles **(b)** (C = 0.23 $g \cdot L^{-1}$) at pH = 7.0 [13]. Numbers at the curves indicate Ru:Yb molar ratio (1:0, 0:1, 2:1, 1:1, 0.9:1, 0.5:1).

Keeping in mind the quenching of Yb-centered luminescence in aqueous solutions due to the ion-pairing between $[Yb(III)\text{-}TCAS]^{4-}$ and $[Ru(dipy)_3]^{2+}$, the obtained data indicate that Ru(II) and Yb(III) complexes within nanoparticles are not in close proximity to each other. It is worth noting that the formation of silica nanoparticles occurs within aqueous nanodroplets of reversed micelles. The conditions of tetraethylorthosilicate hydrolysis should be taken into account to explain the restricted ion-pairing between $[Yb(III)\text{-}TCAS]^{4-}$ and $[Ru(dipy)_3]^{2+}$. The excess amounts of NH_4^+ and OH^- ions, which are weakening the electrostatic attraction between $[Ru(dipy)_3]^{2+}$ and $[Yb(III)\text{-}TCAS]^{4-}$ is the most probable reason of their doping as $[Ru(dipy)_3]X_2$ and $[Yb(III)\text{-}TCAS]Y_4$, where X=Cl or OH and Y=Na or NH_4 [13].

3.3. Covalent surface modification of Tb(III)-TCAS doped silica nanoparticles

The surface modification of the nanoparticles by suitable functional groups is the required task for their binding with biotarget. There are two type of surface modification: covalent and non-covalent. The covalent modification of the silanol groups at the silica interface by the required anchor groups with further bioconjugation with fit biosubstrates is preferable for the application as biomarker, when the efficient binding with biotarget is required.

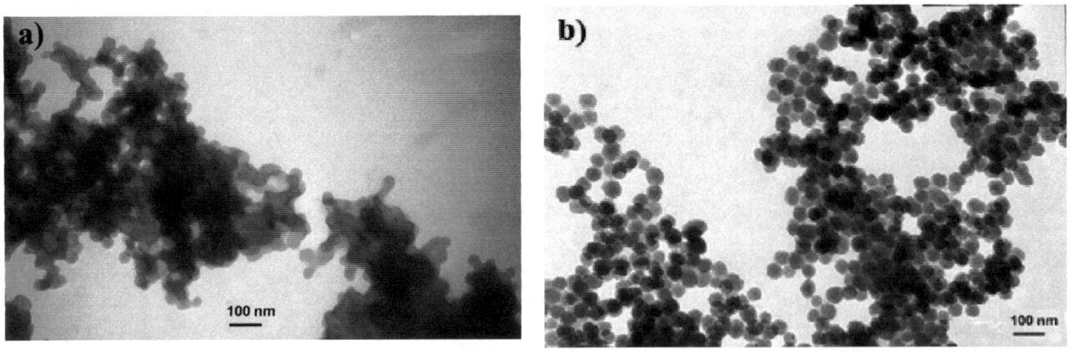

Figure 5. TEM images of amino- **(a)** and carboxy-modified **(b)** silica nanoparticles.

In particular Tb(III)-TCAS doped silica nanoparticles with amino surface groups were obtained by the known procedure[11].TEM data (figure 5) indicate the enhanced aggregation of amino-modified silica nanoparticles, which restricts their size determination. The carboxy-modified nanoparticles, which can be obtained from the amino-modified ones according to scheme 1, exhibit more colloidal stability. Their TEM image indicates more uniform distribution on the copper grids with the size being about 43±5 nm.

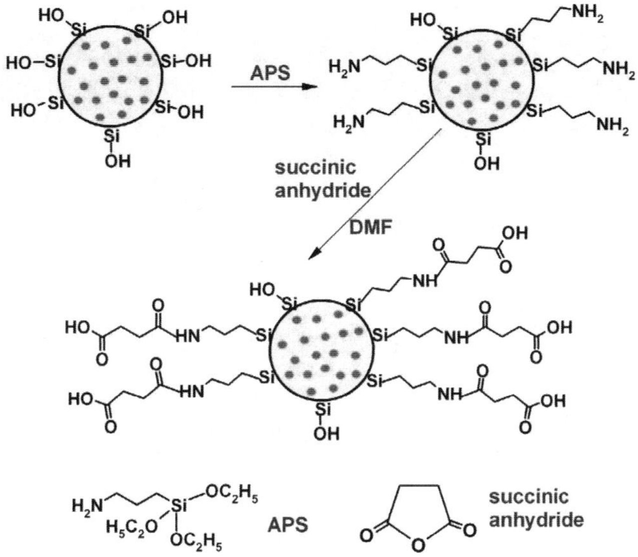

Scheme 1. Synthesis of amino- and carboxy-modified Tb(III)-TCAS doped silica nanoparticles

DLS data confirm low colloidal stability of amino-modified nanoparticles in aqueous dispersions. Silica nanoparticles with amino surface groups are characterized by $\zeta = 0.82$ mV and by the size of about 1300 nm in solution (pH = 8.2) by the DLS method. The surface charge neutralization due to the protonation of amino-groups at the silica surface ($pK_a = 9.0$) is the main reason of the enhanced aggregation of the amino-modified nanoparticles in neutral media. Further modification of such nanoparticles by carboxyl groups results in $\zeta = -37$ mV and in the size of obtained nanoparticles of about 220 nm (pH = 8.2). The appearance of the charged $C(O)O^-$ groups due to the partial dissociation of carboxyl groups C(O)OH ($pK_a = 4.9$) is the main reason of colloidal stability of carboxy-modified nanoparticles in neutral media. High negative surface charge prevents from the aggregation of carboxy-modified silica nanoparticles.

It is worth noting that the covalent surface modification of silica nanoparticles undergoes without any loss of luminescence (figure 6). The enhancement of the Tb-centered luminescence intensity is observed for the modified nanoparticles, which can be explained by the additional shielding of luminescent complexes from the quenching by the aqueous environment.

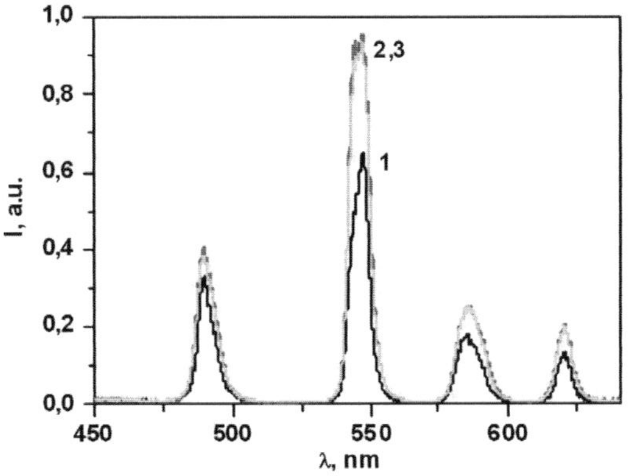

Figure 6. Emission spectra of non-modified Tb(III)-TCAS doped silica nanoparticles **(1)** and modified by amino **(2)** and carboxyl groups **(3)** (C = 0.028 g·L^{-1}) at pH = 7.0.

3.4. Non-covalent surface modification of Tb(III)-TCAS doped silica nanoparticles
However, a development of nanoscale sensors is of the enhanced significance for the bioanalytical application. Unlike biomarkers, biosensors should both bind substrate and report about this binding through "off-on" switching of luminescence. Anions play an important role in living organisms and even phospholipids membranes of cells are negatively charged. Our goal was to obtain highly luminescent, colloidal stable system with sensor function to anions. The main point of our approach is the following. The luminescence of silica nanoparticles should be quenched through the interaction with some quenching anions with further reestablishment of luminescence through the competitive interactions with non-quenching, but biorelevant ones.

Table 2. The averaged hydrodynamic diameters (d), ζ-potential values and polydispersity index (PDI) of Tb(III)-TCAS doped silica nanoparticles (C = 0.028 g·L^{-1}) at various concentrations of $[Co(dipy)_3](ClO_4)_3$ without and in the presence of various concentrations of Na_2HPO_4 at pH = 8.2.

	C_{Co} (mM)	$C_{Na2HPO4}$ (mM)	d (nm)	PDI	ζ (mV)
$[Co(dipy)_3](ClO_4)_3$	$1\cdot10^{-3}$		216,6 ± 1,65	0,201	-26,6
	$2\cdot10^{-3}$		275,6 ± 4,13	0,233	-22,9
	$3\cdot10^{-3}$		724,8 ± 24,09	0,281	-20,3
	$4\cdot10^{-3}$		1105 ± 48,36	0,382	-19,3
	$5\cdot10^{-3}$		1900 ± 101,3	0,319	-17,4
$[Co(dipy)_3](ClO_4)_3$	$3\cdot10^{-3}$	$3\cdot10^{-2}$	153,1 ± 6,71	0,323	-21,7
+	$3\cdot10^{-3}$	$3\cdot10^{-1}$	210,8 ± 4,54	0,230	-27
Na_2HPO_4	$3\cdot10^{-3}$	2,5	223,1 ± 4,48	0,215	-28,1
	$3\cdot10^{-3}$	3	394,2 ± 6,67	0,424	-32,5

It is known that positively charged complexes of cobalt(III) are quenchers of terbium luminescence [14]. Indeed the terbium luminescence of Tb(III)-TCAS doped silica nanoparticles is switched off when cobalt(III) complexes are adsorbed by their surface. More detailed description of these quenching measurements is presented in our previous work [10]. Herein we represent our efforts to develop the luminescent nanoparticles with sensor function using $[Co(dipy)_3^{3+}]$ ions as quenchers. Adsorption of $[Co(dipy)_3^{3+}]$ (C = $3\cdot10^{-3}$ mM) changes the ζ-potential of silica nanoparticles from -33 mV to -20 mV (table 2) and quenches their Tb-centered luminescence (figure 7).

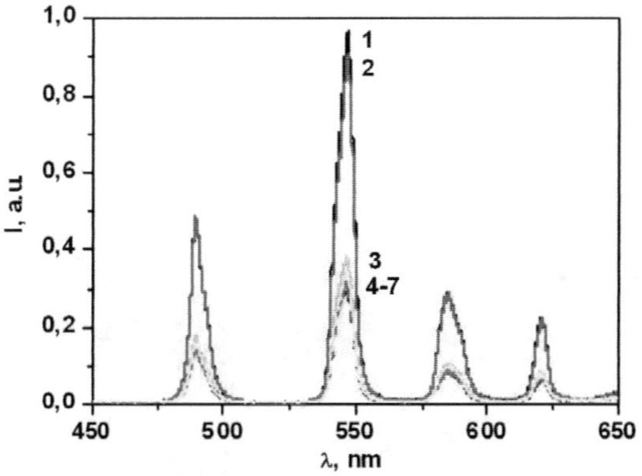

Figure 7. Emission of Tb(III)-TCAS doped silica nanoparticles (0.028 g·L^{-1}) **(1)** in presence of Na_2HPO_4 (C = 1 mM) **(2)** and [$Co(dipy)_3^{3+}$] (C = $3\cdot10^{-3}$ mM) **(3)** at the addition of Na_2HPO_4 : $1\cdot10^{-3}$ mM **(4)**, 0.01 mM **(5)**, 0.1 mM **(6)** and 1 mM **(7)** at pH = 8.2.

It is known that phosphate anions HPO_4^{2-} can form stable ion-pairs with [$Co(dipy)_3^{3+}$] in aqueous solution. But the addition of phosphate anions HPO_4^{2-} to silica nanoparticles with adsorbed [$Co(dipy)_3^{3+}$] does not reestablish Tb-centered luminescence (figure 7). Moreover electrophoretic measurements (table 2) reveal the increased negative charge at the silica/water interface (ζ is about -30 mV) due to the deposition of phosphate anions as the second layer at the [$Co(dipy)_3^{3+}$] ions adsorbed by the silica surface (chart 3). So, the layer by layer deposition of the opposite charged ions restricts the use of unmodified luminescent silica nanoparticles as the basis for the development of the nanoscale sensor to anions.

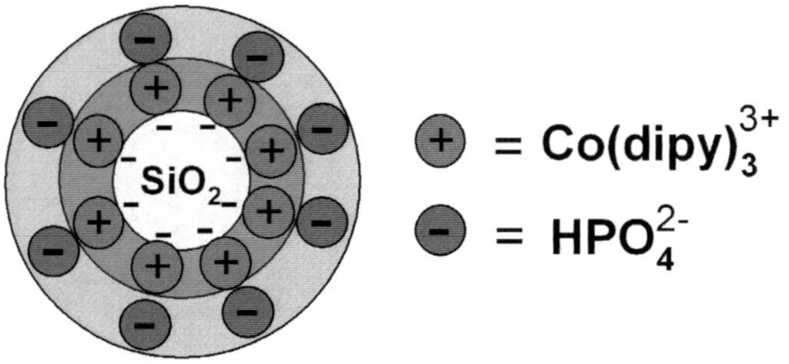

Chart 3. Schematic representation of layer-by-layer deposition of the opposite charged ions on the silica surface.

The decoration of silica nanoparticles with various surfactants is well developed way to modify the aggregation stability and binding ability of nanoparticles [6, 7, 15]. The adsorption of surfactants at a silica/water interface is contributed by van der Waals, electrostatic interactions and multiple hydrogen bonding [16]. Taking into account that in neutral media silica surface is negatively charged due to the dissociation of Si-OH groups (pK_a=8.4) [17], cationic surfactants are most convenient for adsorption at a silica/water interface. The aggregation of cationic surfactants at the silica nanoparticles is exemplified in literature in most details by cetyltrimethylammonium bromide (CTAB) [7, 15], which is monocationic surfactant. Amphiphilic cations cethyltrimethylammonium (CTA^+) form doubly charged bilayer at the silica/water interface, resulting in the recharging of the silica surface from minus to plus [7, 18]. Taking into account that electrostatic interactions play the key role in the formation of the doubly charged bilayer, gemini surfactant with two cationic groups per molecule, namely hexalidene-bis(dimethylammonium) bromide (16-6-16, chart 1) was chosen to recharge silica surface. Though the effects of CTAB and 16-6-16 are similar, there are some differences between them. In particular the change of ξ-values of silica nanoparticles from -33 mV to +22 mV occurs at $3\cdot10^{-3}$ mM of 16-6-16, while higher concentration of CTAB ($3\cdot10^{-2}$ mM) is required for the similar recharging (figure 8) [19]. This tendency is in good agreement with literature data concerning adsorption and aggregation of cationic and dicationic surfactants at a silica surface [16]. The modified nanoparticles possess good colloidal stability and affinity towards anions.

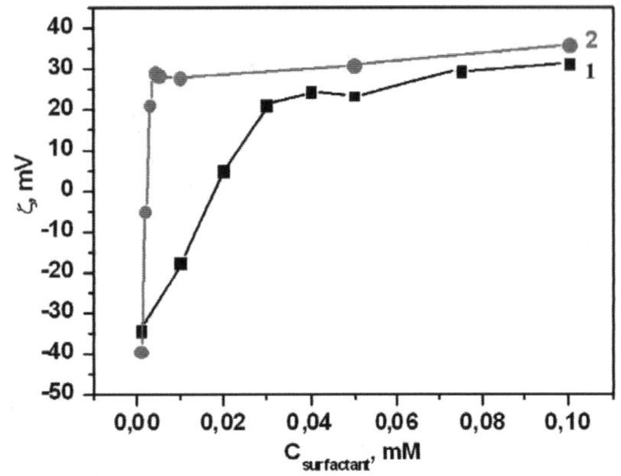

Figure 8. The ζ-potential values of Tb(III)-TCAS doped silica nanoparticles (C=0.028 g·L^{-1}) *versus* concentration of CTAB **(1)** and 16-6-16 **(2)** in aqueous solutions at pH = 9.2.

The quenching effect of dyes on lanthanide centered luminescence is also well known [20] and applied in bioanalysis [21] or pH measurements [22]. According to well-known conceptions the quenching effect results from the intermolecular energy transfer from lanthanides to chromophores according to the so-called Förster mechanism [20]. The distance between luminophoric and quenching molecules greatly affects the intermolecular energy transfer [23], thus luminophoric and quenching molecules should be drawn together by binding or collisional process. So, the detectable quenching of silica coated luminophores is possible as long as the quenching molecules locate at the silica/water interface of nanoparticles.

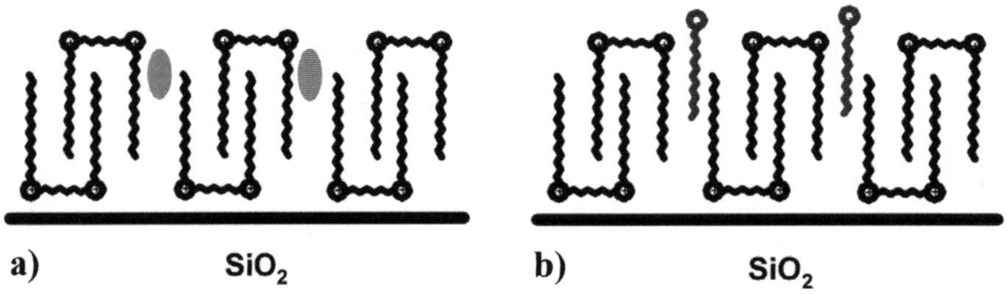

Chart 4. Schematic representation of the packing mode of dicationic surfactant 16-6-16 molecules at the silica/water interface and the penetration of phenol red anions (pink ovals) **(a)** and dodecylsulfate anions (blue molecules) **(b)** into the 16-6-16 bilayer.

The choice of phenol red molecules as quenching ones is determined by several reasons. The first reason is their quenching effect on the luminescence of lanthanide complexes in solutions [20]. The second reason is the efficient penetration of dye molecules into micellar aggregates [7]. The insertion of phenol red anions into surfactant bilayer at silica/water interface (chart 4) results in the quenching of Tb(III)-centered luminescence of the nanoparticles (figure 9). It is also worth noting that the quenching effect of dye anions can be decreased or come to nothing at the competitive binding of gemini modified nanoparticles with various anions.

It is anticipated that a hydrophobicity of anions plays a key role in their localization within micellar layer. Therefore both highly hydrophilic anions, such as HPO_4^{2-} and anionic surfactants, namely sodium dodecylsulfate (SDS) as a source of DS^- anions with high affinity towards cationic micelles were chosen. The widespread use of phosphate buffers in bioanalysis is another reason of the choice of hydrophosphate as an example of hydrophilic anions.

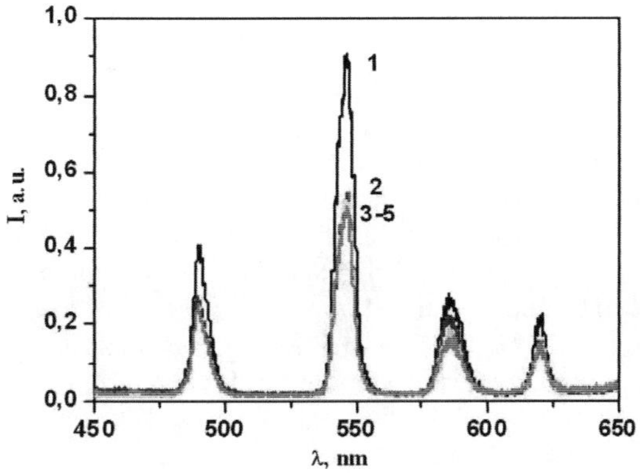

Figure 9. Emission spectra of Tb(III)-TCAS doped silica nanoparticles (0.028 $g \cdot L^{-1}$) in aqueous solution of 16-6-16 ($5 \cdot 10^{-2}$ mM) **(1)**; after quenching by phenol red ($5\ 10^{-3}$ mM) **(2)** and at the addition of Na_2HPO_4 : 0.1 mM **(3)**, 1 mM **(4)** and 10 mM **(5)** at pH = 9.2.

The figures 9 and 10 illustrate the effect of various amounts of Na_2HPO_4 and SDS respectively on the emission spectra of gemini modified Tb(III)-TCAS doped nanoparticles in the presence of quenching anions (C= $5 \cdot 10^{-3}$ mM). The addition of various amounts of HPO_4^{2-} does not reestablish the Tb(III) centered luminescence of the nanoparticles (figure 9), but results in some decrease of the positive surface charge of nanoparticles (table 3). Taking into account high hydrophilicity of phosphate anions they should locate in the Shtern layer of cationic micelles without significant insertion into the micellar core. So, hydrophilic HPO_4^{2-} anions can not substitute hydrophobic anions of phenol red from hydrophobic interior of gemini bilayer at the silica/water interface, therefore no reestablishment of the Tb-centered luminescence of nanoparticles occurs in this case, while detectable decrease of ξ-value is observed.

Table 3. The averaged hydrodynamic diametrs (d), ζ-potential values, and polydispersity index (PDI) of gemini modified Tb(III)-TCAS doped silica nanoparticles (C = 0.028 $g \cdot L^{-1}$) at various concentrations of Na_2HPO_4 and SDS at pH = 9.2.

$C_{Na2HPO4}$ (mM)	C_{SDS} (mM)	d (nm)	PDI	ζ (mV)	Peak Means[a] (nm)
0.1	-	278 ± 2	0.311	27	-
1	-	314 ± 4	0.249	16	-
10	-	-	0.878	19	aggregation[b]
-	0.001	246±4	0.183	35	-
-	0.005	236±3	0.179	42	-
-	0.01	216±3	0.166	34	-
-	0.02	216±2	0.151	43	-
-	0.03	227±4	0.175	46	-
-	0.04	271±4	0.237	46	-
-	0.05	325±4	0.225	47	-
-	0.06	369±5	0.251	47	-
-	0.07	420±4	0.243	46	-
-	0.08	494±6	0.287	43	-
-	0.09	-	1.000	18	aggregation[b]
-	0.1	-	1.000	0.49	aggregation[b]
-	0.15	-	0.474	-25	79±3 (30%), 362±24 (70%)
-	0.2	-	0.385	-27	271±14 (100%)
-	0.3	-	0.411	-29	279±12 (100%)
-	0.4	-	0.473	-32	76±37 (30%), 360±160 (70%)
-	0.5	-	0.462	-33	47±4 (23%), 295±30 (77%)
-	1	-	0.429	-34	45±5 (18%), 256±17 (82%)

[a] it is more correct to use Peak Means instead of averaged hydrodynamic diameters at PDI > 0.300
[b] the individual peaks are indistinguishable

In case of hydrophobic DS^- anions the Tb-centered luminescence is still quenched, when the concentration of the added SDS is less than $5 \cdot 10^{-2}$ mM, while at 10^{-1} mM of SDS the increase of Tb(III) luminescence occurs (figure 10) along with the enhanced aggregation due to the charge neutralization (ξ-value is close to zero) (table 3). The enhanced aggregation of nanoparticles is evident from the visible precipitation. The precipitated aqueous dispersion exhibits very poor emission, therefore the intense mixing is required to observe luminescence. The reestablishment of Tb-centered luminescence is observed at the increase of SDS concentration to $5 \cdot 10^{-1}$ mM (figure 10), followed by the recharging of nanoparticles from positive to negative (table 3). The DLS data (table 3) confirm the enhanced aggregation of nanoparticles at 10^{-1} mM of SDS and reveal the appearance of two sized particles (260±30 and 45±4 nm) at $5 \cdot 10^{-1}$ and 1 mM of SDS. The size being 260 nm should be obviously referred to silica nanoparticles with gemini-SDS aggregates at the surface, while the aggregates being 45 nm concern to mixed gemini-SDS aggregates. Thus, DS^- anions substitute phenol red anions (chart 4) leading to reestablishment of Tb(III)-centered luminescence [19].

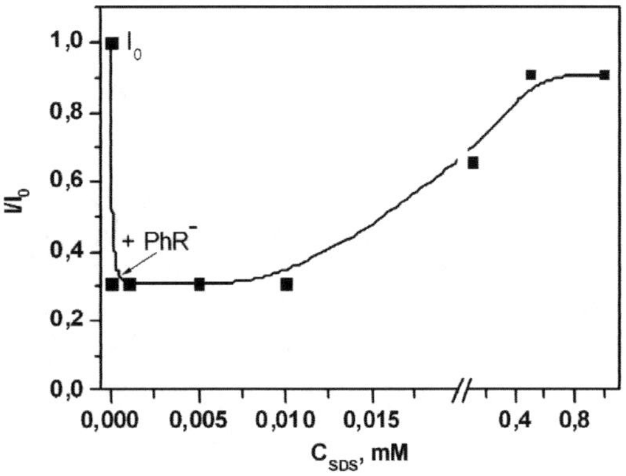

Figure 10. The I/I_0 of Tb(III)-TCAS doped silica nanoparticles (0.028 $g \cdot L^{-1}$) in aqueous solution of 16-6-16 ($5 \cdot 10^{-2}$ mM) in the presence of $5 \cdot 10^{-2}$ mM of phenol red (PhR^-) at the varied concentration of SDS at pH = 9.2. I_0 is the emission intensity without phenol red [18].

So, the "on-off-on" switching luminescent system was obtained. The insertion of hydrophobic dye anions (namely phenol red), which are efficient quenchers of terbium luminescence, into surfactant bilayer at silica/water interface, results in the quenching or "on-off" switching of luminescence of the nanoparticles. The displacement of dye anions by dodecylsulphate anions is manifested by the "off-on" switching of luminescence.

4. Summary

In summary the following points should be mentioned:

- Both Tb-TCAS and Yb-TCAS complexes are enough luminescent and thermodynamically stable to withstand synthetic conditions of water-in-oil microemulsion procedure without significant degradation. Thus the efficient insertion of Tb-TCAS and Yb-TCAS complexes into silica matrix enables to obtain uniformly sized nanoparticles with efficient visible (Tb-TCAS) and near IR (Yb-TCAS) luminescence. The synthesized nanoparticles are characterized by the slowed down leakage, and thus their luminescence is stable in time, which is of particular importance for their practical application. These nanoparticles for the first time exemplify the use of calixarene complexes as luminophoric dopants to silica nanoparticles. The covalent surface modification of these nanoparticles undergoes without any loss of luminescence, which is also of great importance for their bioanalytical application.
- The convenient and reproducible synthetic procedure has been developed to obtain dual NIR-visible luminescent nanoparticles.

- The luminescent nanoparticles, which are able to give luminescent response on the addition of definite substrates, have been developed. The Tb-TCAS doped nanoparticles modified by the dicationic surfactant coating are the basis for the substrate responsible system. The reversible insertion of the quenching anions (namely phenol red) into the surfactant based layer at the surface of luminescent nanoparticles switches off the Tb-centered luminescence. In turn the reversible reestablishment of the luminescence results from the competitive insertion of the non-quenching anions into the surfactant layer at the silica/water interface.

Acknowledgments
We thank RFBR project N 09-03-12260 Ofi_M and N 10-03-00352-a for financial supporting.

References

[1] Overstreet D J, Dhruv H D and Vernon B L 2010 *Biomacromolecules* **11** 1154
Fares M M, Othman A A and Macromol 2010 *J Sci. Part A: Pure and Appl. Chem.* **47** 61
[2] Eastoe J, Hollamby M J and Hudson L 2006 *AdV. Colloid Interface Sci.* **5** 128-130
Yao L, Xu G, Dou W and Bai Y 2008 Colloids Surf. A: Physicochem. Eng. Aspects **316** 8
[3] Wu Ch, Hong J, Guo X, Huang Ch, Lai J, Zheng J, Chen J, Mu X and Zhao Y 2008 *Chem. Commun.* 750–752
Montalti M, Prodi L, Zaccheroni N, Charbonnière L, Douce L and Ziessel R 2001 *J. Am. Chem. Soc.* **123** 12694–695
Wang D-Y, Zhao Y-B, Xu J-G and Guo X-G 1997 *Fresenius J. Anal. Chem.* **358** 514–518
[4] Bunzli J C G and Piguet C 2005 *Chem. Soc. Rev.* **34** 1048–77
[5] Knopp D, Tang D and Niessner R 2009 *Analyt. Chim. Acta* **647** 14–30
Wang L, Zhao W and Tan W 2008 *Nano Res.* **1** 99–115
[6] Wei H, Liu J, Zhou L, Li J, Jiang X, Kang J, Yang X, Dong S and Wang E 2008 *Chem.- A Eur. J.* **14** 3687–93
Mori H, Muller A H E and Klee J E_2003 *J. Am. Chem. Soc.* **125** 3712-13
[7] Bryleva E Yu, Vodolazkaya N A, Mchedlov-Petrossyan N O, Samokhina L V, Matveevskaya N A and Tolmachev A V 2007 *J. Colloid Interface Sci.* **316** 712-722
[8] Iki N, Fujimoto T and Miyano S 1998 *Chem. Letters* **27** 625-626
[9] Mirgorodskaya A, Kudryavtseva L, Pankratov V, Lukashenko S, Rizvanova L, Konovalov A 2006 *J. Gen. Chem.* **76** 1696–1701
[10] Mustafina A R, Fedorenko S V, Konovalova O D, Menshikova A Yu, Shevchenko N N, Soloveva S E, Konovalov A I and Antipin I S 2009 *Langmuir* **25** 3146–3151
[11] Song L, Zhang H-L, Liu T-C, Liu B, Cao Y-Ch, Huang Zh-L, Zhao Y-D and Luo Q-M 2007 *J. biomedical materials research* 752-757
[12] Bagwe R P, Hilliard L R and Tan W 2006 *Langmuir* **22** 4357
[13] Fedorenko S V, Bochkova O D, Mustafina A R, Burilov V A, Kadirov M K, Holin C V, Nizameev I R, Skripacheva V V, Menshikova A Yu, Antipin I S and Konovalov A I 2010 *J. Phys. Chem. C* **114** 6350–55
[14] Skripacheva V, Mustafina A, Rusakova N, Yanilkin V, Nastapova N, Amirov R, Burilov V, Zairov R, Kost S, Solovieva S, Korovin Yu, Antipin I and Konovalov A 2008 *Eur. J. Inorg. Chem.* 3957–63
[15] Wang W, Gu B, Liang L and Hamilton W A 2004 *J. Phys. Chem. B* **108** 17477–83
[16] Chorro C, Chorro M, Dolladile O, Partyka S and Zana R 1998 *J. Colloid. Interface Sci.* **199** 169–176
Parida S K, Dash S, Patel S and Mishra B K 2006 *Adv. Colloid Interface Sci.* **121** 77–110
[17] Kumar R and Milanova D 2009 *Ann. NY Acad. Sci.*, **1161** 472-483
Rimer J D, Lobo R F and Vlachos D G 2005 *Langmuir* **21** 8960–71
[18] Manne S and Warr G G 1999 Supramolecular Structure in Confined Geometries. Ch. 1. Supramolecular Structure of Surfactants Confined to Interfaces (Washington DC: ACS)

[19] Mustafina A R, Elistratova Ju, Bochkova O D, Zakharova L Ya, Kadirov M K, Nizameyev I R, Fedorenko S V, Skripacheva V V, Konovalov A I and Soloveva S E *J. Phys. Chem. B* in press

[20] Lis S, Elbanowski M, Makowska B and Hnatejko Z 2002 *J. Photochem. and Photobiol. A: Chem.* **150** 233–247

[21] Wensel T G and Meares C F 1989 *J. Less-Common Met.* **149** 143–160

[22] Kessler M A 1999 *Anal. Chem.* **71** 1540–43

[23] Van-D Meer B W, Coker G I and Chen S-Y 1994 *Resonance energy transfer. Theory and data* (New York: VCH Publishers)

The formation of nanoparticles, ceramics, and thin films of ZnO in the environment of zinc vapor

A Abduev, A Akhmedov, A Asvarov
Institute of Physics, Dagestan Scientific Center of RAS, Yaragskogo str, 94, Makhachkala, 367003, Russia

E-mail: cht-if-ran@mail.ru

Abstract. The investigations of formation mechanisms of ZnO nanoparticles, ZnO ceramic materials sintering and synthesis of ZnO thin films in zinc vapor-containing atmosphere shown that increase in "zinc/oxygen" relation (P_{Zn}/P_{O2} ratio) leads to a significant change in the structure and formation mechanisms of ZnO ultrafine particles, thin films and bulk materials. We observed that ZnO whiskers growth occurs in the direction of the optimal ratio of partial pressures of Zn and O_2. Growth of ZnO tetrapods during the ejection of Zn vapor in an oxygen atmosphere occurs in two stages: the formation of the ZnO_{1-x} seed and subsequent growth of ZnO whiskers from it. It was also shown that the synthesis of ZnO films by magnetron sputtering in the Zn vapor excess environment can occur by the "vapor-liquid-solid" (VLS) mechanism. Also sintering of Ga-doped ZnO ceramic targets in the atmosphere containing desired pressure of Zn vapor leads to increase in target density and oxygen deficiency. These targets were used for deposit transparent conductive oxide (TCO) thin films onto glass substrate by direct current (dc) magnetron sputtering. The relationship between the structural perfection and electrical characteristics of Ga doped ZnO (GZO) films as a function of synthesis temperature and Ga content (3-6 at.%) were investigated.

1. Introduction

The development of new materials and devices based on zinc oxide is currently experiencing an unprecedented boom in solar energetic, industry of displays and light-emitting diodes. Applied research in developing new ceramic materials for the formation of transparent electrodes and their deposition technologies [1], dye-sensitized photovoltaic cells [2], etc is hence in high demand.
Numerous studies show that an important factor in determining the conditions of nucleation and formation of ZnO-based nanoparticles, thin films and bulk ceramic materials is the ratio of pressure of Zn vapor and partial pressure of oxygen (P_{Zn}/P_{O2}) in the chamber during the synthesis [3].
In this study we attempted to examine the effect of the P_{Zn}/P_{O2} on the structure of ZnO nanoobjects and films deposited by magnetron sputtering, as well as on the processes of sintering of ceramic targets for the formation of transparent electrodes.

2. Experiment and materials

We investigated the influence of P_{Zn}/P_{O2} ratio on the structure of ZnO nanoobjects (whiskers, tetrapods) and ZnO films, as well as on the sintering of TCO ceramic targets.
Synthesis of ZnO nanoobject materials was carried out by various methods:
1) the annealing in the air of the zinc films (thickness ≈ 1000nm), synthesized by thermal evaporation of zinc from a Knudsen cell on the Si or glass substrates;

2) the gas-phase synthesis of ZnO aggregates by means of effusion of the overheated Zn vapor through a thin channels into oxygen atmosphere.

To investigate effect of Zn vapor excess environment on the properties of ZnO films, a series ZnO films were grown by dc magnetron co-sputtering ZnO ceramic target and Zn metal target at different substrate temperatures and various discharge currents of magnetron with Zn target.

Also the synthesis of ceramic targets was carried out by sintering of ZnO-based compacted materials with doping in the atmosphere containing Zn vapor [4,5]. To study the properties of the ZnO-based TCO thin films, ZnO targets with the gallium content of 3 to 6 at% were manufactured. The thin films were synthesized by dc magnetron sputtering at a constant pressure of Ar in the chamber (0.5 Pa), while substrate temperature varied from room temperature to 280 °C.

Morphology of samples surface and film thickness was studied using scanning electron microscope LEO 1450 (Carl Zeiss). The structural properties of the samples were investigated using X-ray difractometer Shimadzu XRD-7000, with CuKα radiation. Precise XRD measurements were performed in order to estimated the full width at half maximum (FWHM) of the (002)ZnO XRD peak. The resistivity, Hall mobility and free carrier concentrations for Ga-doped ZnO TCO thin films were estimated by the four-point probe method and Hall-effect measurement.

3. Results and discussions

3.1. Formation of nanoobjects in the atmosphere of Zn - O_2.

Previous studies [6,7] have shown that the formation of ZnO whiskers at Zn vapor excess occurs by the mechanism of "vapor-liquid-solid" (VLS), moreover "liquid"-forming agent is ZnO_{1-x} phase. The model of zones of the growth of ZnO whiskers and tetrapods is shown in Figure 1a. The nucleation of the ZnO whiskers occur at the edge of the nozzle, and the further formation of whiskers is directed along the axis of the cone, i.e. along the lines of a certain optimal P_{Zn}/P_{O2} ratio.

Figure 1a schematically shows also the change of partial pressure of oxygen and pressure of Zn vapor in the direction from nozzle. During the effusion of zinc in oxygen atmosphere, in addition to the growth of whiskers, also observed the gas-phase formation of ZnO tetrapods (inside the cone generator). ZnO tetrapods, formed in the flow of zinc, shown in Figure 1b.

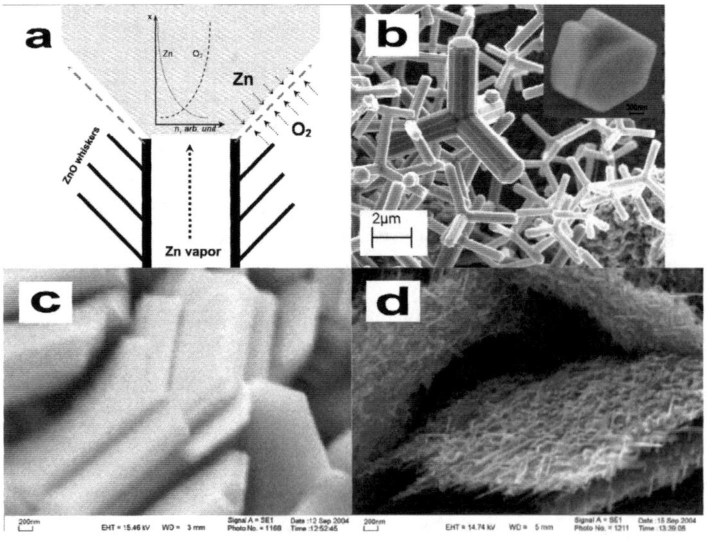

Figure 1. a – the model of zones for the growth of ZnO whiskers and tetrapods; b – SEM photos of ZnO tetrapods; c – SEM photo of as-deposited Zn layer; d – SEM photo of Zn layer after annealing in air.

Numerous observations confirm the fact that the formation of tetrapods occurs during the expansion of zinc in a two-stage scheme.
- A forming of ZnO_{1-x} crystallization seeds into field with relatively high pressures of zinc.
- Subsequent growth of four whisker legs during the expansion of the seed material.

The formation mechanism of whiskers along the lines of a certain optimal P_{Zn}/P_{O2} ratio is also confirmed by our experiment on the oxidation of metallic Zn layers. Figure 1c shows a SEM micrograph of the surface of the as-deposited Zn film.

The annealing of the Zn layer at 380 °C in air results in the formation of a dense web of whiskers on the surface of Zn hexagonal grains (Figure 1d). Analysis of the morphology shows that the whiskers are directed predominantly parallel to the original surfaces of Zn grains, i.e. again along the lines of optimum pressure ratio of oxygen (O_2 comes from the outside) and zinc vapor (Zn comes from the film due to evaporation). The formation of two-dimensional web-like structure of the ZnO dendritic branches was also reported in [8].

However, saturated vapor pressure of elemental zinc at 380 °C corresponds to 10 Pa, i.e. doesn't match conditions for whisker growth. One can propose, that due to autoignition of zinc vapor at 380 °C, near-surface regions of the film reach boiling temperature and higher zinc vapor pressure. This zinc flame front moves deeper within the layer parallel to the surface of crystallites. Thus, whisker growth occurs in zinc flame front plane within the region of optimal ratio of P_{Zn}/P_{O2}.

3.2. DC magnetron synthesis of ZnO films via VLS mechanism.

Previously we have shown that the columnar structure of ZnO thin films was suppressed by dc-magnetron co-sputtering of ZnO and Zn at the substrate temperature 450 °C [9]. In this paper we study the structure of films synthesized at 300, 500 and 800 °C, as well as different flows of Zn vapors. Figure 2a-c shows micrographs of ZnO films cross-section view at different substrate temperature (T_{sub}). At a $T_{sub} < 400$ °C the films have a typical columnar structure (Figure 2a). Figure 2b shows that

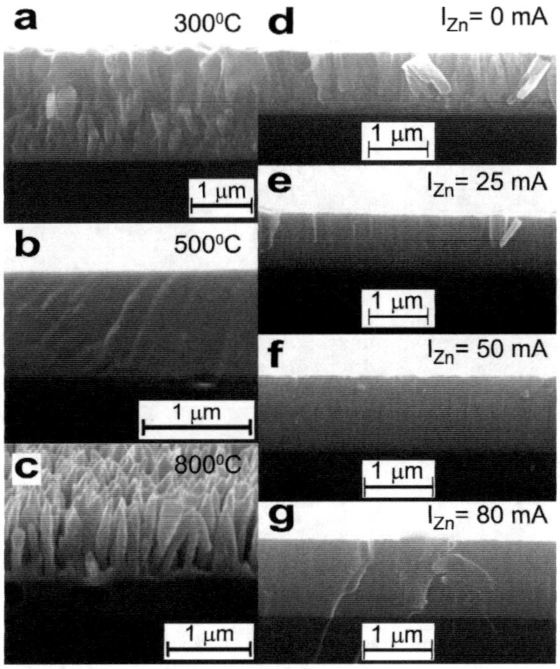

Figure 2. SEM photos of the ZnO films, deposited at different substrate temperatures (a-c) and different discharge currents of Zn target (d-g).

suppression of the formation of columnar structures occurs at $T_{sub} > 400$ °C. This is due to the formation of ZnO_{1-x} low-melting phase on the growing surface under the environment of Zn vapor excess [10,11]. In this case, as in the growth of whiskers, the growth of thin films proceeds in accordance with the VLS model [7]. Further increase in T_{sub} is the result of decrease in the number of crystallization centers on the surface and forming an array of separate columnar structures (Figure 2c). Dependence of the structure on the Zn sputtering intensity is shows on Figure 2d-g. Increasing the flow of Zn does not increase the thickness of the films. This shows that the excess of Zn is not incorporated into the film and only provides a low-melting phase on the growing surface.

3.3. ZnO based TCO ceramic materials.

One of the most important areas of applied research is the development of advanced TCO ceramic targets for magnetron sputtering. It should be noticed that the requirements for the front (*color filter*, C/F) and back (*thin film transistor*, TFT) transparent electrode in the liquid crystal displays (LCD), as well as to the electrodes in organic light emitted diodes (OLED) structures and solar converters, vary significantly. Obviously, adaptation of the target to the specific requirements will significantly improve the performance of devices. In addition, targets must possess high density (not less than 95% of theoretical) to prevent nodules formation. To address these requirements, we developed a method of synthesis of ceramic targets in the atmosphere with excess vapor pressure of zinc [5]. Completed studies show that an increase in shrinkage rate is the result of formation of liquid phase ZnO_{1-x} on the grains surface, leading to a significant increase in diffusion. Synthesizing ceramics have a density of at least 99% of the theoretical value and a deficiency of oxygen in the ceramics of about 8%.

We investigated the relationship between the perfection of the films and electrical characteristics of Ga doped ZnO (GZO) films as a function of synthesis temperature and Ga content (3-6 at.%) in high-density ceramic targets, sintered in an atmosphere containing zinc vapour.

The investigation of the resistivity of GZO films as a function of the Ga content show that the behavior of this dependence with increasing substrate temperature can be divided into three characteristic temperature regions (Figure 3a):

- resistivity decreases monotonically with increased Ga content in temperature range $T_{sub} < 150$ °C;
- the curves reach minimum at Ga content ~ 4-5 at% at 150 °C $\leq T_{sub} \leq$ 200 °C;
- the resistivity increases monotonically in the range of 3 to 6 at% Ga at $T_{sub} \geq 250$ °C.

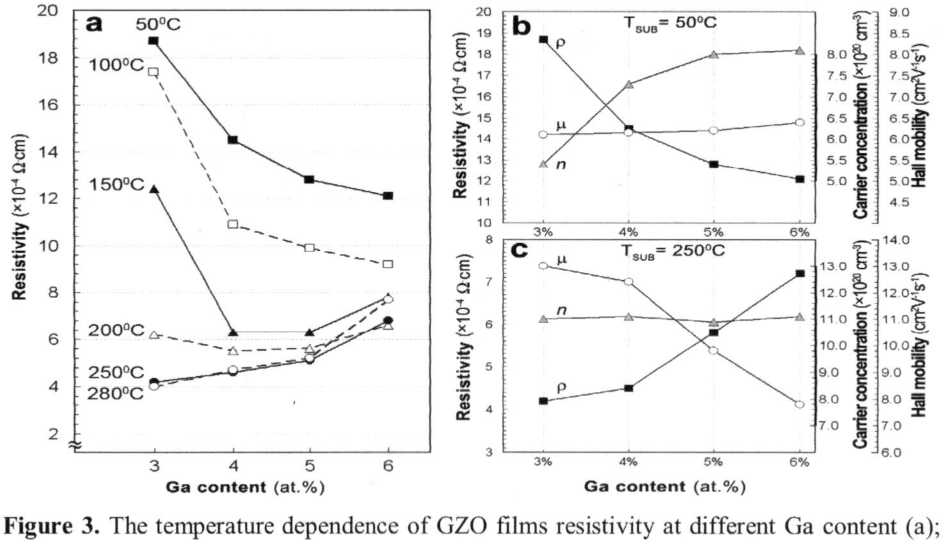

Figure 3. The temperature dependence of GZO films resistivity at different Ga content (a); resistivity (ρ), carrier concentration (n) and Hall mobility (μ) of GZO films vs. gallium content in the target (b - T_{sub} = 50 °C; c – T_{sub} = 250 °C).

These data show that minimum resistivity is reached at low value of Ga contents with increased substrate temperature.

Figure 3b&c shows measurements of Hall parameters of the GZO films synthesized at temperatures of 50 and 250 °C from the targets GZO with 3, 4, 5 and 6 at% Ga. At substrate temperature T_{sub} = 50 °C, carrier concentration increases linearly as Ga content is increased, reaching saturation at 6 at%. At temperature of 250 °C maximum of carrier concentration is achieved when Ga content reaches ~ 3 at%. Further increase of Ga leads to the localization of an excess of Ga at the grain boundaries and reduction of mobility.

These data are consistent with the results of X-ray analysis (Figure 4). It is observed that the GZO films are predominantly in (002) ZnO orientation and do not reveal any other phases.

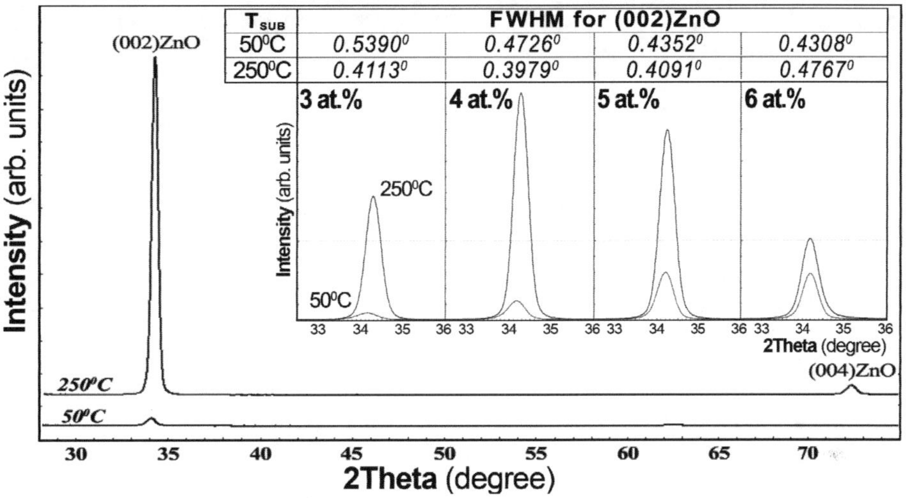

Figure 4. XRD patterns of GZO films with 3at% Ga deposited at 50 and 250 °C. Inset – XRD data for GZO films deposited at same temperatures for different Ga content: a – 3 at%, b – 4 at%, c – 5 at%, and d – 6 at%.

In the insert of Figure 4 shows the profiles reflections (002) ZnO for GZO films with 3-6 at% Ga deposited at substrate temperature of 50 and 250 °C. It is noteworthy that the intensity of the peak (002) at 250 °C has a pronounced maximum at 4 at% Ga, and at 50 °C is a monotonically increasing up to 5-6 at% Ga. FWHM measurements, shown on Figure 4, demonstrate that growth of intensity peaks is coupled with increased grain size. Observed increase in the perfection of GZO films for the temperature of 50 °C may be associated with the reduction of the concentration of native defects due to incorporation of trivalent gallium in ZnO lattice. As concentration continues to increase, gallium excess on the growing surface creates new centers of crystallization, leading to a decrease in grain size.

Next we compared the data of Hall measurements of GZO films (3 at%), synthesized at different temperatures. Resistivity, concentration and mobility of the films are shown in Figure 5a. Of particular interest is the anomalous behavior of the mobility at 100-150 °C. In this case, carrier concentration increases with increasing T_{sub}. At temperatures above 150 °C mobility and concentration steadily increase and show signs of saturation at a temperature of about 250 °C. In the temperature range 50 ÷ 150 °C Hall parameters vary in opposite directions: concentration increases while mobility decreases. With further increase in temperature observed anomalous mobility around 100 ÷ 150 °C can be associated with non-stoichiometry of targets. According to [12], presence of zinc vapor excess in the flow leads to the formation of metal clusters on the surface at temperatures below 150 °C. We propose that sputtering of targets with a deficit of oxygen leads to the presence of zinc vapor in the flow and consequent deterioration of the crystalline perfection of the films. Increase of carrier concentration in the range up to 150 °C is then due to thermal activation process of embedding an impurity in the lattice.

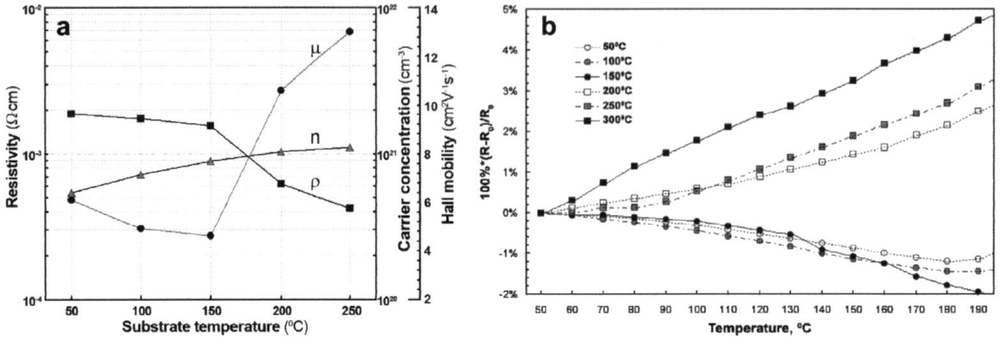

Figure 5. Resistivity, carrier concentration and Hall mobility of GZO films with 3 at% Ga deposited at various substrate temperatures (a); the temperature coefficient of resistance of GZO films with 3 at% Ga deposited at various substrate temperatures (b).

Also, measurements of the temperature coefficient of resistance (TCR) show that the GZO films, synthesized at temperatures below 150 °C, have a negative TCR, whereas the films, deposited at $T_{sub} \geq 200$ °C, TCR becomes positive, i.e. high synthesis temperatures lead to degenerate semiconductor behaviour (Figure 5b).

4. Conclusions

The effect of the P_{Zn}/P_{O2} ratio on the formation and structure of ZnO nanoparticles as well as ZnO-based thin films and ceramics was investigated. The experimental results are summarized in the following.

(1) The ZnO whiskers grow along the lines of a certain optimal P_{Zn}/P_{O2} ratio.
(2) The formation of ZnO tetrapods occurs during the expansion of zinc in a two-stage scheme: the formation of the ZnO_{1-x} seed and subsequent growth of ZnO whiskers from it.
(3) The increase in P_{Zn}/P_{O2} ratio at dc magnetron co-sputtering of Zn and ZnO targets leads to a significant change in the structure of ZnO films at $T_{sub} > 400$ °C according to the VLS mechanism.
(4) The Ga-doped ZnO ceramic targets, sintered in the environment of zinc vapour, have have a density of at least 99% of the theoretical value and a deficiency of oxygen in the ceramics of about 8%. The presence of oxygen deficiency in the sputtered ceramic improves the structural and electrical characteristics of the deposited ZnO-based TCO films. With increasing of substrate temperature the minimal value of resistivity of Ga-doped ZnO thin films is achieved at lower concentrations of gallium.

Acknowledgments

The work is performed using the equipment of the Analytical Center of Collective Usage of the Dagestan Scientific Center of Russian Academy of Sciences. The study was supported by the Rosnauka, Contract no. 02.740.11.0397

References

[1] Minami T 2005 *Semiconductor Science and Technology* **20**(4) S35
[2] Yu K and Chen J 2009 *Nanoscale Research Letters* **4**(1) 1

[3] Ko H-J, Yao T, Chen Y and Hong S-K 2002 *J. Appl. Phys.* **92**(8) 43
[4] Patent RU No. 2280015 (2004); US Pat. Appl. No.US0283802 A1 (2008)
[5] Patent RU No. 2382014 (2008)
[6] Abduev A, Akhmedov A, Baryshnikov V and Shakhshaev Sh 2000 *Tech. Phys. Lett.* **26**(4) 332
[7] Abduev A, Asvarov A, Akhmedov A, Baryshnikov V and Terukov E 2002 *Tech. Phys. Lett.* **28**(11) 952
[8] Fan H J, Scholz R, Kolb F M and Zacharias M 2004 *Appl. Phys. Lett.* **85**(18) 4142
[9] Abduev A, Asvarov A, Akhmedov A, Kamilov I and Sulyanov S 2005 *NATO Science Series II: Mathematics, Physics and Chemistry* **194** 15
[10] Heo Y, Caradarajan V, Kaufman M, Kim K and Norton D 2002 *Appl. Phys. Lett.* **81**(16) 3046
[11] Feng L, Liu A, Ma Y, Liu M and Man B 2010 *Acta Physica Polonica A* **117** (3) 512
[12] Abduev A, Magomedov A and Shakhshaev O 1997 *Inorganic materials* **33**(3) 282

Au/TiO_2 nanocomposites with high concentrated "hot spots" under near IR femtosecond pulsed excitation

A Aiboushev[1], A Astafiev[1], O M Sarkisov[1] and V A Nadtochenko[1,2]
[1] N.N. Semenov Institute of Chemical Physics RAS, Moscow, Russia
[2] Institute of Problems of Chemical Physics RAS, Chernogolovka, Russia

E-mail: aiboosh@gmail.com

Abstract. Luminescence of gold nanoparticles photodeposited on titan dioxide mesoporous films has been studied using multiphoton microscopy. Luminescence was registered under the two-photon excitation by femtosecond pulses of Ti:sapphire laser. It was observed that Au/TiO_2 mesoporous films have high concentration of bright luminescence spots ("hot spots") which reveal stability to degradation under long illumination. The most intense "hot spots" have luminescence enhancement of order of 10^5 therefore Au/TiO_2 nanocomposites can be useful for single molecule spectroscopy and biological objects visualization. Application of Au/TiO_2 mesoporous films for Raman scattering spectroscopy is demonstrated for the case of Rhodamine B. Au/TiO_2 geometry analysed with FDTD to find probable "hot spot" configurations.

1. Introduction

Mesoporous films, composed from TiO_2 nanoparticles can be used as a photocatalyst for conversion of light energy to chemical energy [1]. However the efficiency of photocatalytic reactions is limited by recombination processes and by absorption in visible [2]. Metal nanoparticles deposited on TiO_2 surface can essentially improve catalytic properties of TiO_2 [3]. Moreover enhancement of electromagnetic field near metallic nanoparticles could be the most important reason for dramatic increase of Raman scattering cross section or for luminescence spectroscopy enhancement [4-6]. Thus noble metal nanoparticles are very attractive since their localized surface plasmon resonance (LSPR) in visible.

The frequency of LSPR can be spectrally tuned by changing shape, dielectric environment of the particles and their size [7]. For complex shape particles, there can be several resonances corresponding to each polarization. The width of electromagnetic scattering and absorption wave resonance on metal particle is determined by surface plasmon damping. There are several damping mechanisms: radiative damping, one-electron energy relaxation and Landau damping in finite system [7-8]. These dampings lead to a decrease in the LSPR magnitude and near field intensity as well as to an increase of the LSPR width [9-10].

Clusters of noble metals can reveal photoluminescence in visible [11-12,16]. Surface roughness of clusters and impurities (e.g., oxides) can give a contribution to noble metal photoluminescence.

In the present work we report the effect of enhanced luminescence of gold nanoparticles (photodeposited on TiO_2 mesoporous film) under the two-photon excitation by femtosecond laser

[1] aiboosh@gmail.com

pulses of Ti:sapphire laser. We registered many bright luminescence spots in the film. To interpret experimental results different possible configurations of Au/TiO_2 nanoparticles were analyzed. Application of Au/TiO_2 mesoporous films for Raman scattering spectroscopy is demonstrated for the case of Rhodamine B.

2. Samples

Porous nanocrystalline TiO_2 electrodes were prepared from TiO_2 powder (Degussa P25), using the method similar to that described previously [13]. The powder, to which a small amount of water and acetyl acetone was added was milled for a prolonged time in order to lower the degree of aggregation. A white viscous paste was obtained, to which some nonionic detergent (Triton X-100) was added. Paste was deposited on glass microscope plate surface by "doctor blade" technique. Finally, the samples were heated in a furnace at 450 C (in air) for 2 h.

Au nanoparticles were obtained on the porous TiO_2 by the photodeposition technique: TiO_2 plate dipped in the solution 0.1 mM $HAuCl_4$ in water (ethanol concentration was 5%) and was illuminated by Philips phosphor lamp. The lamp has following characteristics: spectral range between 340 and 380nm, maximum of the spectrum is at 365nm, flux power 3mW/cm^2.

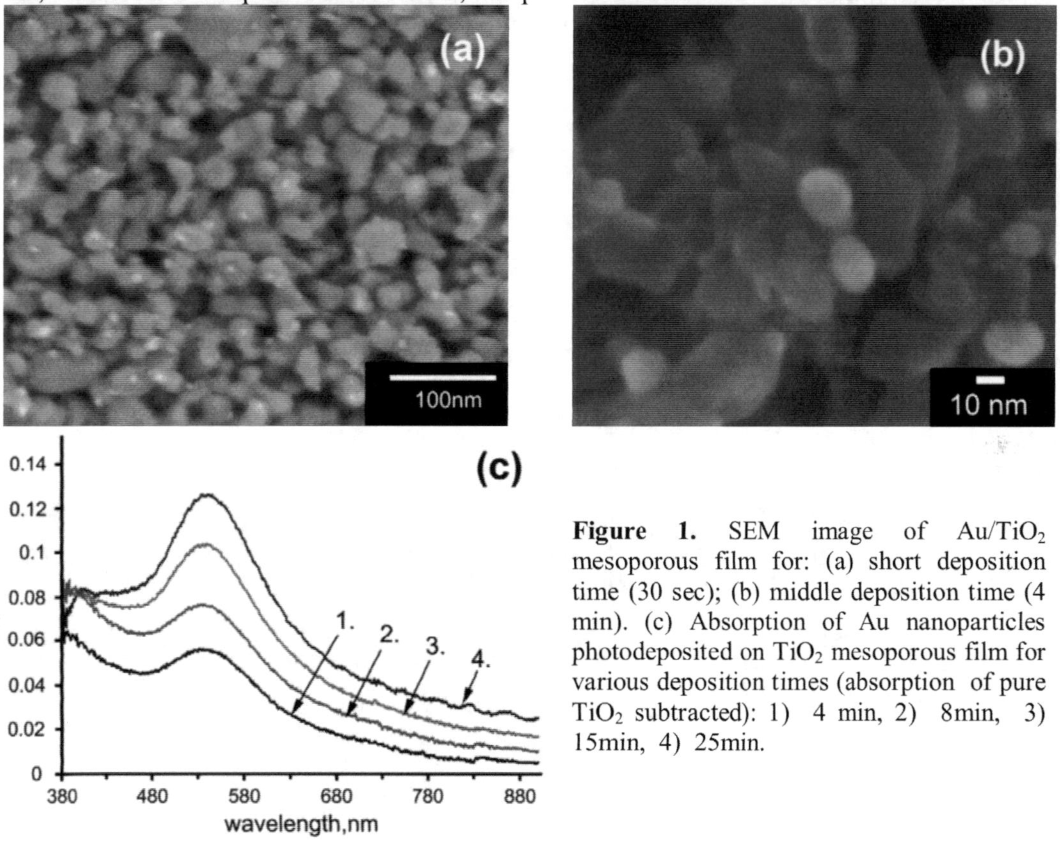

Figure 1. SEM image of Au/TiO_2 mesoporous film for: (a) short deposition time (30 sec); (b) middle deposition time (4 min). (c) Absorption of Au nanoparticles photodeposited on TiO_2 mesoporous film for various deposition times (absorption of pure TiO_2 subtracted): 1) 4 min, 2) 8min, 3) 15min, 4) 25min.

SEM analysis (figure 1) of TiO_2/Au mesoporous film shows that the average size of TiO_2 nanoparticles is 30nm. Average size of Au nanoparticles and their distribution over TiO_2 mesoporous film depend on photodeposition time. For short deposition time (< 1 min) average size of Au nanoparticles was 5 nm. Photodeposition time on figure 1a was 30 sec. Single Au nanoparticles predominated over film surface. For long deposition time average size of Au nanoparticles was 20 nm. In this case both large Au nanoparticles (~50 nm) and nanoparticle agglomerates also were detected.

Au nanoparticles distribution over the film for middle deposition time can be characterized as following. First, single Au nanoparticles predominated over film surface. Second, there are coupled Au nanoparticles (surface concentration ~10^5 cm^{-2}) on the film (figure 1b). In this case distance between nanoparticles is less than nanoparticles size. Such configurations are discussed in section 5.

It is seen from SEM experiments that Au nanoparticles have the shape near to spherical (figure 1a,b). As it was mentioned in section 1 LSPR position (and absorption) of metal nanoparticles depends on nanoparticles shape. To obtain average absorption spectra of gold nanoparticles the same area of TiO_2 film was measured (by Shimadzu spectrophotometer) before photodeposition and after photodeposition of metal nanoparticles. The differential spectra revealed a plasmonic absorption peak, which was at about 540 nm (figure 1c). The magnitude of the peak increases with deposition time, but position and the shape remains approximately constant. According to FDTD analysis (see section 5) such absorption curves correspond to spherical Au nanoparticles with small contact area with TiO_2. It should be noted here that absorption measurements are not microscopic measurements therefore they have statistical sense, i.e. most of Au nanoparticles (but not all) have a small contact with TiO_2. Additionally, solutions absorption was measured after photodeposition of Au nanoparticles. It occurred that for long deposition (~10 min) times there are typical features on absorption curves which correspond to absorption of Au nanoparticles. This means that Au nanoparticles deposited on the film from appropriate solution, but not after photocatalytic process.

In this work middle time of photodeposition studied (figure 1b). The main features of middle time of photodeposition are: 1) average sizes of TiO_2 and Me nanoparticles were 30 nm and 20 nm, respectively; 2) Au nanoparticles on TiO_2 are not the result of deposition from solution but photocatalytic process; 3) Coupled Au nanoparticles detected over the Au/TiO_2 film.

3. Experimental Setup

We used a titanium sapphire Mai Tai HP (Spectra Physics) laser producing 100 fs pulses centered near 800 nm, with a repetition rate of 80 MHz and about 20 mW average power coupled with an inverted microscope (figure 2). The sample was on piezoelectric nano-positioning stage (NT MDT, Zelenograd) which allowed the sample to be scanned over a range of 30 μm in the x and y directions.

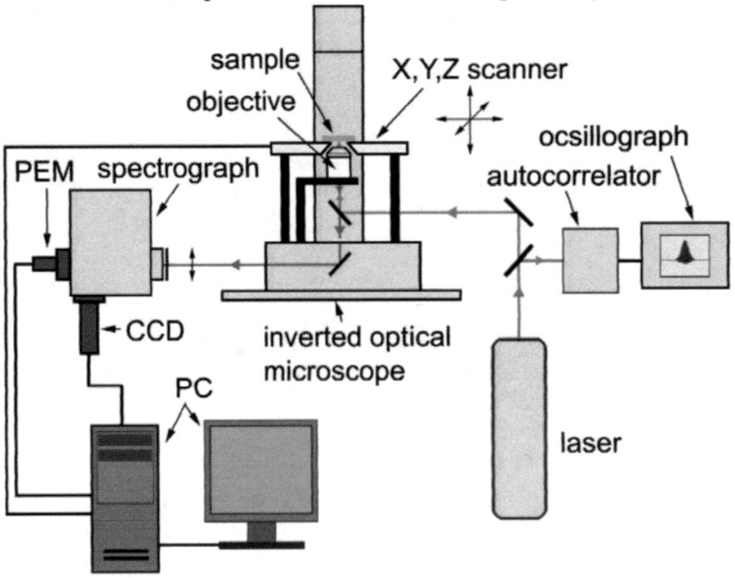

Figure 2. Experimental setup of multi-photon microscope.

All experiments were carried out using Olympus UPLSAPO 100× NA 1.40 apochromatic objective to focus laser beam onto the sample and collect emitted light, which was then imaged after

Journal of Physics: Conference Series **291** (2011) 012040 doi:10.1088/1742-6596/291/1/012040

spectrograph SP-300i (Acton Research) by an electron multiplier CCD camera Newton EM (Andor). Luminescence spectra were measured in every points of scanning field. Laser scattered light in range 700-900 nm was cut by SWF-750-B filter.

Mai Tai HP laser was replaced by He:Ne laser for Raman scattering measurements. He:Ne laser produced continuous-wave radiation at 632.8 nm and 1.5 mW. Such laser wavelength allowed to avoid strong luminescence from Rhodamine B (since low absorption at this wavelength for Rhodamine B).

4. Computational method details

Finite difference time domain method (FDTD) of solving Maxwell equations was used to calculate nanoparticles extinction and electromagnetic field distribution. Quite traditional attributes of FDTD such as PML (perfect match layer), periodic boundary conditions, TFSF method, frequency dispersion of metal dielectric permittivity were implemented in the program code and used in calculations [14-15]. Dielectric functions of TiO_2, Ag and Au were taken from SOPRA web site. Throughout the calculations x, y, z mesh steps were 0.25 nm for precise description of such geometrical features as the gap between nanoparticles etc.

5. Results and discussion

An example of the luminescence scan of Au/TiO_2 mesoporous film represented on figure 3a. Wavelength of excitation λ_e was 825 nm. It is seen that the image consists of a large number of bright spots in which the signal intensity substantially exceeds the background level. The size of such "hot spot" with bright luminescence was ~300 nm (the size of one pixel is 160nm), roughly coinciding with the limit of spatial resolution of the microscope. Spectra in various pixels of the film (circle centers on figure 3a) are shown on figure 3b. Both bright and dark spots of the film have broad luminescence spectra ranging from ~400 to ~800nm. As it was mentioned in section 3 the upper limit (800 nm) is the result of applied filter. But the lower limit coincides with $\lambda_e/2$ (second harmonic). Moreover second harmonic is pronounced for "hot spots" of the film (curves 1, 2, 3 on figure 3b). Investigation of signal level for various laser pulse power also showed second order of non-linearity. Thus a conclusion can be made: the luminescence signal observed in experiments is two-photon luminescence. The intensity ratio between (for example, between curve 2 and curve 5 on figure 3b) the most intense and dark pixels of the film is ~7.

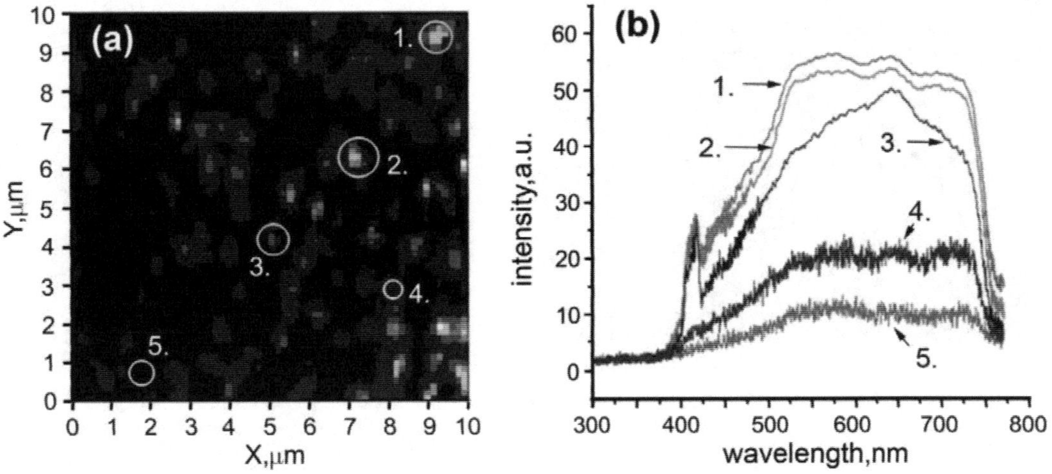

Figure 3. (a) Hot spots distribution in two-photon microscopy scan of Au/TiO_2 mesoporous film, (b) photoluminescence spectra in various pixels of the film. Excitation wavelength $\lambda_e = 825$ nm, scanning area size 10×10 μm.

The concentration of "hot spots" in two-photon experiment (figure 3a) is considerably lower than metal particle concentration on TiO_2 mesoporous film (figure 1a). Therefore some specific configurations of nanoparticles on the film exist which the most effectively enhance electromagnetic field to form the "hot spot" in luminescence experiment. Two types of measurements were made to reveal additional properties of "hot spots" of Au/TiO_2 films.

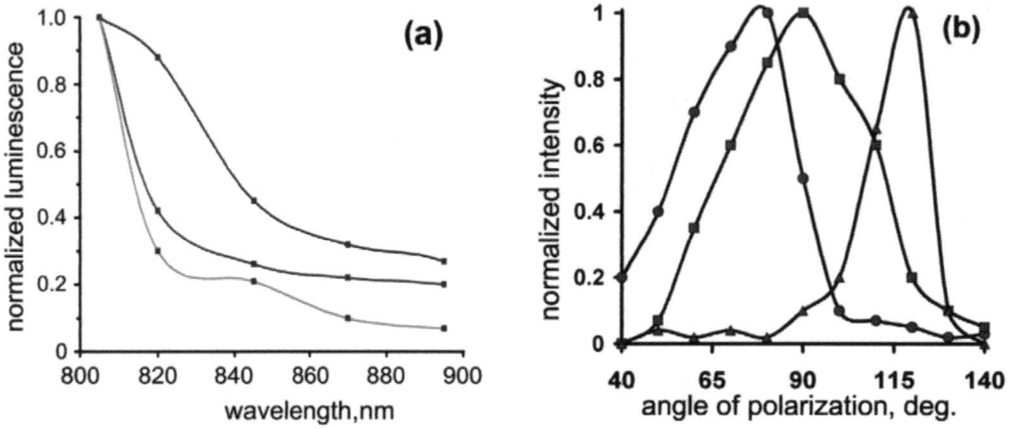

Figure 4. Dependence of integral luminescence intensity of "hot spots" of Au-TiO_2 mesoporous film on: (a) excitation wavelength; (b) direction of linear polarization of excitation laser pulse.

First, the dependence of integral luminescence of "hot spots" on laser excitation wavelength was measured (figure 4a). As can be seen from figure 4a the "hot spots" integral luminescence decreases for longer excitation wavelengths λ_e in experiment. Second, the influence of the direction of linear polarization of incident laser pulse on the integral luminescence in "hot spots" was studied. The dependence of the integral luminescence on various polarization directions (angles with respect to initial direction of pulse polarization) represented on figure 4b. As can seen from figure 4b there is only one pronounced maximum for each "hot spot" of the film.

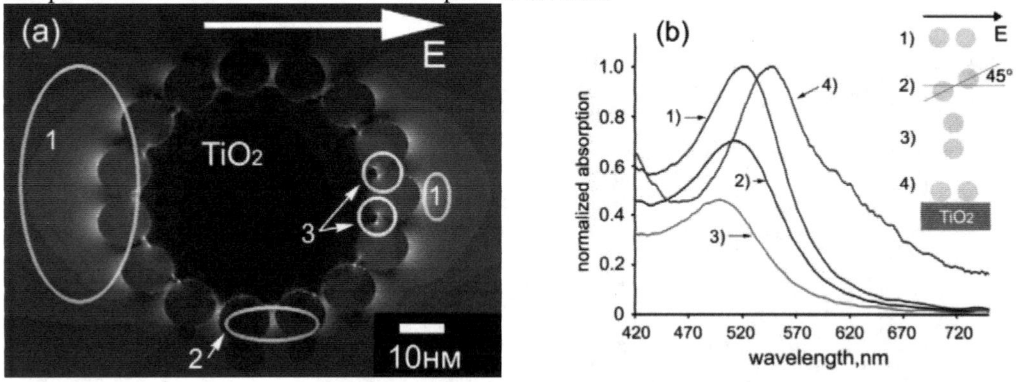

Figure 5. FDTD calculation: (a) electromagnetic field distribution in Au/TiO_2 model system; (b) normalized absorption of two Au spherical nanoparticles in vacuum environment (radius 10 nm, distance between particles 2 nm): 1) p-polarization, 2) 45 degrees between system axis and light polarization, 3) s-polarization, 4) p-polarization, spheres placed on TiO_2 surface.

The electromagnetic field transformation in Au/TiO_2 film could be divided on three main stages. First stage is local electromagnetic field enhancement (EFE) α at λ_e (wavelength of excitation in experiment). Second, two-photon absorption at λ_e. Third, two-photon luminescence of the mesoporous film (figure 3b). Therefore bright spots ("hot spots") with enhanced two-photon luminescence of Au/TiO_2 film should be result of EFE at λ_e. Moreover, two-photon luminescence enhancement is proportional to α^4.

Figure 5a represents the result of FDTD calculation of electromagnetic field distribution for Au spherical nanoparticles on TiO_2 substrate. It is seen several configurations from figure 5a which have enhanced electromagnetic field (lightened areas). Areas labeled by 1 represent typical Mie distribution for single spherical particle. "Hot-spots" of Au/TiO_2 can not be explained by such areas by following reasons. First, in this case "hot-spots" concentration should be approximately the same as particle concentration but it is not. Second, such areas do not have some specific axis to reveal the dependence similar to figure 4b.

Areas labeled by 2 show EFE in the gap between two coupled nanoparticles (dimer). Calculation of absorption spectra for Au coupled nanoparticles (dimer) for various angle between dimer axis and direction of linear polarization represented on figure 5b. It is seen from figure 5b that distance between LSPR position (and absorption peak) and λ_e is minimal for p-polarization. Therefore such a system has maximum EFE α for p-polarization, i.e. single maximum for all range of angles. The same behavior was observed in figure 4b. On the one hand, according to SEM results the shape of Au nanoparticles is close to spherical (see section 2), i.e. figure 4b can not be explained by EFE α of single Au nanoparticles. From the other hand, the dimer of nanoparticles is the simplest configuration which can explain experimental data represented on figure 4a and figure 4b.

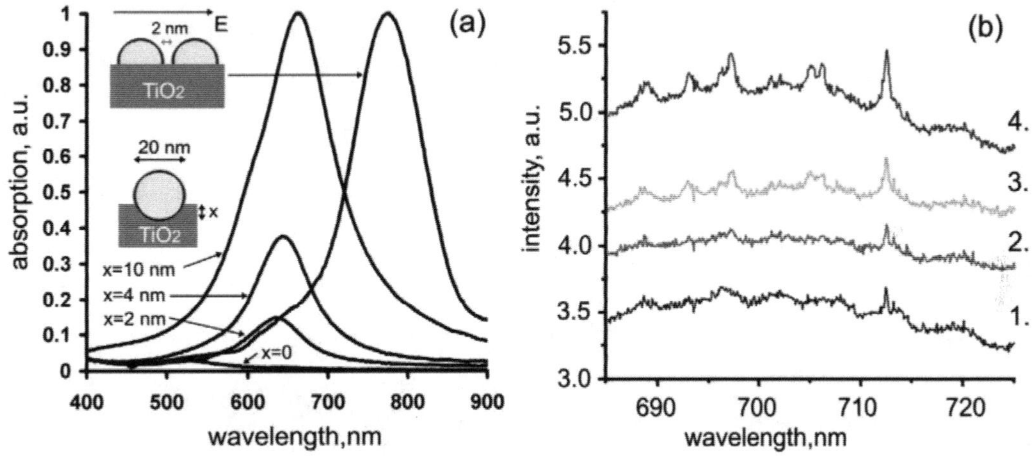

Figure 6. (a) FDTD calculation of absorption spectra of Au/TiO_2 nanocomposites (geometry shown on figure); (b) Intensity of Raman scattering of Rhodamine B on Au/TiO_2 mesoporous film for different photodeposition times : 1 – 4min, 2 – 8 min, 3 – 15 min, 4 – 25min.

Areas labeled by 3 (figure 5a) show EFE in the gap between two coupled nanoparticles near to contact region of Au nanoparticle and TiO_2. It occurred such configurations reveal maximum EFE α. Figure 5b and figure 6a show the influence of TiO_2 substrate on absorption of Au/TiO_2 nanocomposites. It is seen from figure 5b that even small contact with TiO_2 for dimer of Au nanoparticles lead high absorption at λ_e. From the other hand, figure 6a shows that absorption maximum of Au spherical particle on TiO_2 shifts from 550 nm to 665 nm while particle imbedding from 0 nm to 10 nm. Also the imbedding leads to rise of absorption, particularly at λ_e.

It was noted in section 2 that Au nanoparticles formed on TiO_2 surface by photocatalytic process. This means that actually Au nanoparticles are cut by TiO_2 surface. Figure 6a shows absorption spectrum for such cut dimer of Au spherical nanoparticles (r = 10 nm). It occurs that when distance

Journal of Physics: Conference Series **291** (2011) 012040 doi:10.1088/1742-6596/291/1/012040

between nanoparticles is 2nm absorption peak is at 770nm which is very close to λ_e. Moreover, absorption curve has only one pronounced maximum and decreases for wavelength longer than 770 nm. This is also in accordance with figure 4a, i.e. such a system should reveal decrease in integral luminescence for wavelength longer than λ_e (825 nm). Thus, nanoparticles dimer shown on figure 6a should lead near to the highest luminescence enhancement in "hot-spots" of Au/TiO_2 film. Considering this system as ideal configuration let us estimate real "hot-spot" luminescence enhancement. On the one hand, resolution in two-photon experiments was ~300 nm (see above). On the other hand, the most intense electromagnetic field concentrated in the gap between nanoparticles (figure 6a). Taking into account that luminescence enhancement in two-photon experiments was ~7 one can obtain the luminescence enhancement in the gap should be of order of 10^5. Corresponding EFE α should be ≈ 20.

In conclusion a Raman scattering signal was observed for Rhodamine B molecules deposited on Au-TiO_2 film surface. Figure 6b represents a spectrum of signal detected (intensity vs. Raman shift). The position of Raman peaks was in good agreement with the one observed in [17,18]. At the same time no appreciable signal was observed for Rhodamine B molecules on the surface of pure TiO_2 film. Au/TiO_2 nanocomposites show promising application not only for Raman scattering spectroscopy but also for biological objects visualization. This is because of high "hot spots" concentration over the TiO_2 surface (~10^5 cm^{-2}) and large EFE α of "hot spots" in near IR (~λ_e).

6. Acknowledgement

Numerical calculations were carried out on "Chebyshev" supercomputer complex of Moscow State University, Moscow, Russia.

7. References

[1] U Bach, D Lupo, P Comte, J E Moser, F Weissortel, J Salbeck, H Spreitzer and M Gratzel 1998 *Nature* **395** 583
[2] X Baifu 2005 *J.Phys.Chem.*B. **109** 2805
[3] E Hirakawa and P Kamat 2005 *J.Am.Chem.Soc.* **127**(11) 3928
[4] P Johansson 2005 *Phys.Rev.*B. **72** 035427
[5] J Zhang and R Joseph 2005 *J.Phys.Chem.*B. **109** 8701
[6] I Gryczynski, J Malicka and Y Shen 2002 *J. Phys. Chem.*B. **106** 2191
[7] Yu E Lozovik and A V Klyuchnik 1987 *Dielectric susceptibility* ed L V Keldysh (NY: North Holland) chapter 6
[8] M Brongersma and P Kik 2007 *Surface plasmon nanophotonics* (Berlin:Spinger) chapter 1
[9] M Barma and V Subrahmanyam 1989 *J.Phys.Condens.Matter* **1** 7681
[10] A Zayats, 1999 *Opt. Commun.* **161** 197
[11] A Mooradian 1969 *Phy.Rev.Lett.* **22** 185
[12] D L Andrews 2007 *Frontiers in Surface Nanophotonics* (Berlin:Springer)
[13] E Lukashev and V Nadtochenko 2007 *DAN* **415** 696
[14] A Taflove and S Chagness 2000 *The Finite-Difference Time-Domain Method* (Norwood: Artech House)
[15] A Aiboushev and Yu E Lozovik 2007 *Proc.SPIE* **6727** 67272G
[16] A V Aiboushev, A A Astafiev, Yu E Lozovik, S P Merkulova, V A Nadtochenko and O M Sarkisov 2008 *Phys.Lett.*A. **372** 5193
[17] A M Michaels, J Jiang and L E Brus 2000 *J. Phys. Chem.*B. **104** 11965
[18] M Volny, A Sengupta, C B Wilson, B D Swanson, E J Davies and F Turecek 2007 *Anal.Chem.* **79** 4543

Journal of Physics: Conference Series **291** (2011) 012041 doi:10.1088/1742-6596/291/1/012041

Fiber-optic sensor for butylamine

I A Boldov, A S Kuchyanov, A I Plekhanov, N A Orlova[1],I Yu Kargapolova[1] and V V Shelkovnikov[1]

Institute of Automation and Electrometry SB RAS, Novosibirsk, Russia
[1] Novosibirsk Institute of Organic Chemistry SB RAS, Novosibirsk, Russia

E-mail: boldov.ivan@gmail.com

Abstract. Fiber-optic chemical sensor for butylamine was realized. The sensor includes a thin layer nanostructure on the end of optical fiber with a diameter of 600 μm. It consists of luminescent silica nanoparticles modified with pyrylocyanine dye, silver nanoparticles and photonic-crystal film. The sensitivity of the sensor is increased tenfold due to an additional covering of the film with photonic crystal as a porous mirror and the injection of silver nanoparticles with a diameter of 5-7 nm.

1. Introduction

Fiber optics is one of the most dynamically developing field of physics. The application of fiber optics in a range of sensors has several important advantages. Insensitivity to electromagnetic and radiation interference, chemical and thermal stability, remote sensing, relatively low cost makes these types of sensors indispensable in some cases [1]. The coating of the thin-film nanostructures based on silica nanoparticles on the surface of optical fiber allows significantly extend the scope of fiber-optic sensors [2-5]. It is known that amines are usually toxic and in some cases are strong carcinogens. One of the significant shortcomings of existing chemical sensors for amine-type compounds is the lack of selectivity. This paper reports the results of experimental studies of selective sensor for butylamine based on nanostructure thin-film coated at the end of an optical fiber. This film consists of silica nanoparticles of 8-10 nm modified by polymethine dyes [6]. It is well known that pyrylocyanine dyes interact with primary amines to form pyrydocyanines. Pyrydocyanines exhibit luminescence, which is different from the luminescence of the starting pyrylocyanines. This effect was inventive incentive for the synthesis of pyrylocyanine 2 and pyrylocyanine 3 dyes [6], which selectively react with butylamine. It is also well known that the luminescence properties of organic dyes are enhanced in the mesoporous nanostructures of silica and the radiation resistance of the dyes increased [7,8]. Therefore, the binding of luminescent functional dyes with silica nanoparticles is promising for the use of optical chemical sensors. The pyrylocyanine dye may have a covalent bond with the skeleton of silica nanoparticles. The method of chemical modification of silica nanoparticles in a sol-gel synthesis is described in [6].

2. Materials and methods

«Avantes AvaSpec-2048TEC» spectrometer with a spectral resolution of 1 nm was used in the studies of luminescence properties of the thin films, which are consist of modified silica nanoparticles (MSN). «Newport LQC 405 - 85E» semiconductor laser was used as a source of exciting light with a wavelength of 407 nm (figure 1). The thin-film nanostructure is deposited on the end of a freshly cleaved optical fiber with a diameter of 600 μm by dropping into the solution of MSN or deposition and subsequent drying a drop of the solution at the end of the fiber. We have experimentally found that the film of MSN is durable to the various kinds of mechanical effects and demonstrates a high radiation resistance.

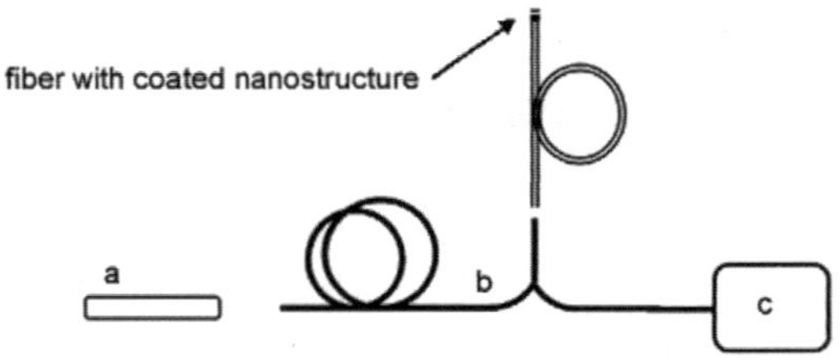

Figure 1. The optical scheme of measuring; a – semiconductor laser, b - reflection probe, c – spectrometer.

A typical luminescence spectrum of the MSN film modified by pyrylocyanine 3 dye has a pronounced peak at 660 nm. The film gives a green-yellow luminescence with a maximum at 560 nm after a treatment with 10% solution of butylamine (figure 2).

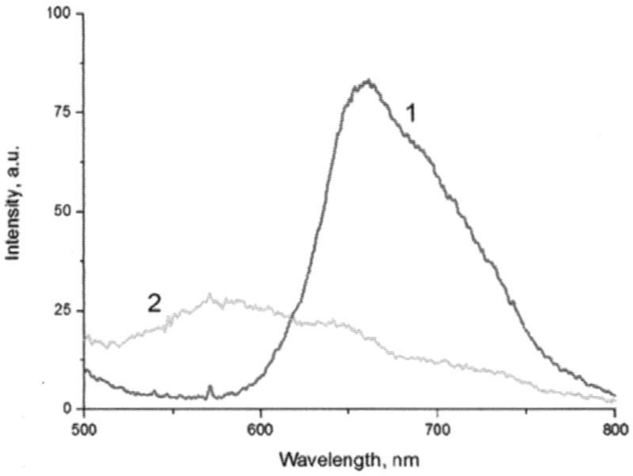

Figure 2. The luminescence spectrum of the MSN thin-film by pyrylocyanine 3 before (1) and after (2) treatment with butylamine solution.

It should be noted that the luminescence effect of the MSN film after the treatment with other amino liquids leads only to the quenching of luminescence in the red spectrum, while the yellow-green luminescence does not appear. The minimum concentration of butylamine in ethanol was 10^{-3} mol/l in which possible to record the luminescent responses from the MSN film modified by pyrylocyanine 3 dye. In contrast to the MSN film modified by pyrylocyanine 3 dye, the films modified by pyrylocyanine 2 do not have the luminescence before the treatment. The maximum of luminescence at 580 nm was exhibited after treatment with butylamine solution (figure 3, curve 1). On exposure to other amino substances, the luminescence peak does not appear which also shows selectivity for the analyte.

To increase the amplitude of the luminescence response from the MSN film coated on the end of optical fiber as a porous selective mirror was applied. The mirror is an artificial opal photonic-crystal (PhC) film. The maximum of reflection of this mirror coincides with the peak of pyrylocyanine luminescence in MSN after the treatment with butylamine solution (figure 3, curve 2).

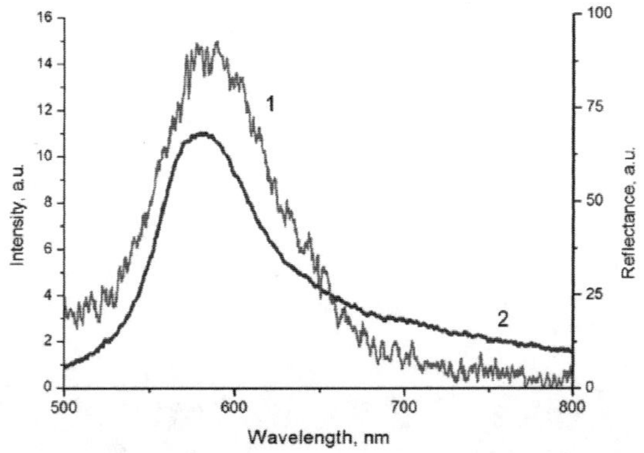

Figure 3. The luminescence spectrum of the MSN thin-film by pyrylocyanine 2 after treatment with butylamine solution (1) and the reflectance spectrum of «PhC mirror» (2).

It was achieved by choosing the diameter of monodisperse silica spheres (MSS) of about 260 nm. The porosity of PhC film allows butylamine and other substances easily penetrate into the MSN film. The PhC opal film was grown by moving meniscus method on the end of the optical fiber [9]. For this purpose optical fiber is fixed in the suspension of MSS on the surface under an angle of 15 degrees. The reflection spectrum of «PhC mirror» deposited on the end of the optical fiber is practically coincided with the luminescence band of pyrylocyanine 2 dye in the region of 580 nm.

3. Results

The experiments have demonstrated that the coating of the «PhC mirrors» on the MSN film leads to an increase in the amplitude of the reflected signal in 3 - 4 times (figure 4). The PhC film was a highly ordered hexagonal close-packed structure of the MSS, and the crystallographic direction (111) was normal to the surface. The presence of «PhC mirror» does not affect the rate of change of the optical response of sensory film.

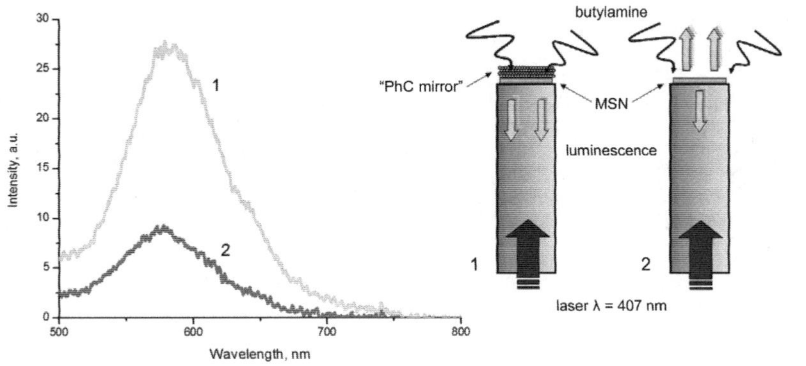

Figure 4. The luminescence spectrum of MSN by pyrylocyanine 2 with (1) and without (2) «PhC mirror».

The further increase in the amplitude of a luminescence of MSN might be enhanced due to the presence of nearby metallic nanoparticles such as Ag or Au [10]. We have found that the addition of

Ag nanoparticles with a diameter of 5 - 7 nm into the MSN film leads to an increase the luminescence in 2 - 3 times (see figure 5).

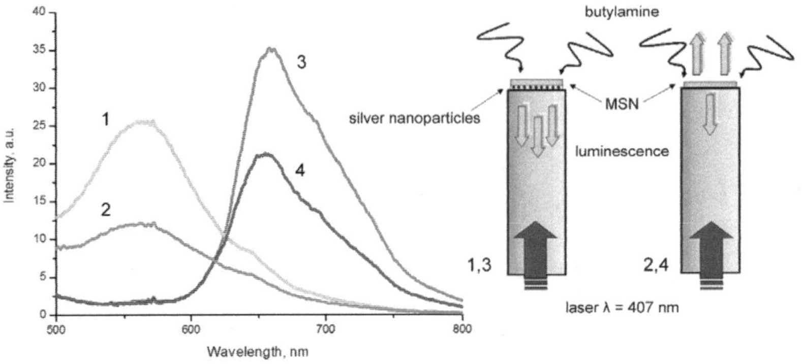

Figure 5. The luminescence spectrum of MSN by pyrylocyanine 3 before (3,4) and after (1,2) treatment with butylamine solution; with (1,3) and without (2,4) injection of silver nanoparticles.

4. Conclusion

This paper proposed an optical fiber butylamine chemical sensor based on selectively altering the luminescent properties of MSN thin-film. The sensor component is a MSN covalently bound with pyrylocyanine dye and deposited at the end of the optical fiber. We confirmed that our sensor has high sensitivity for actual use, and 10% solution of butylamine concentration less than 10^{-3} mol/l can be detected. It was shown that the addition of silver nanoparticles into the film of MSN and the coating it the PhC film as a porous selective mirror on the end of the fiber can increase the sensors sensitivity about 10 times. We assumed that the improving the sensing element by doping it in PhC film with metallic nanoparticles allow us to develop a highly sensitive selective optical chemical sensors remote monitoring.

We are grateful to Dmitry V. Kalinin for providing the suspension of silica particles.

References

[1] E. Udda. 2008. *Fiber-optic sensors.* (Moscow, Technosphere).
[2] J. Li, P.R. Herman, C.E. Valdivia, V. Kitaev, G.A. Ozin. Optics Express **13** (2005) 6454.
[3] H. Yan, M. Wang, Y. Ge, P.Yu. Optical Fiber Technology **15** (2009) 324.
[4] M. Janotta, A. Katzir, B. Mizaikoff. Applied Spectroscopy **57** (2003) 823.
[5] H. Xiao, J. Zhang, J. Dong, M. Luo, R. Lee, V. Romero. Optics Letters **30** (2005) 1270.
[6] N.A. Orlova, I.Yu. Kargapolova, V.V. Shelkovnikov, A.I. Plekhanov. High Energy Chemistry **43** (2009) 602.
[7] I. Sokolov, Y.Y. Kievscy, J.M. Kaszpurenko. Small. **3** (2007) 419.
[8] M.M. Collinson. Trends in analytical chemistry **21** (2002) 30.
[9] A.I. Plekhanov, D.V. Kalinin, V.V. Serdobintseva. Russian Nanotechnology **1** (2006) 245.
[10] C.D. Geddes, H. Cao, I. Gryczynski, Z. Gryczynski, J. Fang, J.R. Lakowicz. Physical Chemistry **107** (2003) 3443.

Journal of Physics: Conference Series **291** (2011) 012042 doi:10.1088/1742-6596/291/1/012042

The nonlinear thermal model of heterojunction light emitting diodes

V A Sergeev[1,2], A M Hodakov[1]
[1] Kotelnikov Institute of Radio Engineering and Electronics of Russian Academy of Sciences, Ulyanovsk Branch, 48/2 Goncharov Str., Ulyanovsk, 432011, Russia;
[2] Ulyanovsk State Technical University, 32 S.Venets Str., Ulyanovsk, 432027, Russia

E-mail: sva@ulstu.ru; ufire@mv.ru

Abstract. Feature of heterojunction light-emitting diodes (LEDs) is that their internal quantum efficiency decreases with the growth of temperature and current density, which leads to additional positive electrothermal feedback. For an estimation of these effects nonlinear electrothermal model of power LED with dependences internal quantum efficiency from temperature and current density and dependence of factor of heat conductivity of a substrate on temperature is offered. It is shown, that action of these electrothermal feedback mechanisms leads to increase in heterogeneity of distribution of temperature and current in structures of LED in comparison with linear model and to nonlinear increase in thermal resistance of LEDs with current growth. Results of measurement of the thermal impedance of serial power LEDs is well described by the offered model.

1. Introduction

It is known, that intensity of radiation and durability heterojunction light emitting diodes (LED) decrease with the increase in the temperature of an overheat of heterojunction [1]. Analysis of thermal properties LED usually assumes, that the power density is uniformly distributed on the heterojunction area [1,2]. However even at uniform distribution of sources of heat on surface of structures as a result of non-uniform heat removal from its various parts distribution of temperature will be non-uniform. Exponential dependence of density of a current on temperature as a result of positive electrothermal feedback (PETF) leads to increase of this non-uniformity, and with increase in a total current the non-uniformity of distribution of density of a current and temperature increases.

The feature of heterojunction LED is that their internal quantum efficiency decreases with the growth of temperature and current density, which leads to additional PETF: full electrical power, transformed to heat, and density of heating power in more warm areas of heterojunction will increase, which leads to stronger increase of heterogeneity.

2. The nonlinear electrothermal model of light emitting diodes

For calculation and the analysis of the specified effects has been considered electrothermal model InGaN/GaN heterojunction structure on the sapphire substrate placed on metal heat-tap (figure 1). As the thickness and depth of placing heterojunction are small, heat sources in this model were considered surface-distributed.

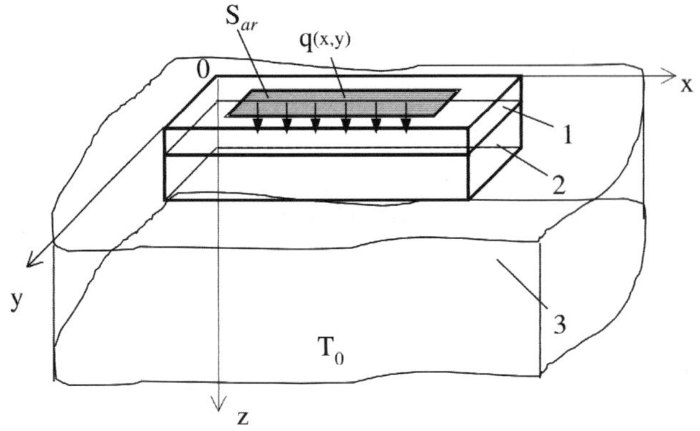

Figure 1. The thermal model of high power LED: 1 – heterostructure; 2 – sapphire substrate; 3 – metal heat sink.

The mathematical model describing thermal properties of considered structure LED, includes system from the stationary equation a heat conductivity and the thermogeneration equations, under adiabatic conditions on lateral borders. Dependence of density of current LED from temperature is defined its *I-V* characteristic:

$$J(x,y) = C \cdot exp\left(-\frac{E_g - e(U_D - \rho_n L_{get} J(T_n))}{pkT_n(x,y)}\right), \quad (1)$$

where C – parameter weakly depending on temperature, U_D – direct voltage on LED, E_g – band energy of the semiconductor, $T_n\,(x,\,y)$ – temperature in a heterojunction plane, $\rho_{n,}\,L_{get}$ – specific resistance and a thickness of heterostructure accordingly, k –Boltzmann's constant, e – an electron charge, p – the parameter, depending on the mechanism of current carrying in heterojunction, $x,\,y$ – coordinates on a surface of heterostructure. The density of electric power, transforming in heat on surface of substrates, is defined by the formula

$$q(x,y) = [1 - \eta(T_n, J)]J(T_n)U_D, \quad (2)$$

where η – the internal quantum efficiency of heterostructure which are function of temperature and density of a current. At the set total current through LED for a finding of value U_D it necessary to known distribution of temperature to solve the such equation

$$\iint_{S_{ar}} J(U_D, T_n)dxdy = I_D, \quad (3)$$

where S_{ar} – the area of active area of structure.

Dependences η from temperature and current density it was described by such function:

$$\eta(T,J) = \eta_0(T_0, J_0)exp(-B\Delta T - D\Delta J),$$

where ΔT and и ΔJ – deviations of temperature and density of a current from initial values T_0 and J_0, $\eta_0(T_0, J_0)$– value of quantum efficiency at T_0 and J_0; values of parameters B, D and η_0 were defined according to [1]. From settlement temperature distributions (figure 2) at different combinations of parameters it is visible that dependence of quantum efficiency on current and temperature density leads to increase in the maximum overheat of structure in 1.2-1.3 times in comparison with linear model $\eta = const$ [3].

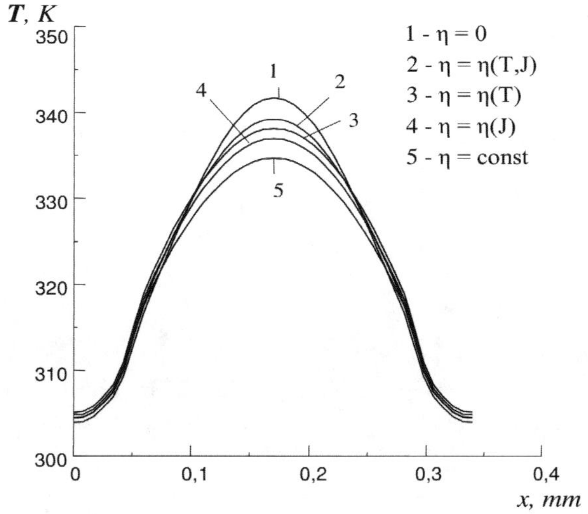

Figure 2. Temperature distributions in the center of a surface of he structure along an axis x at a current 350 mA and various combinations of parameters of model.

Heterogeneity of temperature leads to non-uniform distribution of density of a current for which description use [4] linear approach $J = \bar{J}\left[1 + \xi_T\,(T_n - \bar{T}_n)/T_0\right]$, where ξ_T temperature factor of density of a current under condition of (3). Measuring thermal parameter LED is thermal resistance junction-case, which in one-dimensional model is defined by expression

$$R_{Tj-c} = d/S_{ar}\,\lambda_n\,, \tag{4}$$

where d_n – a thickness of a substrate, S_{ar} – the area of active area of a LED, λ_n – factor of heat conductivity of a substrate material. For more adequate definition of thermal resistance at non-uniform current distribution in expression (4) it is necessary to use the effective area of active area, which is defined as the relation of average density of a current to maximum: $S_{aeff} = S_0\left[1 + \xi_T\left(T_n^{\max} - \bar{T}_n\right)/T_0\right]^{-1}$. As a result of action of considered mechanisms PETF the effective area of LED will decrease, and thermal resistance junction-case of LED increase with full current growth. In linear approach this dependence can be described by the expression $R_{Tj\text{-}c} \approx R_{Tj-c0}(1 + \xi_1 I_D)$. By our estimations $\xi_1 \approx (0{,}3 \div 0{,}4) I_{\max}^{-1}$, where $I_{\max}$ – the maximum working current LED.

Other fundamental reason leading to dependence of thermal resistance from a total current is dependence of factor of heat conductivity of a substrate on temperature. If accept as modeling approximation of this dependence expression $\lambda = \lambda_0 (T_0 / T)$ in linear approach this dependence can be described as the expression $R_{Tj\text{-}c} \approx R_{Tj-c0}(1 + \xi_2 I_D)$. Contribution of this mechanism to the general steepness current dependences is less compared with mechanisms non-uniform current distribution ($\xi_1 > \xi_2$); at the increase in a current from 0 to maximum value thermal resistance to increase by 20-30%.

3. Experimental results

For measurement of parameters of nonlinear thermal models LED the experimental sample is offered of a microcontroller measuring instrument of a thermal impedance of diodes [5]. Results of measurement of frequency dependences of the module of thermal impedance $Z_T(\omega)$ serial high power LED XRC-RED-L1-R2-M2-C-1 ($I_{\max}$ =350 mA), produced by firms Cree, are shown on figure 3.

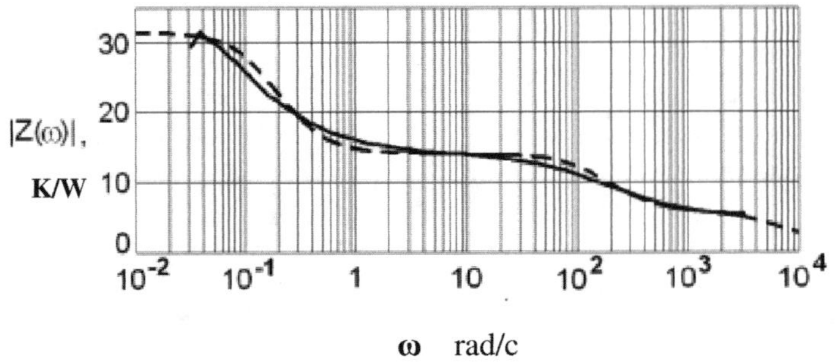

Figure 3. Dependence of the module of a thermal impedance high power LED from frequency of modulation of heating power in an air at a current 350 mA: continuous line – experiment; shaped line – theory.

Values of the module of a thermal impedance are measured on the bottom shelf ($\omega \approx 1500$ rad/s) correspond to junction-case thermal resistance. With increase in a heating current from 200 mA to 550 mA thermal resistance of a substrate increases in 1.6-1.7 times (figure 4) that is well described within the limits of the offered model.

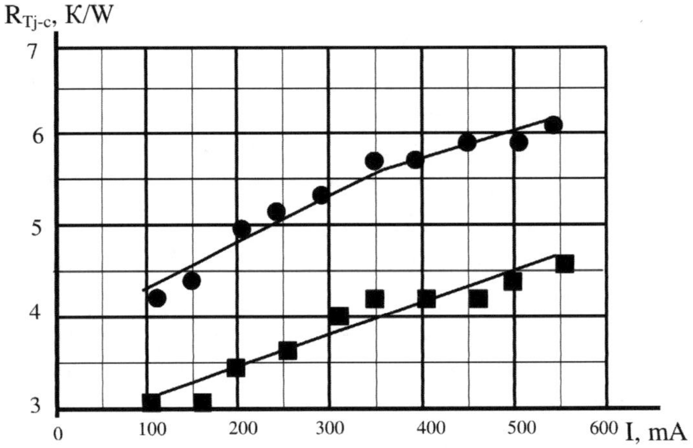

Figure 4. Dependences of junction- case thermal resistance of two high power LEDs from total current.

4. Conclusion

It is shown, that dependence of quantum efficiency on temperature and current density leads to increase in heterogeneity of distribution of temperature and current in structures of power light-emitting diodes in comparison with linear model. As a result of action of this electrothermal feedback mechanism also dependence of the factor of heat conductivity of substrate from temperature effective thermal resistance of LEDs nonlinear increase with current growth.

Acknowledgments

Work is executed in according to the thematic plan of researches of Ulyanovsk branch of Kotel'nikov Institute of Radioegineering and Electronics of RAS and base chair "Radio engineering opto - and nanoelectronics" of Ulyanovsk State Technical University with financial support of the Ministry of

Journal of Physics: Conference Series **291** (2011) 012042 doi:10.1088/1742-6596/291/1/012042

Education and Science of the Russian Federation within the limits of the target program «Development of scientific potential of the higher school (2009-2010)», the project № 2.1.2/4606.

References

[1] Efremov A A, Bochkaryova N I, Gorbunov R I, et al. 2006 *Semiconductors* **40** 605
[2] Rozhansky I V, Zakhejm D A 2006 *Semiconductors* **40** 839
[3] Sergeev V A, Hodakov A M 2010 *Semiconductors* **44** 218
[4] Sergeev V A 2005 *Izvestiya Vuzov. Electronics* **3** 22 (in Russian)
[5] Sergeev V A, Smirnov V I, Gavrikov A A et al. 2010 *Promyshlennye ASU i kontrollery* **3** 45 (in Russian)

Journal of Physics: Conference Series **291** (2011) 012043 doi:10.1088/1742-6596/291/1/012043

Research of Frequency Characteristics and Temperature Drift Power Quantum-Well Lasers

D. F. Zaitsev
Department of Electronics, the Moscow state institute of radio-engineering, electronics and automatics, avenue Vernadskogo 78, Moscow, 119454, Russia

E-mail: zaysev@yandex.ru

Abstract. In the following paper results of AFC (amplitude-frequency characteristics) and PFC (phase-frequency characteristics) and temperature drift studies as well as of time delay of CW and impulse signals in QW InGaAs lasers with CW output power more than 1W at direct modulation are given. It is demonstrated, that power QW lasers can potentially surpass their low-power analogues and DH lasers on one of the major parameters – product of the maximum modulating frequency (bandwidth) and output modulated RF power.The developed analytical mechanism considers QW power lasers' working features in a static and dynamic mode and enables us to calculate AFC, PFC, their temperature drift and delay of modulating signals with a sufficient for engineering calculations accuracy.

1. Introduction

In a modern radar, communication and radio-astronomy it is the equipment on the basis of nanophotonics that is increasing in its application possessing significant advantages in comparison to the traditional electronic one[1]. Nanophotonic equipments usually apply to realization of transferring pathes of the big capacity low-power DH lasers with direct or external modulation with the subsequent strengthening the modulated optical signal by semi-conductor or fiber optical amplifiers. However optical amplifiers have efficiency of approximately 10 % [2]. Meanwhile modern quantum-well (QW) power lasers have efficiency of 70 % and more [3]. In order to realize high-power transmitting paths of nanophotonic equipment the low-power heterolasers - with direct or external modulation and with the following amplification of the modulated optical signal by semiconductor or fiber-optical amplifiers - are utilized.
Therefore it is an issue of great interest to estimate potential frequency characteristics and AFC and PFC temperature drift of QW power lasers with wavelengths of 0,85 - 1,05 μm, which correspond to maximum quantum efficiency of modern photodetectors modulated directly with a radio-frequency signal. However, nowadays only frequency characteristics and their temperature drift of low-power DH lasers of various types have been well-studied under direct modulation with a radio-frequency signal [4-6]. This is due to a much greater complexity in construction of a dynamic model of QW power lasers, as it is necessary to simultaneously take into consideration a lot of parameters, neglecting of which facilitated considerably the analysis of low-power lasers. In addition, experimental check seems problematic, because of significantly greater influences of parasitic elements of the equivalent circuit of power lasers themselves as well as their modulator circuits.

2. AFC and PFC of QW power lasers

By solving total system of the rate equations and implementing the multifactorial analysis of time dependences of QW power lasers' internal parameters from table 1, analytical expressions for calculation AFC and PFC in a wide frequency range have been received in explicit form.
Results of modeling for a power single-mode InGaAs QW SCH "internal" laser at low-signal modulation (up to 70 %) of AFC and PFC at various relations of bias currents to a threshold current are presented in

Fig. 1 (a) and (b) respectively. In order to perform a quantitative check of the calculation method offered the experimental results, received for QW power lasers with similar resonator length L = 3 mm, with the bias current ratio I / I_{th} and working temperature at the continuous operating mode, were used (these results are shown in [7]). The results utilized confirm sufficient correspondence to the theory: the maximal deviation does not exceed 15 %. As these experimental results were received by measuring the level of relative noise of the RIN laser without a modulating signal, they reflect potential dynamic properties of power heterolasers without the influence of the circuit parasitic elements.

Table 1. QW SCH Characteristics of LD structures

Parameter	Symbol	Value	Unit
Forbidden bandwidth	h_v	1.16-active area of a 1.42-wave guide of 1.8-emitters	eV
Active area volume	Va	$2{,}96{*}10^{-15}$	m^3
Resonator length	L	3000	Mm
Resonator width	W	100	μm
Confinement factor	Γ	0.008-0.01	-
Mirror reflectivity 1	R_1	0.05	-
Mirror reflectivity 2	R_2	0.95	-
Intrinsic loss	α_i	0,34	cm^{-1}
Differential gain	g_o	1,32*10	m^3/c
Nonlinear gain suppression	ε	$7.0{*}10^{-23}$	m^3
Effective carrier lifetime	τ_{ef}	$1.5 * 10^{-9}$	S
Photon lifetime	τ_p	$25{,}5 *10^{-12}$	s
Transparency density	N_t	$8{,}0{*}10^{23}$	m^{-3}
Characteristic temperature	T_o	110	°K
Threshold current	I_{th}	430	mA
Drive current at output power 6 W	I_H	7000	mA
Differential quantum efficiency	η_d	1000	mW/mA

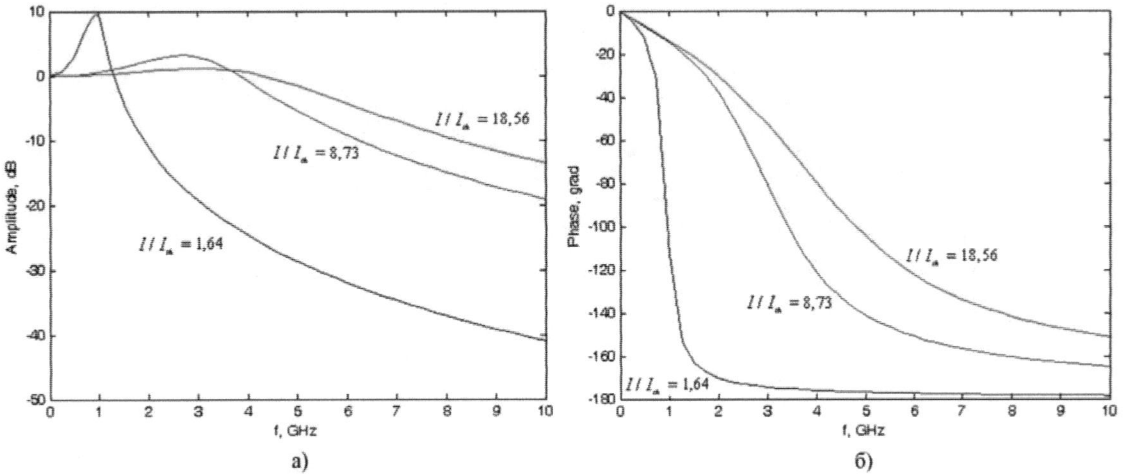

Fig. 1 Analytical solutions of AFC (a) and PFC (b) of QW SCH InGaAs "internal" power laser, without parasitic elements, at various relations of bias currents to a threshold current I / I_{th} = 1.67, 8.73 and18.56 respectively at low-signal (up to 70%) modulation and the temperature of +20°C.

The potential transmission efficiency T_{ef} (that equals to product of target capacity P_o and resonant frequency f_o, i.e. $T_{ef} = P_o * f_o$) comparison of typical QW power lasers and the low-power lasers specially optimized for generation of the maximum value f_o (the maximum operation speed) yields the following results. From [8] for the low-power laser we have: at I/I_{th}= 1.8: T_{ef}= 0.92 * 4.65 ≈ 4.3 GHz * mW. For the power laser [7] at I/I_{th}= 1.7: T_{ef}= 30 * 1.14 ≈ 34.2 GHz * mW. For the low-power laser from [9] at I/I_{th}= 9 we have: T_{ef}≈ 11 * 4.7 ≈ 51.7 GHz * mW, for the power laser [7] at I/I_{th}= 9: T_{ef}= 300 * 2,91 ≈ 873 GHz * mW.

For the power laser with the parameters given in table 1 it is possible to assess: at I/I_{th} ≈ 9 : T_{ef}≈3000 * 3,17 ≈ 9510 GHz * mW, i.e. ~ 10 THz * mW.

Thus, it is possible to draw the following conclusion: QW power lasers, for example InGaAs QW SCH, potentially surpass low-power analogues and DH lasers in one of the major parameters - product of the maximum modulated frequency (bandwidth) output power and radio-frequency of modulation.

3. Temperature Drift Power Quantum-Well Lasers

Owing to the solution of total system of the rate equations and the multifactorial analysis of time and temperature and bias (S_0) dependences of power QW lasers' internal parameters $g_o(T)$, $S_0(T,I)$, $\chi(T)$, $\tau_p(T)$, $\tau_{ef}(T)$, analytical expressions for calculation AFC and PFC and their temperature drift in a wide frequency and temperature range have been received in an explicit form.

For a single mode InGaAs QW SCH "internal" power laser at low-signal modulations with no optical feedback, AFC and PFC depending on temperature, bias current and frequency will be:

$$|A(T,I,\omega)| = \frac{\dfrac{g_o(T)S_0(T,I)}{\chi(T)\tau_p(T)(1+\varepsilon S_0(T,I))}\left(1+\dfrac{\varepsilon}{g_o(T)\tau_{ef}(T)}\right)}{D^{1/2}(T,I)}; \quad (1)$$

$$\varphi(T,I,\omega) = -\mathbf{arccos}\left\{\frac{\left[\dfrac{g_o(T)S_0(T,I)}{\chi(T)\tau_p(T)(1+\varepsilon S_0(T,I))}\left(1+\dfrac{\varepsilon}{g_o(T)\tau_{ef}(T)}\right)-\omega^2\right]}{D^{1/2}(T,I)}\right\}; \quad (2)$$

where:

$$D = \left[\frac{g_o(T)S_0(T,I)}{\chi(T)\tau_p(T)(1+\varepsilon S_0(T,I))}\left(1+\frac{\varepsilon}{g_o(T)\tau_{ef}(T)}\right)-\omega^2\right]^2 +$$

$$+\omega^2\left[\frac{g_o(T)S_0(T,I)}{\chi(T)(1+\varepsilon S_0(T,I))}+\frac{1}{\chi(T)\tau_{ef}(T)}+\frac{\varepsilon S_0(T,I)}{\tau_p(T)(1+\varepsilon S_0(T,I))}\right]^2,$$

$S_0(T,I)$- stationary density of photons , $\chi = 1+\frac{\tau_t}{\tau_e}$, τ_e - escape lifetime, τ_t - transport time.

The size of time drift $\Delta\tau$ PFC can be expressed through the phase drift by using the formula:

$$\Delta\tau = -\varphi(T,I,\omega)/360*f; \qquad (3)$$

where: f is modulation frequency.

Some results of mathematical modelling are given in (fig. 2) and (fig. 3). In fig. 2 there is an example of *3D* amplitude surface $A(\omega,T)$ construction, calculated by the formula (1), as functions of frequency in a range from 0 to 10 GHz and temperatures in range from +20 to +80 °C, for the laser with parameters from tab. 1 and bias current I = 5A and (I/I_{th} =10). In fig.3 there is an example of 3D phase surface φ(ω,T) construction, calculated by the formula (2), as functions of frequency in range from 0 to 10 GHz and temperatures in range from +20 to +80 °C for the similar laser with bias current I = 5A and (I/I_{th} =10).

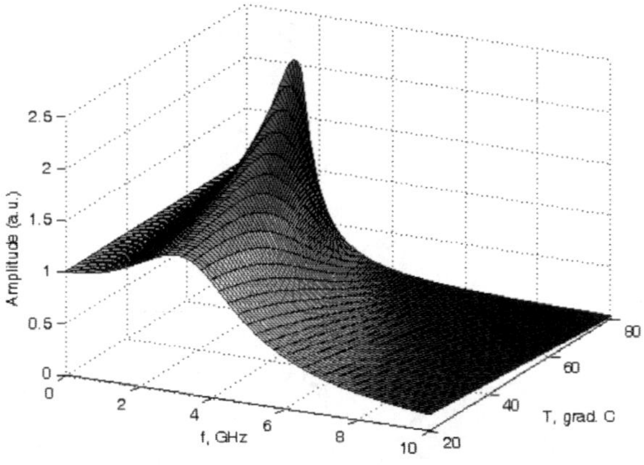

Fig. 2: 3D amplitude surface $A(\omega,T)$ as function of frequency and temperature for QW SCH InGaAs power laser with bias currents I = 5 A. Frequency in GHz is plotted along X-axis, amplitude is plotted along Y- axis, temperature in °C is plotted along Z- axis.

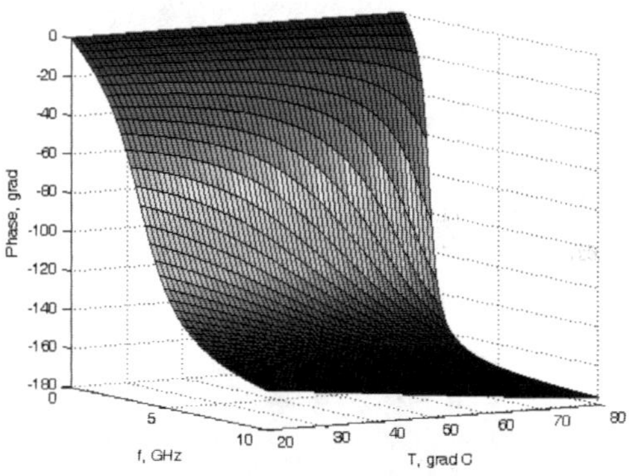

Fig. 3: 3D phase surface $\varphi(\omega,T)$ as function of frequency and temperature for QW SCH InGaAs power laser with bias currents I = 5 A. Frequency in GHz is plotted along X-axis, phase is plotted along Y- axis, temperature in °C is plotted along Z- axis.

4. Conclusions

It has been shown that high-power QW lasers potentially surpass low-power analogues in one of the major parameters - product of the maximum modulating frequency (bandwidth) and output modulated RF power. Also the method of AFC and PFC calculation for QW power lasers as well as of their temperature drift at direct modulation by RF signal has been developed giving sufficient accuracy for engineering calculations. Directly modulated quantum-well high-power lasers can be applied in transmitting paths of nanophotonic transmitters of radio-fibre systems (RoF), including systems of the broadband wireless Internet with large coverage range (WiMAX) and open laser communication systems.

Acknowledgements
The author would like to thank prof. I. S. Tarasov for granting lasers data and fruitful discussions. The work was partially supported by the Russian Ministry of Education and Science Grant No. 2.1.2/6591.

References

[1] Zaitsev D F 2007 Antennas Iss. 9 (124), p 60 (in Russian)
[2] Juodawlkis P, Plant J, Loh W, Missaggia L, Jensen K and O'Donnell F 2009 IEEE Photon. Technol. Let. V 21, No 17, p 1208
[3] Pikhtin N A, Slipchenko S O, Sokolova A Z , Stankevich A L, Vinokurov D A, Tarasov I S, Alferov Zh I 2004 Electron. Let. V 40, No 22, p 1413
[4] Zaitsev D F 1993 IEE Proc. – J. V 140, No. 4 p 227
[5] Zaitsev D F 2003 Antenny Iss. 12 (79), p 33 (in Russian)
[6] Habibullah F, Park J W 2005 Optics Communications **249** p 265
[7] Garbuzov D, Maiorov M, Menna R, Komissarov A, Khalfin V, Kudryashov I, Lunev A, DiMarco L, Connolly J 2002 Proc.of SPIE V 4651
[8] Chen T, Zhao B, Eng L, Zhuang Y, O'Brien J and Yariv A 1993 Electron. Let. V 29, No 17, p 1525
[9] Choe J-S, Kim K, Park S, Kim J, Lee J-M, Kim M-S, Park S and Ju J 2010 Semicond. Sci. Technol. V 25, p 1

The infiltration of nonwetting liquid into nanoporous media and the thermal effect

V. D. Borman[1], **A. A. Belogorlov**[1], **V. A. Byrkin**[1], **G. V. Lisichkin**[2], **V. N. Tronin**[1] **and V. I. Troyan**[1]

[1] National Research Nuclear University MEPhI, Kashirskoe sh. 31, Moscow, 115409 Russia
[2] Lomonosov Moscow State University, GSP-1, Leninskie Gory, Moscow, 119991, Russia

E-mail: vbyrkin@gmail.com

Abstract. In the present work we formulate a model of infiltration nonwetting liquid into the nanoporous medium. This model takes into account the correlation effects in a disordered medium and is based on analytic methods of percolation theory. The infiltration of porous medium is treated as a process of filling pores in an infinite cluster of pores connected with each other. In the framework of the developed model we describe thermal effects that accompany the accumulation of energy by system characterized by different porosity.

1. Introduction

For past few decades the investigation of "nonwetting liquid-nanoporous body" systems behaviour have been drawing increasing attention. For example, Laplace pressure, which for nanometer-sized pores is $10^2 \div 10^3$ atm, must be overcome to make possible the infiltration of nonwetting liquid into the nanoporous medium. Thus, in the transition of bulk liquid into the dispersed state in pores of the nanoporous medium with the specific surface of ~ 1 cm^2/g, both the absorbed energy and the accumulated energy (which returns back when the liquid flows out) can reach a value of $10 \div 100$ kJ/kg. This value is greater than the value of energy in case of widely used materials like polymer composites or shape-memory alloys by an order of magnitude [1–3]. In view of above the use of such "nanoporous medium — nonwetting liquid" systems opens up broad prospects for development of alternative sources of energy and mechanical energy dissipators [1–13].

So far the following systems were investigated: silochromes, zeolites – liquid metals, hydrophobized silica gels, zeolites – water, aqueous solutions of organic compounds and salts [1–49]. In the infiltration-defiltration cycle the hysteresis is observed, i.e. the threshold infiltration pressure is higher than the threshold defiltration pressure. The value of the accumulated energy is determined by the product of the volume of pores filled with the nonwetting liquid and the difference between values of threshold infiltration and defiltration pressures. During infiltration of nonwetting liquid into the porous medium the thermal effects are observed which depend both on the degree of filling (occupancy) of the porous medium and the porosity φ [5,44,47,48]. The integral thermal effect was observed on the infiltration of the modified porous mediums PEP100(C18), PEP300(C18), KSK-G(C16) with water [47,48]. It was established that for these porous media with the porosity $\varphi = 0.35$ and $\varphi = 0.45$ the value $\Delta Q < 0$ decreases with the increase of occupancy degree and is minimal when the occupancy is complete. Therefore, it follows from the comparison of the results obtained in [5,47,48] that the thermal effect depends on the porosity of medium.

On the other hand it was found in [44] that when the hydrophobized silica gel porous medium Sigma-Aldrich with the porosity $\varphi = 0.68$, was filled with water up to 1/3 of its volume the temperature of the system in the heat-insulated chamber did not increase within the measurement error (≤ 0.05 K). But it follows from the estimates made by the authors that the increase in temperature must be equal to $\Delta T = 0.8$ K.

Within the existing theories during the filling of disordered porous medium the absorbed energy can be spent on the formation of liquid-porous medium surface in pores, the formation of meniscus surface in the pore throats,the heat release that accompanies the formation of these surfaces and on the dissipation caused by the motion of liquid in nanopores [5,9,10,20–23,47,48]. Let us note that viscous dissipation could probably be neglected [50]. In case of closed hysteresis loop the change in the internal energy is equal to zero. Because of that the work must be equal to the heat release $\oint dA = \oint dQ$. Under the assumption that wetting angles during infiltration and defiltration are different the work $\oint dA$ is not equal to zero. However, in accordance with [51] the value of thermal effect ($\Delta Q = -\int T\frac{d\sigma}{dT}dS$) is equal to zero for $\Delta S = 0$ in case of a closed cycle.

From the above it follows that the nature of the observed thermal effects is unclear.

For description of infiltration of nonwetting liquids into nanoporous media the percolation theory [7, 8, 52, 53] and lattice models [54–57] are used. In early works [51] the infiltration transition within the framework of the percolation theory was described using the condition of the Laplace with phenomenological wetting angle. To describe the defiltration process the asymmetric pore size distribution as well as the pore throat size distribution were considered. In [7,8] based on percolation theory the infiltration was treated within the statistical fluctuation theory. The probability of filling of a pore was calculated with account of the energy expenditure for the formation of nonwetting liquid-nanoporous medium interface and menisci in the throats between the neighbor pores. The condition of liquid reaching the pore to be filled was provided by the creation of clusters of pores, connected with the surface of the porous medium of a finite size. To describe the infiltration of nanoporous medium with nonwetting liquids in [51,54,55,58] the lattice model treated within a mean-field approach and numerical methods were used. In these works the condition for liquid to reach the pores inside the porous medium was not analyzed. The dependence of infiltration and defiltration pressures on the volume of liquid in pores was determined by phenomenological parameters of model lattice Hamiltonian. Energetics of infiltration-defiltration processes were not considered within lattice models and models based on percolation theory.

In the present work to explain the observed thermal effects we constructed a model of infiltration of nonwetting liquid into nanoporous medium that takes into account the correlation effects in the disordered medium. This model is based on analytic methods of percolation theory [50, 52, 58, 59]. Within this model, the process of infiltration of nonwetting liquid into the porous media was described for media with different porosity. Meanwhile, the infiltration is regarded as the filling of pores of an infinite cluster. Within the model of randomly spaced spheres (RSS) [53] the number of nearest neighbor pores and the square of throats between neighbor pores which analytically depend on porosity were calculated. These results allow to take into account the correlation effect of spatial location and connectivity of pores in medium. Another correlation effect of positional relationship of filled and empty pores on a shell of a cluster of filled pores determines the probability of fluctuation of filling. The expression for this probability is obtained in a closed form.

2. "Nanoporous media — nonwetting liquid" system

Now let us consider the infiltration of nonwetting liquid into a single pore in a partly filled porous medium immersed in a nonwetting liquid and being under external pressure. Depending of the pore radius the pore can be in one of two possible states: it can be able or not able to be filled with a liquid at a given pressure p. The probability of a pore to be in these states can be written as [8]

$$w(p, R, \theta) = \left[1 + \exp\left(\frac{\delta A(p, R, \theta)}{T}\right)\right]^{-1}, \tag{1}$$

where $\delta A(p, R, \theta)$ is the work providing the infiltration of liquid into a pore with the radius R at the pressure p and the temperature T.

If depending on pressure and occupancy of the medium $\delta A(p, R, \theta) < 0$ then the probability $w \sim 1$ and the pore can be filled with liquid. If $\delta A(p, R, \theta) > 0$ then $w = 0$. Consequently, the pore can not be filled with liquid.

Let us calculate the work expended on the fluctuating filling of one pore. To calculate it we assume that each pore in the porous medium has z nearest neighbour pores and pores contact with each other through the throats, each of them of the area S_z. If an empty pore is in contact with a filled pore, a meniscus occurs in the throat. A pore in a porous medium can be filled only if the liquid can reach this pore. The condition that liquid can reach the pore can be provided by means of formation of an infinite cluster of filled pores. In this case only pores that are located on the shell of such a cluster will be filled. It could be shown that the contribution to the filled volume of a finite size cluster which liquid can reach through the filled pores contacting with the surface of the medium is small. At $\theta \sim \theta_c$ the distribution function $f(N)$ with respect to the number of pores in clusters N in finite size clusters near the percolation threshold is determined by scaling dependence $f(N) \sim \frac{1}{N^\tau}$, $\tau = 2.2$ [52]. It follows from this fact that most clusters contain one or several pores which generally are not connected with the porous medium surface. Liquid can not reach these pores, consequently, they can not be filled at $\theta \sim \theta_c$. Taking it into account we could write the work $\delta A(p)$ providing the filling of one spherical pore at the surface of an infinite cluster of filled pores as

$$\delta A_{in}(p, \theta, R) = -pV + \sigma S_m + \delta\sigma(S - S_m) = -p\frac{4\pi R^3}{3} + \delta\sigma(1-\eta)4\pi R^2 + \eta 4\pi R^2 \sigma W(\theta), \tag{2}$$

where V is a pore volume, σ is a liquid surface energy, $\delta\sigma$ is a difference between the surface energies of solid-liquid and solid-gas interfaces, $S = 4\pi R^2$ is a surface area of a pore, R is a pore radius, $S_m = zs_m$ is a total area of menisci per one pore, s_m is area of one meniscus, z is a number of the nearest neighbor pores, $\delta S_m = 4\pi R^2 W(\theta)$ is the change in the surface area of menisci during the filling of a pore with the radius R, $\eta = S_m/S$ is a connectivity coefficient equal to the ratio of surface area of menisci to the pore surface area, $W(\theta)$ is a difference between the number of menisci on the shell of the infinite cluster after and before the filling of the pore per one neighbor pore.

In order to describe the porous medium we use the model of randomly spaced spheres (RSS). Within this model pores represent randomly arranged spherical holes [53]. The RSS model does not take into account the correlations in location of pores of different radii in accordance with the assumption made above that the pore size distribution function is narrow $(\delta R) \ll \bar{R}$.

In accordance with [53] the area of throats which connect neighbour pores s_m and the average number of nearest neighbour pores $\bar{z}$ depend on porosity φ and can be written as

$$s_m = \frac{9\pi^2}{256} R^2,$$
$$z = \bar{z} = -8 \ln(1-\varphi). \tag{3}$$

These allow us to take into account the correlation effect of spatial location and connectivity of pores in the medium. This effect occurs as a result of probabilistic realization of interconnected pore system in the infinite cluster in the medium with the porosity φ. In the porous medium with the porosity $\varphi > \varphi_c$ the infinite cluster of filled pores is formed inside the infinite cluster of empty pores when the degree of occupancy approaches the value $\theta = \theta_c$. It follows from the condition for liquid to reach the cluster of pores that the value of θ increases only if pores are located on the shell of the cluster of filled pores.

Another correlation effect of positional relationship of filled and empty pores on a shell of a cluster of filled pores determines the probability of filling fluctuation. This effect is connected with changes of the surface of menisci during the filling of a pore with radius R.

To calculate $W(\theta)$ let us consider an empty pore located on the shell of the infinite cluster of filled pores. We assume that this pore is in contact with the infinite cluster of filled pores through n throats. In each of n throats the menisci are formed, whereas in other $z-n$ throats menisci are absent. After the filling of the pore under consideration these n menisci disappear and thus the number of menisci becomes equal to $z-n$. In this case $W(\theta)$ could be written as

$$W(\theta) = \sum_{n=1}^{z} (P(\theta))^n (1-\theta)^{z-n+1} \frac{z-2n}{z} \frac{z!}{n!(z-n)!}. \tag{4}$$

Here the first factor determines the probability that empty pore contacts n times with the infinite cluster of filled pores. The second factor corresponds to the probability that empty pore neighbour to the infinite cluster of filled pores if it is surrounded by z empty pores and thus it has $z-n$ throats. The third factor determines the difference between the relative number of menisci after $(z-n)$ and before (n) filling of the pore. Combinatorial factor accounts the number of variants of location of n menisci per the pores neighbouring to the pore under consideration. We have to mention that the expression for $W(\theta)$ obtained above is reduced to the one derived in [58], which coincides with the results of numerical calculations if the third factor is substituted by unity.

The sum in (4) can be performed analytically:

$$W(\theta) = (\theta^2 - 2\theta - P(\theta) + 1 + \theta P(\theta))(P(\theta) - \theta + 1)^{z-1} - (1-\theta)^{z+1}. \tag{5}$$

It follows from (1) and (5) that the work providing infiltration of nonwetting liquid into a single pore in a partly filled porous medium depends on porosity φ and occupancy degree θ of medium with liquid.

3. The description of infiltration process in the "nanoporous medium-nonwetting liquid" system

Consider disordered porous medium with the degree of occupancy θ_0. To fill this medium up to the degree of occupancy θ_1 one has to apply the pressure p_1. If $f(R)$ is the pore size distribution function normalized in accordance with equation $\int R^3 f(R) dR = 1$, then $\theta_1 = \int_0^\infty w(p_1, R, \theta_0) R^3 f(R) dR$. Increasing pressure in this step-by-step manner and taking into account that the probability w changes abruptly from 0 to 1 in a narrow pressure interval one can get the following expression for the occupancy of the porous medium at the n-th step

$$\theta_n = \int_{\frac{F(\theta_{n-1})}{p}}^{\infty} R^3 f(R) dR$$

$$F(\theta_{n-1}) = 3\delta\sigma\{(1-\eta) + \frac{\sigma}{\delta\sigma}\eta W(\theta_{n-1})\}. \quad (6)$$

Equation (6) is equivalent to the following differential equation

$$\frac{d\theta}{dp} = (\frac{F(\theta)}{p})^3 \frac{F(\theta)}{p^2} f(\frac{F(\theta)}{p}) \quad (7)$$

The solution of equation (7) allows us to determine the dependence of occupancy on pressure $\theta(p)$.

Comparison of experimental data [47, 48] and calculation based on the equation (7) for three different porous media: KSK-G(C16), PEP100(C18) and PEP300(C18) is presented in Fig. 1, 2 and 3. Parameters of nanoporous media are presented in Table 1.

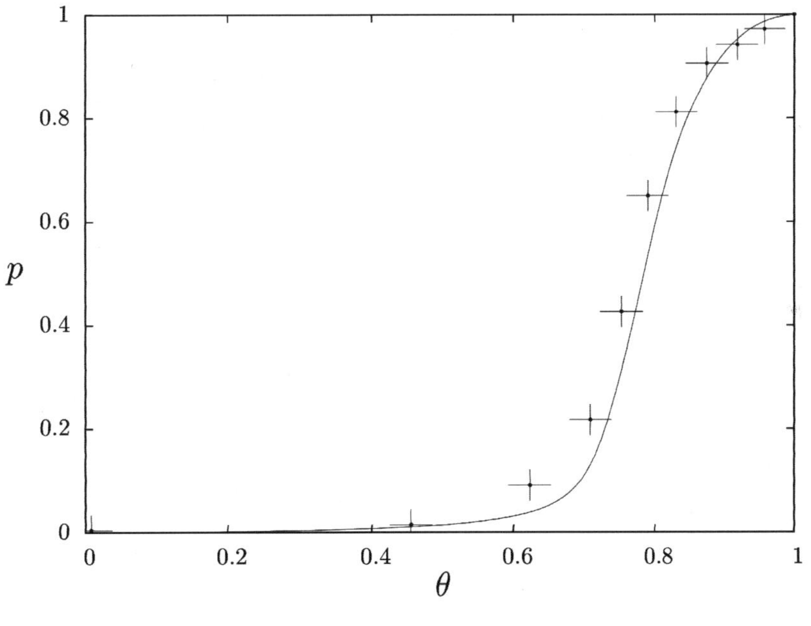

Figure 1. The dependence of occupancy on pressure for system KSK-G — water

Thus taking into account the correlation effects we can qualitatively describe the experimental data on infiltration of nonwetting liquid into the nanoporous medium beyond the percolation threshold. Figures 1-3 show that curves of infiltration depend on the porosity and as it increases, there is less and less deviation of theoretical curves from the experimental data at small degrees of occupancy ($\theta < 0.2$). The deviation of theoretical curves from the experimental data is a consequence of not taking into account in this model the filling of clusters of finite size. It should be mentioned that previously, in [8], curves were described not only qualitatively but

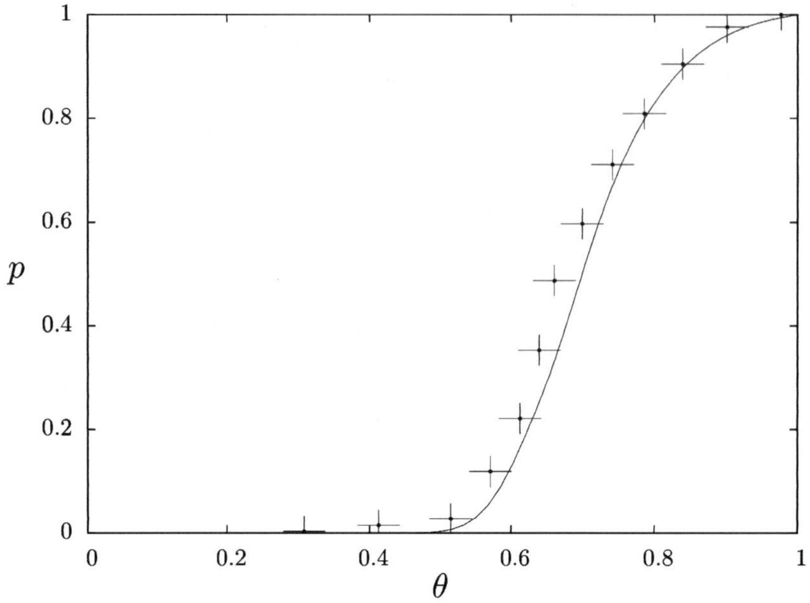

Figure 2. The dependence of occupancy on pressure for system PEP100 — water

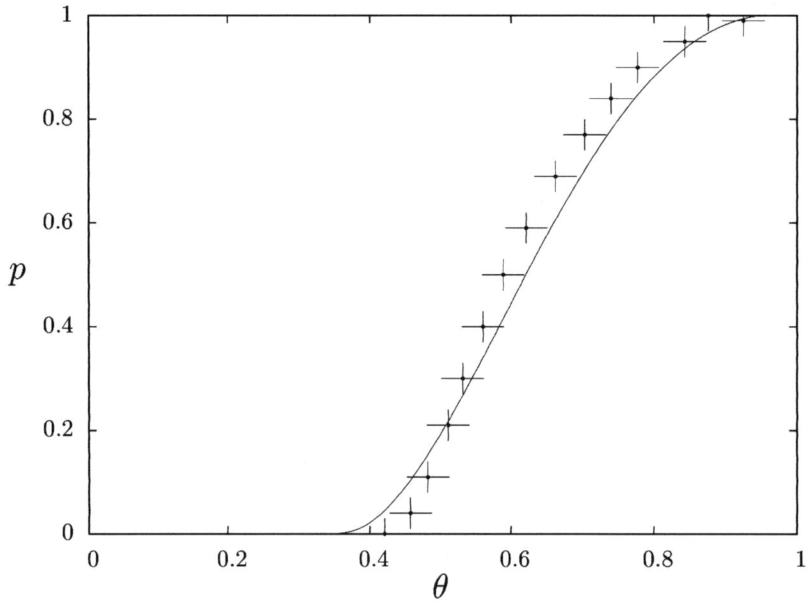

Figure 3. The dependence of occupancy on pressure for system PEP300 — water

also quantitatively using the distribution function of clusters of finite size. This function takes into account the formation of clusters both less and greater than the correlation length. The dependence of the deviation from porosity can be explained by the fact that with increasing of

Table 1. Parameters of nanoporous media

	Mean pore radius, R nm	Standard deviation, ΔR nm	The ratio of surface energies, $\sigma/\delta\sigma$	Porosity, φ
KSK-G	6.5	1.7	1.6	0.23
PEP100	5	2.2	2.4	0.35
PEP300	14.4	0.7	3.1	0.45

porosity contribution to the infiltration from the infinite cluster becomes dominant in comparison with the contribution from clusters of finite size.

Moreover, the function of pore size distribution probably contributes to the difference between the theoretical curves and the experimental data. It is because we used Gaussian distribution as the function of pore size distribution. Actually, the function of pore size distribution in range of large pores is protracted, because large pores are filled first.

4. The thermal effects

During the infiltration of nonwetting liquid into the porous medium the thermal effect ΔQ includes the thermal effect ΔQ_p that takes place due to formation of liquid-solid interface, the thermal effect ΔQ_w caused by the formation-disappearance of menisci and the thermal effect ΔQ_u related to the compressibility of a nonwetting liquid-nanoporous medium system, $\Delta Q = \Delta Q_p + \Delta Q_w + \Delta Q_u$.

The values ΔQ_p and ΔQ_w can be calculated using thermodynamic relation [60], that determines the thermal effect related to the formation of the surface ΔQ_s:

$$\Delta Q_s = -T\frac{d\sigma}{dT}\Delta S, \tag{8}$$

where ΔS is the change in the surface of a system.

To calculate the thermal effect ΔQ_s let us assume that each pore in the porous medium has z nearest neighbours and pores contact with each other by throats, each of them has an area S_z. If an empty pore has contact with a filled one, a meniscus is formed in the throat. Taking into account this fact we can write the thermal effect related to the filling of one pore in the porous medium δQ as

$$\delta Q = -T\frac{d\delta\sigma}{dT}(S - zS_z) - T\frac{d\sigma}{dT}zS_zW(\theta), \tag{9}$$

Considering that infiltration of a porous medium is a result of formation of the infinity cluster of filled pores and taking into account that the probability $P(\theta)$ is normalized, we obtain that the quantity of heat per one pore released during the filling of the porous medium up to the occupancy θ can be written as

$$\Delta Q_p(\theta) = -T\frac{d\delta\sigma}{dT}\int_0^\theta < (S - zS_z) > \theta P(\theta)d\theta,$$

$$< S - zS_z >= \int_0^\infty dR(S - zS_z)f(R), \tag{10}$$

$$\Delta Q_w(\theta) = -T\frac{d\sigma}{dT}\int_0^\theta < zS_z W(\theta) > d\theta$$

where $f(R)$ is the pore size distribution function normalized to unity. Within the RSS model the area of a throat and the average number of nearest neighbours $\bar{z}$ related to the porosity of the medium φ can be written in the form (3). Provided that $(\delta R) \ll \bar{R}$ one can get

$$\Delta Q_p(\theta) = -T\frac{d\delta\sigma}{dT}(1-\eta)4\pi\bar{R}^2\int_0^\theta \theta P(\theta)d\theta,$$

$$\Delta Q_w = -T\eta\frac{d\sigma}{dT}4\pi\bar{R}^2\int_0^\theta W(\theta)d\theta \tag{11}$$

It follows from (11) that the change in thermal energy of the system during the filling of porous medium is determined by the derivatives of specific surface energies $\frac{d\sigma}{dT}$ and $\frac{d\delta\sigma}{dT}$, by geometrical properties of the porous medium and by the evolution of the infinite cluster of filled pores which depends on the properties of the disordered porous medium.

To calculate the thermal effect ΔQ_v that takes place during the defiltration of the liquid out of the porous medium, one should note that the defiltration of the liquid leads to the formation of empty pores in the filled porous medium. They are surrounded by at least one filled pore connected through the other filled pores with the surface of the porous medium. As in case of infiltration the formation of an empty pore corresponds to the change in the energy of a liquid-solid and liquid-gas interfaces related to the formation-disappearance of menisci [8].

Taking it into account when the degree of occupancy decreases from 1 down to θ, the thermal effect during the defiltration of liquid from the porous medium can be written like (11) as

$$\Delta Q_p^v(\theta) = -T\frac{d\delta\sigma}{dT}(1-\eta)4\pi\bar{R}^2\int_0^\theta \theta d\theta,$$

$$\Delta Q_w^v = T\eta\frac{d\sigma}{dT}4\pi\bar{R}^2\int_0^\theta W_1(\theta)d\theta, \tag{12}$$

$$\Delta Q_v = \Delta Q_p^v + \Delta Q_w^v,$$

Relations (11) and (12) differ by the sign of the last term and the functions $W(\theta)$ and $W_1(\theta)$. These functions determine the difference per one nearest neighbor between the number of menisci after and before the infiltration (defiltration) in pores.

Relations (11) and (12) are valid if the process is isothermal. This implies that they can be used for the description of experiments when the characteristic time of heat supply (removal) τ_Q is much less than the characteristic time τ_V of the change in a volume of the nanoporous medium-nonwetting liquid system $\tau_Q \ll \tau_V$. At $\tau_Q \geq \tau_V$, the temperature and, correspondingly, the quantities $\frac{d\sigma}{dT}$, $\frac{d\delta\sigma}{dT}$ become time dependent and thus occupancy θ dependent. In this case they should be placed under the integral sign in equations (11), (12). The inequality $\tau_Q \ll \tau_V$ imposes constraints on the rate of compression of a system when the equilibrium properties of this system are investigated.

Equations (11), (12) allow to describe thermal effects that take place during the filling of porous medium by the nonwetting liquid. It follows from (11) that total thermal effect is determined by parameters ΔQ_p, ΔQ_w. These parameters depend on derivatives $d\sigma/dT$, $d\delta\sigma/dT$ and on integrals of $P(\theta)$ and $[W(\theta) + W_1(\theta)]$. The sign and the value of the thermal effect that occurs due to formation of liquid solid interface ΔQ_p depend on the sign and on the value of parameter $\frac{d\delta\sigma}{dT}$. The sign and on the value of parameter $\frac{d\delta\sigma}{dT}$ could be determined with the help of known dependencies of the threshold pressures of the beginning of the infiltration p_{in} and defiltration p_{out} on temperature.

Experimental data related to the thermal effect measured in three different systems water-silica gel KSK-G(C16) [48], PEP100(C18) and PEP300(C18) [47] are presented in Fig. 4,5. The porosity after the modification was estimated with the help of the dependencies of pressure on volume presented in [47,48]. It was shown that porosity of these systems has the following values: $\varphi = 0.23$ for KSK-G(C16), $\varphi = 0.35$ for PEP100(C18) and $\varphi = 0.45$ for PEP300(C18). Both differential and integral thermal effects were investigated in [48]. In [47] only integral effect was investigated and the experimental data obtained were corrected with account of compressibility of water. The author of the work [48] does not write anything about such correction of his experimental data. However, the description of measurement procedure presented in [61] allows one to suppose that correction of thermal effects related to compressibility of water was done.

It follows from Fig. 4,5 that the value of differential thermal effect (for KSK-G(C16), $\varphi = 0.23$) is negative at $\theta > 0.04$, decreases with occupancy θ, at $\theta \sim 0.8$ it approaches minimum and vanishes at $\theta = 1$. At $\theta < 0.04$ experimental points are systematically located in a region where ΔQ is positive. However, they are close to zero within the measurement error. Integral heat release for the system KSK-G(C16)-water at $\theta > 0.07$ is negative and monotonously decreases approaching minimum at $\theta = 1$. Similarly at $\theta < 0.07$ experimental points corresponding to integral effect are systematically located in a region where ΔQ is positive and close to zero within the measurements error. For systems PEP100(C18), $\varphi = 0.35$ and PEP300(C18), $\varphi = 0.45$ the integral thermal effect is negative and monotonously decreases with increase of occupancy.

Theoretical dependencies, presented in Fig. 4,5 were calculated with the help of equations (11), (4). It was supposed that the pressure in the beginning of infiltration is temperature independent similar to the situation with porous medium with modification C8. Relying on this condition we determined the value of derivative $d\delta\sigma/dT$. It appeared to be equal to $d\delta\sigma/dT = 0.1 \cdot 10^{-4}$ J/m^2 K for KSK-G(C16), $d\delta\sigma/dT = 0.35 \cdot 10^{-4}$ J/m^2 K for PEP300 (C18) and $d\delta\sigma/dT = 0.5 \cdot 10^{-4}$ J/m^2 K for PEP100(C18). One can see that calculated dependencies satisfactory coincide experimental dependencies for PEP100(C18) and PEP300(C18) (Fig. 5). It could be also seen the qualitative agreement of the calculated and experimental dependencies for a KSK-G(C16) (Fig. 4a, 4b). The account of thermal effect related to compressibility of water and partly filled porous medium leads to satisfactory coincidence of experimental and calculated dependencies for KSK-G(C16) (Fig. 4c, 4d). It is worth noting that within the developed model one should expect the occurrence of positive thermal effect (both integral and differential) at small values of occupancy ($\theta < 0.07$)(inserts in Fig. 4a, 4b). Experimental data obtained in [48] are located in this region within the measurements error. Because of that to determine the sign of thermal effect at small values of occupancy additional experiments are needed.

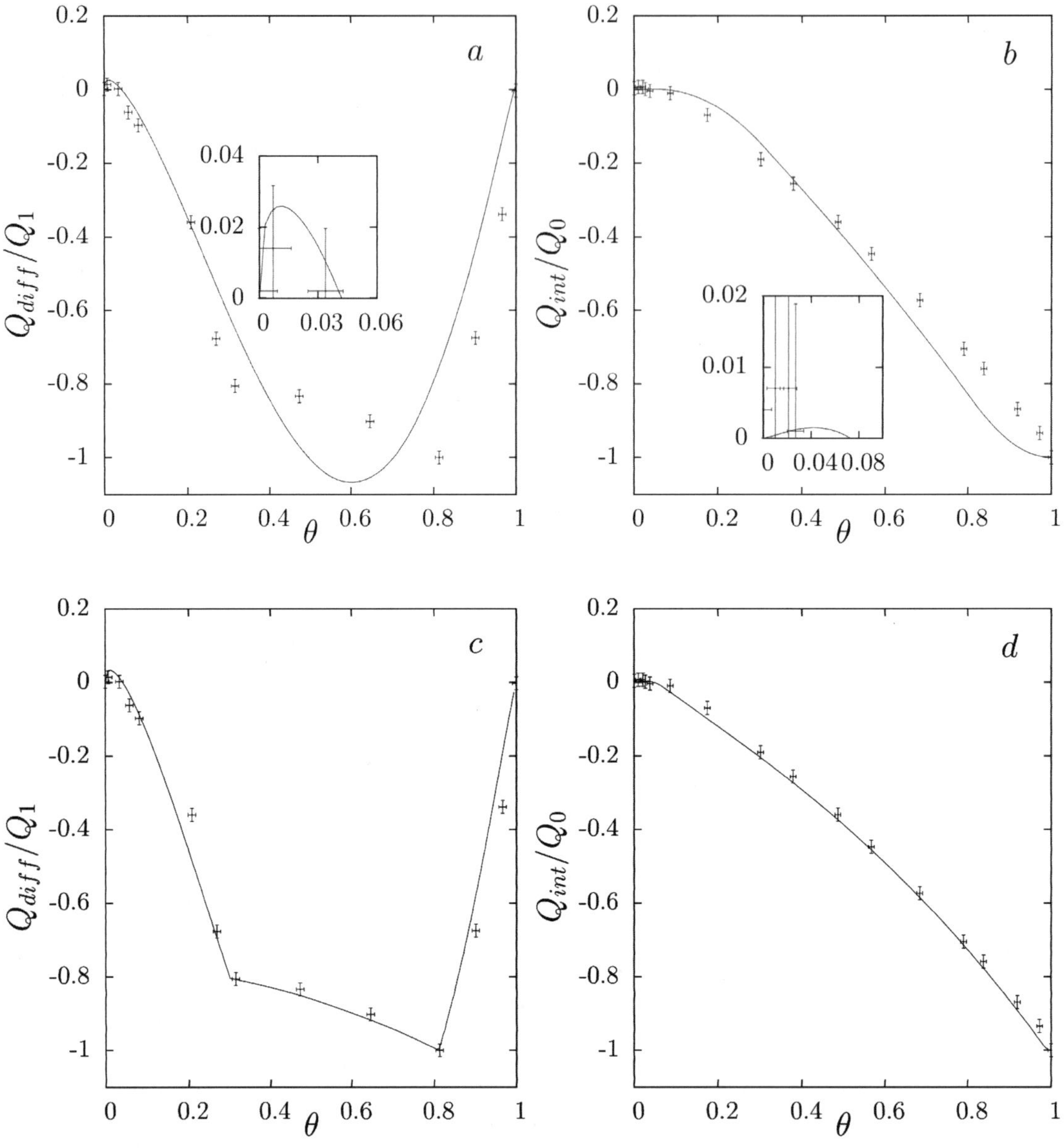

Figure 4. The occupancy dependence of integral thermal effect (b), (d) and differential thermal effect (a), (c). Points correspond to experimental data [48], solid lines correspond to calculation, (a), (b) — without the account of compressibility of water and porous medium during the infiltration, (c), (d) — with account of compressibility of water and porous medium during the infiltration. $Q_0 = 9$ J [48] is maximal absolute value of integral thermal effect, $q_1 = 40$ J/g [48] is maximal absolute value of differential thermal effect

5. Conclusion

In the present work we constructed the model of infiltration of nonwetting liquid into nanoporous medium that takes into account the correlation effects in the disordered medium and is based

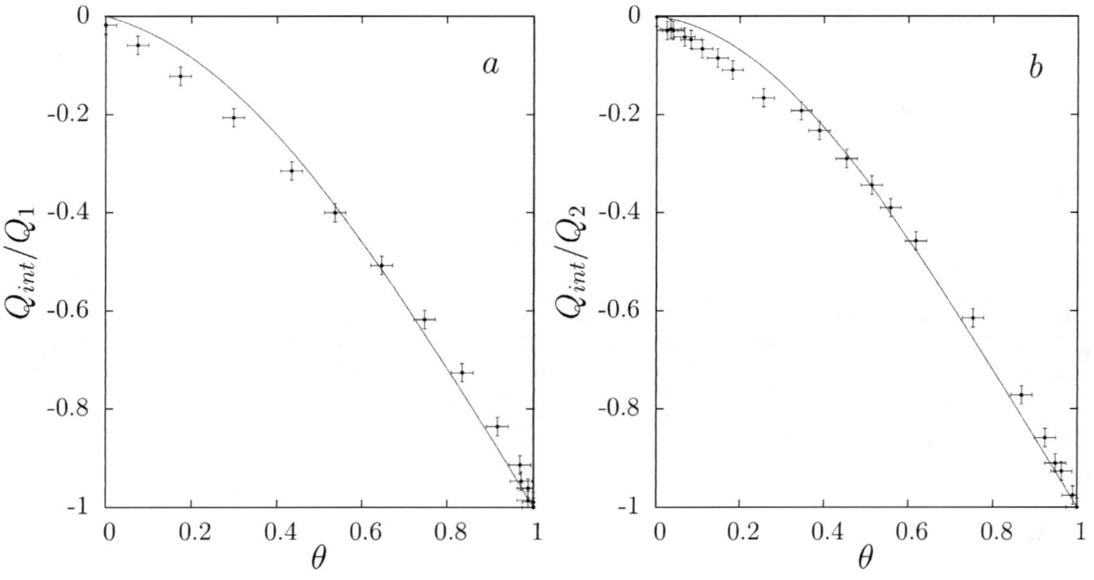

Figure 5. The occupancy dependence of integral thermal effect (per one gram of porous body) for PEP100(C18) (a) and PEP300(C18) (b). Points correspond to experimental data [47], solid lines correspond to calculation. $Q_1 = 5.5$ J/g [47], $Q_2 = 1.6$ J/g [47] are maximal absolute values of integral thermal effect

on analytic methods of percolation theory. This model allows one to find a linkage between the effective parameters of porous medium such as an average number of neighbour pores and an average size of throats and macroscopic characteristics of porous medium such as porosity and specific surface.

Using the proposed model we can describe both the process of filling of nanoporous medium by nonwetting liquid and thermal effects accompanying the energy absorption by systems with different porosity. This allowed us to remove the contradictions in the available experimental data on thermal effects during infiltration of water into modified porous media KSK-G(C16), PEP300 (C18) and PEP100(C18).

Acknowledgments

This work was supported by the Federal target program "Scientific and scientific-educational personnel of innovative Russia" under grant P536.

References

[1] Qiao Y, Punyamurtula V K, Han A, Kong X and Surani F B 2006 *Appl. Phys. Lett.* **89** 251905
[2] Han A and Qiao Y 2007 *Appl. Phys. Lett.* **91** 173123
[3] Han A and Qiao Y 2008 *Chem. Phys. Lett.* **454** 294
[4] Bogomolov V N 1995 *Phys. Rev. B* **51** 17040
[5] Erochenko V A 2002 *Rus.Chem.Journal* **XLVI** 31
[6] Fadeev U and Erochenko V A 1997 *J. of Coll. and Interface Sci.* **187** 275
[7] Borman V D, Grekhov A M and Troyan V I 2000 *JETP* **91** 170
[8] Borman V D, Belogorlov A A, Grekhov A M, Lisichkin G V, Tronin V N and Troyan V I 2005 *JETP* **100** 385

[9] Lefevre B, Saugey A, Barrat J L, Boequet L, Charlaix E, Gobin P F and Vigier G 2004 *Coll. and Surf. A* **241** 265–272
[10] Lefevre B, Gobin P F, Martin T, Saugey A, Barrat J L, Bocquet L, Charlaix L and Vigier G 2004 *J. Chem. Phys.* **120** 4927
[11] Iwatsubo T, Suciu C V, Ikenagao M and Yaguchio K 2007 *Journal of Sound and Vibration* **308** 579
[12] Chen X, Surani F B, Kong X, Punyamurtula V K and Qiao Y 2006 *Appl. Phys. Lett.* **89** 241918
[13] Erochenko V A, Piatiletov I, Coiffard L and Stoudenets V 2007 *Proc.I MechE, Part D* **221** 301–312
[14] Sebastian I and Halasz I 1974 *Chromotographia* **7** 371
[15] Unger K K 1979 *Porous silica, its properties and use as suppert in column liquid chromatography. (J.Chromotogr.Libr. V.16)* (Elsevier, Amsterdam)
[16] Bokasanyi L, Liardon O and Kovats E 1976 *Adv.Colloid Interface Sci.* **6** 95
[17] Lisichkin G V and Fadeev A Y 1996 *Rus.Chem.Journal* **XL** 65
[18] Fadeev U and Erochenko V A 1996 *J. Phys.Chem.* **70** 1482
[19] Fadeev U and Erochenko V A 1995 *Colloid. Journal.* **57** 480
[20] Suciu C V, Iwatsubo T and Deki S 2003 *J. of Coll. and Interface Sci.* **259** 62
[21] Suciu C V, Iwatsubo T, K Y and M I 2005 *J. of Coll. and Interface Sci.* **283** 196
[22] Coiffard L and Eroshenko V A 2006 *J. of Coll. and Interface Sci.* **300** 304–309
[23] Denoyel R, Beurroies I and Lefevre B 2004 *J. of Petroleum Science and Engineering* **45** 203
[24] Han A, Kong X and Qiao Y 2006 *J. Appl. Phys.* **100** 014308
[25] Kong X and Qiao Y 2005 *Philosophical Magazine Letters* **85** 331
[26] Qiao Y, Cao G and Chen X 2007 *J. Am. Chem. Soc.* **129** 2355
[27] Kong X, Surani F B and Qiao Y 2005 *J. Mater. Res.* **20** 1042
[28] Kong X and Qiao Y 2005 *Appl. Phys. Lett.* **86** 151919
[29] Surani F and Qiao Y 2006 *J. Appl. Phys.* **100** 034311
[30] Surani F, Qiao Y and Kong X 2005 *Appl. Phys. Lett.* **87** 251906
[31] Surani F B, Han A and Qiao Y 2006 *Appl. Phys. Lett.* **89** 093108
[32] Han A, Lu W, Punyamurtula V K, Kim T and Qiao Y 2009 *J. Appl. Phys.* **105** 024309
[33] Kim T, Han A and Qiao Y 2008 *J. Appl. Phys.* **104** 034304
[34] Han A, Punyamurtula V K and Qiao Y 2008 *Appl. Phys. Lett.* **92** 153117
[35] Eroshenko V A, Regis R Cand Soulard M and Patarin J 2001 *J. Am. Chem. Soc.* **123** 8129–8130
[36] Eroshenko V A, Regis R Cand Soulard M and Patarin J 2002 *C.R. Physique* **3** 111–119
[37] Han A, Lu W, Chen X and Kim T 2008 *Phys. Rev. E.* **78** 031408
[38] Lu W, Han A, Kim T, Punyamrtula V K, Chen X and Qiao Y 2009 *App. Phys. Lett.* **94** 023106
[39] Qiao Y, Liu L and Chen X 2009 *Nanolett.* **9** 984–988
[40] Liu L, Chen X, Lu W, Han A and Qiao Y 2009 *Phys. Rev.Lett.* **1-2** 184501
[41] Kim T, Lu W, Han A, Punyamrtula V K and Qiao Y 2009 *App. Phys. Lett.* **94** 013105
[42] Chen X, Cao G, Han A, Punyamurtula V K, Liu L, Culligan P, Kim T and Qiao Y 2009 *Nanolett.* **8** 2988–2992
[43] Liu L, Chen X and Qiao Y 2008 *App. Phys.Lett.* **92** 101927
[44] Qiao Y, Punyamurtula V K, Xian G, Karbhari V M and Han A 2008 *Appl. Phys. Lett.* **92** 063109
[45] Han A, Lu W, Punyamurtula V K and Qiao Y 2008 *App. Phys. Lett.* **103** 084318
[46] Han A, Lu W, Punyamurtula V K, Chen X, Surani F B, Kim T and Qiao Y 2008 *J. Appl. Phys.* **104** 124908
[47] Gomez F, Denoyel R and Rouquerol J 2000 *Langmuir* **16** 4374–4379
[48] Gusev V Y 1994 *Langmuir* **10** 235–240
[49] Surani F, Kong X, Panchal D and Qiao Y 2005 *Appl. Phys. Lett.* **87** 163111
[50] Borman V D, Belogorlov A A, Lisichkin G V, Tronin V N and Troyan V I 2009 *JETP* **108** 389–410
[51] Sashimi M 1993 *Rev.Mod. Phys.* **65** 1393
[52] Isichenko M B 1992 *Rev. of Mod. Phys* **64** 961
[53] Heifetz L I and Neimark A V 1982 *Multiphase processes in porous media [Russian translation]* (Chemistry,Moscow)
[54] Woo H J and Monson P A 2003 *Phys.Rev. E* **67** 041207
[55] Surani F 2006 Ph.D. thesis Science University of Akron
[56] Porcheron F and Monson P A 2005 *Langmuir* **21** 3179
[57] Porcheron F, Thommes M and Ahmad R 2007 *Langmuir* **23** 3372–3380
[58] Grinchuk P S and Rabinovich O S 2003 *JETP* **96** 301
[59] Abrikosov A A 1979 *ZhETF* **29** 72
[60] Landau L D and Lifschitz E M 1980 *Statistical Physics, Part 1, 3rd ed* (Oxford, England: Pergamon Press)
[61] Gusev V Y and Fomkin A A 1994 *J. of Coll. and Interface Sci.* **21** 4567

SYNTHESIS FUNCTIONAL AND CONSTRUCTIONAL NANOMATERIALS ON A BASIS CARBIDE TUNGSTEN, MOLYBDENUM AND METALS OF A TRIAD OF IRON IN IONIC MELTS

H. B. Kushkhov, M. N. Adamokova, V. A. Kvashin, and A. L. Kardanov
Kabardino-Balkar State University, ul. Chernyshevskogo 173,Nalchik, 360004 Russia
e-mail: hasbikushchov@yahoo.com

Abstract-Single and cyclic voltammetry is used to study the electrode processes that occur during electrochemical synthesis of hard-alloy compositions based on tungsten carbide and an iron triad metal in tungstate and tungstate-carbonate Na_2WO_4–Li_2WO_4–Li_2CO_3 (5.0-22.0 wt %) melts. The conditions of bringing the electroprecipitation potentials of tungsten, carbon, and an iron triad metal into coincidence are determined.

INTRODUCTION

Methods for producing fine-grained alloys in a nanocrystalline state, where the particle sizes do not exceed 40 nm, are being actively developed [1–4]. The most promising method of forming nanostructured materials with a low sintering temperature and high strength properties (due to a fine-grained structure) is the use of nano- and submicrocrystalline carbides as alloy components [5, 6]. Plasmachemical synthesis is one of the most widely used chemical methods for producing highly dispersed nitride and carbide powders [7, 8]. The synthesis of a fine-grained tungsten carbide powder using an arc plasma generator was described in [9]. The authors of [10] considered a low-temperature gas-phase synthesis of a WC nanopowder; this is a twostep process, which includes tungsten reduction in the form of a powder from tungstic acid salts and subsequent carbonization of tungsten in a CO-containing gas mixture. A simpler one-step process of synthesizing a nanocrystalline disperse WC carbide was described in [11]. The modern technologies of producing nanocomposite WC–Co mixtures include the following sequential operations: precipitation from colloidal solutions, cryogenic spraying and drying, and carbothermic reduction. For example, an effective method for producing nanocrystalline tungsten carbide and cobalt compositions used to manufacture hard alloys is based on the coprecipitation of WC and Co precursors from colloidal solutions [12, 13]. This technology was patented as a spray conversion process (SCP) [14]. An almost ideal version of a mechanochemical synthesis, which is combined with the formation of a nanocomposite mixture, was descried in [15]. An analysis of available data demonstrates that nanostructured hard-alloy WC–Co compositions can be synthesized by various methods and that the most efficient methods are based on the use of a WC–Co nanomixture rather than only nanostructured tungsten carbide.

One of these promising methods is high-temperature electrochemical synthesis of hard-alloy compositions based on tungsten carbide and an iron triad metal. At present, high-temperature electrochemical synthesis of tungsten carbides is performed in the following two fields: in the first field, a carbonate ion is used as a carbon source; in the second field, carbon dioxide is used. The contribution of the Ukrainian school of Shapoval and Kushkhov to the development of electrochemical synthesis of carbides should be emphasized in designing more efficient electrolytes for the manufacture of carbides having no hard-to-dissolve borates and fluorides of alkali metals [16–19]. High-temperature electrochemical synthesis is based on the multielectron electrochemical reactions of precipitation of hard-alloy components from an ionic melt. A necessary synthesis condition consists in bringing the electroprecipitation potentials of tungsten, carbon, and an iron triad metal into coincidence. The purpose of this work is to study the electrochemical behavior of a Na_2WO_4–Li_2WO_4–MWO_4 (M = Ni, Co, Fe) melt and the possibility of combined electroprecipitation of an iron triad metal, tungsten, and carbon from oxide melts.

EXPERIMENTAL

We studied the I–V characteristics and the combined electroprecipitation of tungsten, carbon, and an iron triad metal by single and cyclic voltammetry using a PARSTAT 2273 potentiostat–galvanostat with a computer control system. As base electrolytes, we used tungstate Na_2WO_4 and tungstate–carbonate Na_2WO_4–Li_2WO_4–Li_2CO_3 (5.0–22.0 wt %) melts. Individual salts were preliminarily resolidified and then melted in an atmosphere of purified and dried argon. The sodium and lithium tungstates used as a base electrolyte had special purity grade and were preliminarily dried at a temperature of 573 K. The quality of the base electrolyte was determined by measuring the residual polarographic currents. The experiments were carried out in a three-electrode cell. As an indicator electrode, we applied a half submerged needle platinum electrode; as an anode, we used a platinum plate. A quasi-reversible platinum electrode served as a reference electrode. X-ray diffraction analysis of cathode precipitates was performed on a DRON-6 X-ray diffractometer using CuK_α radiation at a wavelength of 1.54051 A. X-ray diffraction patterns were recorded in the angular range $\Theta = 10°–90°$ at a step of 2° per minute with a diffraction angle measurement accuracy of 0.005°. To refine the lattice parameters, we repeatedly recorded some diffraction lines at a rate of 0.5°/min. The exposure time was 1 s. An elemental analysis of cathode precipitates was performed on a MAKS-GV X-ray fluorescence elemental analyzer. To determine the free carbon content in samples, we used a Multi EA 2000 CS carbon analyzer.

Journal of Physics: Conference Series **291** (2011) 012045 doi:10.1088/1742-6596/291/1/012045

RESULTS AND DISCUSSION

Figure 1a shows the I–V characteristic of a Na_2WO_4–Li_2WO_4–$NiWO_4$ melt. The introduction of the first nickel tungstate portions into the base electrolyte of sodium tungstate leads to the appearance of a small peak of a nickel ion reduction wave at a potential of –(0.5–0.6) V with respect to the platinum–oxygen reference electrode (Fig. 1a, curves 2–7). As is seen from Fig. 2 (curve 1), an increase in the nickel tungstate concentration leads to a linear increase in the current density to a concentration of 3.0×10^{-4} mol/cm^3. As the nickel tungstate concentration increases, the concentration dependence of the current density of the $NiWO_4$ electroreduction peak ceases to be linear, which is explained by reaching the limiting nickel tungstate solubility in sodium tungstate and reaching the equilibrium

$$NiWO_4 + Na_2WO_4 \leftrightarrow NiO + Na_2W_2O_7. \quad (1)$$

As follows from the studies of the solubility of nickel tungstate in sodium tungstate, the maximum nickel tungstate concentration achievable in the given melt is about 3.0 $\times 10^{-4}$ mol/cm^3, which corresponds to 2.5-3.0 wt %. X-ray diffraction analysis of the cathode precipitates of potentiostatic electrolysis at the potential of the end of the Ni ion electroreduction wave (–0.7 V) demonstrates the presence of only metallic nickel (Fig. 3a). We then introduced lithium tungstate into the melt to study electrochemically active W particles, namely, ions cationized by lithium ions. As is seen from Fig. 1b (curves 5, 6), the introduction of lithium tungstate leads to the appearance of a W electroreduction wave at a potential of –(1.6–1.8) V with respect to the platinum–oxygen reference electrode. The difference in the nickel and tungsten precipitation potentials in this melt is about 1.0 V. An increase in the lithium tungstate concentration to 40 mol % does not decrease the difference between the Ni and W precipitation potentials; however, an apparent shift in the entire I–V characteristic toward a positive potential region is observed, which is caused by a change in the potential of the platinum–oxygen reference electrode. X-ray diffraction analysis of the cathode precipitates formed at a potential of –(1.6–1.8) V reveals the presence of metallic Ni, W, and NiW alloy (Fig. 3b). Similarly to the study of the combined electroreduction of tungsten and nickel ions, we should first determine the solubility of cobalt tungstate in sodium tungstate in order to analyze the mechanism of combined W and Co electroprecipitation. To determine the solubility of cobalt tungstate in sodium tungstate, we also used cyclic voltammetry. The dependence of the current density of the $CoWO_4$ electroreduction on the $CoWO_4$ concentration is linear up to a concentration of $3.0 \, . \, 10^{-4}$ mol/cm^3 (2.5–3.0 wt %; see Fig. 2, curve 2). The electroreduction potential of cobalt ions against the background of sodium tungstate is –(0.5–0.8) V. As is seen from the I–V characteristics (Fig. 1c), the mechanism of cobalt electroreduction against the background of sodium tungstate is irreversible, which is indicated by the shape of the I–V characteristics (which are extended along the potential axis) and by the difference between the cathode and anode peak potentials (200–300 mV). X-ray diffraction analysis of the products of potentiostatic electrolysis at a potential of –0.7 V demonstrates the presence of metallic cobalt (Fig. 3c). An addition of lithium tungstate leads to the appearance of a tungsten ion reduction wave as a result of the formation of electrically active particles at a potential of –(1.5–1.7) V with respect to the platinum–oxygen reference electrode. Thus, excess lithium ions lead to both a shift in the I–V characteristics

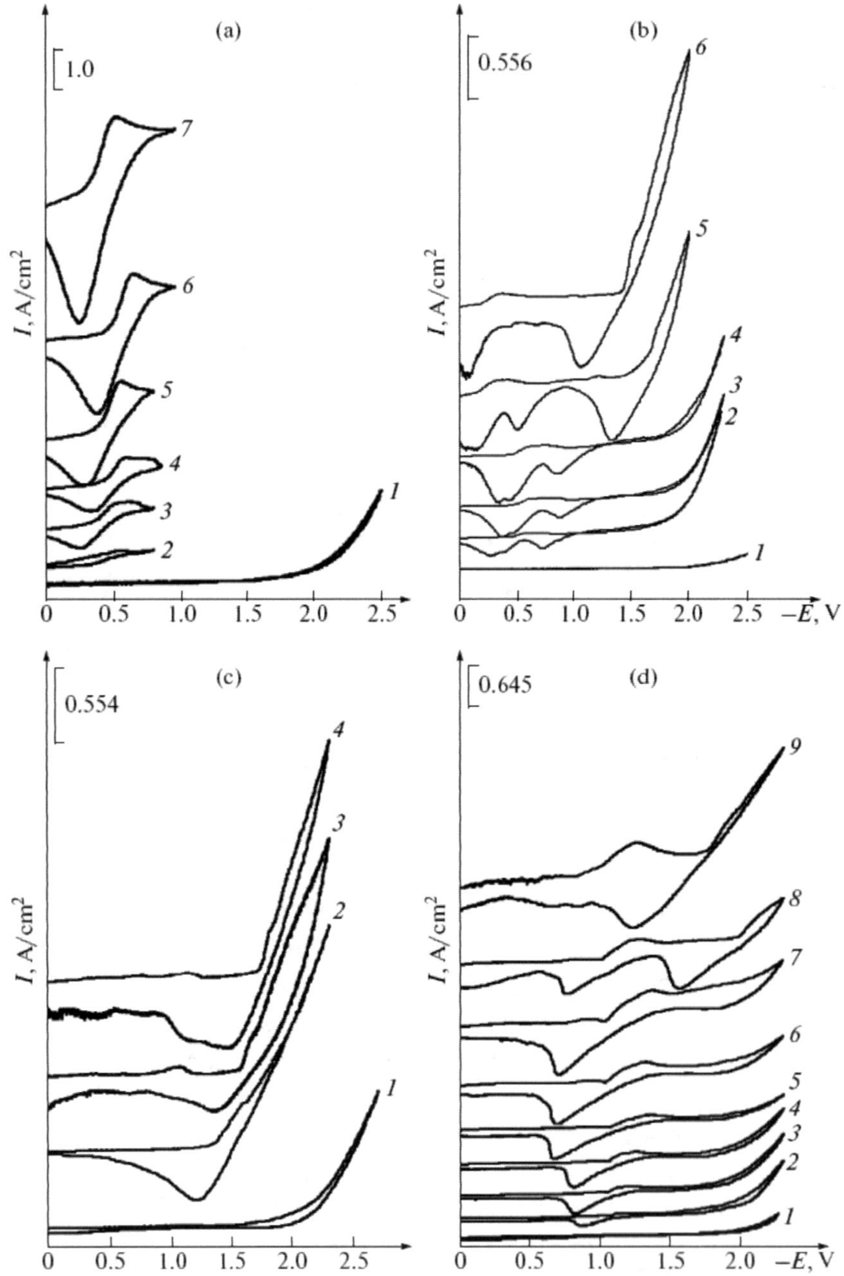

Fig. 1. I–V characteristics of (a) Na_2WO_4–Li_2WO_4 melt in which $NiWO_4$ is gradually added and a Na_2WO_4 melt containing (b) $NiWO_4$ and Li_2WO_4, (c) $CoWO_4$ and Li_2WO_4, and (d) $FeWO_4$ and Li_2WO_4. Pt cathode, the polarization rate is 0.1 V/s, T = 1173 K. (a) C($NiWO_4$) . 10^{-4}, mol/cm^3: (1) 0, (2) 0.5, (3) 1.5, (4) 2.0, (5) 3.0, (6) 3.5, and (7) 4.0. (b) C($NiWO_4$) . 10^{-4}, mol/cm^3: (1) base curve, (2) 2.0, (3) 3.0, and (4–6) 4.0; C(Li_2WO_4), mol/cm3: (1–4) 0, (5) 20.0, and (6) 40.0. (c) C($CoWO_4$) . 10^{-4}, mol/cm^3: (1) base curve, (2) 0, (3) 1.0, and (4) 2.0; C(Li_2WO_4), mol/cm^3: (1) 0 and (2–4) 40.0. (d) C($FeWO_4$) . 10^{-4}, mol/cm^3: (1) base curve, (2) 1.0, (3) 2.0, (4) 3.0, (5) 4.0, (6) 6.0, and (7) 8.0; C(Li_2WO_4), mol/cm^3: (1–7) 0, (8) 20.0, and (9) 40.0.

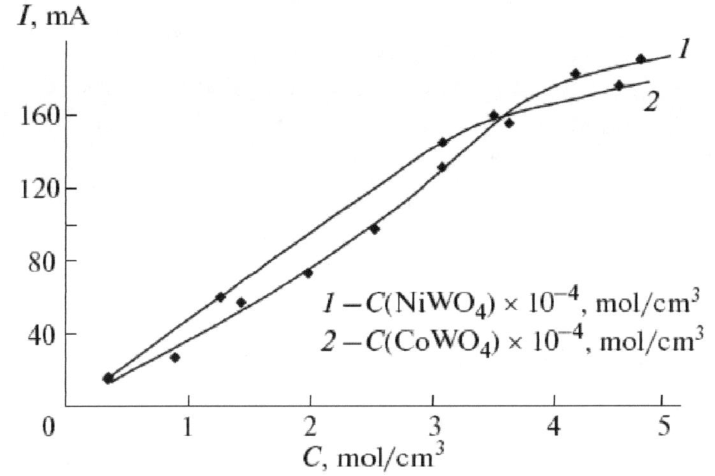

Fig. 2. Electroreduction peak current of (1) $NiWO_4$ or (2) $CoWO_4$ vs. the $NiWO_4$ or $CoWO_4$ concentration, respectively.

toward a positive potential range with respect to the base curve and a small (~100 mV) shift in the tungsten and cobalt precipitation potentials toward each other (Fig. 1c). The tungsten reduction current height increases with the lithium ion concentration, which is related to an increase in the concentration of electrically active particles, namely, tungsten ions cationized by lithium ions $[LixWO_4]^{x-2}$ (where $x \geq 2$). X-ray diffraction analysis of the cathode precipitates formed at a potential of –(1.5–1.7) V demon strates the presence of metallic W and Co_7W_6 and Co_3W intermetallics (Fig. 3d). The cyclic voltammetry study of the solubility of iron tungstate in a sodium tungstate melt demon strates a higher solubility of iron tungstate in this melt (up to 1.0 . 10^{-3} mol/cm^3, or up to 8 wt %) as compared to nickel or cobalt tungstate (Fig. 5a). The iron ion electroreduction potential in a sodium tungstate melt is –(1.1–1.2) V with respect to the platinum–oxygen reference electrode. However, as the $FeWO_4$ concentration increases above 3.0×10^{-4} mol/cm^3, the second wave appears in I–V characteristics at a potential of –(1.2–1.4) V. As follows from the electroreduction potential of this peak, it can belong to electrically active $W_2O_7^{2-}$ particles (Fig. 5a, curves 6–8). The effect of excess lithium tungstate on the tungsten electroreduction processes in this melt is similarto those in the electrolytic systems with nickel and cobalt tungstates: the I–V characteristics shift toward a positive potential range by 250–300 mV, and the electroreduction potentials of iron ions and electrically active tungsten particles move toward each other (Fig. 1d; curves 8, 9). The dependence of the current density of the $FeWO_4$ electroreduction peak on the $FeWO_4$ concentration is linear up to a concentration of 3.0×10^{-4} mol/cm^3 (2.5–3.0 wt %). X-ray diffraction analysis of the cathode precipitates formed at a potential of –(1.5–1.7) V demonstrates the presence of metallic W and Fe. Based on an analysis of the I–V characteristics of a tungstate melt containing the tungstate of an iron triad metal, we assume that the equilibrium in the reactions

$$Na_2WO_4 + NiWO_4 \leftrightarrow Na_2W_2O_7 + NiO, \quad (2)$$
$$Na_2WO_4 + CoWO_4 \leftrightarrow Na_2W_2O_7 + CoO, \quad (3)$$
$$Na_2WO_4 + FeWO_4 \leftrightarrow Na_2W_2O_7 + FeO \quad (4)$$

shifts toward the formation of ditungstate $W_2O_7^{2-}$ ions when going from nickel tungstate to iron tungstate. $W_2O_7^{2-}$ ions are reduced at potentials close to the reduction potential of iron ions (–(1.5–1.7) V), which explains the appearance of metallic tungsten in the cathode product during the electrolysis of a Na_2WO_4-$FeWO_4$ melt, in contrast to the Na_2WO_4-$NiWO_4$ and Na_2WO_4–$CoWO_4$ systems. As shown above, the cathode precipitate in the last case consists of metallic Ni and Co. Since the tungstates of the iron triad metals are relatively expensive and their synthesis is difficult, it is interesting to study the electrochemical behavior of the chlorides of Fe, Co, and Ni, which are available and cheaper as compared to their tungstates, in a tungstate melt. The results obtained demonstrate that the I–V characteristics of the chlorides and tungstates of Fe, Co, and Ni in a tungstate melt are similar, which is seen from the I–V characteristics shown in Figs. 4a, 4b, and 5b. Based on these data, we applied the chlorides of Fe, Co, and Ni for electrochemical synthesis of hard-alloy compositions based on tungsten carbide and an iron triad metal. Using the results of studying combined electroreduction of tungsten, carbon, and iron (cobalt, nickel) ions, we tried to perform electrochemical synthesis of double W–Fe(Co, Ni)–C carbides in molten systems (mol %) 51.0 Na_2WO_4–34.0 Li_2WO_4–15.0Li_2CO_3–MCl_2 (1.5×10^{-4} mol/cm^3), 48.9 Na_2WO_4–32.6 Li_2WO_4–18.5 Li_2CO_3–MCl_2 (1.5×10^{-4} mol/cm^3), and 46.8 Na_2WO_4–31.2 Li_2WO_4–22.0 Li_2CO_3–MCl_2 (1.5×10^{-4} mol/cm^3). Electrolysis was carried out in a potentiostatic mode at low (1.0–2.0 A/cm^2), growing (1.0–8.0 A/cm^2), and high (8.0–12.0 A/cm^2) current densities. Table 1 presents the results of X-ray diffraction analysis and chemical analysis performed to determine the free carbon content. It follows from the data in Table 1 that 15.0 mol % is a sufficient Li_2CO_3 concentration in the 51.0 Na_2WO_4–34.0 Li_2WO_4–Li_2CO_3–MCl_2 (1.5×10^{-4} mol/cm^3)

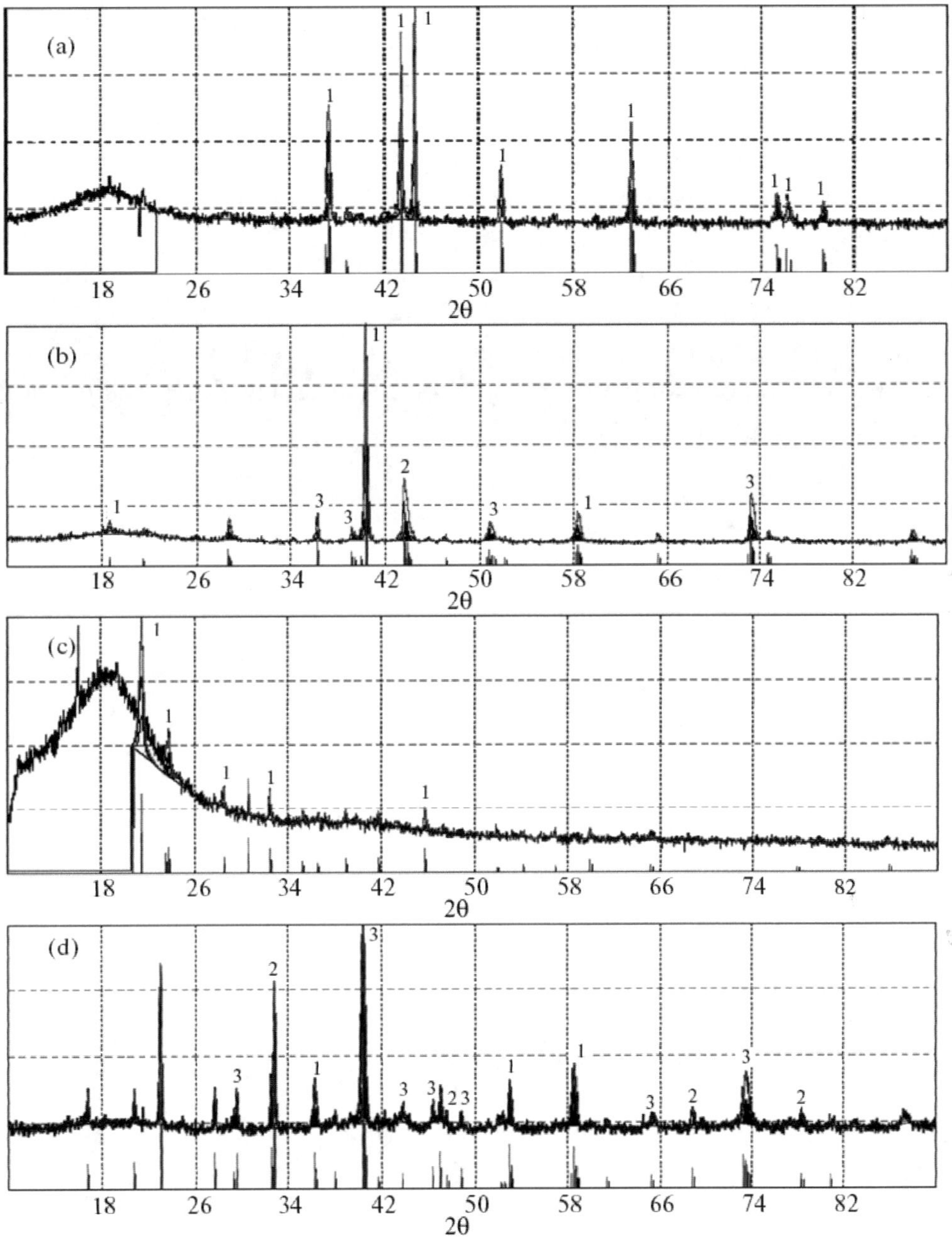

Fig. 3. X-ray diffraction patterns of the cathode precipitate formed upon potentiostatic electrolysis from a (a, b) Na_2WO_4–$NiWO_4$ or (c, d) Na_2WO_4–$CoWO_4$ melt. The cathode is made of a platinum plate, S = 2.7 cm^2, T = 1173 K, and the potential is –0.7 V. Diffraction lines (a) (1) Ni; (b) (1) W, (2) Ni, and (3) NiW; (c) (1) Co; and (d) (1) W, (2) Co_7W_6, and (3) Co_3W.

(M = Fe, Co, Ni) system at a current density of 1.0-2.0 A/cm^2, 18.5 mol % is a sufficient Li_2CO_3 concentration in the 48.9 Na_2WO_4–32.6 Li_2WO_4–Li_2CO_3–MCl_2

(1.5×10^{-4} mol/cm^3) system at a current density of 1.0–8.0 A/cm^2, and 22.0 mol % is a sufficient Li_2CO_3 concentration in the 46.8 Na_2WO_4–31.2 Li_2WO_4–Li_2CO_3–MCl_2 (1.5×10^{-4} mol/cm^3) system at a current density of 8.0–12.0 A/cm^2. These results are supported by X-ray diffraction analysis (DRON-6 diffractometer) and an analysis for a total carbon content (multi EA 2000 CS). Using an elemental fluorescence Spektroscan MAKS-GV analyzer, we analyzed the W and M (M=Fe, Co, Ni) contents in binary tungsten and iron triad metal carbides in the samples produced by potentiostatic electrolysis in molten systems (mol %) 51.0 Na_2WO_4–34.0 Li_2WO_4–15.0 Li_2CO_3–MCl_2 (1.5×10^{-4} mol/cm^3), 48.9 Na_2WO_4–32.6 Li_2WO_4–18.5 Li_2CO_3–MCl_2 (1.5×10^{-4} mol/cm^3), and 46.8 Na_2WO_4–31.2 Li_2WO_4–22.0 Li_2CO_3–MCl_2 (1.5×10^{-4} mol/cm^3). The results of electron-probe fluorescence analysis of the binary tungsten carbides are given in Table 2.

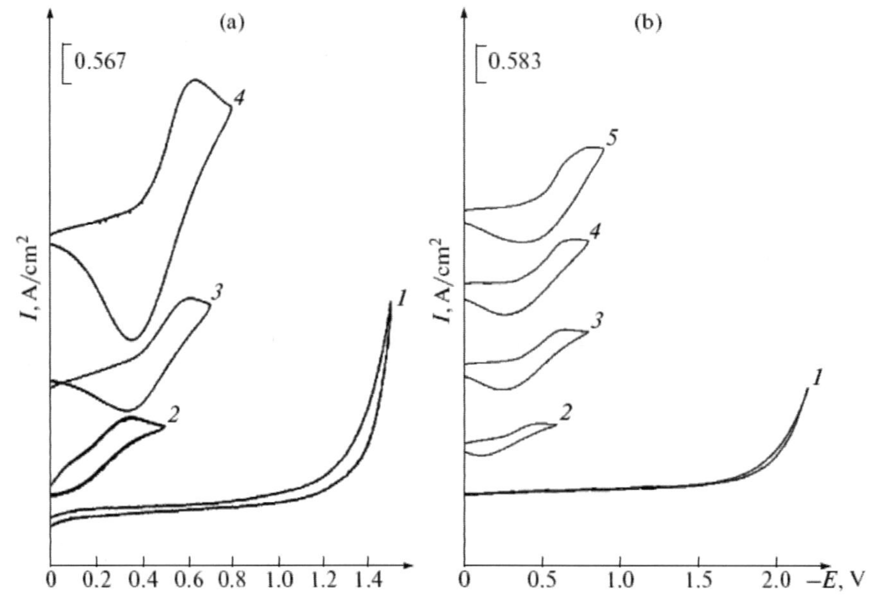

Fig. 4. I–V characteristics of a Na_2WO_4 melt containing (a) $NiCl_2$ or (b) $CoCl_2$. Pt cathode, polarization rate is 0.1 V/s, and T =1173 K. (a) C($NiCl_2$) . 10^{-4}, mol/cm^3: (1) base curve, (2) 1.0, (3) 2.0, and (4) 4.0. (b) C($CoCl_2$) . 10^{-4}, mol/cm^3: (1) base curve, (2) 2.0, (3) 4.0, and (4) 6.0, and (5) 8.0.

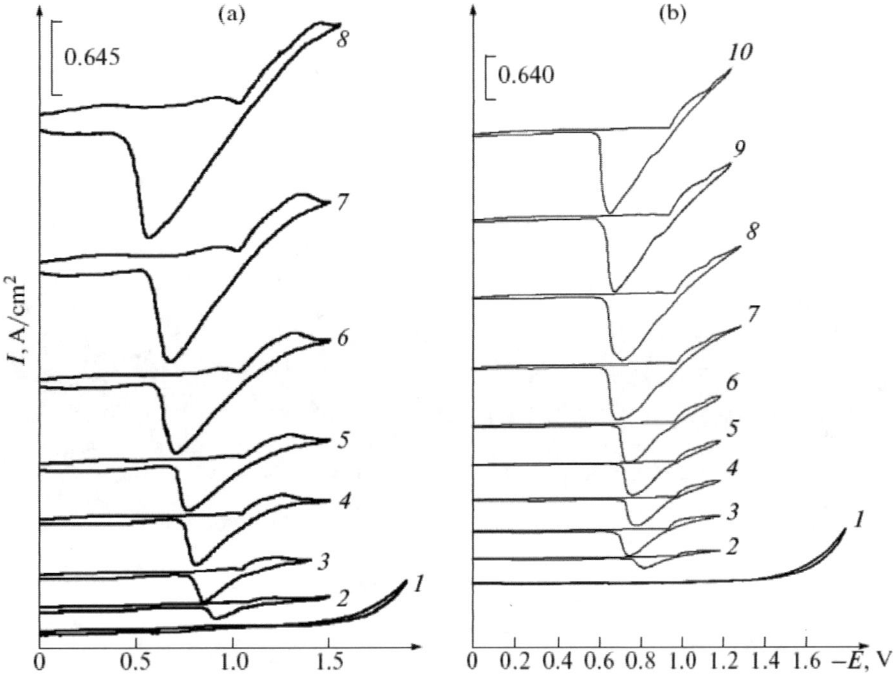

Fig. 5. I–V characteristics of a Na_2WO_4 melt (a) added gradually by $FeWO_4$ and containing (b) $FeCl_2$. Pt cathode, polarization rate is 0.1 V/s, and T = 1173 K. (a) C($FeWO_4$) . 10^{-4}, mol/cm^3: (1) base curve, (2) 1.0, (3) 2.0, (4) 3.0, (5) 4.0, (6) 6.0, (7) 8.0, and (8) 10.0. (b) C($FeCl_2$) $\times10^{-4}$, mol/cm^3: (1) base curve, (2) 1.0, (3) 2.0, (4) 3.0, (5) 4.0, (6) 5.0, (7) 6.0, (8) 7.0, (9) 9.0, and (10) 1.3×10^{-3} mol/cm^3.

Table 1. Results of X-ray diffraction analysis and analysis for the free carbon content in the binary tungsten carbides produced by potentiostatic electrolysis of a Na_2WO_4–Li_2WO_4–Li_2CO_3 (15.0, 18.5, 22.0 mol %)–MCl_2 (1.5×10^{-4} mol/cm^3; M = Ni, Co, Fe) in the current density range 1.0–12.0 A/cm^2

Electrolyte composition, mol %	Current density, A/cm^2	Bath voltage, V	Total carbon, %	Phase composition
51.0 Na_2WO_4–34.0 Li_2WO_4–15.0 Li_2CO_3–$FeCl_2$ (1.5×10^{-4} mol/cm^3)	1.0–2.0	2.0–4.5	9.283	WC, Fe_6W_6C
51.0 Na_2WO_4–34.0 Li_2WO_4–15.0 Li_2CO_3–$CoCl_2$ (1.5×10^{-4} mol/cm^3)	1.0–2.0	2.0–4.5	7.136	WC, Co_6W_6C
51.0 Na_2WO_4–34.0 Li_2WO_4–15.0 Li_2CO_3–$NiCl_2$ (1.5×10^{-4} mol/cm^3)	1.0–2.0	2.0–4.5	6.025	WC, NiC_x, NiW
48.9 Na_2WO_4–32.6 Li_2WO_4–18.5 Li_2CO_3–$FeCl_2$ (1.5×10^{-4} mol/cm^3)	1.0–8.0	2.0–4.5	10.16	WC, Fe_6W_6C
48.9 Na_2WO_4–32.6 Li_2WO_4–18.5 Li_2CO_3–$CoCl_2$ (1.5×10^{-4} mol/cm^3)	1.0–8.0	2.0–4.5	6.834	WC, Co_6W_6C
48.9 Na_2WO_4–32.6 Li_2WO_4–18.5 Li_2CO_3–$NiCl_2$ (1.5×10^{-4} mol/cm^3)	1.0–8.0	2.0–4.5	8.874	WC, NiC_x, NiW
48.9 Na_2WO_4–32.6 Li_2WO_4–18.5 Li_2CO_3–$CoCl_2$ (1.5×10^{-4} mol/cm^3)	8.0–12.0	2.0–4.5	3.748	W, WC, W_2C, Co_6W_6C
48.9 Na_2WO_4–32.6 Li_2WO_4–18.5 Li_2CO_3–$FeCl_2$ (1.5×10^{-4} mol/cm^3)	8.0–12.0	2.0–4.5	4.662	WC, W_2C, $W_6C_{2.54}$, Fe_6W_6C
48.9 Na_2WO_4–32.6 Li_2WO_4–18.5 Li_2CO_3–$NiCl_2$ (1.5×10^{-4} mol/cm^3)	8.0–12.0	2.0–4.5	3.645	WC, W_2C, NiC, NiW
46.8 Na_2WO_4–31.2 Li_2WO_4–22.0 Li_2CO_3–$FeCl_2$ (1.5×10^{-4} mol/cm^3)	8.0–12.0	2.0–4.5	5.1	WC, Fe_6W_6C
46.8 Na_2WO_4–31.2 Li_2WO_4–22.0 Li_2CO_3–$CoCl_2$ (1.5×10^{-4} mol/cm^3)	8.0–12.0	2.0–4.5	5.4	WC, Co_6W_6C, Co_4W_2C
46.8 Na_2WO_4–31.2 Li_2WO_4–22.0 Li_2CO_3–$NiCl_2$ (1.5×10^{-4} mol/cm^3)	8.0–12.0	2.0–4.5	8.255	WC, NiC_x, NiW

Table 2. Results of electron-probe fluorescence analysis of the binary tungsten carbides produced by potentiostatic electrolysis of a Na_2WO_4–Li_2WO_4–Li_2CO_3 (15.0, 18.5, 22.0 mol %)–MCl_2 (1.5×10^{-4} mol/cm^3; M = Fe, Co, Ni) in the current density range 1.0–12.0 A/cm^2

No.	Hard-alloy composition based on the carbide of	Current density, A/cm^2	W, %	*M* (Fe, Co, Ni), %	C, %
1	tungsten and iron	1.0–2.0	86.335	0.038	9.283
2		1.0–8.0	89.798	0.042	10.16
3		8.0–12.0	94.87	0.040	5.10
4	tungsten and cobalt	1.0–2.0	91.576	1.288	7.136
5		1.0–8.0	90.834	2.332	6.834
6		8.0–12.0	84.190	2.320	13.49
7	tungsten and nickel	1.0–2.0	90.787	3.188	6.025
8		1.0–8.0	88.763	2.363	8.874
9		8.0–12.0	86.006	5.739	8.255

As it has been shown in the work [20] lithium carbonate in a tungstate melt is reduced at potential - (1.0-1.3) V. Therefore, for electrochemical synthesis of double molybdenum and tungsten carbides, management of the factors influencing for speed of allocation of carbon is required. In the researched system the concentration of Li_2CO_3 is a such factor.

At addition of lithium carbonate (fig. 6, a curve 2 and fig. 7) in tungstate-molybdate melts Na_2WO_4-Li_2WO_4 (20.0 wt %)-Li_2MoO_4 (5.0 wt %) a wave of carbonate-ions reduction coordinated with cationic of lithium is appear in the field of positive potentials - (1.1-1.3) V(fig. 7).

With increasing of lithium carbonate concentration (fig. 6, a curve 3) these waves merge in one joint carbon, molybdenum and tungsten deposition wave stretched on an axis of potentials.

At addition of lithium carbonate in tungstate-molybdate melts the form of the anodic dissolution of cathode product wave changes too. A wave of carbon dissolution is observed at potentials - (0.250-0) V with the wave of metal phase dissolution.

Proceeding from our researches on joint electroreduction of tungsten, molybdenum and carbon ions we carried out the electrochemical synthesis of double tungsten and molybdenum carbides from tungstate-molybdate-carbonate melts (a mol %) (45.0-55.0) Na_2WO_4-(30.0-37.0) Li_2WO_4-(1.0-5.0) Li_2MoO_4-(7.0-21.0) Li_2CO_3. The electrolysis was carried out at cathode current density (1.0-2.5) A/sm^2 with nickel cathode in graphite crucible (MPG-7) at temperature 1173 K.

The results of the electrolysis products analysis are given in Table. 3. The basic opportunity of double molybdenum and tungsten carbides synthesis by electrolysis of melted Na_2WO_4-Li_2WO_4-Li_2MoO_4-Li_2CO_3 system is follow.

Thus, the concentration of molybdate and a lithium carbonate in melts are important, but their parity is more important. Besides that, the phase structure of cathode deposits are changes with increase in electricity quantity.

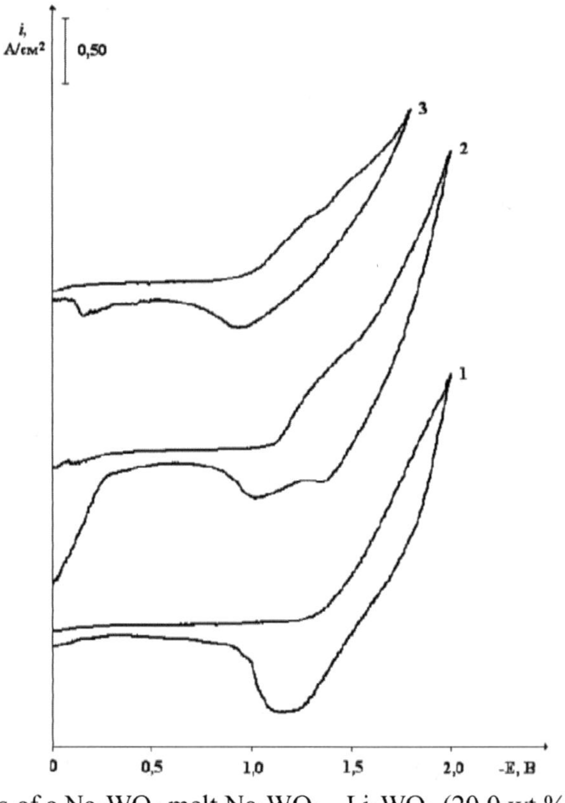

Fig. 6. I–V characteristics of a Na_2WO_4 melt Na_2WO_4 - Li_2WO_4 (20.0 wt %) - Li_2MoO_4 (20.0 wt %) at consecutive addition Li_2CO_3, mol %: 1 - 0, 2 - 5.0; 3 - 20.0. Pt cathode, polarization rate is 0.1 V/s, and T = 1173 K.

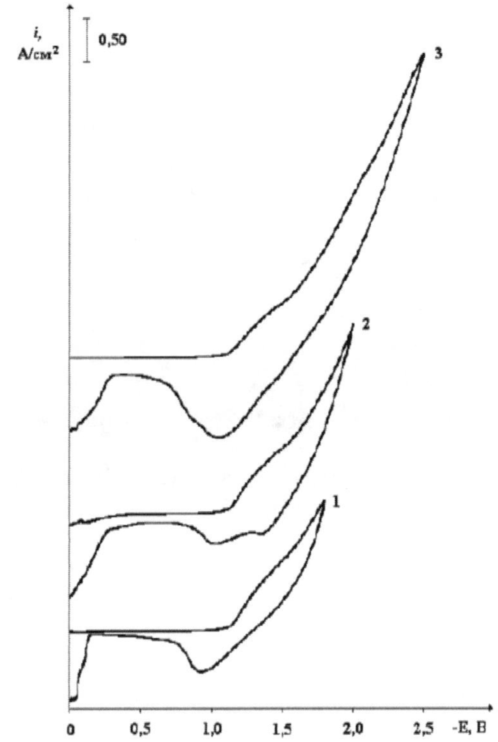

Fig. 7. I–V characteristics of melt Na_2WO_4 - Li_2WO_4 (20.0 wt %) - Li_2MoO_4 (20.0 wt%) - Li_2CO_3 (5.0 wt%) at various potentials, 1 - 1.8; 2 - 2.0; 3 - 2.3. Pt cathode, and T = 1173 K.

Table 3

Results of X-ray diffraction analysis of products electrolysis fused systems Na_2WO_4-Li_2WO_4-Li_2MoO_4-Li_2CO_3 at 1173 K

Electrolyte composition, mol %	Current density, A/cm^2	Duration electrolysis, min	Phase composition
55.2 Na_2WO_4 36.8 Li_2WO_4 1.0 Li_2MoO_4 7.0 Li_2CO_3	2.4	10	W_2C, Mo_2C
	2.3	20	WC, MoC, W_2C, Mo_2C
	2.0	30	W_2C, Mo_2C, W, Mo
50.4 Na_2WO_4 33.6 Li_2WO_4 2.0 Li_2MoO_4 14.0 Li_2CO_3	1.9	30	W_2C, Mo_2C
	1.8	60	WC, MoC, W_2C, Mo_2C
	2.0	70	W_2C, Mo_2C, W, Mo
	2.0	80	W, Mo
45.6 Na_2WO_4 30.4 Li_2WO_4 3.0 Li_2MoO_4 21.0 Li_2CO_3	2.0	30	W_2C, Mo_2C
	2.0	50	W_2C, Mo_2C
	1.6	70	W_2C, Mo_2C

CONCLUSIONS

By electrolysis of tungstate-carbonate melt containing iron triad metal ions, we carried out the electrochemical synthesis of hard-alloy compositions based on tungsten carbide. The processes of joint electrodeposition of tungsten, molybdenum, carbon ions and iron triad metal ions are carried out in tungstate-molybdate-carbonate melts. The structures of cathode deposits are established too.

REFERENCES

1. A. I. Gusev, “Nanocrystalline State Effects in Compact metals and Compounds,” in Nanocrystalline Materials: Production Methods and Properties (UrO RAN, Yekaterinburg, 1998), pp. 5–9.

2. A. I. Gusev and A. A. Rempel, in Nanocrystalline Materials (Nauka, Moscow, 2000), pp. 13–19.

3. A. I. Gusev and A. A. Rempel, “Nonstoichiometric Transition Metal Carbides as Alternative Hard Materials,” in Nanocrystalline Materials (Cambridge Intern. Science Publ., Cambridge, 2004), pp. 256–263.

4. A. I. Gusev, in Nanomaterials, Nanostructures, and Nanotechnologies (Fizmatlit, Moscow, 2005), pp. 5–9.

5. S. Berger, R. Porat, and R. Rosen, “Nanocrystalline Materials: A Study of WC-based Hard Metals,” Prog. Mater. Sci. 42, 311–320 (1997).

6. A. I. Gusev and A. S. Kurlov, “Hard Alloys Today and Tomorrow,” Metally Evrazii 2, 42–45 (2005).

7. T. N. Miller, Neorg. Mater. 15, 557 (1979).

8. T. Ya. Kosolapova, G. N. Makarenko, and D. P. Zyatkevich, “Plasmachemical Synthesis of RefractoryCompounds,” Zh. VKhO im. Mendeleeva 24, 328–333 (1979).

9. Yu. V. Blagoveshschenskii, E. A. Danilkin, T. P. Egorikhina, and V. I. Terekhov, in Proceedings of the IV All_Russia Conference on Physical Chemistry of Ultradispersed Systems, Moscow, Russia (MIFI, Moscow, 1998), p. 274.

10. L. Gao and B. H. Kear, "Nanostructure Tungsten Carbide Catalysts for Polymer Electrolyte Fuel Cells," Nanostruct. Mater. 5, 555–563 (1995).

11. L. Gao and B. H. Kear, "Formation of Fine Intragranular Ferrite in Cast Plain Carbon Steel Inoculated by Titanium Oxide Nanopowder," Nanomech. Material 9, 205–211 (1997).

12. L. E. McCandlish, B. H. Kear, and B. K. Kim, "Processing and Properties of Nanostructured WC–Co," Nanostruct. Mater. 1, 119–123 (1992).

13. L. Wu, J. Lin, B. K. Kim, et al., in Proceedings of the 13th International Plansee Seminar, Ed. by H. Bildstein and R. Eck (Metallwerk Plansee, Reutte, 1993), Vol. 3, pp. 667–678.

14. L. E. McCandlish, B. H. Kear, and S. J. Bhatia, US Patent 5352269, 1994.

15. M. A. Xueming and J. I. Gang, "Relative Effects of All Chemical Elements on the Electrical Conductivity of Metals and Alloys: An Alternative to Norbury–Linde Rule," Alloys J. Compd., 245–247 (1996).

16. V. I. Shapoval, Yu. K. Delimarskii, and Kh. B. Kushkhov, "Kinetic Electroreduction Currents in KCl–NaCl Melts," Elektrokhimiya 14 (8), 1141–1145 (1978).

17. V. I. Shapoval and Kh. B. Kushkhov, "Effect of the Li^+ Cation on the WO_4^{2-} Electroreduction Kinetics in an Equimolar KCl–NaCl Melt," Ukr. Khim. Zh. 45 (8), 698–701 (1979).

18. V. I. Shapoval and Kh. B. Kushkhov, "Mechanism of Formation of Electrochemically Active Particles during the Electroreduction of WO_4^{2-} in a KCl–NaCl Melt," in Electrode Processes and Investigation Methods (Naukova Dumka, Kiev, 1978), pp. 165–169.

19. Kh. B. Kushkhov, L. M. Beroeva, M. N. Adamokova, and S. Dzh. Gonov, "Combined Electroprecipitation of Tungsten, Molybdenum, and Carbon from a Sodium Tungstate Melt at an Excess Carbon Dioxide Pressure," in Proceedings of the XII Russian Conference on the Physical Chemistry and Electrochemistry of Ionic

Journal of Physics: Conference Series **291** (2011) 012045 doi:10.1088/1742-6596/291/1/012045

Melts and Solid Electrolytes, Nal'chik, Russia (Kabard.-Balkar Gos. Univ., Nal'chik, 2001), Vol. 1, pp. 249–252.

20. Kushhov H.B., Malyshev V.V., Shapoval V. I. Electroreduction of carbonate-ions in tungsten melts. (Ukr. Khim. Zh. 54, №11, 1988), pp. 1155-1158.

Superconducting spintronic devices based on nanostructures ferromagnet/superconductor

M G Khusainov[1,2], **Yu Y Petrushenko**[1], **V L Matukhin**[1], **Yu A Sakhratov**[1] **and Yu N Proshin**[2]

[1]Kazan State Power Engineering University, Kazan 420066, Russia
[2]Kazan State University, Kazan 420008, Russia

E-mail: sakhratov@gmail.com

Abstract. The layered nanostructures ferromagnet/superconductor (F/S) due to combination of incompatible in homogeneous materials properties are the most perspective materials for use in the new field of electronics – a superconducting spintronics. A new type of logical devices based on the layered F/S nanostructures and combining the advantages of the superconducting and magnetic recording channels in one sample is offered. Each channel can be separately controlled by weak magnetic field or current pulse and the switching time is of order of 10^{-10} - 10^{-11} s. The implementation of such devices on base of high-temperature superconductors will allow using nitrogen instead of expensive helium for cooling.

1. Introduction

The coexistence of the superconducting and ferromagnetic order parameters (OPs) is unlikely in a uniform system, although it is easily achievable in artificially prepared layered F/S systems consisting of alternating ferromagnetic (F) and superconducting (S) layers. Owing to the proximity effect, a superconducting order parameter can be induced in the F layer; on the other hand, the neighboring pair of F layers can interact with one another via the S layer. Such systems exhibit rich physics, which can be controlled by varying the thicknesses of the F and S layers or by placing the F/S structure in an external magnetic field [1].

The modern technologies of production of layered structures, such as molecular-beam epitaxy, permit one to apply layers of atomic thicknesses and study the properties of layered F/S systems as function of the thickness of the ferromagnetic (d_f) and superconducting (d_s) layer. Numerous experiments on the F/S structures (junctions and superlattices) revealed nontrivial dependences of the temperature of the superconducting transition T_c on the thickness of the ferromagnetic layer. Of special interest is the study of multilayered F/S structures, in which various types of magnetic order can arise in F layers due to their indirect interaction via S layers. Recently, logical elements of a new type (spin switches) were suggested based on the interrelation between the superconducting and magnetic OPs in three-layer F/S/F and four-layer F/S/F/S structures. The general theoretical interest in the problem of the mutual influence of superconductivity and magnetism in F/S structures and possible engineering applications make the problem discussed quite topical.

2. Results and Discussion

The three-dimensional model of proximity effect for layered F/S structures is expounded in [2]. In the framework of this model the phase diagrams T_c versus d_f and d_s have been calculated. On the base of these results we propose logical elements/control devices of a new type (spin switches).

Let us consider the superconducting spin valve, cryotron, based on the three-layered F/S/F nanostructure (figure 1). The calculated phase diagram of this structure is depicted on figure 2. The state of F/S structure can be managed by a weak magnetic field H, which affects the phase diagram slightly. In this case there are three characteristic quantities of this magnetic field: coercive field H_{coer}, critical field H_c and pinning field H_p. The value of critical field H_c is determined by the difference between critical and operating temperatures (T_c - T), which can be negligibly small.

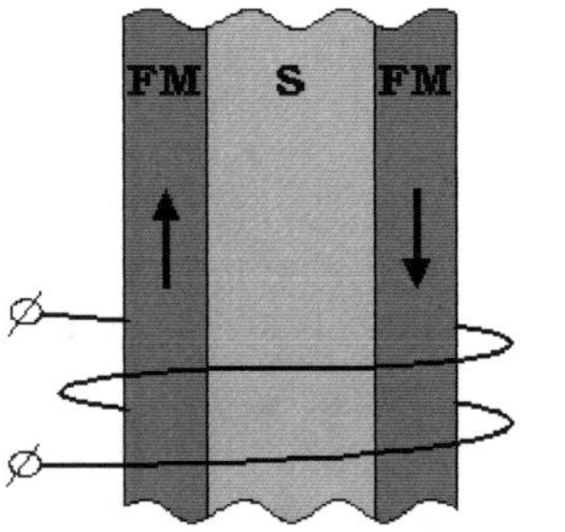

Figure 1. The sketch of a superconducting spin valve with two channels, separately controlled by an external magnetic field.

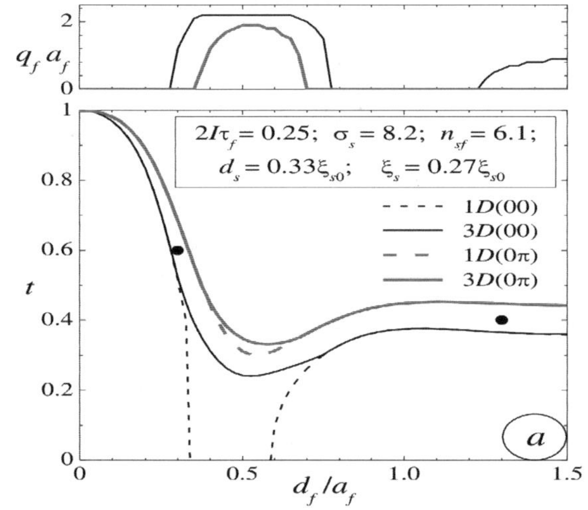

Figure 2. Reduced critical temperature t (down) and pair's momentum q_f (up) versus reduced thickness of the FM layer d_f/a_f. The thin line corresponds to AFS state, the thick dashed line corresponds to FS state. The solid and dashed lines correspond to 3D ($q_f \neq 0$) and 1D ($q_f = 0$) cases, respectively.

Let the system be in one of the state depicted on figure 2 by points. In zero magnetic field the system is in the base antiferromagnetic superconducting (AFS) state. Let us suppose, that the magnetic fields are related as follows: $H_{coer} < H_c < H_p$. When we apply an external magnetic field at the direction of pinning field $\mathbf{H}_p$, transition from the base AFS state to the ferromagnetic normal state (FN) occurs at $H \approx H_{coer}$. If the orientation of magnetization of the left F-layer, for example, up (↑), then the transition can be represented as ↑S↓ → ↑N↑. Applying of an external magnetic field of opposite direction makes possible two other transitions: first form the base AFS state to antiferromagnetic normal state (AFN) at $H \approx H_c$ (↑S↓ → ↑N↓), and then to the FN state at $H \approx H_p$ (↑N↓ → ↓N↓).

Thus, for a given choice of operating points, we have up to 4 possible different logical states (one superconducting state ↑S↓ and three normal states ↑N↑, ↑N↓, ↓N↓), and we can choose the recording of information on the superconducting current (the first channel) and the relative orientation of magnetization (the second channel). Note that you can get up to five different states (↑S↓, ↑S↑, ↑N↑, ↑N↓, ↓N↓), if slightly shift the operating points below the 3D (FS) curve on figure 2.

Let us consider the concept of cryotron based on the four-layered F/S/F′/S′ nanostructure. The phase diagram of this structure is depicted on figure 3. Add an additional layer of a magnetic dielectric to the left of the outer layer with a purpose to fix the direction of magnetization in the F layer. One possible consequence is the return of our system to its initial state after turning the magnetic field off. Thus, formally our system becomes MI/F/S/F′/S′. Changing an external magnetic field **H**, directed along the pinning field $\mathbf{H}_p$, which sets the direction of magnetization in the outer left F layer (call it a direction “up”), we can call the system transitions between the ground antiferromagnetic superconductor (AFMS) state, mixed state FM with an unrelated superconductivity and ferromagnetic normal (FMN) state. An external magnetic field of opposite orientation can cause the other three transitions between the ground state AFMS, the mixed state AFM, AFMN state, and, finally, FMN state. Fix the orientation of the magnetization of the outer layer of F, for example, up (↑). At $H = 0$ the system is in AFMS state. If we apply a weak external magnetic field **H**, greater than the coercive field ($H_{coer} < H < H_c$), in the direction of magnetization **M** of the F layer and the pinning field ($\mathbf{H}\uparrow\uparrow\mathbf{H}_p$), the direction of magnetization **M′** in the F′ layer reversed. The system goes into a state with ferromagnetic magnetizations. Thus, we can say that as a result of the transition $\uparrow S \downarrow S \xrightarrow{H_{coer}} \uparrow N \uparrow S$ analysed above, the information recorded on the superconducting properties of the layer S′, has survived. Information recorded on the relative orientation of magnetizations of the layers F and F′ and superconducting current of the inner layer of S, has changed. If we now act on the system by a field, greater then critical field for the outer layer of S′ ($H > H_c'$), then the superconductivity in this layer also disappears and the whole system goes into the FMN state. In other words, in this transition changes the information recorded on the conductive properties of the layer S′. Thus, we have the following chain of transitions: $\uparrow S \downarrow S \xrightarrow{H_{coer}} \uparrow N \uparrow S \xrightarrow{H_c'} \uparrow N \uparrow N$. An external magnetic field of opposite orientation ($\mathbf{H}\downarrow\uparrow\mathbf{H}_p$) can cause three other transitions in the system.

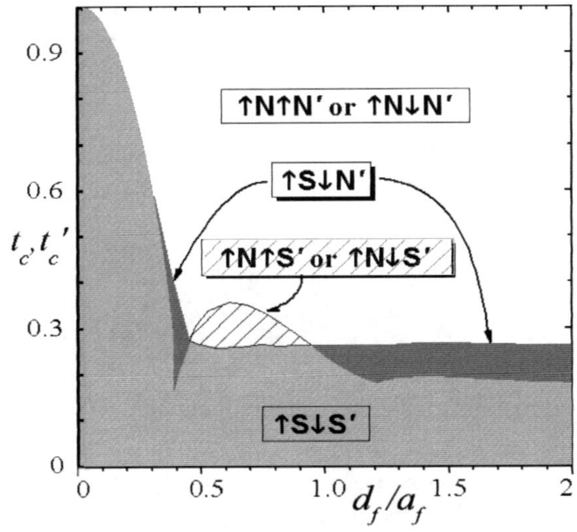

Figure 3. The generalized phase diagram of four-layered F/S/F′/S′ system. The vertical arrows indicate the direction of magnetization in corresponding ferromagnetic layer. The letters S and N are used to denote the superconducting and normal state of superconducting layers respectively.

If H is a little more than H_c, but less than H_c', then the system undergoes a transition from the AFMS state into a mixed AFM state $\uparrow S \downarrow S \xrightarrow{H_c} \uparrow N \downarrow S$. In this case, only information recorded using the conductive properties of S-layer is changed, while the information recorded on the relative orientation of magnetizations of the layers F and F′ and superconducting current layer S′, remains unchanged.

The following transitions in this series are observed with further increase of the magnetic field in the same direction "down" (from $H_c' < H < H_p$ to $H > H_p$). Thus, the system F/S/F′/S′ has six logically different states: ↑S↓S, ↑N↑S, ↑N↑N, ↑N↓S, ↑N↓N, ↓N↓N.

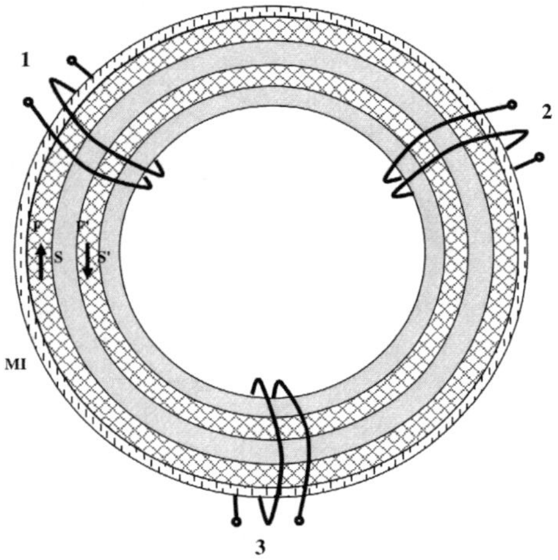

Figure 4. The sketch of a memory cell with two separately controlled channels. 1 - recording coil, 2 - interrogate coil, 3 - reading coil.

A possible physical realization is depicted on figure 4. In the absence of the magnetic field (current in the coil 1) the system is in the ground state ↑S↓S. Magnetization reversal in the ferromagnetic layers is achieved by applying a current in the coil 1 of an appropriate direction and magnitude. For reading information a current pulse is passed through the coil 2, which creates a magnetic field in the direction of carrying the system to its original state. If the information was recorded, the respondent pulse creates alternating magnetic flux, inducing a current pulse in the reading coil 3. In the opposite case, the reversal does not occur and the signal does not appear in the reading coil. When the magnetic field is turned off the system is restored to the ground state ↑S↓S.

3. Conclusion

The three-layered F/S/F and four-layered F/S/F/S nanostructures are the most perspective candidates for the use in the superconducting spin nanoelectronics (spintronics). These structures possess two data-record channels, namely, on the superconducting current and the magnetic order of the FM layer magnetizations. We can separately manage these channels by weak magnetic field or current pulse. The advantages of such spin devices linked to small size (thickness $d_f \sim 0.5$–5 nm, thickness $d_s \sim 25$–80 nm), high switching speed ($\sim 10^{-10} - 10^{-11}$ c), high enough critical currents. Combination of both channels in a single sample would significantly increase the information recording density. Currently, an implementation of such nanodevices on base of high-temperature superconductors (HTSC) is considered, which would allow using cheap liquid nitrogen rather than expensive helium.

References

[1] Y A Izyumov, Yu N Proshin, and M G Khusainov 2002 *Physics Uspekhi* **45** (2) 109.
[2] Y N Proshin, A Zimin, N Fazleev, and M G Khusainov 2006 *Phys. Rev. B* **73** 184514.

Journal of Physics: Conference Series **291** (2011) 012047 doi:10.1088/1742-6596/291/1/012047

Development and realization methods for the study of local magnetic and transport characteristics of modern nanostructured superconducting materials

S. V. Pokrovskiy, I. A. Rudnev, A. I. Podlivaev

National Nuclear Research University "MEPHI", Kashirskoe shosse 31, Moscow, Russia

sergeypokrovskii@gmail.com, iarudnev@mephi.ru

Abstract. It is well known that the Hall probe magnetic imaging represents a powerful technique for the investigation of the magnetic field distribution in nanostructured superconducting materials of many different types and geometries. We present a new method of the evaluation of current distributions in thin nanostructured superconducting tapes and films from the data of two-dimensional magnetic field profiles measured over the sample surface. We have analyzed the different conditions of magnetic flux trap of external magnetic field and the magnetic field distribution of a superconductor at the flow of transport current. Based on the standard algorithm was developed original method of non-contact investigation of local transport properties of modern nanostructured superconducting materials.

1. Methods

The superconducting tapes were investigated in two modes: the penetration and trap of an external magnetic field, the flow of transport current. Using method of local Hall magnetometry were obtained the 2D spatial distribution of trapped magnetic flux or field component of the transport current perpendicular to the surface of the tape $B_z(x,y)$. On the basis of these data is possible to calculate the spatial distribution of currents corresponding to the magnetic field.

The solving of this problem is resulted in the inversion of Biot–Savart law. We have analysed standard approaches of solving the problem [1] and have developed a new modified calculation algorithm 2D distribution of supercurrents in thin superconducting tapes and films [2].

The new algorithm of calculation allows correct solve inversion problem using experimental data of the magnetic field only over the sample, whereas standard algorithm requires that the scanning area of the field was several times larger than the size of the sample. In addition, optimal filter input data allowes to detect the inhomogeneities of current flow which typical dimensions on the order of magnitude lower step of the experimental grid scanning the magnetic field.

2. Materials

Have been investigated various modern high-temperature superconducting tapes based on nanostructured thin films of superconducting $YBa_2Cu_3O_{7-\delta}$ and $Bi_2Sr_2Ca_2Cu_3O_{10-\delta}$.

3. Rezults

Tapes of various architecture and characteristics, both new and earlier used have been investigated. In this paper we demonstrate only the most typical observed experimental data and the corresponding calculations.

Figure 1 show the distribution of trap magnetic flux in HTS 2G tape and restored surfaces of currents (across and along the tape see figure. 2,3). Architecture of the tape: substrate Ni-5%W alloy (deformation texturing), buffer stack Y_2O_3/YSZ/CeO_2 (~75 nm, high rate reactive sputtering), YBCO (~1 μm, metal organic deposition of TFA), Ag (DC sputtering).

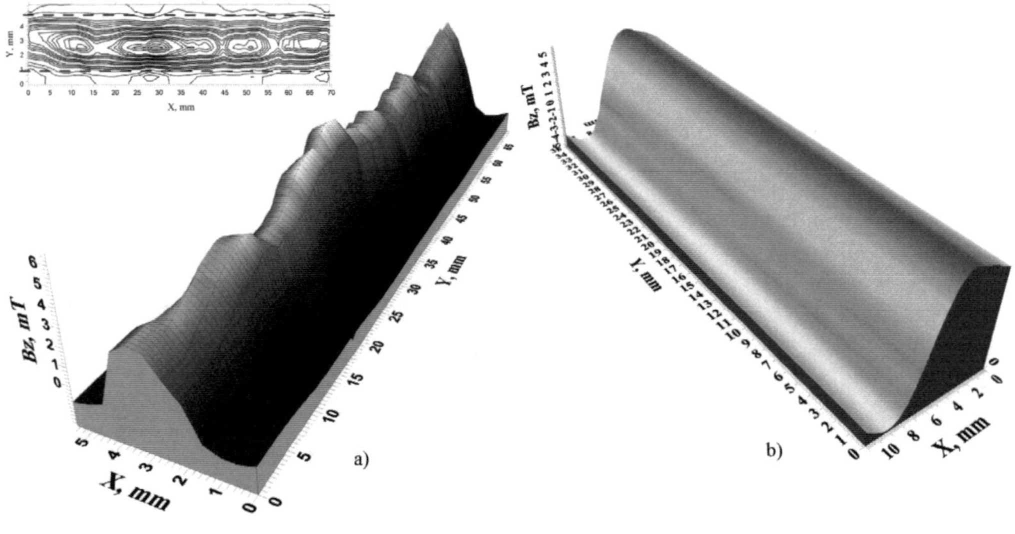

Figure 1. a) distribution of trap magnetic field in high-temperature superconducting tape b) distribution of magnetic field generated by transport current in high-temperature superconducting tape

The magnetization surface profile corresponds to the Bean model. Note that magnetic flux distribution displays moderate inhomogeneity which corresponds to the inhomogeneity of the current distribution in the tape. Surfaces of currents distribution testify to the circular course of the induced currents.

Let's consider separately components of the current along ($J_y(x,y)$) and across the tape ($J_x(x,y)$) (figures 2). Distribution $J_y(x,y)$ non-uniform, that in turn results in nonzero distribution $J_x(x,y)$. That is owing to imperfection of the tape, non-homogeneity of its properties, arise overflows of currents, flows (appearance of nonzero component of the current across the tape). Similar results have been received for other tapes. Also, results were obtained in the mode of transport current flow (figure 3).

Analysis of the results allows us not only describe the homogeneity of the magnetic and transport properties that determine the quality of the finished tape or tapes on the stage of production, but also to determine the fault location and area of redistribution and the flow of current (current domains) in the superconducting tape, by which one can conclude about the defectiveness of tapes, and also to find a region with small current-carrying characteristics. These regions lead to the decreasing of the critical current of the tape, and modern data falling current-carrying capacity can reach 30% [3].

In addition, our method allows to study a local characteristics of current-carrying superconducting tapes during the flow of transport current. Similarly carry out the spatial distribution of the magnetic flux, but induced by passing transport current through the tape. Such measurements provide a clear way to see the current path, identify weak areas of tapes, to evaluate the homogeneity of current flow and efficiency of the superconducting tape.

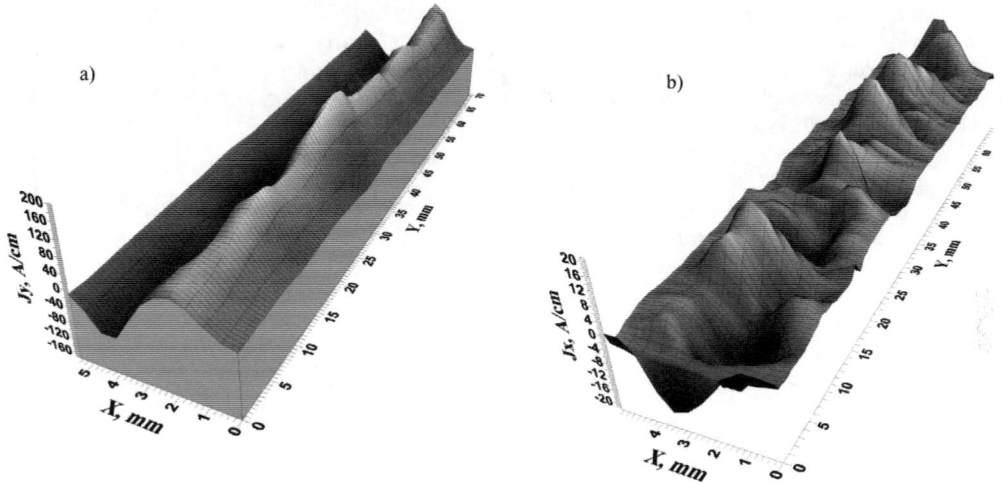

Figure 2. Spatial distribution of the current components a) Jy(x,y) – along the magnetized superconducting tape, b) Jx(x,y) – across the magnetized superconducting tape

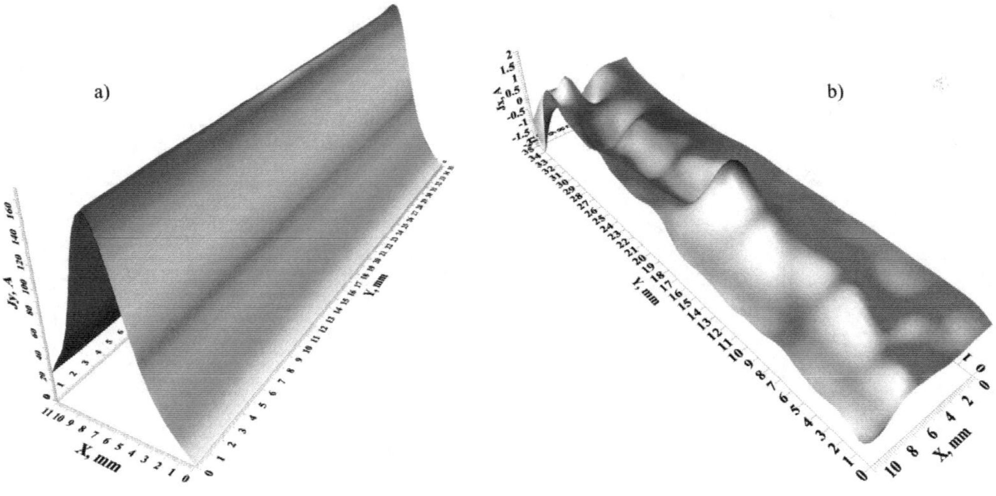

Figure 3. Spatial distribution of the transport current components a) Jy(x,y) – along the superconducting tape, b) Jx(x,y) – across the superconducting tape

The proposed method has no specific binding to a model, and the spatial resolution is limited only by the experimental method, which is measured spatial distribution of the magnetic field component perpendicular to the plane of current flow. There is only one restriction – it can be studied objects with a two-dimensional current flow. Such a limitation follows from the possibility of a correct solution of the inversion problem Biot-Savart-Laplas equation. For today we carried out measurements using the Hall probe as a sensor of magnetic field. As a result of accurate testing and work through the entire procedure it can be possible to detect the region of current flow (defect HTSC film) an order of magnitude smaller then the sensitive area of the Hall probe.

Our approach can be implemented not only with respect to HTS films, but also to other nanoobjects with two-dimensional current flow. On the redistribution of current flow one can conclude about, for example, the presence of defects or damage to the object of research. Moreover, using magnetic atomic force microscope will allow to explore the processes of percolation and redistribution of power in the objects of nanoscale.

4. Conclusions

It was developed a noncontact method for investigation of local transport characteristics of modern nanostructured superconductors.

The calculation method allows us investigate any nanoobjects with two-dimensional current flow.

5. Acknowledgements

Work was performed during realization of the federal target program "Scientific and scientific-pedagogical personnel of innovation Russia 2009-2013"

References

[1] Jooss Ch, Albrecht J, Kuhn H, Leonhardt S and Kronmüller H 2002 Rep. Prog. Phys. 65 651
[2] Pokrovskiy S.V., Rudnev I.A., Podlivaev A.I. // J. Phys. Conf. Ser. 2009. V. 150. P. 052211
[3] Selvamanickam V. et al // IEEE Tran. on Appl. Superconductivity. 2009. V. 19. P. 3225

Journal of Physics: Conference Series **291** (2011) 012048 doi:10.1088/1742-6596/291/1/012048

Peculiarities of magnetization of second generation high-temperature superconducting tapes in a wide temperature range

I Rudnev[1,2], A Komarova[1], P Bobin[1]

[1] National Research Nuclear University "MEPHI" . Moscow Russia

[2] International Laboratory of High Magnetic Fields and Low Temperatures. Wroclaw Poland

E-mail: iarudnev@mephi.ru

Abstract. We present the results of study of magnetization and critical current of coated conductors with magnetic and nonmagnetic substrates. The measurements of magnetization curves were done in a wide temperature range from 4,2 to 100 K and magnetic field up to 14 T. To determine the dependence of transport critical current on the magnetic field we measured a set of current-voltage characteristics in the range of magnetic field from 0 to 8 T at T = 77 K with perpendicular to the tape field orientation. It was obtained that the substrates magnetism dramatically changes the form of magnetization curves but not influence the value of critical current. Comparison of field dependence of critical current, obtained by contact and contactless method at T = 77 K shows that for both samples is observed coincidence of the curves at low fields and a strong divergence at H> 1 Tesla.

1. Introduction

Recently, considerable attention is paid to the development of the second-generation current-carrying tapes based on high-temperature superconductor (HTSC) $YBa_2Cu_3O_{7-x}$ with critical temperature 92 K. Tapes are thin (less than 1 micron) layers of HTSC, deposited in various ways to the metal substrate alloy Hastelloy (Ni + W). Depending on the W concentration, which ranged from 0 to 5%, these substrates have different magnetic properties, which may influence the integral magnetization of HTS tapes. Since the substrates are weak ferromagnetic but high-temperature superconductors - diamagnetic, the total magnetization of the tape will depend on the relative contribution of the substrate and the HTSC magnetism. At the same time in different magnetic fields and at different temperatures, the contribution of various components of the magnetic tape can be different. In this report we present results of studies of the magnetization of HTSC tapes with different magnetic characteristics of metal substrates.

2. Materials and methods

Industrial tape manufactured by AMSC (magnetic substrate), and Super Power (non-magnetic substrate) were selected for study. Magnetization measurements were performed using vibration magnetometer [1] on samples of size 4x4x0, 1 mm. The temperature varied in the range 4,2-100 K, the magnetic field was varied from 0 to 14 T, the orientation of the field was perpendicular to the plane of the tape.

Determination of the magnetization of the substrate was carried out in two ways. The first - the magnetization of the samples was measured at T = 95 K, i.e. when the superconductor is in normal state. However, it remained unknown temperature dependence of magnetic susceptibility of the substrate and it is not clear how the values of magnetization and susceptibility, obtained at T = 95 K correspond to the values at lower temperatures. Therefore, for the sample with magnetic substrate we used a different procedure. After all measurements of the integral of the magnetization from 4,2 K to 77 K, the sample was split accurately on the HTS layer into two parts, a thin layer of HTSC was scraped carefully, and two half have been stuck together. The result was a new sample, which had a geometric dimensions and weight of almost the same as the original, but without the superconducting layer. It thus becomes possible to measure the magnetization of the substrate at temperatures ranging from 4.2 to 95 K. These measurements showed that the intrinsic magnetization of the substrate changes with temperature from 95 K to 4,2 K is not more than 3% (Figure 1).

3. Results and discussion

Measurements of superconducting samples both at T = 77 K and at lower temperatures showed a significant difference in the form of curves M(H) due to influence of the magnetic properties of the substrates. Moreover, the influence of ferromagnetism of the substrate AMSC tape is so strong that the curve M(H) was absolutely no resemblance to the classical curve M(H) for type-II superconductors, such as, for SP tape (Figure 2, Figure 3). In fact, the diamagnetic HTS layer veiled completely the influence of the substrate. However, if the from total magnetization of tapes AMSC deduct the magnetization of the substrate, measured at the same temperature, the magnetization of the HTSC layer is obtained, analogous to the magnetization of SP tape. Obviously, small differences remain due to the difference in the properties of HTSC materials.

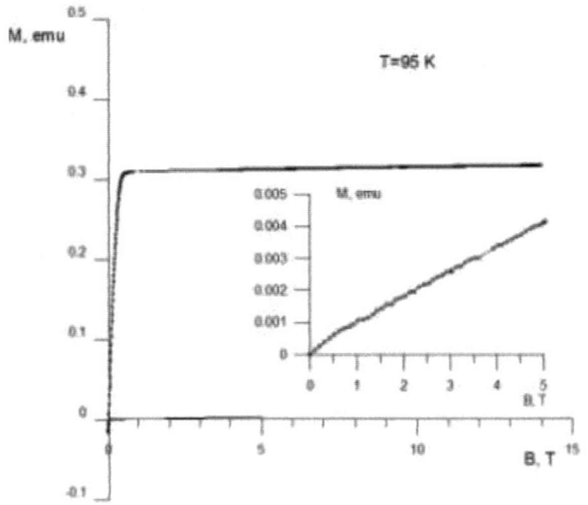

Figure 1: Magnetization of 2G sample with magnetic substrate and nonmagnetic substrate (inset) at T=95 K

The significant difference between the curves of the magnetization of HTSC films on magnetic and nonmagnetic substrates is manifested in the whole temperature range from 4.2 to 77 K.

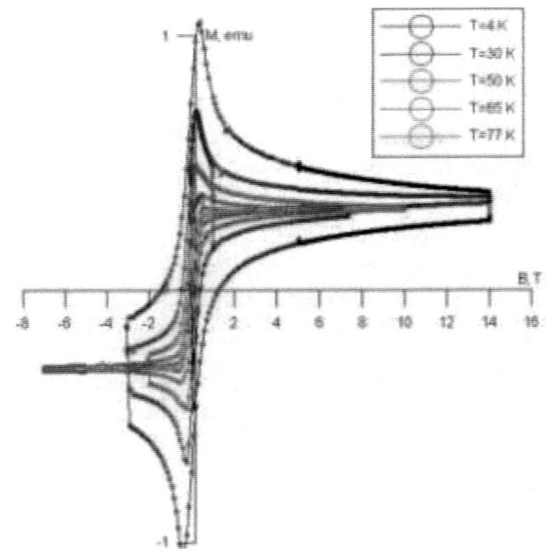

Figure 2: M(H) curves for 2G tape with *magnetic* substrate

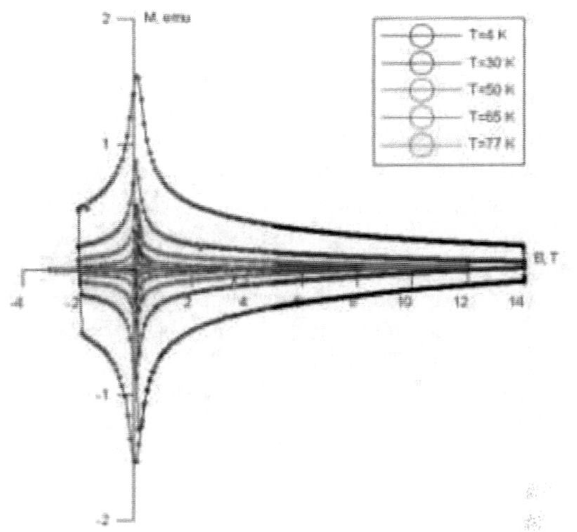

Figure 3: M(H) curves for 2G tape with *nonmagnetic* substrate

Thus, it is experimentally shown that the ferromagnetism of the metallic substrate significantly distorts the shape of magnetization curves of HTSC tape. However, after taking into account the magnetic susceptibility of the substrate, the magnetic characteristics of HTS layers are similar (Figure 4). Since the hysteresis of M(H) curves for 2G tape with magnetic substrate shown on Figure 2 is absent, we can expect that the magnetism of the substrate will not affect the magnitude of energy loss in alternating electromagnetic field. Analysis of the curves M (H) can give detailed information about the hysteresis losses in 2G HTS tapes AMSC and SP in wide temperature range from 4.2 to 77 K in magnetic fields up to 14 Tesla.

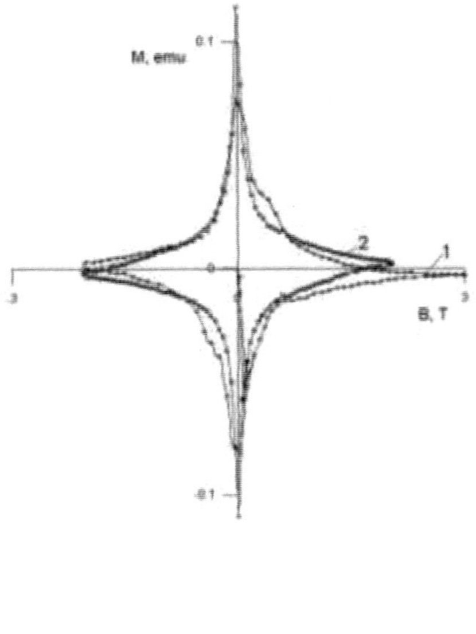

Figure 4: M(H) curves for superconducting layers AMS (1) and SP (2) at T=77 K. Curve (2) is obtained by *subtraction* from magnetization of the entire tape magnetization of the substrate

High values of dissipationless flow of transport current in modern high-tapes of the second generation led to methodological difficulties in measuring the critical current, which are associated with the overheating of the current supply wires and electrical contacts. It is known that alternative transport measurements are various contactless techniques, in particular, the measurement of magnetization, from which, based on the critical-state model [2,3], we can determine the values of the critical current. The unquestionable advantages of the magnetization measurements should include the possibility of measurements in a wide range of temperatures and the possibility of using relatively small samples of superconducting tapes. The dependence of $I_c(H)$ can be relatively easily verified by direct transport measurements at the boiling point of liquid nitrogen and determine the range of magnetic fields, which applies non-contact method for measuring the critical current.

We compared the normalized dependences of the critical current, obtained by contact and contactless method. For both samples the coincidence of the curves at low fields and strong divergence at H> 1 Tesla. The field at which the data obtained by different methods are roughly the same limited by the value 0.4-0.6 Tesla depends on sample types. Meanwhile, it should be noted that this range of fields is of practical importance, since the fall of the critical current below the level of 0,1 I_{c0} makes bleak possibility of real devices.

4. Conclusion

In this report it is experimentally shown that the ferromagnetism of the metallic substrate significantly changes the shape of magnetization curves of HTSC second generation tape. The strong discrepancy between the results of transport and magnetic measurements of the critical current in fields above 1 T seems to indicate non-applicability of the critical state model for calculating the critical current of HTS tapes in high magnetic fields.

Acknowledgments

The work was done within the framework of the Russian Federal Program "Research and scientific-pedagogical personnel of Innovative Russia in 2009 - 2013 years".

References

[1] Nizhankovskii V I, Lugansky L B 2007 *Measurement Science and Technology* **18** 1533
[2] Bean C P 1962 *Physical Review Letters* **6** 250
[3] Bean C P 1964 *Review of Modern Physics* **36** 31

Journal of Physics: Conference Series **291** (2011) 012049 doi:10.1088/1742-6596/291/1/012049

Thin film solar cells based on CdTe and $Cu(In,Ga)Se_2$ (CIGS) compounds

P P Gladyshev[1,2], S V Filin[2,3], A I Puzynin[2,3], I A Tanachev[2,3], A V Rybakova[2,3], V V Tuzova[2,3], S A Kozlovskiy[2,3], V F Gremenok[4], A V Mudryi[4], E P Zaretskaya[4], V B Zalesskiy[5], V M Kravchenko[5], U R Leonova[5], A A Khodin[5], V A Pilipovich[5], A M Polikanin[5], G S Khrypunov[6], E P Chernyh[6], N A Kovtun[6], E K Belonogov[7], V M Ievlev[7], M B Dergacheva[8], V N Stacuk[8], L A Fogel[8]

[1]International University of Nature, Society and Man "Dubna", Dubna, Russia
[2]Center of High Technologies of FSUE "Applied Acoustics Research Institute", Dubna, Russia
[3]Co Ltd "Scientific and Technical Evaluation Center "Nanoteh-Dubna", Dubna, Russia
[4]State Scientific and Production Association «Scientific-Practical Materials, Researcher center of National Academy of Sciences of Belarus», Minsk, Belarus
[5]Institute of Physics of National Academy of Sciences of Belarus, Minsk, Belarus
[6]National Technical University "Kharkov Politechnical Institute", Kharkov, Ukraine
[7]Voronej State Technical University, Voronej, Russia
[8]Institute of Organic Catalysis and Electrochemistry by D.V. Socolskii, Alma Ata, Kazakhstan

E-mail: pavel.gladyshev@niipa.ru

Abstract. We are publishing recent results in chalcogenide photoelectric convertors fabrication, which are efforts of many scientific teams from Russia, Belarus, Ukraine, and Kazakhstan. Competitively high efficiency of photoelectric convertors (11.4% for CdTe and 11% for CIGS) was achieved in the process of our work. Furthermore, luminescent filters for improvement of spectral response of such chalcogenide solar cells in a short wavelengths region were also developed and investigated here.

1. Introduction

A market of thin film chalcogenide photoelectric convertors (PhEC) on the base of CdTe and $Cu\{In,Ga\}Se_2$ (CIGS) are growing extensively due to their lower cost per watt of installed capacity in comparison of silicon solar cells [1]. An efficiency of thin film solar cells is 12-14%, which allows to obtain about 100 watts of electric power from one square meter of photovoltaic surface. Unlike silicon, chalcogenide materials have an optimal band gap, better light absorption and good radiation resistance. Moreover, current efficiency of such solar cells does not reach a half of theoretical limit (24-29%). Thus, they have large potential for efficiency growth in the future. Another advantage of chalcogenide systems is the ability of manufacture of thin, light and flexible cells on the base of polymer [2-3] and metal foil [4] substrates.
In this paper we represent different scientific centers, which joined their efforts in solar energy conversion area for research and developing of PhEC.

Active layer formation for CdTe and CIGS solar cells (see fig.1) is usually realized by variety of vacuum techniques. At this moment such methods are well known and applied for layer sputtering widely.
Also, there are so-called "wet-methods" for the layer deposition. Wet methods are usually based on deposition from different solutions of appropriate chemicals or on the modern printing technologies (such as ink jet, etc.) Moreover, we have to note that wet methods can give some advantage over standard vacuum sputtering such as lower price and higher speed of layer deposition. Obviously, that is very important in the case of mass manufacture.

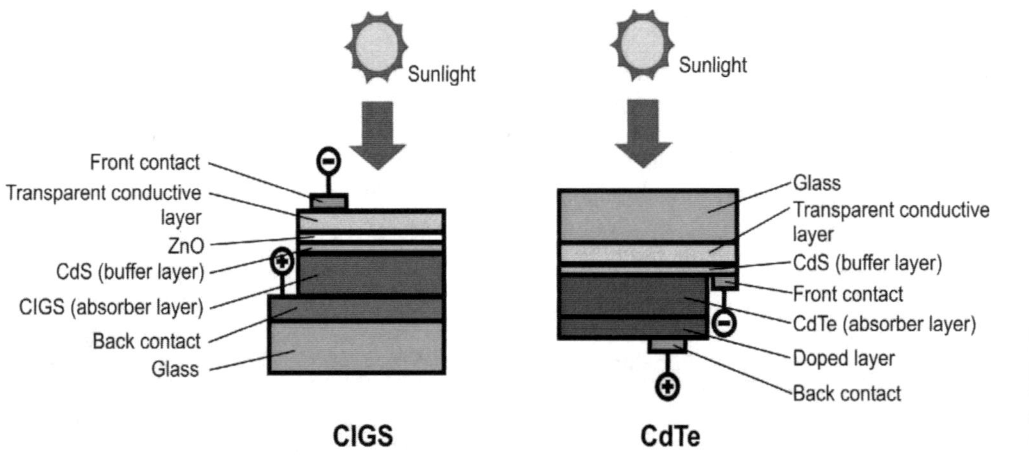

Figure 1. Simplified diagram of solar cells on the base of CdTe and CIGS semiconductor layers.

2. Experimental

Well known that different schemes for CdTe and CIGS PhEC fabrication can be applied. The difference between them is mostly in the order of layer deposition and nature of some working layers. Assembling of CdTe PhEC usually starts from glass or polymer substrate with transparent conductor layers (such as ITO, etc.). This approach has been called "substrate method". Opposite scheme of assembling is called "superstrate method" and it is usually applied in the case of CIGS (see fig. 1).

2.1. Vacuum technologies for thin film structures formation

2.1.1. Thermal vacuum technologies and close space sublimation. Thermal vacuum technologies were used for thin film solar cells ITO/CdS/CdTe/Cu/Au formation on the base of flexible polyimide films [2].

2.1.2. Ion-beam sputtering. Ion-beam sputtering was used for transparent electroconductive SnO_2 layer forming on glass substrate.

2.1.3. Magnetron sputtering. Magnetron was used for sputtering of ITO, CdS and CdTe layers, as well as for metal layer of back contact in CdTe PhEC.

2.1.4. Ion-plasma sputtering. Ion-plasma method was used for metal precursors (Cu-In-Ga) sputtering with 10 nm/min rate.

2.1.5. Metal precursor selenization. Metal precursor selenization is very important stage of CIGS PhEC production. The selenization was carried out in two stages. First stage is the reaction between metals and selenium at reasonably low temperature (250 ^{0}C). Second stage is slow recrystallization of semiconductor layer at the temperature region from 250 to 540 ºC.

2.2. Wet technologies of thin film structure formation

2.2.1. Chemical deposition methods. Chemical bath deposition is used for buffer CdS layer deposition in different PhEC constructions [6]. This process is based on thiourea decomposition reaction in presence of cadmium salt water solution (see equations 1-3).

$$4NH_3 + Cd^{2+} \leftrightarrow [Cd(NH_3)_4]^{2+} \quad (1)$$

$$SC(NH_2)_2 + 2OH^- \rightarrow S^{2-} + CH_2N_2 + 2H_20 \quad (2)$$

$$[Cd(NH_3)_4]^{2+} + S^{2-} \rightarrow CdS + 4NH_3 \quad (3)$$

2.2.2. Pulverization method. Pulverization method was used for TCO and CdTe layers formation. For this aim we had synthesized special colloidal CdTe dispersion that can be applied for such process.

2.2.3. Electric deposition. The advantage of electric deposition process is in ability of preparation of high homogenous films for large areas, which is important for mass production. Electrochemical behavior of Se (IV), Cu(II), In(III), Ga(III) ions is studied in different electrolytes on the base of sulfuric, sulfanic, citric and sulfosalicylic acids at pH (0,5÷3) on glass-carbonic and molybdenum electrodes [7]. Process can be described with following reactions:

$$Cu^{2+} + In^{3+} + 2H_2SeO_3 + 8H+ + 13e^- \rightarrow CuInSe_2 + 6H_2O \quad (4)$$

$$Cu^{+2} + xH_2SeO_3 + 4xH+ + (2+4x)\bar{e} \rightarrow CuSe_x + 3xH2O \quad (5)$$

$$2CuSe_x+2xH++2x\bar{e} \rightarrow Cu_2Se+xH_2Se \quad (6)$$

$$Se^{-2} + In^{+3} = In_2Se_3 \quad (7)$$

3. Results

3.1. Development of CIGS PhEC

For production of CIGS PhEC with Mo/Cu(In,Ga)Se2/CdS/ZnO/(Al-Ni) structure on glass substrate there were tested various combinations of wet and vacuum technologies. In the basic technology films were sputtered onto glass substrates with proportions 7.0 x 2.5 cm^2. Metal precursors (Cu-In-Ga) were sputtered with ion-plasma sputtering with the speed of 10 nm/min. After that heat treatment of the initial Cu-In-Ga layers in an inert atmosphere in the presence of Se and S vapors to self-organized synthesis of homogeneous films $Cu(In,Ga)(S,Se)_2$ solid solutions was held. At the first stage selenium including into precursor and binary phase formation was held under temperature of 250 ^{o}C. It supports slow reaction of Se with binary phase, which leads to better quality of crystal film. It also allows to include Se into precursors before recrystallization process, which supports better p-type conductivity. Second stage of CIGS films formation is held with 250-540°C. Further temperature rising did not lead to structure properties enhancement. Ratio of Cu:In:Ga:Se elements in film was 25,24:20,47:5,17:49,12, respectively. Best laboratory CIGS samples achieved efficiency 11%. Optimization of Ga concentration in $Cu(In,Ga)Se_2$ is expected to improve PhEC parameters and develop stable technologies of production of solar cells with efficiency higher than 15% at glass substrate and 10-12% at metal substrate.

Comparative analysis of In_2S_3 layer usage instead of toxic CdS was also held. An analysis of the charge-transport mechanism showed that replacement of the CdS barrier layer by In_2S_3 in the obtained thin-film heterostructures (TFHSs) does not result in significant changes in the electrical properties. Observed broadband photoconversion and induced photopleochroism are indicative of interference antireflection of TFHSs. Thus, it was established that In_2S_3 films can be introduced into TFHSs as barriers. This technology can be used for the fabrication of ecologically safe cadmium-free next generation of TFHSs, while the use of polarization photoelectric spectroscopy can ensure the monitoring of the widerange antireflection of these photoconverters.
The electrochemical behavior of Se (IV), Cu(II), In(III), Ga(III) ions in different electrolytes (Sulfuric, sulfomat, citric, sulfosalicylic acid-based electrolytes for pH (0,5÷3) on glass carbon and molybdenum electrodes is studied. The cyclic sweep method is used to research the characteristics of reduction these ions in details. The new electrolyte on the base of sulfosalicylic acid allows to electrodeposite the thin films of compounds $CuInSe_2$ and $CuInGaSe_2$ [7].

3.2. Development of CdTe PhEC
Various combinations of vacuum and wet technologies were used for CdTe PhEC producing. CdTe contact with metal forms Schottky barrier. So there was formed an intermediate system of layers between CdTe and metal electrode. Best efficiency for CdTe PhEC was achieved for thin film solar cells ITO/CdS/CdTe/Cu/Au. They were formed at polyimide films by vacuum methods [2]. Technological conditions, output parameters and light diode characteristics of this thin film solar cells were investigated.
Experimental research demonstrated that optimal thickness of CdS in flexible ITO/CdS/CdTe/Cu/Au solar cells (fig. 2) comes to 0.5 μ. It depends on two competitive physical processes. First is diode saturation current density and the second is photocurrent density.
It was revealed that polyimide films with TCO layers preliminary annealing with 400 °C temperature allows to improve solar cell efficiency up to 11.4%. It improves because of open circuit voltage growth from 733 mV to 765 mV, fill factor growth from 0.59 to 0.71 and short circuit current growth from 19.3 to 20.9 mA/cm^2. Light diode characteristics analysis demonstrates that efficiency improvement occurs because of serial resistance decrease to $R_s = 2.1\ \Omega/cm^2$, diode current saturation decrease to $J_0 = 1.0 \cdot 10^{-8}$ A/cm^2, photocurrent increase up to $J_F = 21$ mA/cm^2, and shunting resistance increase up to $R_{sh} = 6400\ \Omega \cdot cm^2$.

Figure 2. Flexible CdS/CdTe based solar module sample.

For PhEC module fabrication there was used corresponding metal mask tooling. This method is more cost effective then photolithograthy. Technological innovation of new module (fig. 2) construction consists in fabrication of high-resistance CdS and CdTe layers without chloride activation as dielectric layer for elements separating in the module. For the first time there was fabricated flexible CdTe micromodule at polyimide films with 5.4% efficiency, composed of five consecutive elements, and with 4.5%, composed of four parallel elements. Analysis shows that cause of micromodule lower efficiency is in decreasing of all output parameters (fig.3). It is connected with active area enlarging, which causes higher probability of macrodefects formation.

Methods of thin film CdTe PhEC producing with ion-beam deposition of SnO_2 at glass substrate, chemical deposition of CdS and electrochemical CdTe deposition were also developed. Speed of electrochemical deposition is limited by tellurium diffusion to electrode. Optimal electrochemical deposition potential satisfies requirements of tellurium (IV) reduction-limit current, but does not reach redox of cadmium potential.

The morphology of surface had been investigated by AFM microscope. It showed the column structure. The sizes of the particles as-deposited CdTe film on CdS were 50-100nm and width of the forbidden zone of CdTe layers was 1,5 eV. It is more than 1,44 eV for CdTe monocrystal.

Such methods result in high fill factor 71-75%. It allows to predict 10% efficiency. [7].

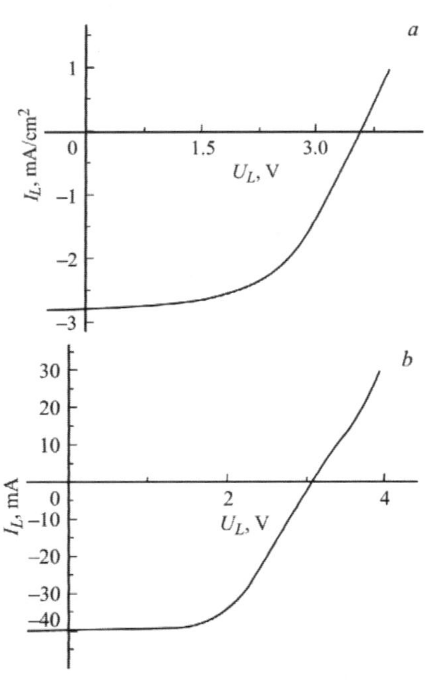

Figure 3. J-V curves of flexible micromodule (a) and full module (b).

Also methods of formation thin film CdTe PhEC based on magnetron and “wet” technologies are developed. They show encouraging results. In particular PhEC with 3.5% efficiency was obtained at balanced magnetron with using cold glass substrate and 6.5% at balanced magnetron with hot glass substrate. Thicknesses of layers were optimized. Optimal materials and technique of back contact deposition were selected. Also active layers are supposed to be deposited by unbalanced magnetron at polyimide films as substrate. Modules would be produced by laser scribing.

3.3. Luminescent filters (LF) for PhEC efficiency increasing

Common disadvantage of chalcogenide PhEC is low efficiency of photoelectrical conversion in spectral region below 500 nm. The problem of loses in short wavelengths region is connected with presence of CdS buffer layer, which is good absorber of sun light in blue spectral area. Furthermore, these losses increase under polyimide film usage as a substrate instead glass. One of the ways of this problem overcoming is down-conversion of solar radiation by LF. Such filters contain organic luminophores (OL) or colloidal quantum dots (CQD) [9,10] (see fig. 4). Undoubted advantage of CQD relatively to OL is in high radiation resistance and stable optical properties during long operation under solar radiation treatment. State-of-the-art synthetic capabilities allow to modify full width half maximum and wavelength of CQD fluorescence in wide spectral area. That is good for optimal spectral response of solar cells.

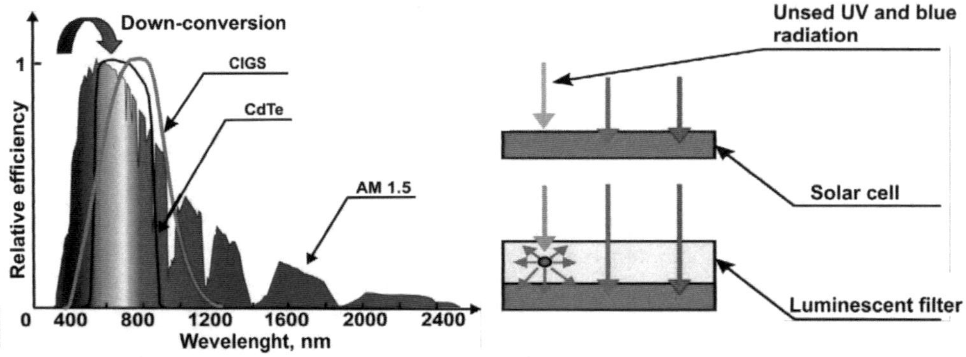

Figure 4. Concept to improving of efficiency of CdTe and CIGS solar cells with luminescent filter.

Figure 5 shows spectral characteristics of CdTe solar cells without LF and with LF. It demonstrates efficiency increasing due to down conversion of some amount of sun power from blue region to the reddish (useful) one. OL which have been used in our work are depicted in table 1. Comparative investigation showed that both CQD and OL can be used in fabrication of LF.

Photopolymerization of polymethylmethacrylate (PMMA) and other acrylic monomers was mostly used for LF production in our work. Acrylic copolymers are sufficient transparent polymeric matrix for OL, as well as for CQD. The set of experimental investigations helped us to reveal an optimal concentration of OL and thickness of LF. As an example, we found that OL MNBI-4 has limited solubility in acrylic monomers. The maximum concentration of MNBI-4 luminophore can be only about 0.5 g/L. Definitely, this fact influents on down conversion efficiency of LF. Coumarin-6 is good soluble, so its concentration varied from 1 to 2 g/L. The higher concentration of luminophore leads to the higher efficiency of down conversion. Also, to get the best results we played with thickness of LF, so it was varied from 0.08 to 0.94 mm.

Journal of Physics: Conference Series **291** (2011) 012049 doi:10.1088/1742-6596/291/1/012049

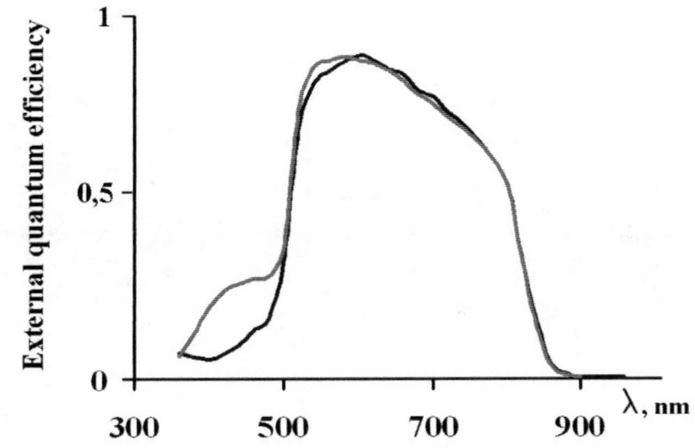

Figure 5. PhEC spectral response with LF and without LF.

In the set of experiments we used LF with different thicknesses and different concentration of luminophores. Under increasing of thickness with the same concentration of luminophores short circuit photocurrent saturates and even tends to go down as you can see from figure 6. It allows to find an optimal thickness of LF and concentration of OL for the best sun light conversion.

Table 1. OL used in our work.

Luminophor	Thickness (mg cm^{-2})	Fluorescence quantum yield, %
4-MNBI 4-morpholine-1,8-naphteline-1',2'-benzoimidazole	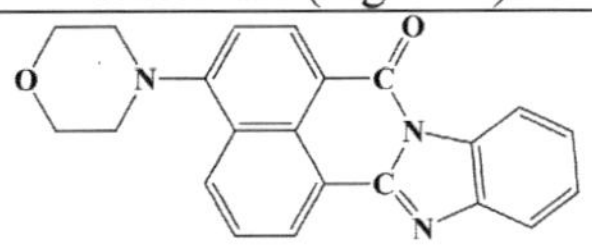	40 – 50
Coumarin 6 3-(2-benzothiazolyn)-7-(diethylamino) coumarin	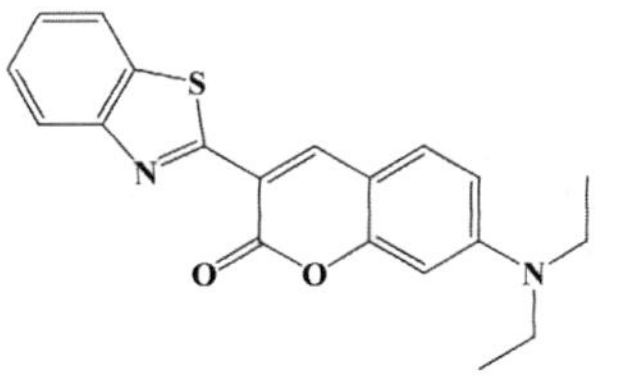	78 (in ethanol)

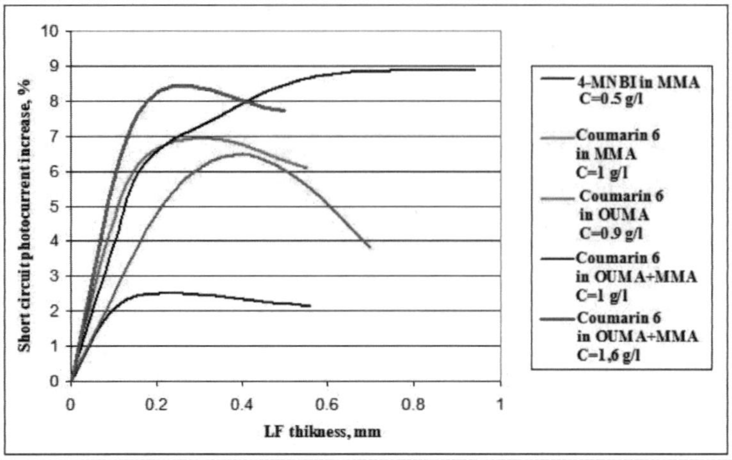

Figure 6. Short circuit photocurrent increasing for CdTe PhEC in the dependence of luminophores concentration and LF thickness.

Application of LF gave the relative increasing of efficiency for CdTe (10%) and for CIGS (9%). Thus, summary efficiency of CdTe and CIGS PhECs can be lift up to 11%.

4. Conclusion

Various methods of production of chalcogenide PhECs were investigated. Maximal efficiencies 11.4% for CdTe PhEC and 10.7% for CIGS PhEC are achieved. Application of LF allows to increase efficiency of chalcogenide PhECs up to 10% additionally to the maximal efficiency. Joint efforts of scientific teams from Russia, Belarus, Ukraine, and Kazakhstan allow to solve problems of production of thin film chalcogenide solar cells with competitive technical and economical properties.

5. References

[1] Gremenok V F, Tivanov M S, Zalesskii V B 2007 Solar cells consisting of semiconductor materials (Minsk: BSU Center) p 222

[2] Khripunov G S, Chernyh E P, Kovtun N A, Belonogov E K 2009 *J. Semiconductor physics and technics* 43 1084-1089

[3] Belonogov E K, Khripunov G S 2009 *J. Alternative energetic and ecology* 2 76

[4] Mudryi A V, Gremenok V F, Zachmann H, Braun A, Heinker S, Yakushev M V 2009 *J. Thin Solid Films* 517 2209-2212

[5] Zalesskii V, Gremenok V 2007 *J. Alternative enrgetics* 507 14-17

[6] Rybakova A V, Filin S V, Gladyshev P P 2010 Regional scientific conference with international participation “Chemistry-2010. Nanochemistry”, 3-4 June 2010, Thesis reports, Dubna 2010, p 31

[7] Dergacheva M B, Statsyuk V N, Fogel L A 2004 *J. Journal of Applied Chemistry* 77 No 2.

[8] Filin S V, Puzynin A I 2010 Regional scientific conference with international participation “Chemistry-2010. Nanochemistry”, 3-4 June 2010, Thesis reports, Dubna p 37

[9] Gladyshev P P, Filin S V, Wakstein M S. et. 2009 Second International Nanotechnological Forum Rusnanotech 2009 (6-8 october 2009, Moscow) pp 61-62

[10] Gladyshev P P, Filin S V, Wakstein M S and et. 2010 The 22nd ISTC Korea Workshop, May 11, 2010. Gumi Electronics & Information Institute, Korea, pp 23-29

Interaction of silver nanoparticles with biological objects: antimicrobial properties and toxicity for the other living organisms

E M Egorova
[1]Laboratory of nanopathology,
Institute of General Pathology and Patophysiology of RAMS,
Baltijskaya st., 8, 125315 Moscow, Russia
[2]Science-technology Company "Nanomet", Moscow, Russia

E-mail: emenano@mail.ru

Abstract. This paper presents several examples of the biological effects of small-sized silver nanoparticles (10.5±3.5nm) observed in experiments on bacteria, slim mold, unicellular alga and plant seeds. The nanoparticles were prepared by the biochemical synthesis, based on the reduction of metal ions in reverse vicelles by biological reductants - natural plant pigments (flavonoids). It is found that, except for the plant seeds, silver nanoparticles (SNP) act as a strong toxic agent, both in water solution and as part of liquid-phase material. It is shown also that the biological action of silver nanoparticles can not be reduced to the toxic action of silver ions in equivalent concentrations or to that of the surfactant (the SNP stabilizer) present in the SNP water solution. Possible SNP applications are suggested.

1. Introduction

Silver nanoparticles (SNP) are widely used now for the creation and production of modified materials with special properties. However, the mode of their action on the living organisms and the conditions providing the safety of their applications remain poorly understood. Therefore, of particular interest are studies allowing to reveal the effects of nanoparticles on various functions of the biological systems of various levels of organization. In particular, it is important to test the SNP antimicrobial activity, in order to define their minimal concentrations providing the 100% death of a given microorganism and thus to estimate their effectiveness for the use as disinfectant. At the same time, it is clear that it is desirable to avoid the pathological changes in the plant, animal or human organisms caused by their contact with SNP present in air, water or on various surfaces after treatment with SNP-containing disinfectants. Hence it is reasonable to assume that studies of the antimicrobial properties of silver nanoparticles should be combined with the tests of their biological action allowing to determine the concentration limits within which the nanoparticles are not toxic for a given higher organism subjected to their influence.

It is well-known today that size, form, stability and properties of metal nanoparticles depend on the method used for their synthesis. So it seems justified to test the action of silver nanoparticles on various biological objects using the SNP preparations obtained by one and the same method. Such an attempt was made by us in studies of the biological effects of SNP obtained by the original method of biochemical synthesis [1,2], based on the reduction of metal ions by natural plant pigments (quercetin, rutin, morin) in reverse micelles formed by the anionic surfactant AOT. As shown earlier [2-4], the method allows to obtain SNP small in size (below 25 nm) and stable on air in micellar solution for a long time (up to several years). For the studies of the nanoparticles interaction with biological objects we used water solutions of SNP obtained by their transfer from micellar solution into the water phase.

In this paper we briefly describe the results of studies of the SNP effects on bacteria species in water solution and in the biodegradable polymer films, as well as of the toxicity exhibited by these nanoparticles towards some other biological systems. The experiments were carried out on small nanoparticles (about 10 nm in size) for which one could expect that specific properties of the nanosized state are clearly expressed

2. Materials and Methods

Silver nanoparticles were obtained by the biochemical synthesis in AOT reverse micelles. Here the reduction of metal ions to atoms and subsequent formation of nanoparticles takes place in the micelle water core at small hydration extent w = $[H_2O]/[AOT] < 4$. From micellar solutions, water dispersions of SNP were prepared by the specially developed procedure [5]. The SNP concentration in micellar or water solutions was found from optical absorption spectra recorded on spectrophotometer Helios-α (Thermo Electronics, GB) in 1 mm quartz cell. Particle sizes in micellar or water solution were determined by transmission electron microscopy (TEM) on LEO912 AB OMEGA microscope (Carl Zeiss, Germany). From electron micrographs particle size distributions were found for no less than 300 particles using Gauss approximation.

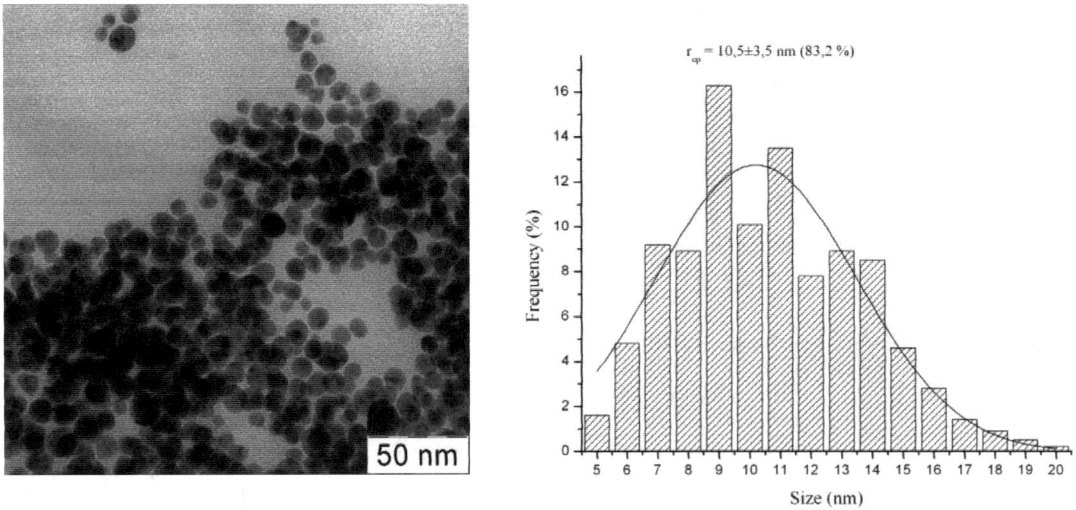

Figure 1. TEM image and size distribution of silver nanoparticles in standard water solution.

The spherical SNP 10.5 ± 3.5 nm in size (see Figure 1) were used in studies of the biological effects both as water solutions and as a part of liquid-phase (e.g. paints or polymer films) or solid materials (activated carbon, cloths, polyamide filtering membranes covered with SNP). The SNP concentration in solution is expressed either in the corresponding molar concentration of silver salt or as equivalent weight content of metallic silver (in mg/liter or in μg/ml). The details of experimental procedures in various biological tests are given in [6].

3. Results and discussion

We present here three examples of the antimicrobial activity of SNP and three examples of their effect on the other biological objects – acellular slim mold, unicellular alga and plant seeds.

Antimicrobial activity is illustrated by the tests with (1) SNP water solution and (2) biodegradable polymer film containing SNP.

(1) In the tests fulfilled on pathogen bacteria species, Esherichia coli and Staphyllococcus aureus, bacteria cell suspensions with $2*10^5$ CFU/ml were incubated with various SNP concentrations obtained by dilution of the initial SNP water solution (Table 1). The SNP concentration in cell suspension, C(SNP) lied in the range 0.27 - 6.5 μg/ml. The initial SNP concentration was 216 μg/ml. It was found that both bacteria species were fully destroyed at C(SNP) ≥2.88 μg/l SNP (≥1.5% of initial concentration) after 30 min of incubation.

Table 1. Action of SNP water solution on the bacteria species Esherichia coli 1257 and Staphyllococcus aureus 906 in aqueous medium. Tests of the antibacterial activity were carried out by means of incubation of bacteria with silver nanoparticles at various dilutions of SNP initial water solution. (-) – absence and (+) – presence of bacterial growth. The data obtained by Moscow Disinfection Center.

Microorganism	Exposition (hours)	Bacteria growth at various C(SNP). Upper row - % of initial SNP concentration, lower row – SNP concentration in the medium (μg/ml)							
		3 (6,5)	2 (4,32)	1,5 (2,88)	1.0 (2,16)	0,75 (1,44)	0,5 (1,08)	0,25 (0,54)	0,125 (0,27)
E.coli 1257	0,5	-	-	-	+	+	+	+	+
	1,0	-	-	-	+	+	+	+	+
	2,0	-	-	-	+	+	+	+	+
	24,0	-	-	-	-	-	+	+	+
S.aureus 906	0,5	-	-	-	+	+	+	+	+
	1,0	-	-	-	+	+	+	+	+
	2,0	-	-	-	+	+	+	+	+
	24,0	-	-	-	-	-	+	+	+

(2) Figure 2 shows the results observed in the tests on E.coli in water suspension. It is seen that, at the high initial concentration of bacteria cells ($3*10^8$ CFU/ml) the high level of inactivation (> 90%) is achieved after incubation for 30 min in the whole range of SNP concentrations (3 – 30 μg/ml). Beginning from 5 μg/ml, total death of bacteria (100 % inactivation) is registered.

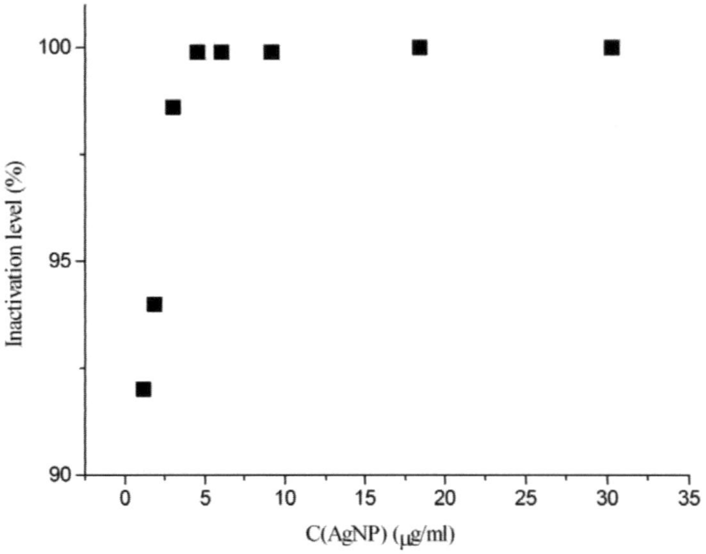

Figure 2. The effect of SNP on inactivation level of E.coli cells in water suspension. The SNP concentration in suspension was varied by means of dilution of the initial SNP water solution. Initial concentrations of SNP – 60.5 μg/ml, AOT – 20 mM. At each SNP concentration the suspension was incubated for 30 min before quenching the SNP activity. Initial concentration of bacteria cells – $3^* 10^8$ CFU/ml. Data from the Institute for Genetics and Selection of Industrial Microorganisms of the State Scientific Center (Moscow).

(3) Polymer films with SNP were obtained by mixing the SNP water solution with that of biodegradable polymer (caboxymethyl chitin); details of the film preparation are given in [7]. Such films were tested on antimicrobial activity against Staphyllococcus aureus and Salmonella typhimurium (Table 2). It was found that the films containing small SNP additions (0.03-0.06 wt%) possess a high antimicrobial activity, at high concentrations of bacteria cells in the films ($\sim 4^*10^7$ CFU/cm^3).The results obtained testify to the strong bactericidal effect of the films with SNP. It is supposed that such films can be applied in medicine, for example, in the treatment of skin injuries.

Table 2. Dynamics of the interaction of bacteria strains with polymer films made from carboxymethyl chitin with or without SNP. Bacterial culture was introduced into the segment (1/4) of polymer film (d=50 mm, thickness 50 μ). After incubation for a given time films were resuspended in the nutrient broth and the CFU number was calculated after cultivation in the standard conditions. Data obtained from the Gamaleya Institute of epidemiology and microbiology (RAMS).

Bacteria species, Dose	Films without (control) and with various concentrations of SNP	Log of the number of living cells in the films after the time periods (hours)			
		1	3,0	6,0	24,0
Salmonella typhimurium	Initial bacteria culture	6,0	6,0	6,0	6,0
TMLR66	Chit - 10. A (control)	6,0	5,8	5,5	3,0

10^6 KOE	Chit – 10. B (0,03wt% Ag)	5,1	4,2	3,0	0
	Chit – 10.C (0,06 wt% Ag)	0	0	0	0
Staphylococcus aureus	Initial bacteria culture	6,0	6,0	6,0	6,0
10^6 KOE	Chit - 10. A (control)	6,0	5,7	5,3	3,5
	Chit – 10. B (0,03wt% Ag)	5,2	4,4	3,2	0
	Chit – 10.C (0,06 wt% Ag)	0	0	0	0

From the results of experiments described above and other similar tests (see [6]) it was concluded that SNP water solutions possess a high antimicrobial activity and can be considered as basic solutions for the creation of disinfectants of a new type, effective and less dangerous for users than those containing chlorine and its derivatives or amino compounds.

The effect on living organisms of the higher levels of organization was studied on the objects of various types, including plasmodium of the slim mold, unicellular algae, plant seeds, animal cells in vitro and animal organisms. In the control experiments, SNP action was compared to that of Ag^+ ions in equivalent concetrations and of the AOT water solution added to the concentration equal to that introduced with SNP water solution. Here we give three examples of the in vitro studies of the SNP effect.

(1) Plasmodium of the slim mold *Physarum polycephalum* is a popular object used in studies of chemotactic phenomena. It is a multinuclear protoplasm surrounded by the mutual membrane capable of growth and amoeboid movement. The aim of our experiments (made in collaboration with Institute of theoretical and experimental biophysics of RAS) was to determine the effect of SNP on the Plasmodium motive activity and growth. For the description of experimental procedure see [8]. Example of the result obtained on Petry dishes showing the SNP effect on the Plasmodium growth is presented in Figure 3.

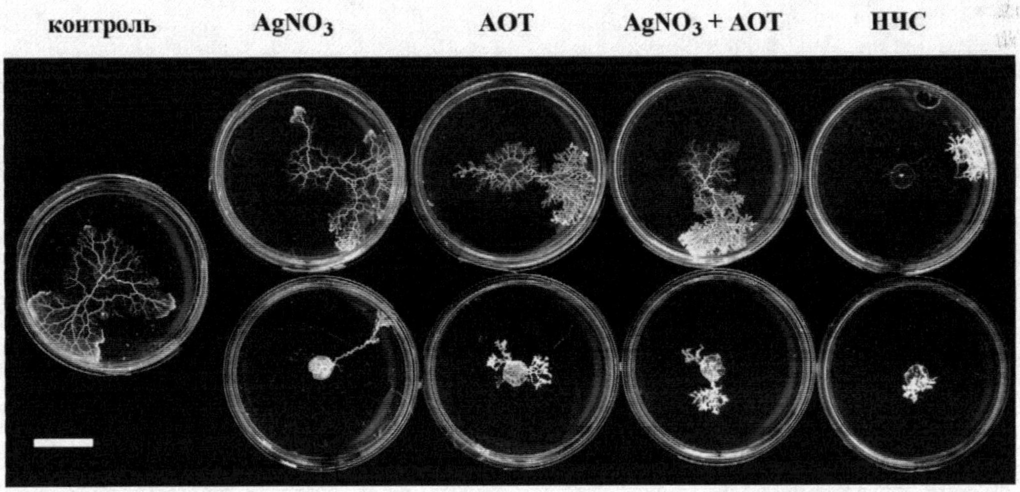

Figure 3. Spreading of Plasmodia on agar substrate in 6h. SNP (НЧС), AOT and Ag^+ concentrations are, respectively, 10^{-5} g-ion/l (1,08 μgAg/ml), $2*10^{-3}$ M and 10^{-5} M (upper row) and ten times higher (lower row). Scale bar, 1 cm. Reproduced from [8] with permission from Springer.

It is seen that, at C(SNP) = 10^{-4} M no growth is observed in all cases.At C(SNP) = 10^{-5} M, there is a marked difference in growth in the medium containing SNP and in three media containing $AgNO_3$, AOT and AOT + $AgNO_3$ mixture. The data obtained allowed to work out the row of effectiveness: $AgNO_3$ << AOT< $AgNO_3$ +AOT << SNP. These results indicate that, in case of SNP, toxic action is strongly amplified, and this effect can not be reduced to the destabilizing action of AOT, or to the release of Ag^+ ions from the surface of nanoparticles. Similar conclusion was made in our studies of the SNP effect on mice *in vivo* and (for Ag^+ ions) in studies of SNP antimicrobial activity on E.coli [6]

(2) Toxic effects of AgNP on microalga *Chlorella vulgaris* were studied by means of microelectrophoresis in collaboration with laboratory for biotesting of the Ministry of Natural Resources (Izhevsk, RF). The use of microelectrophoresis technique for the estimation of toxicity of various agents towards algae is based on the fact that the viability of living cells may be characterized by the surface charge of their plasma membrane, namely, the decrease of surface charge reflects the decrease of cell viability. Hence the toxicity of a given agent may be estimated from the decrease in membrane surface charge which is determined by measuring the changes in the amplitude of cell vibration in the alternative electric field.

Toxic effect is characterized by the toxicity index $T = (A_c - A_{cell})/A_c$ where A_c and A_{cell} are the average amplitudes of cell vibration found for the control and experimental cell suspension, respectively. Apart from SNP, toxic effects of Ag^+ ions and mixture AOT+Ag^+ were estimated independently.

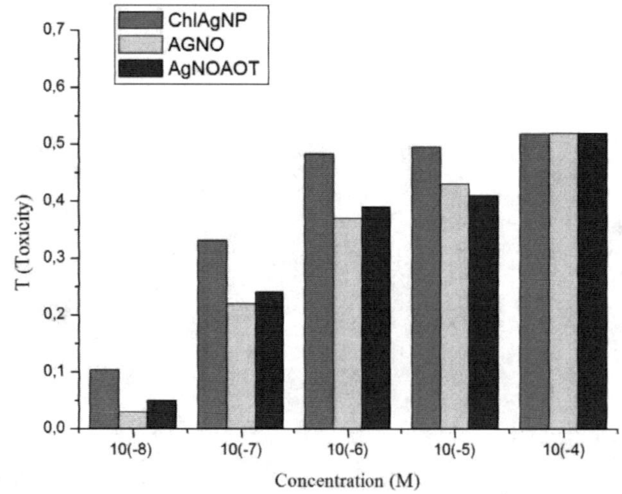

Figure 4. The effect of SNP(AgNP), Ag^+ ions and mixture Ag^+ + AOT on the viability of *Chlorella vulgaris*. The toxicity index T was estimated after 25 min incubation of alga cells in distilled water with the agents studied added to concentrations indicated.

The results are shown in Figure 4. It is seen that the toxicity of SNP for chlorella increases with SNP concentration and then remains constant at C(SNP) from 0.108 μg/ml (10^{-6} M) to 10.8 μg/ml (10^{-4} M), the achieved level of toxicity being equal to approximately half of the maximal (T = 1). At almost all concentrations studied the SNP are more toxic for chlorella than Ag^+ ions or Ag^+ + AOT mixture added to the same total concentrations as those introduced with nanoparticles. Hence here the toxic action of AgNP is unlikely to be caused by only Ag^+ ions released from the nanoparticles surface or by the combined effect of Ag^+ ions and AOT molecules present in the nanoparticles solution. This is in accordance with observations made on Plasmodium Physarum. At the highest concentration (10^{-4} M), the equal toxicity of all three agents studied is, most likely, the result of the cell death or near death state, similar to that observed for Plasmodium at the same concentration (Figure 3).

(3) The effect of SNP on the germination of plant seeds (arabidopsis and soy-bean) and on the lifetime of mice was studied in Vavilov Institute of General Genetics RAS. The influence of SNP water solution was compared to that of AOT and AgNO3 water solutions. The required concentrations of reagents were achieved by dilution of the initial solutions, for details see [6]. The results for arabidopsis are presented in Figure 5.

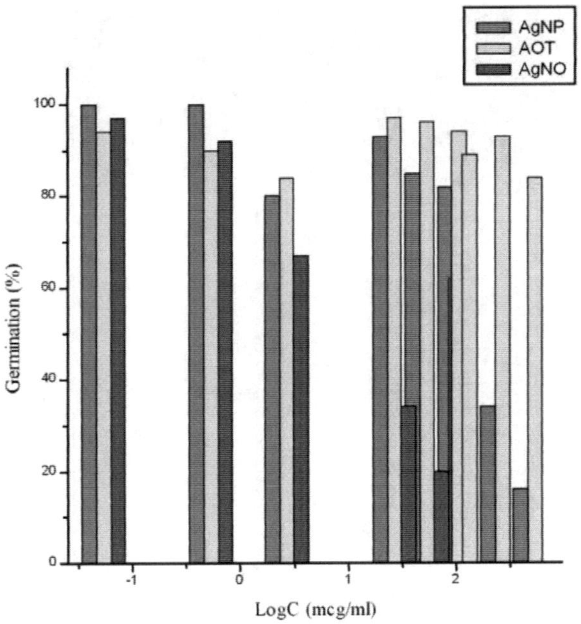

Figure 5. The effect of AgNP, Ag^+ ions and AOT water solutions on the germination of plant seeds (*Arabidopsis thaliana*). The seeds were soaked in distilled water for 24 hours, then washed and placed into Petry dishes with SNP(AgNP), AOT or $AgNO_3$ water solution for another 24 hours. Afterwards the seeds were washed with distilled water and placed for germination in the standard conditions.

As issues from these data, in the wide range of concentrations ($5*10^{-7}$ - $1*10^{-3}$ M) the nanoparticles in water solution are not toxic for the seeds studied. This shows that SNP solution may be used for the protection of seeds from infection by the pathogen bacteria. Unexpectedly, a strong toxic effect was observed for Ag^+ ions, contrary to what was found in experiments with Plasmodium, alga and animal organisms [6]. The origin of this effect remains to be elucidated.

Journal of Physics: Conference Series **291** (2011) 012050 doi:10.1088/1742-6596/291/1/012050

4. Conclusions

1. Silver nanoparticles 10.5±3.5 nm in size act as a strong toxic agent towards both microorganisms and some other living systems, excluding plant seeds studied. With microorganisms, the SNP effect is registered both in the water medium and for the SNP-modified liquid-phase polymer films;
2. In most cases studied, the effect of silver nanoparticles exceeds significantly that of Ag+ ions in equivalent concentrations. This allows to suggest that the effect of silver nanoparticles can not be regarded as only the consequence of the action of Ag+ ions released from the nanoparticle surface.

References

[1] Egorova E M, Revina A A, Kondratieva V S *Patent RF N 2147487* priority from 01.07.1999
[2] Egorova E M, Revina A A 2000 *Colloids Surf. A* **168** 87
[3] Egorova E M, Revina A A 2002 *Kolloidn. Zh.* **64** 334 (*Colloid. J.* **64** 301)
[4] Egorova E M 2010 *Zh. Fiz. Khim.* **4** 713 (*Russ. J. Phys. Chem. A* **84 (4)** 629)
[5] Egorova E M, Revina A A, Rumyantsev B V, et al. 2002 *Zh. Prikl. Khim.* **75** 1620 (*Russ. J. Appl. Chem.* **75** 1585)
[6] Egorova E M. 2010 *In: Silver nanoparticles: properties, characterization and application* (New York: Nova Science Publishers), p.221-258.
[7] Shirokova L N, Alexandrova V A, Egorova E M, Vihoreva G A 2009 *Applied biochemistry and microbiology* **45** 380
[8] Matveeva N B, Egorova E M, Beylina S I, Lednev V V 2006 *Biofizika* **51** 859 (*Biophysics* **51** 758)

GaAs P-I-N structures as detectors of x-ray radiation

L. M. Fedorov, D. I. Mikulik, T.A.Orlova, N.K. Panteleev, N.A. Poletaev, C. A. Snytkina and Yu.V.Zhilyaev

A.F.Ioffe Physico-Technical Institute, RAS , 194021, Russia, Saint-Petersburg, Polytekhnicheskaya, 26

dmitry.mikulik@gmail.com

Abstract. Novel epitaxial growth regimes have been identified to grow thick layers of pure GaAs (up to 250 mcm) of large area in a single growth experiments by gas phase epitaxy. Characteristics of semiconductor epitaxial p-i-n structures based on GaAs have been studied. The obtained layers of i-GaAs have residual impurity concentration < 10^{12} cm^{-3}. Pilot samples of x-ray radiation detector have been fabricated and energy resolution of the devices was about 600 eV at the absorbed photons energy of 60 keV and 200 eV at 5.9 keV, respectively.

1. Introduction

There are two types of semiconductor devices for detection of x-ray radiation: p-i-n diodes and Schottky diodes. Also, photovoltaic detector needing no reverse bias, in contrast to the above two device types, had been reported [1]. At present, silicon, germanium and gallium arsenide are mainly used as semiconductor detector materials for the energy range of 10 – 100 keV. The choice of semiconductor detector material determines the specific efficiency of x-ray quanta registration. This parameter is much higher for GaAs and Ge compared with Si. Device characteristics for GaAs and Ge are close; however, gallium arsenide detectors can be operated at room temperature without cooling. Special mention should be made of the potential of A_2B_6 semiconductor compounds for fabrication of uncooled semiconductor detectors of X-ray, γ-ray and other types of nuclear radiation. Single crystal CdTe-detectors proved their advantages compared to other semiconductor detectors in efficiency of registration of x-ray quanta; however the presence in them of large concentrations of various impurities significantly affects characteristics of these devices. Besides, GaAs is superior to wide-gap A_2B_6 materials regarding charge carriers kinetic parameters (mobility and lifetime), as well as the perfection of many techniques of planar integral technologies [2].

On the strength of the above considerations we have chosen GaAs p-i-n diodes as the most promising and competitive devices from the viewpoint of eventual commercial production. Currently, the main objectives in the fabrication technology of GaAs p-i-n diodes are: achievement of high structure quality, growth of large area structures and approach to ideal device characteristics. The carrier concentration in the pure epitaxial layer should be as low as possible in order to allow the use of the entire thickness of the epitaxial layer as the sensitive detector area. One more problem of the technology is the drastic increase of the reverse current in detectors of large active area (greater than 6 mm^2) [2]. Thickness of the detector active area determines the energy range in which effective registration of x-ray quanta takes place. For our purposes the required active area thickness should be at least 200 mcm, which is a serious challenge for the current growth technology of high-quality gallium arsenide layers.

This study aimed at fabrication of high-quality gallium arsenide p-i-n structures suitable for subsequent manufacture of detectors of x-ray radiation in the range energy 10-100 keV.

2. Experiment

All the semiconductor structures have been grown in a custom made quartz reactor by the method of Chloride Vapor Phase Epitaxy (CVPE) in its standard form [3]. CVPE is a quite well developed technology for fabrication of GaAs layers enabling to achieve high growth rates and good layer quality. In our experiments we used a horizontal flow reactor with resistive heating system capable of maintaining the required temperature regime both in the source zone (800°C) and in the growth зоне (730 – 760°C). $AsCl_3$ was used as arsenic source carried to the reactor with a H_2 flow. In the source zone a boat with metallic Ga was placed.

Epi-ready substrates of n^+-type GaAs doped with Si to 10^{18} cm^{-3} had dimensions of 5x10 cm^2 (rectangular) and a thickness of 300 mcm. The growth process in the reactor was a two-stage one: 1) growth of the thick (up to 200 mcm) high-resistivity basic layer of GaAs with majority carriers concentration of 10^{11} - 10^{13} cm^{-3}; 2) growth of p^+-GaAs layer doped with Zn of a thickness 1-3 mcm and concentration in the range of 10^{18} cm^{-3}. Doping with acceptor was accomplished as follows – metallic Zn was heated to T=500 °C and carried with H_2 flow to the growth zone.

For elimination of longitudinal thickness wedge occurring in the growth in a horizontal reactor an additional flow of $AsCl_3$ was applied, which at high temperature is interacting with H_2 producing hydrogen chloride shifting the useful gas mixture toward the end of the growth zone. This process serves to equalize deposition at the beginning and the end of the growth zone.

For characterization of the obtained layers and structures various methods have been used requiring application of ohmic contracts. The contacts were formed by vacuum deposition simultaneously onto both sides of the structures: on n^+-layer Ge/Au film was deposited, on p^+-layer – Zn/Au film. Then a top layer of Ni was deposited to strengthen the Au films.

3. Result and discussion

In a series of experiments structures with specular surface have been grown. The thickness of the samples was measured on the cleaved edge using optical microscope. The thickness of the active i-layer varied from 150 to 250 mcm depending of the growth conditions, the thickness of p^+-layer of GaAs was on the order of 1 mcm. The use of additional flow of $AsCl_3$ allowed to practically completely eliminate the longitudinal thickness wedge: the thickness reduced by 10 – 20 mcm over 10 cm of the wafer length.

For controlling the electrophysical parameters of pure GaAs layers p^+-layer was etched away for subsequent photoluminescence and voltage-capacitance measurements.

Photoluminescence from the i-layer was measured at 2 K using He-Ne laser (see fig. 1). This diagnostic technique proved useful in previous studies [4]. The measurement data provide information on the purity of i-layers and the presence of acceptors and donors there. D^0x broadening, as a rule, is caused by layer nonhomogeneity and high impurity content. Intensity ratio of the spectral lines makes it possible to estimate residual impurity concentration in the material [4] and gives a value of N_D-N_A = 10^{11} - 10^{12} cm^{-3}.

For estimating the free carrier concentration in the layer of n^0-GaAs voltage-capacitance measurements were performed. The values obtained by this method were $<10^{12}$ $см^{-3}$, which agrees with photoluminescence measurements and meets requirements to the active layer purity of the chosen x-ray detector type.

Current-voltage characteristics of the obtained structures are similar to those of typical diode structures. Of particular interest are reverse branch of CVC because semiconductor detector predominantly operate at reverse bias. In all samples breakdown occurred at reverse voltage not lower than 500 V, which is an evidence of the high purity of the i-layers.

From electroluminescence intensity distribution in the direction perpendicular to the p-n-junction plane the charge carrier diffusion length can be determined. Fig. 2 shows typical electroluminescence pattern from the p-n junctions. From the fact that the electrical field created at the p-n junction boundary (at zero point on the Ox axis) exists throughout the low-doped i-layer thickness it can be concluded that the space charge region extends over the entire i-layer. Thus, the entire n^0-GaAs layer represents the detector active layer, in which efficient registration of x-ray radiation quanta takes place. The calculated diffusion lengths of charge carriers were in the range from 70 to 100 mcm.

One of the main problems in the fabrication of x-ray detectors based on p-i-n structures is the growing of a thick (about 200 mcm) pure high-resistivity GaAs layer of uniform thickness. Insufficient thickness of the n^0-GaAs layer reduces detector sensitivity, negatively affecting the x-ray detector signal. A number of researchers reported [5] that in GaAs layers of a thickness exceeding 100 mcm the amount of defects in the form of growth pyramids is increasing. We were able to eliminate these negative phenomena by optimizing the growth conditions and the fabrication process as a whole, including pre-growth preparation of the substrate material. Also, the possibility of obtaining large area (about 4000 mm^2) structures of uniform thickness and parameters in a single technological process considerably reduces expenses and lowers the costs of commercial production of the detector structures.

The obtained gallium arsenide p-i-n structures have been used to fabricate pilot samples of x-ray radiation detector. Energy resolution of the devices was about 600 eV at the absorbed photons energy of 60 keV and 200 eV at 5.9 keV, respectively.

Besides, prospects exist for manufacture of pixel matrices based on p-i-n structures described above. X-ray radiation detectors have been fabricated with matrices of 256x256 pixels, the pixel size being 55x55 mcm.

An important outcome of this research is that due to the very high quality of the obtained structures in some areas of scientific research it will be possible to operate x-ray detectors without any external power supply, as mentioned above, making use of the photovoltaic effect – separation of electron-hole pairs generated by x-ray photons in the built-in electrical field.

This study has been supported by Russian Foundation for Fundamental Research. (Grant ,№ 10-08-00782)

References

[1] R.A. Akhmadullin, V.F. Dvoryankin, G.G. Dvoryankina, Yu.M. Dikaev et al., Letters to ZhTF, 2002, v. 28, no.1, p.p.34-38

[2] V.M, Zaletin, Yu.V. Tuzov. Uncooled semiconductor radiation detector on epitaxial gallium arsenide for spectral and radiographic analysis, Electronic scientific publication "GEORAZREZ", 2008, Issue #2

[3] M.L. Cardwell, Journal of Crystal Growth 70 (1984) 97

[4] Yu. V. Zhilyaev, N. K. Poletaev, V. M. Botnaryuk T.A. Orlova, L. M. Fedorov, Sh. A. Yusupova, A. Owens, M. Bavdaz, A. Peacock, B. O' Meara, H. Helava. Optical characterization of ultra-pure GaAs. - Phys. Stat. Sol., (c), v. 0(3), p.1024–1027 (2003)

[5] V.F. Dvoryankin, G.G. Dvoryankina, Yu.M. Dikaev et al., ZhTF, 2007, v. 77, no..10, p.p.121-124

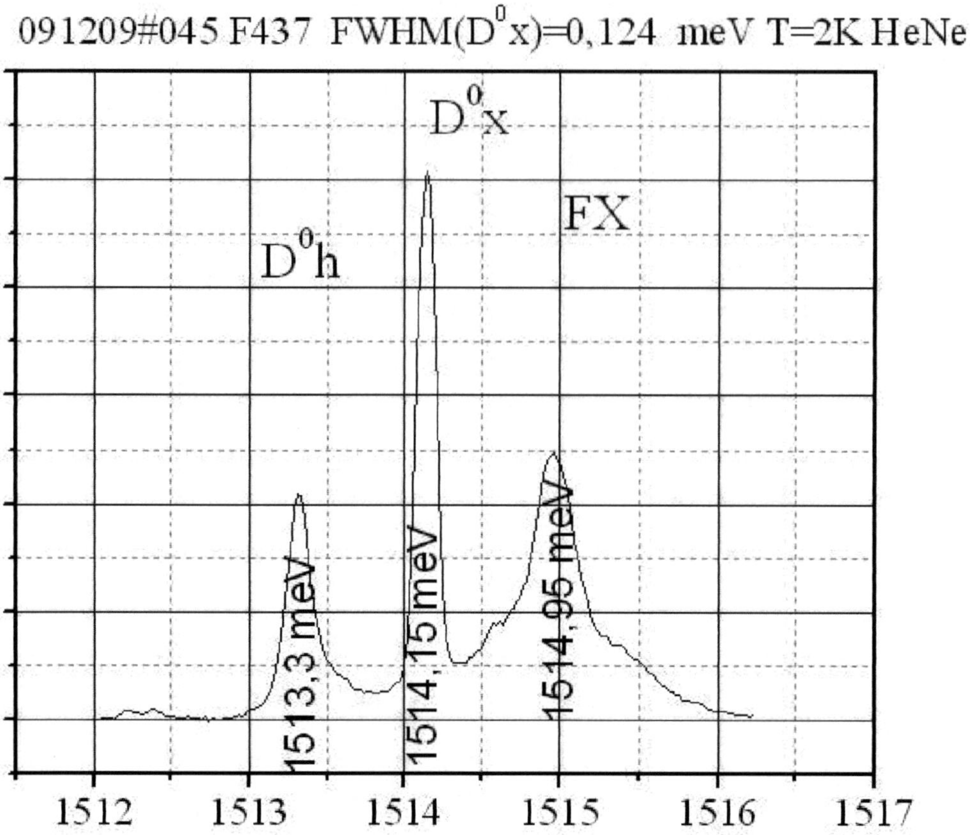

Figure 1. Photoluminescence spectrum of a pure layer of GaAs , concentration N_D-N_A ~ 5×10^{11}cм$^{-3}$

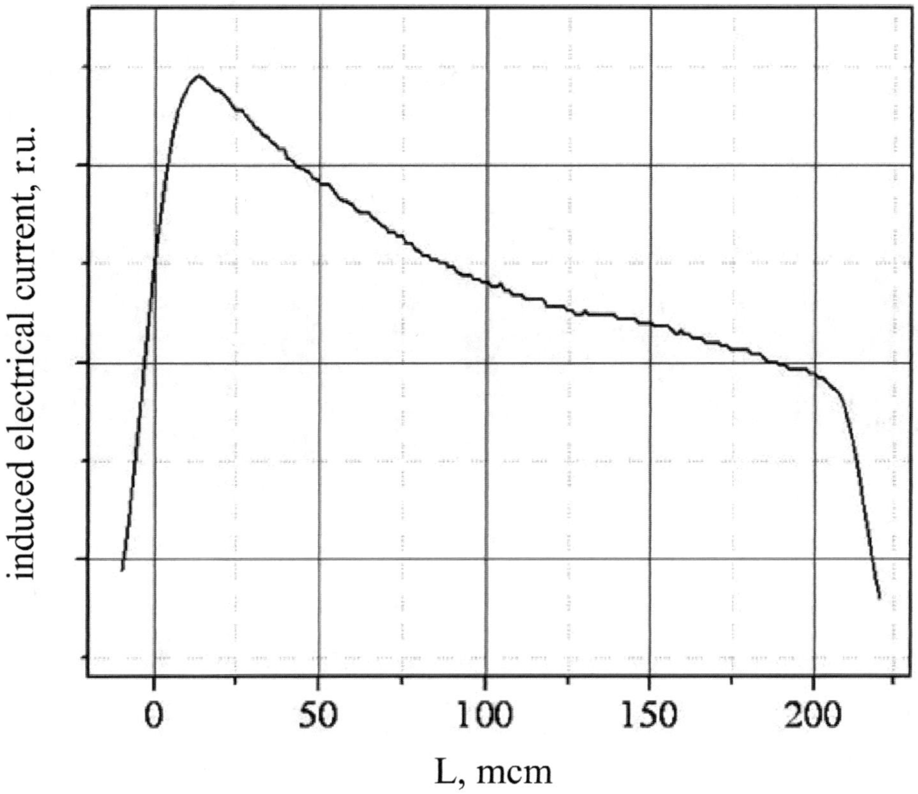

Figure 2. Electrical field distribution in an asymmetric p-n junction obtained from induced electrical current, $d_{n0\text{- GaAs}}$ = 213 mcm.

Journal of Physics: Conference Series **291** (2011) 012052 doi:10.1088/1742-6596/291/1/012052

Reproductive toxicity of carbon nanostructured material - a promising carrier of drugs in laboratory mice

A.A. Gusev[1], E.A. Snegin[2], I.A. Polyakova[1], E.B. Gorsheneva[1], A.G. Tkachev[3], A.V. Emeliyanov[1], S.V. Shutova[1], O.N. Zayceva[1], A.V. Shuklinov[1], A.V. Fedorov[4], T.V. Vasilieva[4], E.A. Smirnova[4], E.M. Lazareva[4],G.E. Onishenko[4]

[1]Tambov State University of the name of G.R. Derzhavin, 392000, Russia, Tambov, Internacionalnaya str., 33,
e-mail: nanosecurity@mail.ru
[2]Belgorod State University, 308015, Russia, Belgorod, Pobedy str., 85
[3]Tambov State Technical University, 392000, Russia, Tambov, str. Sovietskaya, 106
[4]Moscow State University of the name of M.V. Lomonosov, Biological faculty, 119991, Russia, Moscow,
Leninskie gory, 1/12, MSU

INTRODUCTION

Currently, to the prospects of nanotechnology is given an increasing worldwide attention. However, nanomaterials might have very different physical and chemical properties and biological (including toxic) effect than substance in the usual physical and chemical condition that is why they are refered to new types of materials and products. Among the actual and potential areas of their application - microelectronics, micromechanics, energy, environment, agriculture, etc. Especially we focus on medical applications – such as carbon nanotubes, due to its unique properties which is a high strength, hydrophobicity, chemical inertness, high penetrating power, are ideal carriers for drug delivery, providing targeted, the preservation of the active components in the internal environment of an organism , prolongation of action [1]. In a short while a close contact of humans and other biological objects are expected and this problem is one of the most considered issues of the science nowadays. At the same time, the question of the toxicity of carbon nanotubes is the subject of ongoing debate [2]. There are number of papers which demonstrate the negative impact of these structures on living organisms [3,4,5]. More over, the issues of reproductive toxicity of carbon nanotubes are studied improperly. At the same time, the purpose of this study was to evaluate the reaction of the reproductive system on the carbon nanostructured material "Taunit (multi-walled carbon nanotubes), that was injected to laboratory mice. Nutritional path was selected as a way of injection as it is the most

Journal of Physics: Conference Series **291** (2011) 012052 doi:10.1088/1742-6596/291/1/012052

possible method of proceeding nanomaterials into a body [6].

OBJECTS AND METHODS

The object of research - a carbon nanomaterial "Taunit" (multiwalled carbon nanotubes), which is produced on an industrial scale LLC NanoTechCenter (Tambov). This material is one-dimensional, nanoscale, filamentous formation of polycrystalline graphite, a cylindrical shape with an internal channel in the form of loose powder black. Hydrophobic, chemically inert, purity - more than 98%. Granules of nanomaterial of micrometric dimensions have a structure of a tangled bundles of multiwalled tubes (Fig. 1), method of obtaining - chemical vapor deposition on a metal catalyst or CVD-process. "Taunit" is a promising material for the aerospace Industry, nuclear, space technology, medicine, pharmaceutics, production of supercomputers, video equipment, flat screen monitors, filters, multi-purpose. Addition of "Taunit" improves the quality of lubricants, structural composites and building materials. Granules "Taunita" can serve as carriers of catalysts or drugs, as well as adsorbents, cold electron emission sources [7].

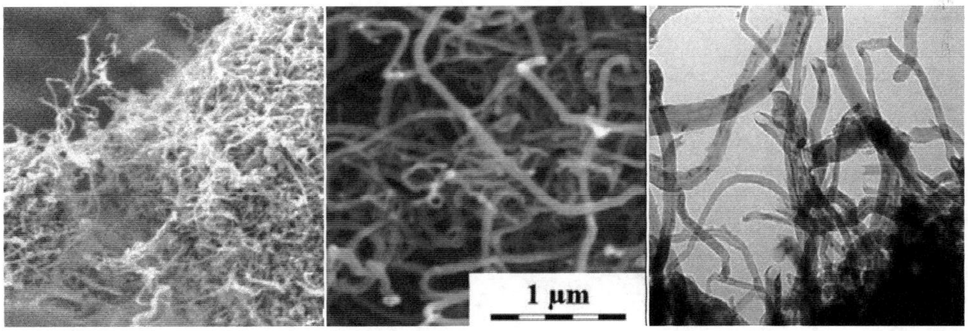

Fig. 1. Electron micrographs of nanostructured carbon material "Taunit". Photos courtesy LLC "NanoTechCenter", Tambov

The experiment consisted of two parts: the first assessed the physiological effect of the investigated nanomaterials in males, in the second - on female laboratory mice.

1. To study the effects of nanomaterials on male reproductive function formed by two coeval group of animals (experimental and control groups) consisting of 10

adult male non-linear laboratory mice each. Throughout the experiment, all animals were kept in standard conditions. Quarantined for 10 days.

Within 30 days of the experimental group was administered orally by UNM "Taunit" by replacing the drinking water in colloidal aqueous solution. The solution was prepared in distilled water using ultrasonic installation. The sample initial CNM prepared for TEM as follows. 25 mg of nanomaterial placed on a surface of liquid pitch (Epon 812), poured in a test tube, centrifuged 3 min at 6000 ×g and polymerized by a standard technique. Ultrathin cuts filled in Epon investigated samples methods TEM and diffractions of electrons by means of appearing through electronic microscopes JEM-1011 (JEOL) with digital camera GATAN ES500W and LEO 912AB (Carl Zeiss).

The average daily dose of the material for each male in the experimental group was 30 mg / kg. Assess the overall condition of the animals.

The control group consisted of mice injected with distilled water in similar quantities.

Upon expiration of the exposure to each male were sat down to 3 Virginy unexposed females. Noted the presence of sexual behavior. Replanting of females to male guinea was performed weekly for three weeks. Separated females were dissected after 15-17 days. Estimated number of pregnant females, females with living fetus, with a dead fetus, the number of living and dead fetuses per litter.

2. To assess the actions of nanomaterial on the reproductive function of females formed two coeval group of animals (experimental and control groups) consisting of 10 adult females, non-linear laboratory mice each. Throughout the experiment, all animals were kept in standard conditions. Quarantined for 10 days. Colloidal aqueous solution of nanomaterials for watering animals prepared as above. After several days of sharing content with the mature unexposed males, females were separated by individual cells and from that moment passed by watering test fluid. The experiment was completed after the end of the nursing offspring. Assessed the following indicators:% of female deaths (overall and for each day of the experiment),% mortality of young during the first three days of life;% not given birth

to females, the number of dead embryos in females, fertility, and general condition of the females.

At each stage, in addition to the experimental group, and contained a similarly studied control group of mice, which is used for drinking distilled water.

RESULTS AND DISCUSSION

1. Preliminary by us it has been spent characterization of structure of nanomaterial. On images structure of CNM (fig. 2A) - conglomerates of MWCNT is visible. Morphological features MWCNT are well visible in drawing 2B. The site with MWCNT, not containing graphitized carbon (fig. 2C), has been characterised by the method of electrons diffraction. As is shown in fig. 2D, MWCNT thanks to the ordered periodic packing of atoms of carbon form on electronogram rings, characteristic for polycrystalline samples.

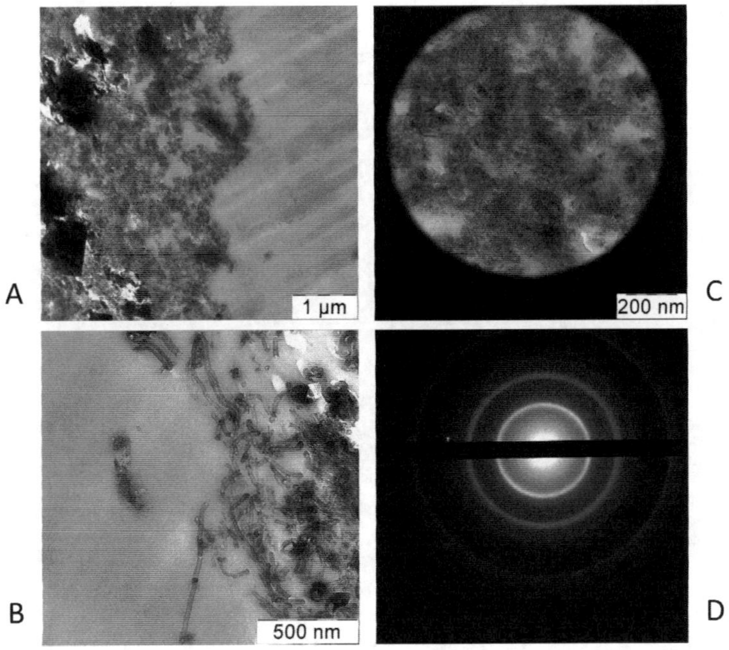

Fig. 2. The structural characteristic of nanomaterial. Explanatories in the text

2.According to the method of study to males from the experimental and control groups during the 3 weeks of sit down Virginy unexposed females of reproductive age. In both groups, there were the presence of normal sexual behavior. External lesions in the experimental group was noted. Follow-up for females revealed that the carbon nanomaterial, introduced orally in a dosage study, causes complete male sterility in laboratory mice (Fig. 3).

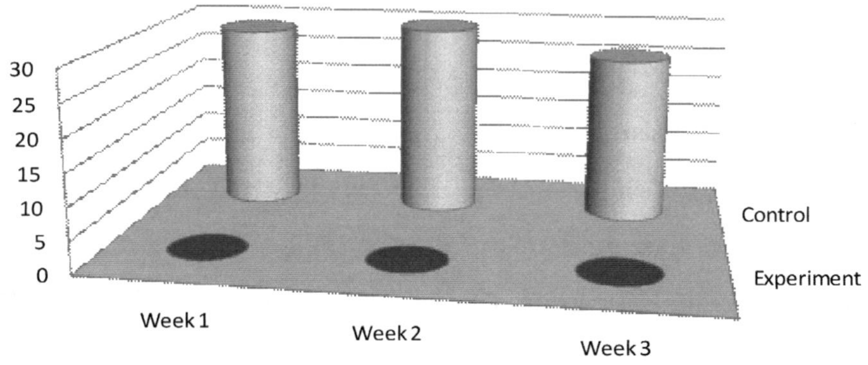

3.Fig. 3. The number of successfully inseminated females during the experiment

Study of histological sections of testes of experimental animals showed significant structural damage in comparison with the control group (Fig. 4). The result can indicate that, along with bodies such as the liver, spleen and lungs testes are target organs for the nanoparticles.

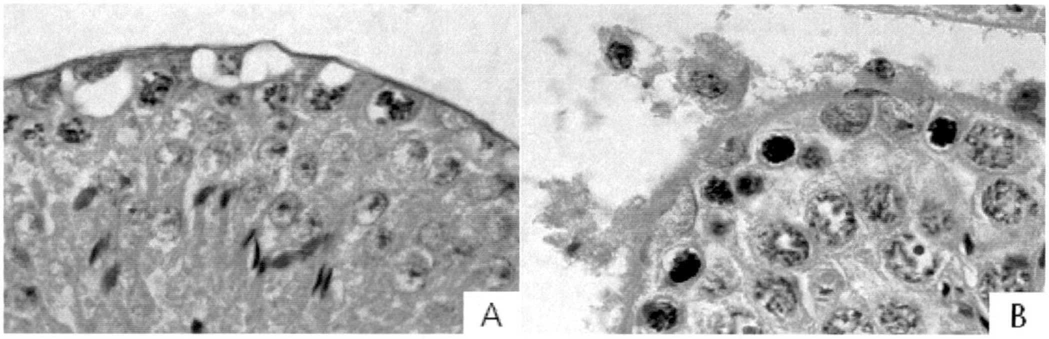

Fig. 4. Histological sections of testes of mice from the control (A) and experimental (B) groups. The right image shows disturbance seminiferous tubules, numerous necrotic cells. The

photos were made members of the Biological Faculty of Moscow State University. MV Lomonosov Moscow State University under the guidance of professor of cell biology and histology, Dr. G. Onishchenko

A possible mechanism of toxicity of carbon nanotubes is oxidative stress. Carbon nanotubes are cylindrical in shape. Presence in the tissue long enough and hard (Multilayer) of nanotubes causes a "failure" in the behavior of macrophages [8]. These cells are unable to swallow "such a large particle, and only emit toxic environment oxygen species (ROS) and hydrolytic enzymes, causing an inflammatory reaction [8]. Oxidative stress can cause infertility in males. Hyperproduction of reactive oxygen species leads to a decrease in cell viability in semen that leads to an inability to fertilize [9]. In the course of nanomaterial toxicity may cause chromosomal aberrations. Most of them occur in the chromosomes derived from the father [10]. Perhaps it is these violations occurred during the experiment, which ultimately led to reduced viability or sperm, or cause damage to spermatogenic tissue, which prevented the normal formation of germ cells. As a result, males of both experimental groups could lose the ability to fertilize.

Stress resulting from mechanical injury or poisoning of nanomaterials may be accompanied by increased levels of stress hormones - corticosteroids and catecholamines. Hormonal changes that accompany stress can cause pathological or adaptive changes at the level of the CNS, particularly in the hypothalamus, resulting in possible violation of the secretion of the hormone GnRH in it, it reduces the release of luteinizing hormone, acting in the testis are Leydig cells, which in response secrete the hormone testosterone. This may subsequently cause dysfunction of the reproductive system [11]. Thus, hormonal imbalance under the influence of nanomaterials could lead to suppression of reproductive function at the level of humoral regulation of spermatogenesis.

2. To study the effect of nanomaterials on female reproductive function, after the joint maintenance of a mature unexposed males and females during pregnancy

and nursing were treated with test fluid. Pathological abnormalities in the appearance and behavior of experimental females have been recorded, on the contrary, they are more heavily recruited body weight than females of the control group. The positive trend was observed in the analysis of the studied indicators of reproductive activity. It was found that when exposed to the organism UNM "Town, administered orally in a dosage of the study is an increase in female fertility by 30% reduction in mortality of young animals by 46% and a decrease in the proportion of individuals not given birth by 17% (Fig. 5).

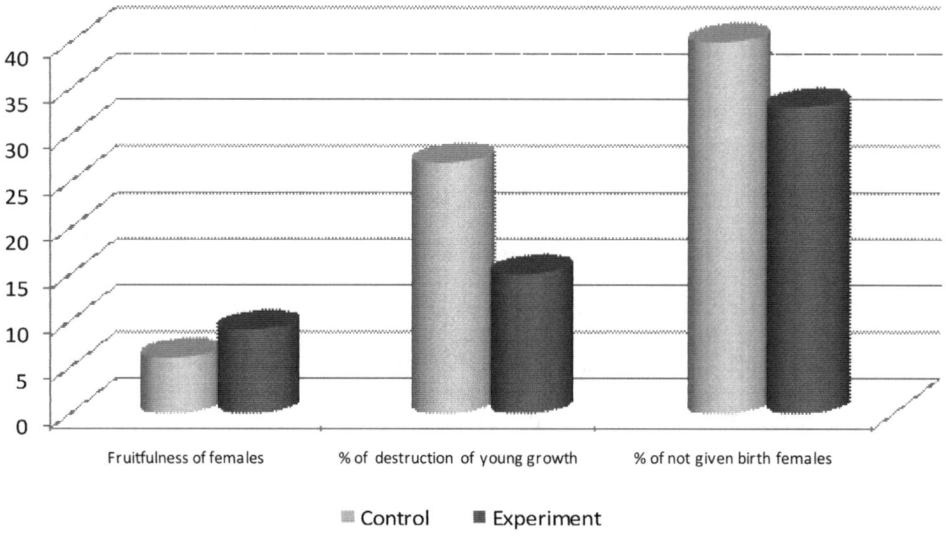

Fig. 5. Comparison srednegruppovyh indicators of female fertility, the proportion who died young, none gave birth to female mice in the control and experimental groups.

A similar effect can be explained as follows. Different reactions to the same effect in the first and second experiments, possibly connected with the peculiarities of physiology of male and female organisms, which are expressed in dissimilar norms of reaction, specific to each gender [12]. For example, the above-described possible hormonal imbalance in the male body can cause spermatoksichesky effect, while in the female body similar hormonal pattern may stimulate the activity of the reproductive system. Anyway, the mechanisms of observed changes we require deeper investigation.

Journal of Physics: Conference Series **291** (2011) 012052 doi:10.1088/1742-6596/291/1/012052

CONCLUSIONS

Thus, study the effect of industrial carbon produced by the nanostructured material on the reproductive function of laboratory mice. Shown that oral administration of colloidal aqueous solution of carbon nanomaterial "Taunit" at a dosage of 30 mg / kg of body weight is a cause of infertility in male mice. At the opening of potentially pregnant females, none in the uterine horns, there were no signs of pregnancy, which constitutes a serious violation of the reproductive function of males under the influence of nanomaterials. At the same time, studied the dosage of nanomaterial stimulates increase fertility in females, leads to a reduction in mortality of young animals, as well as a decrease in the proportion of individuals not given birth.

The results can be taken into account when developing content standards of carbon nanomaterials in the environment, contact with animals and humans, as well as the creation based on multiwalled carbon nanotubes, next-generation drugs.

REFERENCES

1. Foster L. Nanotechnologies. A science, innovations, possibilities. Moscow: Technosfera, 2008. – 352 p.

2. Zhao Yuliang, Xing Genmei, Chai Zhifang. Are carbon nanotubes safe? // Nature Nanotechnology. №4, March 2008. P. 191-192.

3. Kostarelos K. The long and short of carbon nanotube toxicity // Nature Biotechnology, №7, July 2008. V. 26. P. 774-776.

4. Miyawaki J., Yudasaka M., Azami T. et al. Toxicity of Single-Walled Carbon Nanohorns // Acsnano, №2, 2008. V. 2. P. 213–226.

5. Poland C. A., Duffin R., Kinloch I. et al. Carbon nanotubes introduced into the abdominal cavity of mice show asbestoslike pathogenicity in a pilot study // Nature Nanotechnology. July 2008. V. 3. P. 423-428.

6. Onishenko G.G. The decision of the Main state health officer of the Russian Federation

from October, 31st, 2007 №79 Moscow «About the statement of the Concept of toxicological researches, methodology of an estimation of risk, methods of identification and quantitative definition nanomaterial» [Electronic resource]. URL: http://www.rospotrebnadzor.ru/documents/postanov/1344/ (Reference date: 12.02.2010).

7. Tkachev A.G., Zolotuhin I.V. Equipment and methods of synthesis solid-state nanostructure: the monography / – M.: " Publishing Mashinostroenie-1", 2007. – 316 p.

8. Nasibullaev S.K. Fulleren their derivatives and nanotubes [Electronic resource]. URL: http://originweb.info/education/chemistry/fullerenes6.html (Reference date: 22.02.2010).

9. Bikova M.V. Redoks-balance infringement spermatozoids and seed plasma of men at patospermii. The dissertation author's abstract on competition of a scientific degree of Cand.Biol.Sci. Krasnoyarsk, 2008.

10. Uiba V.V. Working out Medico - biological safety in the field of nanotechnologies [Electronic resource]. URL: http://rusnanotech08.rusnanoforum.ru/sadm_files/disk/Docs/1/7/14.pdf (Reference date: 10.03.2010).

11. Kettal V.M, Arki R.U. Pathology of endocrine system. The lane with English - SPb. - M: «The Neva dialect» - «Publishing house BINOM», 2001. - 336 p., ill.12. Geodakyan V.A. Differential death rate and norm of reaction man's and female. Jornal of the general biology. 35 № 3, 1974. P. 376—385.

Research is executed at partial support of the state contracts № 16 24.09.2010 and № P 208 23.04.2010

Journal of Physics: Conference Series **291** (2011) 012053 doi:10.1088/1742-6596/291/1/012053

Structure and properties of titanium β-alloys Ti-Nb-Mo-Zr of medical purpose

O A Golosova[1], T N Vershinina[1], M B Ivanov[1], Yu R Kolobov[1], E S Pigorev[1] and A A Zisman[2]

[1]Research Education and Innovative Centre "Nanostructured Materials and Nanotechnologies", Belgorod State University, Belgorod, Russia

[2]CRISM "Prometey", St. Peterburg, Russia

E-mail: Ollilac@rambler.ru

Abstract. Low-modulus titanium β-alloy Ti-26Nb-7Mo-12Zr for medical application, composed of nontoxic elements V and Al was considered. The influence of deformation in combination with heat treatment on the structure and properties of β-alloys such as elastic modulus, yield strength and tensile strength, ductility was investigated.

1. Introduction

There is an increasing demand in implants made of metallic materials for various purposes in modern medicine in recent years. At the implantation of any biomaterial there is its direct contact with the tissues and media of the human body. Because of this such type of materials should have certain properties. They have to provide not only mechanical and chemical biocompatibility, but wear and corrosion resistance, strength, the ability of germination and integration with the biological environment [1]. One of the most important characteristics of medical alloys is the elastic modulus, which determines the functional reliability of implants in an actual operating conditions in a human body. The value of the elastic modulus should be close to the elastic modulus of bone (30 GPa). It allows to redistribute a significant part of loads on the bone. This corresponds to normal physiological conditions and prevents premature degradation of the bone material [2].

It is believed that the best material for the production of implants are light and durable titanium alloys because of their unique corrosion resistance and biocompatibility [1, 3]. When using titanium alloys for the manufacture of implants and prostheses the extremely favorable factor is their low (approximately two times less than steel) modulus of elasticity [4]. Titanium β-alloys with elastic modulus 60-80 GPa are the most promising from this viewpoint. Elastic modulus of β-alloys is noticeably smaller then that of α-titanium (109 GPa) and most widespread in medicine two-phase (α+β)-type Ti-6Al-4V alloy. Modulus of elasticity of the most rapidly decreasing with correct selection of alloying elements, but they should not contain toxic elements (V, Al, Ni, Co et. al. [5 - 7]). These elements render allergic effect on living tissue or general toxic effect on the human body. Almost ideally biocompatible metal elements are as follows: Pt, Ta, Nb, Ti and Zr [7 - 9]. Alloying of these elements can lead to a decrease in elastic modulus and an increase in the strength of titanium, which provides the best connection to bone and minimal damage at the junction of the bone-implant [4].

In this regard, formerly authors have chosen a system of alloying of niobium, molybdenum and zirconium for the new biomedical β-type titanium alloys.

2. Materials and experimental procedures

In the present work titanium β-alloy Ti-26Nb-7Mo-12Zr in the initial state with an average size of β-grains of 280 microns in the form of a sheet sample of thickness 6.7 mm was investigated. Melting of titanium β-alloy Ti-26Nb-7Mo-12Zr was carried out by triple vacuum arc remelting in Corporation VSMPO-AVISMA (Verkhnyaya Salda, Russia).

Low-modulus titanium β-alloy was subjected to sheet rolling for fine-grained structure formation. Sheet rolling was carried out without heating at room temperature with a reduction of 100-200 microns in one pass. As a result of rolling 3 states with a total reduction of 30, 50 and 70% were obtained. Deformed specimens were annealed at 850°C after rolling and then quenched in water.

Structural studies of titanium β-alloys were carried out using optical microscope Olympus 71 and scanning electron microscope Quanta 600 FEG with a field emission gun. Investigation of the structural-phase composition and the crystallographic texture was also performed using the methods of automatic analysis of the electron back-scattered diffraction patterns (EBSD) on scanning electron microscope Quanta 600 FEG at an accelerating voltage of 20 kV, beam currents of 13 nA, and software TexSEM Lab (TSL). X-ray studies performed on a universal X-ray diffractometer, ARL X`tra.

Specimens for the above methods were cut using electrical discharge machine, and then subjected to mechanical grinding to install TegraPol-31 firm "Struers". Preparation of the sample surface for optical metallography and X-ray diffraction (XRD) analysis included chemical etching in a solution of 4 ml HF, 6 ml HNO_3 and 190 ml H_2O, for scanning electron microscopy using electrolytic polishing at the facility LectroPol-5 (Struers) in a solution of 60 ml of $HClO_4$, 600 ml CH_3OH and 360 ml CH_3 (CH_2) $CH_2OCH_2CH_2OH$ at voltage U = 50 V.

Mechanical tensile tests were carried out on a universal floor electromechanical testing machine Instron 5882 with a strain rate of 1.5 mm / min. The test samples were cut at electrosparking machine in the form of double blades with working part 2.88 x4,08x38 mm^3. Further, the surface of samples subjected to mechanical grinding. According to the results of tests the elastic modulus, yield strength and tensile strength, ductility were evaluated. Modulus of elasticity was evaluated from the slope of the elastic part of stress-strain curve.

3. Results and discussion

The microstructure of the cold-rolled samples of the alloy Ti-26Nb-7Mo-12Zr with a total deformation of 30 and 60% after annealing is represented by partly recrystallized grains, the volume fraction of which is 43 and 94, respectively (figure 1a, b). According to the analysis of electrons back-scattered diffraction patterns (figure 2) the formation and growth of recrystallization nuclei occurs at high-angle boundaries of the original deformed grains. As a result of cold rolling and subsequent annealing with quenching the average size of grains of the alloy decreased by an order of magnitude in compare with the initial state and makes 20,5 ± 0,9 μm (30% deformation and annealing followed by quenching) and 10,1 ± 0,3 μm (60% deformation and annealing followed by quenching).

In the alloy Ti-26Nb-7Mo-12Zr after rolling with a maximum degree of deformation (90%) in contrast to the previous state (with a lesser deformation degree), recrystallization was complete. It led to the formation of a homogeneous globular structure (figure 1c, 3). As can be seen from the presented in Figure 3 histogram of grain size defined by a secant average grain size is 9 μm. As it can be seen from images of the microstructure, there are observed areas with large and small grains. This is probably due to the fact that the nucleation of grains during the initial stages occurs at the grain boundaries, and further in the volume of deformed grains.

According to the analysis of EBSD patterns it was found that in the studied alloy with fully recrystallized structure most of the boundaries represent the high-angle grain boundaries. The volume fraction of such grains is ~ 80% (figure 3a).

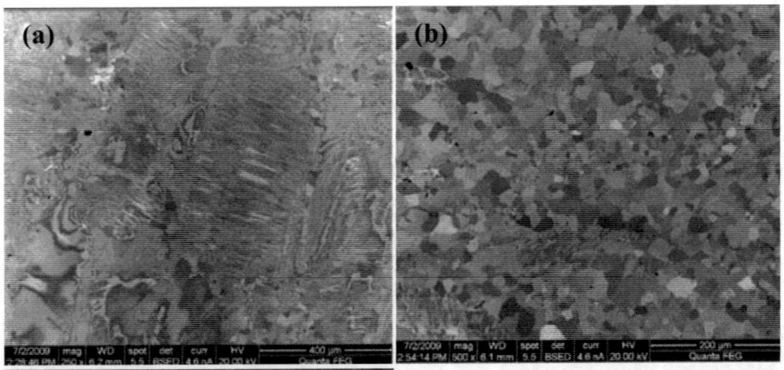

Figure 1: The image of the microstructure of the alloy Ti-26Nb-7Mo-12Zr after rolling at room temperature with the degree of deformation a) 30% b) 60%, c) 90% and subsequent annealing at T = 850°C followed by quenching (scanning electron microscopy).

Figure 2: Map of the distribution of crystallographic orientations and map of grain boundaries misorientations (green - low-angle, blue - high-angle grain boundaries) for alloy Ti-26Nb-7Mo-12Zr after rolling with a deformation degree 30% and subsequent annealing followed by quenching.

Measurement of the elastic modulus showed that its value in the initial state is 84 GPa. As a result of cold rolling and subsequent annealing is a reduction of the elastic modulus. With increasing degree of preliminary cold deformation this value decreases (table 1). As can be seen from the data in the table, strength characteristics following thermomechanical treatment changed slightly. In particular the plasticity increased in 1.5 times compared with the initial state.

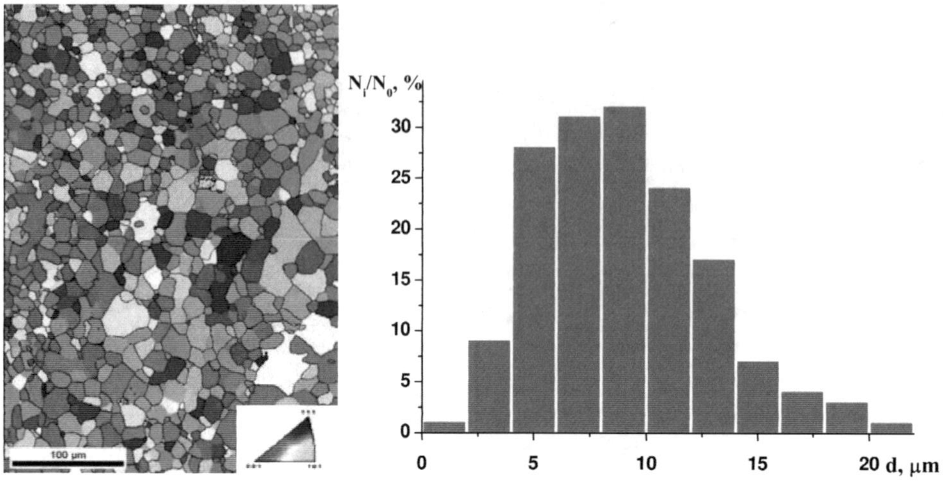

Figure 3: Map of the distribution of crystallographic orientations (high-angle grain boundaries are marked) and histogram of the distribution of grains size for alloy Ti-26Nb-7Mo-12Zr after rolling with a deformation degree 90% and subsequent annealing followed by quenching.

Table 1. The mechanical properties of the alloy Ti-26Nb-7Mo-12Zr after preliminary cold deformation and annealing followed by quenching at T = 850°C.

Preliminary cold deformation	Elastic Modulus, GPa	Yield strength, MPa	Ultimate strength, MПa	Flexibility, %
-	83,7	793	800	6,3
30%	70,5	755	757	5,5
60%	69,2	767	768	7,1
90%	66,6	754	759	8,5

4. Conclusion

The formation of a homogeneous globular structure of titanium β-alloys Ti-26Nb-7Mo-12Zr leads to a significant decrease in the modulus compared to the coarse-grained (initial) state and noticeable increase of plasticity without significant changing of the mechanical properties.

Acknowledgment

The work was carried out under a state contract "Biomedical research structural changes of organs and tissues during stent implantation of new generation» № 14.740.11.0182.

References

[1] Kolobov Yu R 2009 *Nanotechnologies in Russia* **11-12** 758

[2] Leyens C and Peter M 2003 *Titanium and Titanium Alloys. Fundamentals and Applications* (Weinheim)

[3] Hanawa T, Hiromoto S and Yamamoto A 2001 *Structural biomaterials for the 21 century* 145

[4] Geetha M, Singh A K, Asokamani R and Gogia A K 2009 *Progress in materials science* **54** 397

[5] Yoshimitsu Okazaki, Sethumadhvan Rao, Yoshimasa Ito and Tetsuya Tateishi 1998 *Biomaterials* **19** 1197
[6] Nakai M, Niinomi M and Akahori T 2008 *Advances in Materials Research* **10** 167
[7] Hench L and Jones D 2007 *Biomaterials, artificial organs and tissue engineering* (Moscow: Technosphere)
[8] Niinomi M, Kuroda D and Fukunaga K 1999 *Titanium 99: Science and technology* ed Froes F H and Caplan I (The Minerals, Metals&Materials Society) pp 1195-11201
[9] Kawahara H 1992 *Bulletin of the Japan Institute of Metals* **3**1 1033-1039

Functionalization of single-walled carbon nanotubes regulates their effect on hemostasis

A.V. Sokolov[1,2], A.V. Aseychev[1], V.A. Kostevich[1,2], A.A. Gusev[1], S.A. Gusev[1] and I.I. Vlasova[1]

[1] Research Institute for Physico-Chemical Medicine, FMBA, M. Pirogovskaya Str. 1a, Moscow, Russia

[2] Research Institute for Experimental Medicine, RAMS, Acad. Pavlov Str. 12, Saint Petersburg, Russia

E-mail: irina.vlasova@yahoo.com

Abstract. Applications of single-walled carbon nanotubes (SWNTs) in medical field imply the use of drug-coupled carbon nanotubes as well as carbon nanotubes functionalized with different chemical groups that change nanotube surface properties and interactions between nanotubes and cells. Covalent attachment of polyethylene glycol (PEG) to carboxylated single-walled carbon nanotubes (c-SWNT) is known to prevent the nanotubes from interaction with macrophages. Here we characterized nanotube's ability to stimulate coagulation processes in platelet-poor plasma (PPP), and evaluated the effect of SWNTs on platelet aggregation in platelet-rich plasma (PRP). Our study showed that PEG-SWNT did not affect the rate of clotting in PPP, while c-SWNT shortened the clot formation time five times compared to the control PPP. Since c-SWNT failed to accelerate coagulation in plasma lacking coagulation factor XI, it may be suggested that c-SWNT affects the contact activation pathway. In PRP, platelets responded to both SWNT types with irreversible aggregation, as evidenced by changes in the aggregate mean radius. However, the rate of aggregation induced by c-SWNT was two times higher than it was with PEG-SWNT. Cytological analysis also showed that c-SWNT was two times more efficient when compared to PEG-SWNT in aggregating platelets in PRP. Taken together, our results show that functionalization of nanoparticles can diminish their negative influence on blood cells. As seen from our data, modification of c-SWNT with PEG, when only a one percent of carbon atoms is bound to polymer (70 wt %), decreased the nanotube-induced coagulation in PRP and repelled the accelerating effect on the coagulation in PPP. Thus, when functionalized SWNTs are used for administration into bloodstream of laboratory animals, their possible pro-coagulant and pro-aggregating properties must be taken into account.

Prospects for using single-walled carbon nanotubes (SWNTs) in targeted drug delivery and tumor treatment have raised concerns about their biocompatibility and possible adverse effects on human health. In recent years, many studies have been devoted to in vitro investigation of the interaction between carbon nanotubes and blood proteins and cells [1]. In particular, it was shown that SWNTs added either to isolated platelets or platelet-rich plasma (PRP) stimulate platelets to aggregate [2, 3, 4]. However SWNTs may interact simultaneously with many factors of hemostasis, and their effect on blood coagulation will not be limited to induced platelet aggregation. Therefore, in this study we focused on the effect of SWNTs on coagulation not only in human PRP, but also in platelet-poor plasma (PPP). Furthermore, applications of SWNTs in medical field imply the use of drug-coupled, functionalized carbon nanotubes as well as carbon nanotubes functionalized with different chemical groups, leading to a significant change in nanotube surface properties and interaction between nanotubes and cells. In order to increase the dispersability of carbon nanoparticles, they are modified by the addition of charged groups on the surface. For example, the oxidation of SWNTs generates carboxyl groups on the nanotube surface. The further covalent attachment of polyethylene glycol (PEG) to the produced carboxylated single-walled carbon nanotubes (c-SWNTs) can prevent the interaction of the nanotubes with macrophages. Intravenously injected PEG functionalized SWNTs (PEG-SWNTs) conjugated with a cancer chemotherapy drug paclitaxel was shown to suppress the breast tumor growth in mice with a higher efficacy than clinical paclitaxel [5].

Here we evaluated the interaction of functionalized nanotubes c-SWNTs and PEG-SWNTs (Carbon Solutions Inc., USA, [6]) with blood plasma proteins in PPP, characterized nanotube's ability to stimulate coagulation processes in PPP, and evaluated the effect of SWNTs on platelet aggregation in PRP.

The nanotube's capacity to bind human plasma proteins appeared to be significantly different between c-SWNTs and PEG-SWNTs. c-SWNTs efficiently and non-selectively adsorbed proteins with a concomitant sedimentation. PEG-SWNTs adsorbed by an order lower amount of proteins and displayed no adsorption preference for major plasma protein albumin (Fig. 1).

The pro-coagulant property of nanotubes was assessed by their effect on the coagulation cascade. The latter involves circulating coagulation factors (which act as enzymes, which require activation, and cofactors), as well as calcium and platelets and a binding surface upon which the coagulation cascade proceeds. The coagulation cascade has the tissue factor pathway (or the extrinsic pathway) and the contact activation pathway (or the intrinsic pathway) that converge at the generation of thrombin from pothrombin, the cleavage of fibrinogen, by thrombin, and the formation of fibrin fibers which form blood clots. We characterized blood coagulation in PPP by (1) the clot formation time, i.e. the amount of time required for stable clot formation, and (2) the presence of fibrinopeptides and absence of prothrombin in serum (after clotting). The clot formation was stated if the meniscus at the top of plasma in a reaction tube remained parallel to the bottom of the tube even when we inclined the tube at 60 degrees. PEG-SWNTs did not affect the rate of clotting, while c-SWNTs shortened the clot formation time five times compared to the control PPP. Probably, the difference in the affinity of plasma proteins to different nanotubes underlies the latter discrepancy between the two types of nanotubes. In all samples after clotting, fibrinopeptides and no prothrombin were detected in serum. Since c-SWNTs failed to accelerate coagulation in plasma lacking coagulation factor XI, it may be suggested that c-SWNTs affects the contact activation pathway. Considering that factor XII autoactivation, which readily occurs in contact with a variety of negatively charged surfaces, is enough for initiation of the contact activation pathway, we may suggest that c-SWNTs provides a surface that is suitable for contact activation of blood coagulation.

At the same time, both SWNTs types nearly equally quicken the clotting in PRP. The addition of c-SWNTs or PEG-SWNTs to PRP shortened the clot formation time 4 and 3 times, respectively. Since platelet participation in blood coagulation hastens the process 15 times, it was necessarily to examine the effect of SWNTs on platelets. Platelet aggregation was assessed by measuring the aggregate mean radius [7]. The results of our experiments show that in PRP, platelets responded to both SWNTs types with irreversible aggregation. However, the rate of c-SWNTs induced aggregation was 2 times higher than it was for PEG-SWNTs (Fig. 2A). The addition of EDTA completely inhibited platelet aggregation, which indicates calcium ion (Ca^{2+}) participation in the SWNT-induced platelet aggregation. Calcium ions can act as a bridge between negatively charged groups on the cell surface and negatively charged groups (carboxyl or hydroxyl) on the nanotube surface. Calcium is also required to stabilize the binding of coagulation factors V and VIII to platelets. A slight accelerating effect, similar for both c-SWNTs and PEG-SWNTs, on ADP-induced platelet aggregation in PRP (Fig. 2B) can be attributed mainly to pre-activation of platelets by the nanotubes before the addition of ADP.

These phenomena have been also observed for non-functionalized carbon nanotubes [2, 3, 4]. Cytological analysis of PRP demonstrated as well that platelets are aggregated by c-SWNTs and PEG-SWNTs (Fig. 3). The number of small aggregates (2-4 platelets per aggregate) and the number of big aggregates (5-10 platelets per aggregate) was 4- and 11-times, respectively, higher than that in the control blood plasma sample. A similar effect was seen with PEG-SWNTs, though it was significantly less. The number of small aggregates increased 2.6-times, and the number of big ones increased 4-times. Thus, despite significant differences in the binding of plasma proteins to c-SWNT and PEG-SWNTs, these nanotube types are only slightly different in respect of the interaction with platelet membrane and induction of platelet aggregation.

Taken together, our results show that even functionalized carbon nanotubes can cause irreversible alterations in hemostasis: both SWNT types used in our work induce platelet aggregation and blood coagulation. However, the functionalization of nanoparticles can diminish their negative influence on blood cells. As seen from our data, modification of c-SWNTs with PEG, when only a one percent of carbon atoms is bound to polymer (70 wt % [6]), decreased the nanotube-induced coagulation in PRP and repelled the accelerating effect on the coagulation in PPP. Thus, when functionalized SWNTs are used for administration into bloodstream of laboratory animals, their possible pro-coagulant and pro-aggregating properties must be taken into account. The simple methods used in our work to evaluate the effect of SWNTs on blood coagulation can be useful for preliminary evaluation of the effect of functionalized nanoparticles on hemostasis. Prospects for the use of SWNTs in nanopharmacology call for additional approaches to nanotube functionalization to prevent thrombogenesis.

This work was supported by the Russian Foundation for Basic Research, grant 09-04-01043-a.

References

[1] Laserda L, Bianco A, Prato M and Kostarelos K 2006 Carbon nanotubes as nanomedicines: from toxicology to pharmacology. *Adv. Drug Deliv. Rev.* **58** 1460-70.

[2] Radomski A, Jurasz P, Alonso-Escolano D, Drews M, Morandi M, Malinski T and Radomski M W 2005 Nanoparticle-induced platelet aggregation and vascular thrombosis. *Brit. J. Pharmac.* **146** 882-93.

[3] Bihari P, Holzer M, Praetner M, Fent J, Lerchenberger M, Reichel C A, Rehberg M, Lakatos S and Krombach F 2010 Single-walled carbon nanotubes activate platelets and accelerate thrombus formation in the microcirculation. *Toxicology* **268** 148-54.

[4] Semberova J, De Paoli Lacerda S H, Simakova O, Holada K, Gelderman M P and Simak J 2009 Carbon nanotubes activate blood platelets by inducing extracellular Ca2+ influx sensitive to calcium entry inhibitors. Nano lett. 9 3312-17.

[5] Liu Z, Chen K, Davis C, Sherlock S, Cao Q, Chen X and Dai H 2008 Drug delivery with carbon nanotubes for in vivo cancer treatment. *Cancer Res.* **68** 6652-60.

[6] Zhao B, Hu H, Yu A, Perea D and Haddon R C 2005 Synthesis and characterization of water soluble single-walled carbon nanotube craft copolymers. *J. Am. Chem. Soc.* **27** 8197-203.

[7] Gabbasov Z A, Popov E G, Gavrilov I Yu and Pozin E Ya 1989 Platelet aggregation: the use of optical density fluctuations to study microaggregate formation in platelet suspension. *Thromb. Res.* **54** 215-23.

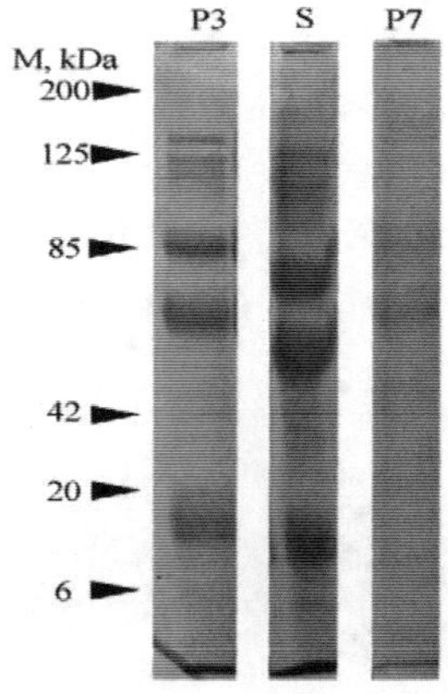

Figure 1. SDS-PAGE of plasma proteins (middle line) and proteins adsorbed on c-SWNTs (P3, left line) and PEG-SWNTs (P7, right line). The molecular masses (kDa) are indicated on the left.

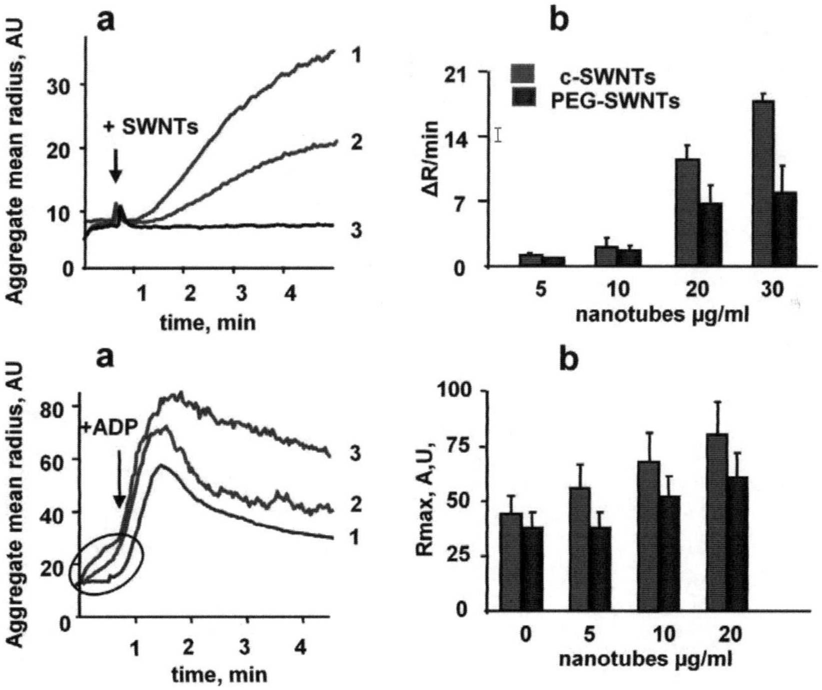

Figure 2. (A) SWNTs induce platelet aggregation in PRP. a) Aggregate mean radius in PRP as function of time: (1) c-SWNTs, (2) PEG-SWNTs, (3) PRP incubated 2 min with 100 μM EDTA before SWNTs addition. SWNTs =20 μg/ml. (b) The rate of aggregate mean radius formation increases on nanotubes concentration.
(B) SWNTs activate ADP-induced platelet aggregation in PRP. a) Aggregate mean radius was measured as function of time after addition of 1 μM ADP to control plasma (1) or plasma pre-incubated with nanotubes: (2) PEG-SWNTs or (3) c-SWNTs. SWNTs =20 μg/ml. b) Maximum radius of aggregates in PRP on nanotubes concentration.

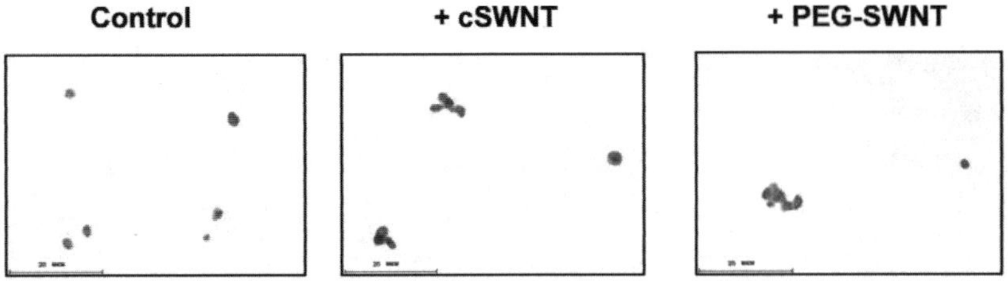

Figure 3. Images of platelets in PRP. Smears were stained by Romanovsky-Giemsa method. The single platelets of various shape and size are present in control plasma, formation of platelet aggregates were detected after nanotubes addition (20 μg/ml, 10 min - incubation time).

Journal of Physics: Conference Series **291** (2011) 012055 doi:10.1088/1742-6596/291/1/012055

Aqueous dispersions of oxide nanoparticles as a treatment for pyoinflammatory diseases with chronic component

Ph Rutberg[1], V Kolikov[1], A Moshkin[2], V Snetov[1], A Stogov[1] and M Khalilov[2]
[1] Institute for Electrophysics and Electric Power Russian Academy of Sciences, 18 Dvortsovaya nab., St.-Petersburg, 191186, Russia
[2] Oryol State University, Medical Institute, October st. 25, Oryol, 302028, Russia

E-mail: Stogov2007@yandex.ru

Abstract. Promising direction of surgery related to the treatment of acute purulent wounds with chronic component could be utilization of aqueous dispersions of nanostructures (ADN) produced by pulsed electric discharge in water. The investigation is addressed to finding out the opportunity of usage of an ADN for treatment of purulent wounds with a chronic component and comparison of its efficiency with the widespread antiseptics. For realization of investigation was used ADN, which has maximal share of "small" nanostructures (<100 nm) with the greatest surface electric charge. High activity of reparative processes is established at use of ADN and subsequent moderate changes of the further healing. The attributes of cellular atypia and preternatural representations about inflammatory reactions are not revealed at local use of ADN.

1. Introduction

Promising direction of surgery related to the treatment of acute purulent wounds with chronic component could be utilization of aqueous dispersions of nanostructures (ADN) produced by pulsed electric discharge in water [3, 6]. The group of patients with such wounds are of large number and differ on etiology of diseases but their mutual feature is long treatment without marked positive changes [5]. Thus long application of antibiotics leads to abnormality of immune processes and antibacterial resistance of microbial flora [1]. Moreover, local antiseptics are frequently toxic and one can oppress processes of reparation in a wound [2].
The investigation is addressed to finding out the opportunity of usage of an ADN for treatment of purulent wounds with a chronic component and comparison of its efficiency with the widespread antiseptics.

2. Materials and procedure

One of effective ways of oxide metal nanostructures manufacture is processing of water by pulsed electrical discharge. In this case, ADN formed by PED contains both nanostructures of electrode metal or its oxides and metal's ions [4, 7]. ADN were produced by electrical discharge setup with voltage 60 kV, current 40 A, current pulse duration of 5 - 10 microseconds and pulse repetition rate of 50 Hz.

Distribution of a share of "small" nanostructures subfractions (≤ 100 nm) and a surface electric charge of nanostructures depending on pulse duration had been determined by laser correlation spectrometer "LCS-03" and the device capillary electrophoresis "NANOPHORE 01" (Figure 1).

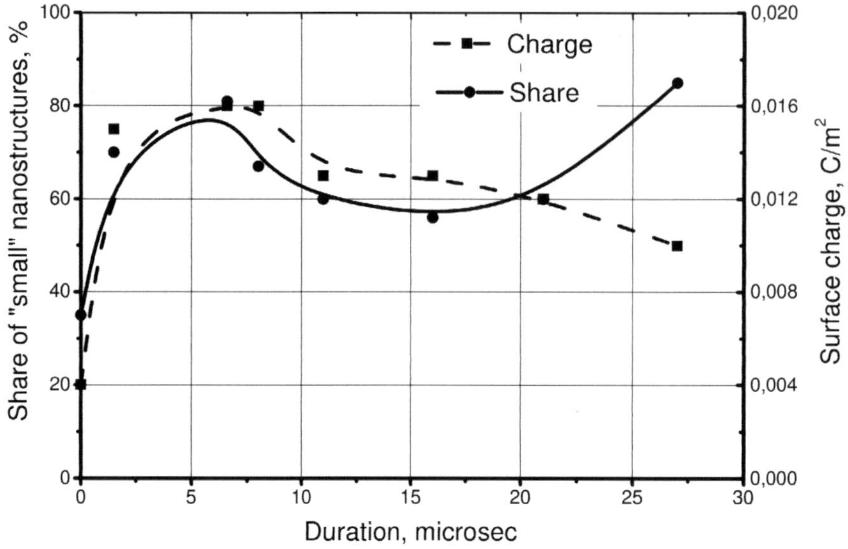

Figure 1. Share and charge of "small" (<100 nm) nanostructures.

For the clinical researches, the ADN with maximal share of "small" nanostructures and maximal surface electric charge has been chosen. Such nanostructures at pulse duration of 5 - 10 μs are formed.

Clinical research on 133 rats of both sexes of line "VISTAR" of weight 170 ± 30 g was carried out. For treatment of wounds were used the ADN of silver alloy of 5 mg/l mass concentration in 2 months after its preparation, solution of sodium hypochlorite of 600 mg/l mass concentration and 0.05% solution of chlorhexidine bigluconate. The animals were divided in four groups: one control group of 31 animals and three test groups of 51 animals each.

Purulent wounds were made by means of sewing in inter shoulder-blade areas of a gauze napkin impregnated with a solution of low viral strain of Staphylococcus aureus. Lancing and surgical sanitation of an abscess on 7th day after infecting were carried out. Initial wound surface area was 150 ± 50 mm^2.

3. Results and discussion

The wound treatment was carried out by daily overlaying of gauze napkin moistened with solutions being compared. Figure 2 presents a graph to reduce the area of wounds for 14 days for the three solutions used.

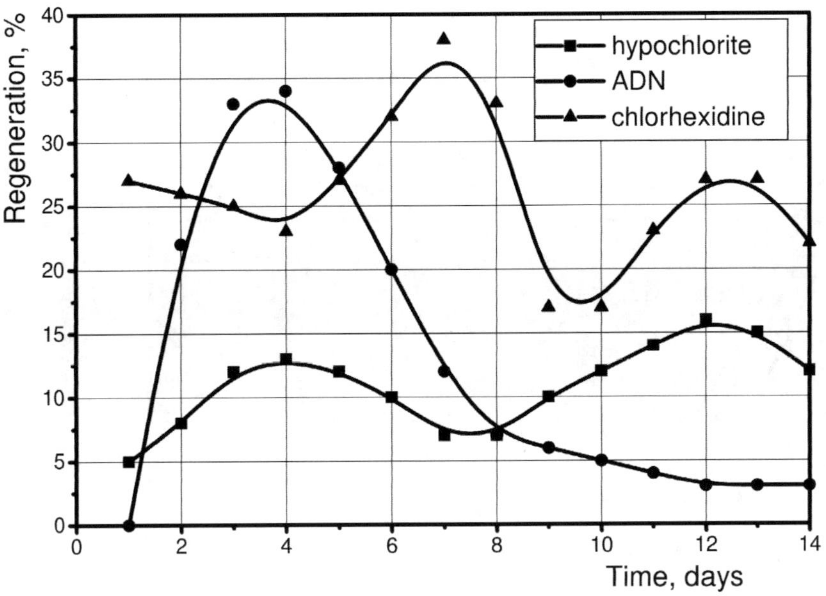

Figure 2. Reduction of the area of wounds at processing by antiseptics.

The plot shows the expressed tendency to reduction of the area of wounds in all three groups of animals with the common law of waviness of the healing current, caused by as conversion processes, and local inflammatory reactions. It is necessary to note high activity conversion processes at use ADN during the first 5 days from the beginning of treatment with a maximum on 3 - 4 days and subsequent moderate changes of the further healing.

The important factor distinguishing action ADN from two other antiseptics is absence secondary infection, observed on 4 and 9 days (chlorhexidine) and on 7 day (hypochlorite sodium).

The plot of changes of bacterial seeding (number of bacteria per gramme of weight of a wound tissue) shows oppression of tissue flora in all groups of animals. Approach of wound sterility occurred on 5 - 7 days, that is related to the activity of antiseptics and the factors of local immunity (Figure 3).

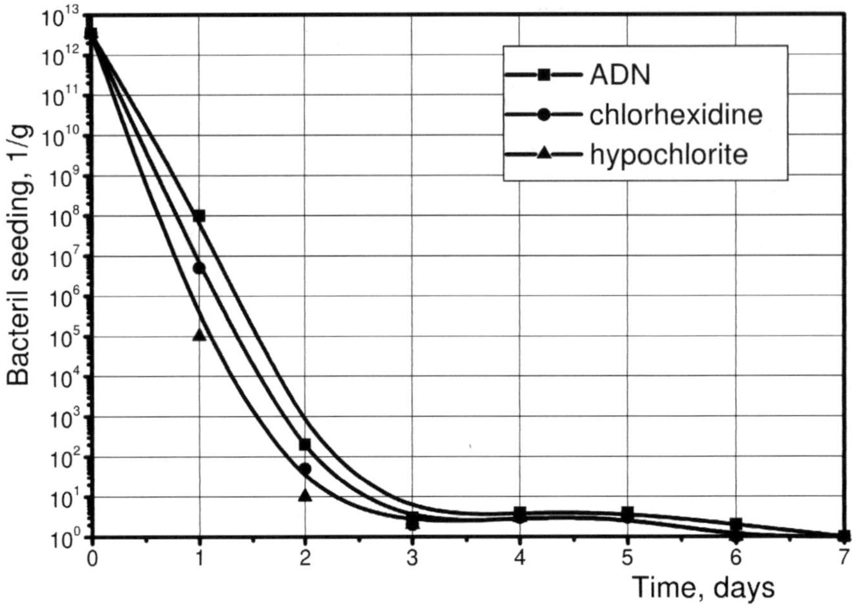

Figure 3. Changes of bacterial seeding.

The average weight of animals in test groups decreased due to long proceeding inflammatory reaction (Figure 4).

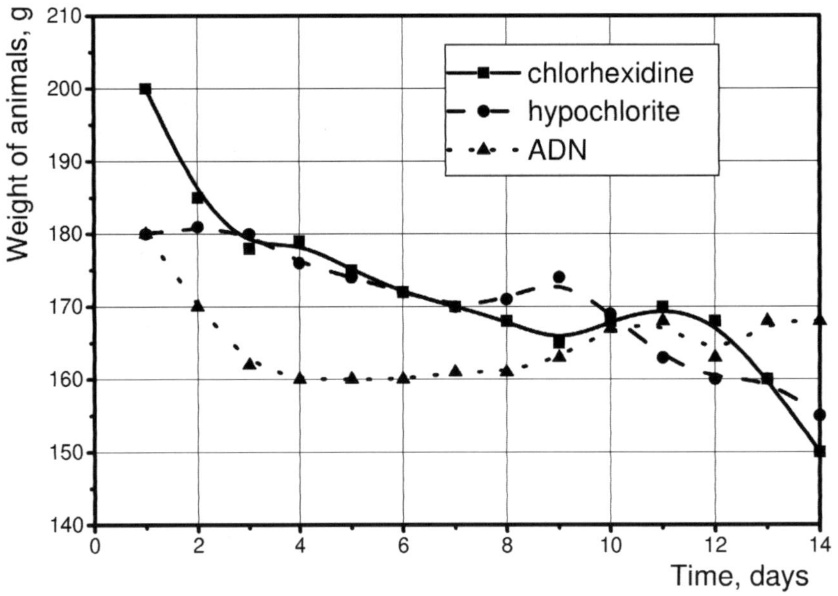

Figure 4. Change of average weight of animals during experiment.

At use of ADN average absolute rate of healing (mm^2/day) is a little bit lower, than in percentage expression that is related, apparently, with intensive wounds healing in the initial days and with the subsequent decrease of healing rate in conditions when the wound is disinfected and has the smaller area (Figure 5).

It is important to note that at use of ADN the number of animals with full wound healing by the end of 14 day appeared a little bit more than in two other groups.

Conformity of a microscopic picture to a visual estimation of wound process has been established at research of histologic cuts. At local use of ADN there were no revealed a attributes of cellular atypia which are beyond of preternatural representations about inflammatory reactions.

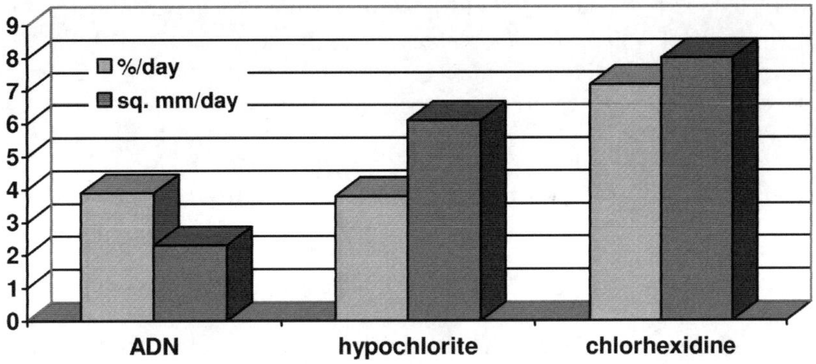

Figure 5. Average for 14 day parameters of reduction of the wound areas concerning its initial size.

Current practice of treatment of chronic purulent wounds on the background of the damaged tissue metabolism and local pathomorphism shows that there is no reason to rely on their quick healing, and therefore, the aim of treatment is to prevent complications.

In addition, the reactivation of an existing microflora when adapting it to antibiotics and the accession of additional associations of pathogens in the nutrient medium of wounds result to considerable adverse effects. Antibiotic resistance is becoming every year more and more pressing problem. Traditionally used antiseptics can inhibit regeneration in the wound and sometimes they lead to total intoxication patient with their long-term use in large-affected areas.

The investigation showed that the ADN have not of systemic toxicity, which may open new possibilities to cure a broad category of patients.

Another significant positive quality of ADN is the relative ease and cheapness of their manufacturing, as well as the fact that their storage does not require special conditions and they retain their antibacterial properties for a long time.

4. Conclusion

For realization of investigation was used ADN, which has maximal share of "small" nanostructures (<100 nm) with the greatest surface electric charge.

High activity of reparative processes is established at use of ADN during the first 5 days from the beginning of treatment with a maximum on 3 - 4 days and subsequent moderate changes of the further healing.

At use of ADN the number of animals with full wound healing by the end of 14 day appeared a little bit more than in groups where were used chlorhexidine and hypochlorite sodium.

The attributes of cellular atypia and preternatural representations about inflammatory reactions are not revealed at local use of ADN.

It is possible to assume that use of ADN will allow to correct a little in the long term practice of out-patient therapy of chronic and slow pyoinflammatory diseases.

Acknowledgements
Work was supported in part by the Russian Foundation for basic research, projects №№ 08-08-00477-a, 10-08-00239-a.

References
[1] Volenko A V, Menshikov A A, Titova G P and Kuprikov S V 2004 *Surgery* Preventive maintenance wound infections by immobilised antibacterial preparations **10** 54-58.
[2] Ljapunov A N, Datsenko B M and Mohert N A 1995 *The problem of medicinal therapy* Theory and practice of local treatment of purulent wounds Kiev 132-196
[3] Kolikov V A, Kurochkin V E, Panina L K and Rutberg F G 2005 *Doklady Biological Sciences,* Pulse electric discharges and prolonged microbial resistance of water **403** 279-281
[4] Kolikov V A, Kurochkin V E, Panina L K, Rutberg A F, Rutberg F G, Snetov V N and Stogov A J 2007 *Technical Physics* Prolonged microbial resistance of water treated by a pulsed electric discharges **52**, no.2 263-70
[5] Fedorov D N, Ivashkin A N, Shinin V V, Vasiljev A V and Ivanov A A 2002 *Archive of a pathology* Morphological and immunological characteristic conversion processes in long not healing wounds **64** № 1 8-11
[6] Rutberg Ph G, Kolikov V A, Kurochkin V E, Panina L K and Rutberg A Ph 2007 *IEEE Transactions on Plasma Science* Electric Discharges and the Prolonged Microbial Resistance of Water **35** no.4 1111-18
[7] Rutberg Ph, Kolikov V, Snetov V, Stogov A, Noskin L, Landa S and Arutunjan A 2007 *IEEE Pulsed Power and Plasma Science Conf. (PPPS-2007)* Pulsed electric discharges in water and oxide nanostructures 1244-47

Journal of Physics: Conference Series **291** (2011) 012056 doi:10.1088/1742-6596/291/1/012056

Peroxidase-induced degradation of single-walled carbon nanotubes: hypochlorite is a major oxidant capable of *in vivo* degradation of carbon nanotubes

I.I. Vlasova[1], T.V. Vakhrusheva[1], A.V. Sokolov[1,2], V.A. Kostevich[1,2] and A.A. Ragimov[3]

[1] Research Institute for Physico-Chemical Medicine, FMBA, M. Pirogovskaya Str. 1a, Moscow, Russia

[2] Research Institute for Experimental Medicine, RAMS, Acad. Pavlov Str. 12, Saint Petersburg, Russia

[3] National Research Centre of Surgery, RAMS, Abrikosovskiy per. 2, Moscow, Russia

E-mail: irina.vlasova@yahoo.com

Abstract. Due to their extraordinary properties, single-walled carbon nanotubes (SWNTs) have a tremendous potential for medical applications such as clinical diagnostics, targeted drug (or gene) delivery and cancer therapy. Hence, effects of SWNTs on living systems as well as mechanisms for biodegradation of SWTNs are of great importance and must be studied before starting to explore SWNTs for medical use. This study was undertaken to compare the potential of different peroxidases in degrading carboxylated SWNT (c-SWNT) and to elucidate the role of peroxidase-generated reactive products in this process. A detailed study showed that neither reactive intermediate products nor free radicals generated via peroxidase cycle can considerably oxidize c-SWNT. Biodegradation of c-SWNT in model system can be induced by free radicals generated as a result of heme degradation. The latter explains why hemoglobin, which is a pseudo-peroxidase possessing low peroxidase activity, is able to oxidize carbon nanotubes with a higher efficiency than horseradish peroxidase. However, c-SWNT in the presence of blood plasma (15 vol %) demonstrated no degradation even at high concentrations of hemoglobin and H_2O_2. The comparison of the ability of various peroxidases to degrade SWNTs *in vitro* revealed that MPO, due to its ability to produce hypochlorite, and lactoperoxidase, due to its ability to produce hypobromite, are extremely efficient in degrading carbon nanotubes. Since neutrophils are a main source of human MPO, we tested the effect of SWNTs on these cells. SWNTs were unable to stimulate neutrophils. On the other hand, they dose-dependently enhanced opsonized zymosan-induced cell stimulation as detected by measuring the amount of hypochlorite produced. This finding may be relevant to the *in vivo* situation, for example, at inflammatory sites. In order to imitate conditions characteristic of phagosomes and inflammatory sites, we titrated the suspension of c-SWNT in the presence of diluted blood plasma at pH 5.8 with high concentrations of (MPO + H_2O_2) or hypochlorite and found significant degradation of nanotubes. Collectively, our results indicate that hypochlorite is a main candidate for oxidative degradation of carbon nanotubes *in vivo*.

The extraordinary properties of single-walled carbon nanotubes (SWNTs) make them potentially useful in a wide variety of applications not only in hi-tech industry, but also in biotechnologies and medicine, in particular, for clinical diagnostics, targeted drug (or gene) delivery, cancer therapy [1, 2]. Effects of SWNTs on living systems as well as mechanisms for biodegradation of SWTNs and ways for their clearance from the body are of great importance

Journal of Physics: Conference Series **291** (2011) 012056 doi:10.1088/1742-6596/291/1/012056

and must be studied before starting to explore SWNTs for medical use. It was shown previously that free radicals and oxidants generated in the reactions mediated by horseradish peroxidase (HRP), by myeloperoxidase (MPO), and by free iron ions can oxidize and destruct carboxylated SWNT (c-SWNT) [3, 4]. Moreover, degradation of nanotubes was induced not only by their incubation with MPO [4, 5] but also by the incubation with neutrophils [4], the cells containing MPO and releasing it upon stimulation.

This study was undertaken to compare the potential of different peroxidases in degrading c-SWNT and to elucidate the role of peroxidase-generated reactive products in this process. We employed HRP, hemoglobin, MPO, and lactoperoxidase (LPO). Among them, MPO and LPO exhibit, in addition to peroxidase activity, halogenation activity, oxidizing halide ions (e.g., Cl^- and Br^-) to the corresponding hypohalogenites [6]. SWNT degradation was assessed spectrophotometrically by measuring the absorption near-infrared spectra of nanotubes pretreated by sonication. Decrease in the absorption correlates well with nanotube degradation (Fig. 1A). Such approach was confirmed by other methods [3, 4].

In the absence of halide ions, the addition of the enzymes and H_2O_2 to nanotube suspension led to c-SWNT degradation, though the extent of degradation was low (Fig. 1B). A more detailed study showed that neither reactive intermediate products nor free radicals generated via peroxidase cycle can considerably oxidize c-SWNT. The destruction of the active site in the presence of H_2O_2 and further free radical formation mediated by heme, Fe^{2+}, Fe^{3+} and H_2O_2 could account for peroxidase-induced nanotube degradation [3]. So, hemoglobin, which is a pseudo-peroxidase possessing low peroxidase activity, is able to oxidize carbon nanotubes with a high efficiency (Fig. 1B). The hemoglobin heme group is indeed noncovalently inserted in the heme pocket and, hence, is not stable. However, it is unlikely that this pathway for nanotube degradation will be realized in the body, since antioxidant systems, among which the mechanisms preventing the appearance of heme and free iron ions in the bloodstream, impede the accumulation of free radicals in the amount needed to degrade resistant structures such as SWNTs. In line with this assumption, c-SWNT demonstrated no degradation in the presence of blood plasma (15 vol %) even at high concentrations of hemoglobin and H_2O_2 (Fig. 2A).

The addition of chloride ions to the incubation medium caused an increase in peroxidase-induced degradation of c-SWNT only when the latter were exposed to MPO (Fig. 1B). Since MPO, in the presence of chloride ions, generates hypochlorite, this result provides compelling evidence that enzymatically produced hypochlorite contributes considerably to the degradation of c-SWNT with MPO. It was also shown previously that hypochlorite added as reagent can oxidize carbon nanotubes [4, 5, 7]. Our experimental protocol of repeated sequential additions of the MPO and H_2O_2 aliquots to nanotube suspension (see legend for Fig. 1) precluded the participation of iron ions in nanotube degradation and proved that hypochlorite generated in MPO-catalysis provides for efficient degradation of c-SWNT. Similarly, when c-SWNT suspension was titrated with LPO and H_2O_2 in the presence of bromide ions, the nanotube degradation increased twice compared to the Br^--free medium (Fig. 1,). This can be again attributed to enzymatically produced hypohalogenite, namely, hypobromite.

Chemical modification of carbon nanotubes is used to increase their efficacy in the medical field. Functionalization of SWNT by polyethylene glycol (PEG) increases nanotube solubility and prevents their interaction with macrophages [2, 8]. In our work, we revealed that functionalization of SWNT by PEG had no effect on the pattern of peroxidase-induced degradation of carbon nanotubes. The best degradation of PEG-functionalized SWNT (PEG-SWNT) was observed with the $MPO/Cl^-/H_2O_2$ system or the $LPO/Br^-/H_2O_2$ system.

Thus, our experiments demonstrated that biodegradation of SWNTs can occur under action of MPO due to its exclusive ability to produce hypochlorite. Can this mechanism be realized *in vivo*? In blood, carbon nanotubes adsorb plasma proteins. Since the reactions of hypochlorite with protein amino acids are several orders of magnitude faster than with carbon nanotubes, hypochlorite will react first of all with proteins, in particular with amino groups. Modification of an amino group with hypochlorite produces chloroamine, a strong and stable oxidant, which could oxidize carbon nanotubes. We used a model N-chloramine, chloramine T, and showed that it induces effective degradation of c-SWNT comparable to this obtained with hypochlorite. In an attempt to better simulate the *in vivo* conditions, we added blood plasma (15 vol %) to c-SWNT suspension. We titrated the mixture with reagent hypochlorite. As a result, the optical absorbance of nanotubes decreased significantly, indicating their degradation (Fig. 2A). This marks a promising possibility for SWNTs to be biodegraded in the body. We hypothesize that nanotube degradation should preferably occur at sites of inflammation and in phagosomes, where high concentrations of MPO (0.5–2 mM) [2], catalytically active NADPH oxidase and the acidic pH levels provide for steady-state high levels of hypochlorite.

Since neutrophils are a main source of human MPO, we tested the effect of SWNTs on these cells. SWNTs were unable to stimulate neutrophils. On the other hand, they do not prevent the interaction of a cell stimulator, opsonized zymosan, with neutrophils but even dose-dependently enhanced stimulator-induced activation of the cells as detected by measuring the amount of hypochlorite produced (Fig. 2B). This finding may be relevant to the *in vivo* situation, for example, at inflammatory sites. It is difficult to examine the mechanisms for SWNT degradation in neutrophil suspension, since isolated neutrophils are alive only within a few hours. Moreover, loss of cell integrity is accompanied by the appearance of strong antioxidants, taurine and ascorbate, in the medium [6]. For this reason, in order to imitate conditions characteristic of phagosomes and inflammatory sites, we placed c-SWNT in the diluted blood plasma at pH 5.8. Titration of this mixture with MPO and H_2O_2 resulted in the effective degradation of nanotubes (Fig. 2A). On the contrary, at pH 7.4, we failed to observe c-SWNT degradation. This can be explained by the higher pH value that, along with certain experimental conditions, is less favorable for chlorinating activity of MPO.

Collectively, our results indicate that hypochlorite is a main candidate for oxidative degradation of carbon nanotubes *in vivo*. Potentially, structural and chemical modification resulting from degradation should decrease cytotoxicity and other negative effects of nanotubes on the body. The work was supported by the grant No. 09-04-01043 from the Russian Foundation for Fundamental Research.

References

[1] Foldvari M and Bagonluri M 2008 Carbon nanotubes as functional excipients for nanomedicines. *I. Pharmaceutical properties. Nanomedicine* **4** 183-200.

[2] Liu Z, Chen K, Davis C, Sherlock S, Cao Q, Chen X and Dai H 2008 Drug delivery with carbon nanotubes for in vivo cancer treatment. *Cancer Res.* **68** 6652-60.

[3] Allen B L, Kotchey G P, Chen Y, Yanamala N V, Klein-Seetharaman J, Kagan V E and Star A 2009 Mechanistic investigations of horseradish peroxidase-catalyzed degradation of single-walled carbon nanotubes. *J. Am. Chem. Soc.* **131(47)** 17194-205.

[4] Kagan V E, Konduru N V, Feng W, Allen B L, Conroy J, Volkov Y, Vlasova I I, Belikova N A, Yanamala N, Kapralov A, Tyurina Y Y, Shi J, Kisin E R, Murray A R, Franks J, Stolz D, Gou P, Klein-Seetharaman J, Fadeel B, Star A and Shvedova A A 2010 Carbon nanotubes degraded by neutrophil myeloperoxidase induce less pulmonary inflammation. *Nat Nanotechnol.* **5(5)** 354-9.

[5] Vlasova I I, Chekanov A V, Matskevich V A and Sokolov A V 2009 Biodegradation of single-walled carbon nanotubes by leucocytic myeloperoxidase. *Abstract book of II Nanotechnology International Forum, Moscow* 603-5.

[6] Klebanoff S J 2005 Myeloperoxidase: friend and foe. *J. Leukoc. Biol.* **77** 598-625.

[7] Wu C H 2007 Studies of the equilibrium and thermodynamics of the absorption of Cu2+ onto as-produced and modified carbon nanotubes. *J. Colloid. Interface Sci.* **311(2)** 338-46.

[8] Zhao B, Hu H, Yu A, Perea D and Haddon R C 2005 Synthesis and characterization of water soluble single-walled carbon nanotube craft copolymers. *J. Am. Chem. Soc.* **127** 8197-203.

[9] Vlasova I I Arnhold J, Osipov A N and Panasenko O M 2006 pH-dependent regulation of myeloperoxidase activity. *Biochemistry* (Moscow) **71** 825-37.

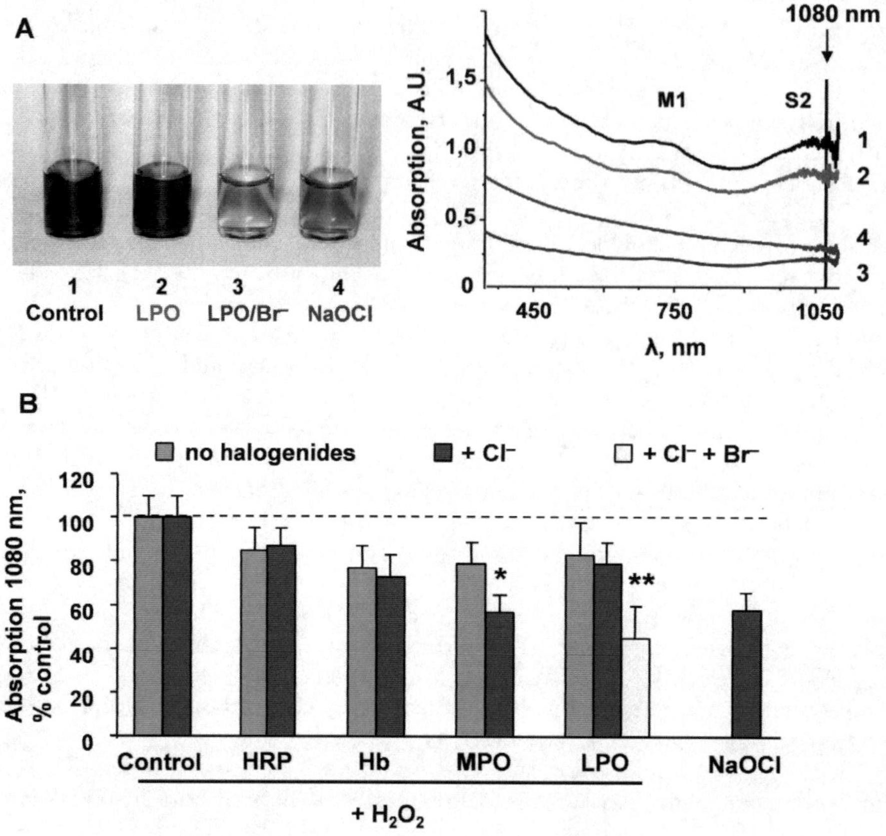

Figure 1. (A) Suspensions of c-SWNT and their spectra after repeated additions of: (control) - water; (LPO) - 400 nM LPO and 200 μM H_2O_2 twice (15 min interval); (LPO/Br^-) - the same but plus 6 mM Br^-, (NaOCl) - 1 mM hypochlorite. The additions were made twice a day; 6 additions totally. 100 mM Na-phosphate buffer, 140 mM NaCl, 100 μM DTPA, pH 7.4. M1 and S2- characteristic absorption bands in the spectra of SWNTs.
(B) Degradation of c-SWNT by peroxidase/H_2O_2 and by hypochlorite. The percent degradation was measured by the change in peak absorbance at 1080 nm. The total number of additions – 4. Final concentrations after each addition: 400 nM horseradish peroxidase (HRP); 100 nM hemoglobin (Hb); 200 nM MPO; 400 nM LPO and 2 × 200 μM H_2O_2; 1 μM NaOCl. $^{*}p < 0.05$ versus Cl^--free buffer; $^{**}p < 0.05$ versus Br^--free buffer.

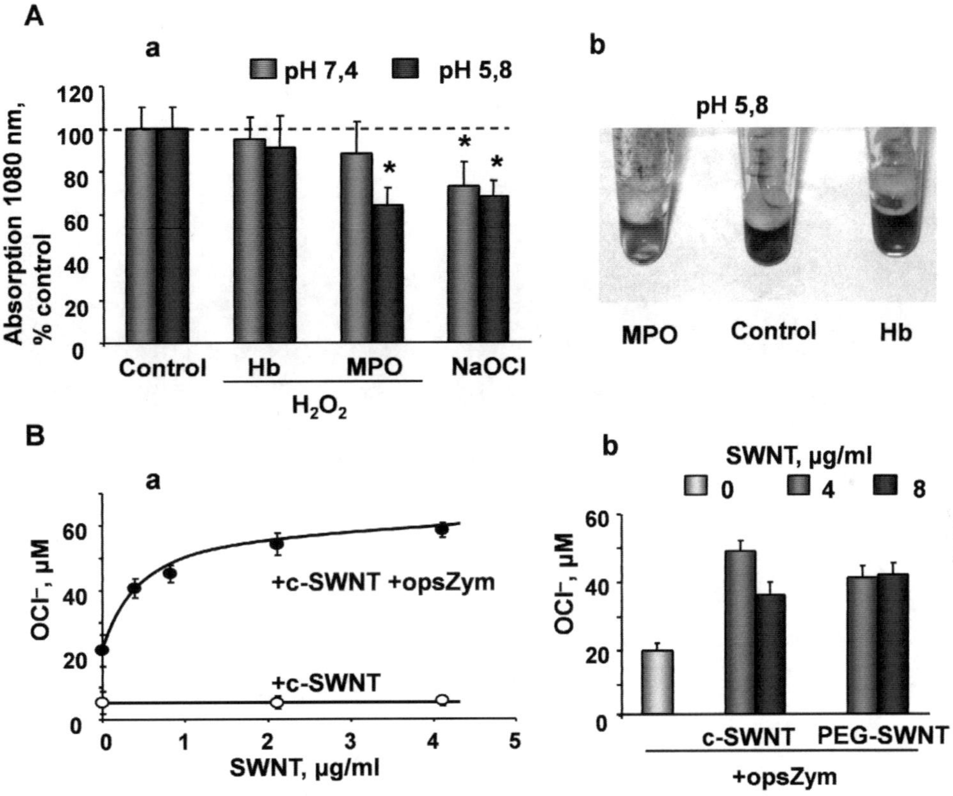

Figure 2. (A) Degradation of c-SWNT in the presence of blood plasma. (a) Absorption of c-SWNT (% control) after incubation with hemoglobin/H_2O_2, MPO/H_2O_2 or hypochlorite at different pH values. Plasma 15 % by volume. The total number of additions - 6. Final concentrations after each addition: 2 μM Hb, 2 μM MPO and 5x(500) μM hydrogen peroxide; 10 mM hypochlorite. $^*p < 0.05$ versus control. (b) Nanotube suspensions after incubation.

(B) Concentration dependence of the effect of SWNTs on hypochlorite generation by neutrophils. (a) Hypochlorite generation by neutrophils (1 million cells per ml) in the presence of c-SWNT alone and in the presence of both c-SWNT and opsonized zymosan after incubation for 30 minutes at 37. (b) The effect of c-SWNT and PEG-SWNT on hypochlorite generation by neutrophils stimulated with opsonized zymosan. Hypochlorite was measured with the taurine chloramine assay.

AUTHOR INDEX

AUTHOR INDEX

AUTHOR INDEX

Institute of Physics
Dirac House, Temple Back
Bristol BS1 6BE UK

CURRAN ASSOCIATES INC
proceedings
.com